Proceedings in Life Sciences

Vertebrate Circadian Systems

Structure and Physiology

Edited by
J. Aschoff S. Daan G.A. Groos

With 154 Figures

Springer-Verlag
Berlin Heidelberg New York 1982

Professor Dr. Jürgen Aschoff
Max-Planck-Institut für Verhaltensphysiologie,
D-8138 Andechs

Dr. Serge Daan
Zoologisch Laboratorium, Rijksuniversiteit Groningen,
Kerklaan 30, NL-Haren

Dr. Gerard A. Groos
National Institute of Mental Health,
Bethesda, Maryland 20205/USA

ISBN 3-540-11664-8 Springer-Verlag Berlin Heidelberg New York
ISBN 0-387-11664-8 Springer-Verlag New York Heidelberg Berlin

Library of Congress Cataloging in Publication Data. Main entry under title: Vertebrate circadian system. (Proceedings in life sciences) Papers presented at a meeting held in October 1980 at Schloss Ringberg. Bibliography: p. Includes index. 1. Circadian rhythms--Congresses. 2. Vertebrates--Physiology--Congresses. I. Aschoff, Jürgen. II. Daan, S. (Serge), 1940- . III. Groos, G. (Gerard), 1951- . IV. Series. [DNLM: 1. Circadian rhythm--Congress. QP 84.6 V567 1980] QP84.6.V48 1982 596'.01'882 82-10249

Typesetting and printing: Beltz, Offsetdruck, Hemsbach/Bergstr.
Bookbinding: Brühlsche Universitätsdruckerei, Giessen

2131-3130-543210

Dedicated to
Curt P. Richter

Preface

By evolutionary adaptation to the perpetual day-night changes in environmental conditions, eukaryotic organisms have acquired an endogenous programme. This mechanism exhibits the characteristics of a self-sustaining oscillation the period of which approximates that of the earth's rotation. For animals such a property was first clearly demonstrated by Maynard S. Johnson (1939) who recorded, in constant conditions, free-running activity rhythms of white-footed mice (*Peromyscus leucopus*). Johnson concluded from his observations that "this animal has an exceptionally substantial and durable self-winding and self-regulating clock, the mechanism of which remains to be worked out". Twenty years later, the formal properties of this "circadian" clock and its use by organisms as a time-keeping device were summarized at the Cold Spring Harbor Symposium in 1960 (Chovnick 1961).

During the following two decades, investigations have turned towards an analysis of the physiological mechanisms involved in and the search for a central masterclock. These efforts led to the discovery that the pineal organ of submammalian vertebrates and the suprachiasmatic nuclei of birds and mammals are major candidates for a role as central circadian pacemakers. At the same time the neural pathways through which these structures are coupled to the light-dark cycle were identified. Furthermore, it was established that the pineal gland and the suprachiasmatic nuclei are closely related structures that integrate the functions of circadian timekeeping and photoperiodic time measurement. Both functions are crucial for an animal's adaptation to the daily and seasonally changing environment. The concept of a single central circadian clock, located in the suprachiasmatic nuclei or the pineal gland, is attractive in its simplicity, but does not account for all the phenomena. It became increasingly obvious from lesioning experiments, particularly in mammals, that some oscillating capacity must exist elsewhere. This supports the older notion that circadian systems comprise a multiplicity of oscillating elements. Yet the localisation of pacemaker structures attracted the active interest of investigators with diverse backgrounds in biology and inspired physiological, anatomical and behavioural research on the mechanism of identified circadian clocks.

These new developments in the circadian rhythm field were the theme of a meeting in October 1980 at Schloss Ringberg. Generously supported by the Deutsche Forschungsgemeinschaft and the Max-Planck-Gesellschaft, the symposium brought together anatomists, physiologists, endocrinologists and pharmacologists to share their experiences with biological clocks. This volume presents most of the papers read at Schloss Ringberg, supplemented by a number of invited contributions. It aims at portraying recent progress made in the analysis of the circadian organization of higher vertebrates, but does not claim to give a complete survey on the structure and physiology of circadian pacemakers; it rather complements two other publications which appeared within the last 3 years (Suda et al. 1979, Aschoff 1981).

Of major importance on our way from Johnson's self-winding clock to the localization of circadian pacemakers is the work of the outstanding psychobiologist Curt P. Richter. Until the publication in 1965 of his monograph *Biological Clocks in Medicine and Psychiatry*, no one else had been equally successful in recording free-running rhythms over many months, and no one else had made attempts, in a similar straightforward and skillful manner, to localize the clock "per exclusionem" i.e. by the removal of organs and by lesioning parts of the brain. Richter summarized his findings in three sentences: "The fact that the various endocrine glands can be removed without affecting the clock demonstrates effectively that the clock cannot reside in any of them. Exhaustive experimentation has convinced us that it must be located in the brain. In our attempts to stop or alter the functioning of the clock we have produced literally hundreds of lesions in all parts of the brain of rats and have narrowed down the site to a small area in the hypothalamus but we are not satisfied as to its exact boundaries." The discovery, made about 7 years later, of the role played by the suprachiasmatic nuclei presumably would have been delayed if there had not been the pioneering work of Richter and his elegant encirclement of the clock's locus. It is for this reason that the contributors unanimously decided to dedicate this volume to Curt P. Richter in honor of his many and important accomplishments.

Andechs, July 1982 J. Aschoff · S. Daan · G.A. Groos

Aschoff J (Ed) (1981) Biological rhythms. Handbook of behavioral neurobiology, Vol 4 (F King, Edt). Plenum New York

Chovnik A (Ed) (1960) Biological Clocks. Cold Spring Harbor Symp quant Biol 25

Johnson MS (1939) Effect of continuous light on periodic spontaneous activity of white-footed mice (Peromyscus) J exp Zool 82:315-328

Richter CP (1965) Biological clocks in medicine and psychiatry. Charles C Thomas Publ, Springfield, Illinois

Suda M, Hayaishi O, Nakagawa H (Eds) (1979) Biological rhythms and their central mechanism. Elsevier North Holland Biomedical Press, Amsterdam New York Oxford

Contents

Contributors

You will find the addresses at the beginning of the respective contribution

1.1 The Search for Principles of Physiological Organization in Vertebrate Circadian Systems

M. Menaker[1]

It is only within the last 15 years that a few favorable experimental situations have been identified that have made it possible to study the physiology of circadian systems in some vertebrates (Aschoff 1981). Given the complexity of the systems under study, the long time constants of experimentation, and the relatively small number of scientists engaged in the work, it is not surprising that I must address the *search* for principles rather than the principles themselves. Insofar as documented principles exist, they are disturbingly vague. However, in discussing them it becomes clear that at least we know where we should be looking further. A few principles, proto-principles and pseudo-principles are discussed below.

1 Vertebrate Circadian Systems Are Organized in Physiological Hierarchies

It is difficult to imagine that anything else could be the case, but direct evidence has only recently become available. Much of this volume is devoted to studies that support this principle either by direct experimental test or by strong inference. Indeed, the specific demonstration of particular circadian hierarchies in several vertebrate species is one of the most exciting aspects of recent research in the field.

When the principle of physiological hierarchies is taken together with the fact that large numbers of physiological and biochemical processes express circadian rhythmicity, a potentially useful classification emerges. It seems likely that the many circadian processes that can be measured will fall into one of three categories.

1.1 Slaves

Those that are driven by a hierarchically superior central rhythm generator (CRG) or pacemaker onto which they have little or no feedback. I will call them slave processes to distinguish them from Pittendrigh's slave oscillators (Pittendrigh, 1981). Probable examples include circadian rhythms of liver enzyme activity, heart rate, skin cell mytosis, etc. For many of these rhythms, feedback to a CRG seems unlikely on physiological grounds (e.g., strong coupling pathways are not known to exist between mitotic

[1] Science III, Institute of Neuroscience, University of Oregon, Eugene, Oregon 97405, USA

Vertebrate Circadian Systems
Ed. by J. Aschoff, S. Daan, and G. Groos

activity in the skin and central nervous system function). In fact, the well-known resistance of circadian pacemakers to fluctuations in physiological variables such as metabolic rate is demanded in a system that must keep accurate time in spite of the stresses to which its „owner" is subjected. Many, perhaps most, overt rhythms are likely to be in this category and it is important to ask what can and cannot be learned from their study.

1.1.1 Markers of the State of the System

Those that can be easily recorded make good markers of the period of the circadian system or its phase relative to an external temporal referent. Unless the dynamics of their coupling to the CRG are well understood, they will function as markers better when the entire system is in steady state than during transients.

1.1.2 Indicators of Internal Desynchronization

If more than one such rhythm can be recorded simultaneously, the phase relationships among them may indicate the state of internal temporal organization. One would expect to measure abnormal phase relationships among some slaves for several cycles after large phase shifts of the entire system, such as those which produce jet-lag.

1.1.3 Not a Pathway to the Pacemaker Mechanism

If a physiological or biochemical process is a circadian slave, knowledge of that fact and a reasonable degree of information concerning the form of its oscillation will be critical to an understanding of its specific role in the system of which it is a part. Thus, endocrinologists need complete descriptions of slave rhythms in the endocrine system and physiologists must understand slave metabolic rhythms in order to understand the overall organization of energy utilization. *On the other hand, because this is the level at which most physiological research on circadian rhythms is done, it is important to recognize that characterization of slave processes is unlikely to provide insight into the mechanism of the CRG.* Indiscriminate attempts to render slave processes in concrete physiological detail seem likely to provide many facts but little understanding. In fact, descriptive research of this kind is of little direct interest to the biologist concerned with the physiology of circadian organization per se, precisely because each of the large number of slave rhythms must by definition be connected to the CRG by a unidirectional pathway of indeterminate length. There must be many such pathways with little in common except great length and complexity. The likelihood of either uncovering important generalizations about them or working back along them to the CRG appears small.

1.2 Rhythmic Processes with Significant Feedback onto the CRG (Feedback-Only Processes) and Rhythmic Processes that are Actual Components of the CRG (Component Processes)

At present, in the vertebrates there are no strong candidates for specific processes that fall into either category. Because estradiol is known to change the free-running period of locomotor rhythmicity in hamsters (Morin et al. 1977), a rhythm in its secretion might be a prime candidate for inclusion in the feedback-only category. If the *rhythmicity* of its secretion were important to its effect on the CRG, then one would expect the response to exogeneous estradiol to vary with the circadian time of its administration (i.e., there should be a phase response curve to estradiol pulses). Somewhat surprisingly, this does not appear to be the case (Fig. 1). If feedback-only processes can be found, it will be useful to study them not only for the insight they will provide into the overall organization of the circadian system, but also because their outputs will become tools with which to probe the CRG itself.

Clearly, the distinction between feedback-only and component processes is only quantitative and disappears at some arbitrary point as the importance of the feedback increases. If its feedback onto the CRG is great enough, a rhythmic process becomes a part of the CRG. It is easy to imagine increased complexity in CRG mechanisms evolving as a result of strengthening of feedback relationships (i.e., feedback-only processes becoming component processes). It may be useful to draw a somewhat arbitrary line between these two classes of process on the basis of the expected effects of their removal on the overall circadian system: removal of a feedback-only process might be expected to change the numerical value of one or more circadian parameters (e.g., phase, period, α/ρ ratio, or shape of the PRC); removal of a component process might be expected to abolish rhythmicity completely.

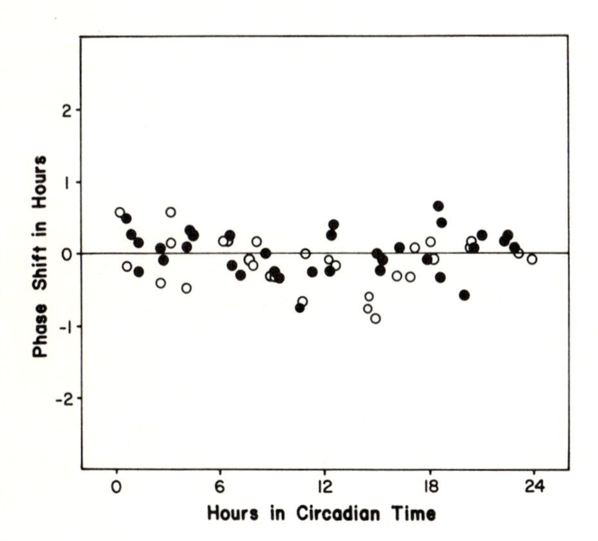

Fig. 1. Response of ovariectomized female hamsters to subcutaneous injections of estradiol benzoate (10 μg in oil). *Filled circles* indicate responses of individual animals injected with hormone; *open circles* indicate responses of control animals injected with vehicle only. Note the lack of phasic response. Higher doses (100 μg) produce slightly large phase shifts but, as with the dosage shown here, the magnitude and direction of the shift are independent of the time of administration. (Unpublished data of Takahashi, Liepe, Hudson and Menaker)

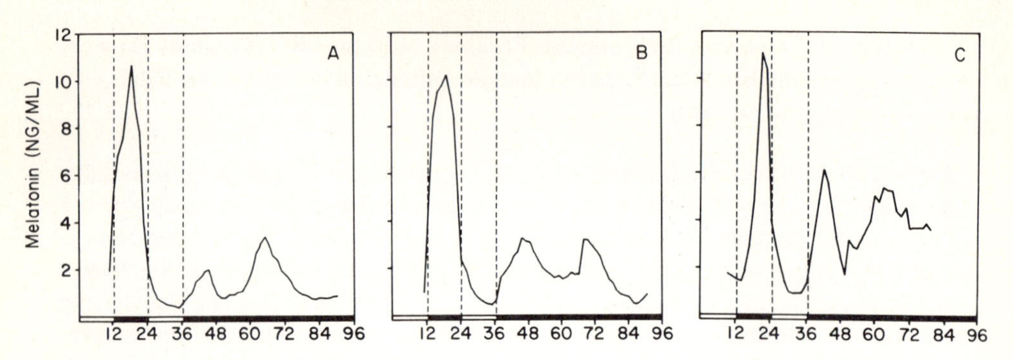

Fig. 2. Rhythms of avian pineal RIA assayable melatonin production in vitro, **A** House sparrow *(Passer domesticus)* half-gland. **B** Starling *(Sturnus vulgaris)* half-gland. **C** Chicken *(Gallus domesticus)* quarter-gland. Note the similarity of the rhythmic melatonin output of the isolated pineals of these three species. (Takahashi, 1981)

2 The Core of the Vertebrate Circadian System Probably Resides in Three Structures of Diencephalic Origin: the Pineal Organ, the Suprachiasmatic Nuclei of the Hypothalamus (SCN), and the Retinae

The role played by each of these structures varies somewhat from one vertebrate species to another. This produces variability in important physiological detail that is puzzling and challenging. Before we can claim to understand vertebrate circadian organization, we must be able to explain the fact that pinealectomy abolishes circadian rhythmicity in most species of passerine birds (Takahashi and Menaker 1979) [starlings are a partial exception (Gwinner 1978)]; affects it very little or not at all in chickens or Japanese quail (Simpson and Follett 1981); and has a variety of effects among lizards (Underwood 1977, 1981). It seems likely that as the effects of SCN lesions are further explored, especially in non-rodent mammals and in non-mammalian vertebrates, similar variability will be uncovered (Reppert et al. 1981).

The evident physiological variability among vertebrates should not distract us from an appreciation of an emerging overall pattern. It has recently begun to seem as if the same structures, perhaps doing the same kind of thing, have simply been coupled together in different ways in the different vertebrate groups. An example can be drawn from the circadian behavior of isolated avian pineals. The rhythms of melatonin output from isolated pineals of house sparrows, starlings, and chickens behave similarly in organ culture (Takahashi et al. 1980) (Fig. 2), although these three species represent the entire range of pineal involvement in avian circadian systems. In vitro, the pineals of all three species entrain to light-dark cycles and free-run in constant darkness at reduced amplitudes for from two to four cycles before becoming arrhythmic. The similarities in the in vitro behavior of their pineals suggest that the differences among the three species probably do not lie at the level of the pineal itself. Pinealectomy fails to abolish circadian rhythmicity in chickens and Japanese quail because the pineal's circadian output has not been incorporated into the CRG (i.e., it is a slave or a weakly coupled feedback-only process in the terms discussed above) and not because the pineal is incapable of circadian oscillation (it clearly *is* capable of oscillation in chickens, as the

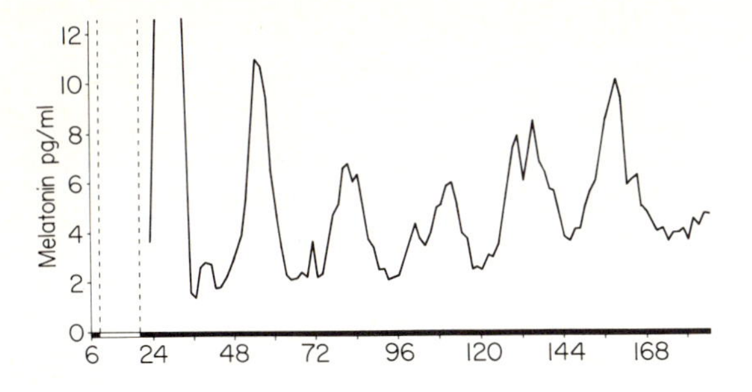

Fig. 3. Free-running rhythm in DD of RIA-assayable melatonin from the isolated pineal of the iguanid lizard *Anolis carolinensis* (male) cultured at 30 ° C. The culture medium and flow-through apparatus used were identical to those employed by Takahashi et al. 1981. Six peaks were recorded; the period of the rhythm is clearly circadian; it varies between 25 and 27 h and appears to be shortening. (Unpublished data of Menaker and Wisner)

in vitro data demonstrate). The sparrow pineal, which may have no greater inherent capacity for oscillation than does that of the chicken, has been incorporated into the CRG to the extent that its removal abolishes overt rhythmicity. In starlings, the pineal occupies an intermediate position (strongly coupled feedback-only or a weak component process).

We have discovered recently that the pineal of the lizard *Anolis carolinensis* also oscillates in vitro, more robustly in fact than do any of the avian pineals so far examined (Fig. 3). We do not yet know what role the pineal of this lizard plays in its overall circadian organization but it is clear from Underwood's work (Underwood 1977, 1981) that the variability in pineal role among lizard species is as great as it is among birds. As in birds, the variability among lizards may result from differences in strength of feedback and in level of incorporation into the CRG rather than from inherent pineal differences.

Everything that we know about the mammalian pineal (most of it derived from work with rodents) suggests that it has lost the capacity for independent oscillation as well as its position in or near the CRG (see Chap. 5 in Suda et al. 1979). The relegation of the mammalian pineal to slave status and its loss of the capacity for independent oscillation may well have been the most profound changes in circadian organization to have occurred during vertebrate evolution. They certainly merit further comparative study.

3 Hormonal Links Among Components of Vertebrate Circadian Systems Employ Melatonin or Related Compounds[2]

Melatonin is synthesized rhythmically by the pineals and retinae of many vertebrates. Exogeneously administered melatonin has large effects on the free-running period of

---

[2] Recent melatonin measurements have been made almost exclusively by radioimmunoassay (RIA) and are subject to the difficulties as well as the advantages of that technique. As an example, our sparrow, starling and lizard data shown in Figs. 2 and 3 are expressed as „RIA-assayable melatonin" because validation of the assay for these tissues is as yet incomplete. We have carried the validation procedure far enough to be certain that at least the first peak from both sparrow and lizard (but not chicken) pineals contains, in addition to melatonin, large amounts of a non-melatonin substance(s) of pineal origin that cross-reacts in the melatonin RIA (Cassone and Menaker, unpublished data). In the speculative discussion that follows, „melatonin" is used as shorthand for „Melatonin-like compounds synthesized rhythmically by the retina and pineal gland".

circadian locomotor rhythms in some birds and lizards. Daily melatonin injections entrain pinealectomized starlings. Melatonin, given by injection or by implant, has complex effects on the reproductive systems of mammals, especially those with photoperiodically regulated (and therefore circadian clock-dependent) seasonal reproductive cycles. The circadian rhythm of melatonin level, whether in blood, pineal organ or retinae, always bears the same general phase relationship to an environmental light cycle regardless of the phase of the rest of the organism's circadian system, and is the only circadian rhythm (with the exception of the phase response curve) of which that is true. In both diurnal and nocturnal vertebrates, high levels of melatonin occur only at night. (References for the above statements can be found in Sect. 4 of this volume.) Because it is produced in photoreceptive organs in response to environmental light conditions, melatonin is an hormonal analogue to light-evoked electrical activity in the nervous system with, however, the important difference that its oscillation persists in the absence of light. Melatonin is available within the organism as a symbolic light cycle – the subjective day is characterized by low levels of melatonin, the subjective night by high levels – and may thus have become an internal temporal bench-mark in general use among vertebrates.

It may be instructive to speculate on the evolution of this special role of melatonin. I would like to suggest that melatonin's original function was to drive the daily rhythms of photoreceptor and/or pigment cell movement hat underlie light and dark adaptation in lower vertebrate retinae (Levinson and Burnside, 1981; Pang and Yew, 1979). In order for melatonin to fill this role, its synthesis would have to be coupled to the light cycle; synthesized at night retinal melatonin would cause movement to the dark-adapted state; its absence in the light would allow these movements to be reversed. Since melatonin is known to bind to microtubules and inhibit their assembly in several systems (Banerjee and Margulis 1973, Cardinali and Freire 1975), it is reasonable to imagine that it might regulate cellular and intracellular movements. Melatonin synthesis in photoreceptive structures may have become circadian as a result of selection pressure for phase control of the adaptation process (to anticipate environmental changes in light level). The circadian rhythm of melatonin synthesis would then be available to time other events within the retina, such as photoreceptor disc shedding, that might operate more efficiently temporally restricted.

In early pineal structures, melatonin may also have been involved in photoreceptor adaptations to the light cycle but, in addition, a new evolutionary direction was taken. Circulatory system adjustments were made and the melatonin-synthesizing capacity of the structures was amplified, enabling a mechanism evolved for local control to become a true endocrine process. Perhaps one of the first functions served by the new hormone was the regulation of pigment granule aggregation in the melanocytes of the skin in adaptive phase relationship to the environmental light cycle. It is not difficult to imagine that in this system, as in the retina, the major site of melatonin action might be the microtubules. By acting on microtubules, melatonin could in principle regulate not only retinomotor movements and pigment granule aggregation but also cell division, axoplasmic flow and neurosecretion.

At this hypothetical stage in the evolution of the pineal complex, there was available to natural selection a rhythmically circulating hormone phased by the environmental light cycle, with the capacity to regulate a large number of important physiological processes through a common pathway. Given such a „constitutional opportunity", it would hardly be surprising if the pineal became involved in the control of a wide variety of adaptive responses to daily and annual environmental cycles. Because such responses must be keyed to the specialized niche occupied by each species and because selection on such adaptations would be intense, one might expect the rapid evolution of pineal involvement in diverse processes related to each other only by a common necessity for adaptive phase relationship to some environmental cycle. Such an evolutionary history might well have produced the heterogeneity in pineal function that has been uncovered by recent research (Reiter 1981).

As long as the synthesis of melatonin were under the direct control of the external light cycle, its usefulness as a signal would be restricted by its lack of phase flexibility – in such an arrangement, the phase of the melatonin cycle is inexorably locked to the phase of the light cycle. One can imagine three evolutionary changes from this primitive state, each of which would introduce new degrees of flexibility in phase control: the first is coupling of the synthesis to a circadian clock (phase control achieved in this way is extensively discussed in Pittendrigh 1981); the second is phase modulation by an additional input that is not directly driven by the light cycle [perhaps this is the function of the neural (sympathetic) input to the pineals of lower vertebrates]; finally, maximal phase control could be achieved if both responsiveness to light and circadian rhythmicity were physically separated from melatonin synthesis but remained linked to it through neural structures capable of complex integration.

In fact, mammals have evolved to this third stage; they have lost pineal photoreception and shifted the locus of circadian control of the pineal away from the gland itself. The independence that this organization confers is illustrated by Reiter's new data (Reiter 1982) on melatonin rhythmicity in the pineal of the Syrian hamster on different photoperiods (Fig. 4). Note that the phase of the rhythm is related not to light or darkness but to a specific feature of the LD cycle – dawn. Furthermore, neither the phase, the amplitude nor the peak width („α/ρ ratio") is strongly dependent on photoperiod. This system would serve admirably as *part* of a photoperiodic time measurement device. To measure day length, it would have to be coupled to another system that was phase-related to dusk. Perhaps organisms find it more efficient or more reliable to use the phase angle between two rhythmic processes than the duration of a single process to measure the passage of time (see Pittendrigh 1981, for a related discussion). Mammals do use their pineals as part of a photoperiod-measuring system while birds, which can measure photoperiod with equal precision, do not use their pineals for this purpose. Perhaps the avian pineal, when employed to produce a general internal signal indicating the alternation of day and night, is unable to support the finer discriminations involved in day-length measurements.

4 Photoreceptive Input to Vertebrate Circadian Systems is Specialized, and Distinct from that Involved in Vision

Of the principles discussed here, this is the least vague and best documented (it is also true for most invertebrates). Since the discovery of extraretinal input to the circadian system of the house sparrow (Menaker 1968), extensive work by many investigators has uncovered similar phenomena in virtually every vertebrate group examined except

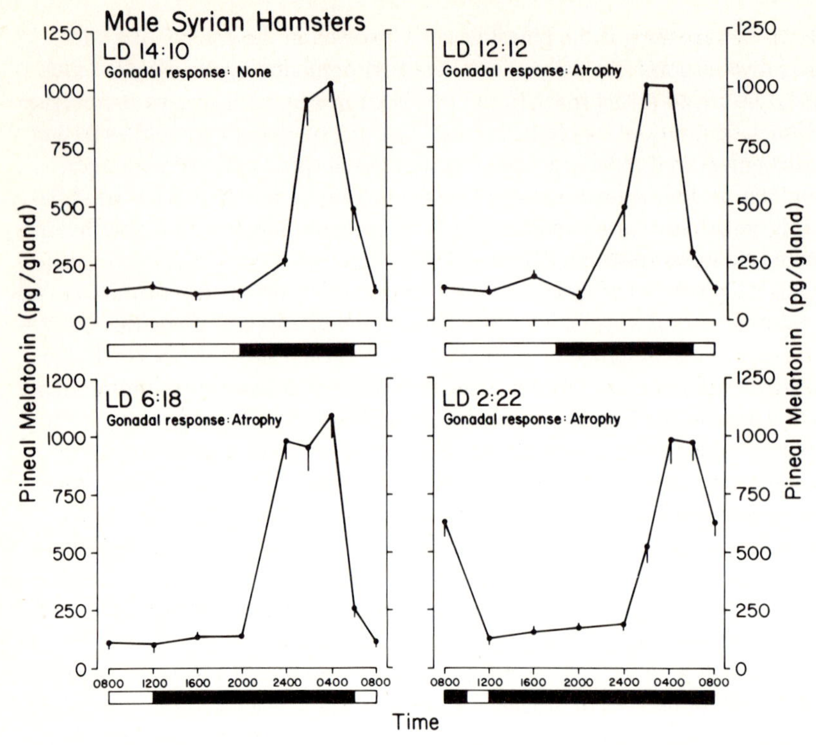

Fig. 4. Rhythms of pineal melatonin content in hamsters held on different photoperiods. Note that the phase of the peak varies little with respect to the dark-to-light transition (dawn) (Reiter, 1982)

the mammals (Underwood and Menaker 1976). Non-mammalian vertebrates have multiple photic inputs to their clocks: the retinae, pineal and parapineal structures and hypothalamus may all be involved in circadian photoreception and there is reason to believe that they are not simply performing redundant roles (McMillan et al. 1975). Reflection on this large and consistent literature suggests that extraretinal photoreceptive input may be a functional necessity. Perhaps the circadian system needs to have its information about the environmental light cycle in a form that is not consistent with the needs of an image-processing visual system. But if that were true, what of the mammals which repeatedly have been shown to use only their eyes (Zucker and Nelson 1981)?

Although mammals do not have extraretinal photic input to their clocks, it has been known for some time that the major pathway for entrainment of their circadian rhythms by light cycles is the specialized, non-visual retino-hypothalamic tract which terminates in the suprachiasmatic nuclei (Moore 1979). This anatomical fact in itself strongly suggests that the mammalian clock, like the clocks of other vertebrates, requires non-visual light information. Recently, we have investigated the physiology of clock-related photoreception in hamsters and mice with surprising results.[3]

[3] A series of experiments has been performed over the past 18 months by (in alphabetical order) L. Bauman, P. DeCoursey, D. Hudson, M. Menaker and J. Takahashi using the phase shift produced by a single light pulse as a measure of the effect of light on the circadian system. The extensive data are summarized very briefly here. Several manuscripts are in preparation.

In these nocturnal rodents, 97 %-98 % of the retinal photoreceptors are rods (Blanks et al. 1974). The remainder are scattered cones about which we know very little (e.g., it is not yet known what photopigment they contain). It is therefore not unexpected that our data indicate that the action spectrum for phase-shifting the hamster circadian clock is similar to the absorption spectrum for rhodopsin (visual pigment 502). More surprisingly, the threshold fluence required to produce a phase shift is six to seven orders of magnitude higher than the threshold for human scotopic vision. This makes a certain amount of ecological sense; the high threshold would probably prevent phase-shifting by moonlight in the wild but is physiologically unusual since the animal is held in complete darkness for 7 days before the light pulse and is therefore maximally dark-adapted.

In further experiments, we measured the temporal range over which intensity x duration reciprocity holds [i.e., if the product of intensity (I) of the light pulse and its duration (D) = K, and K is held constant at a value that produces a half-maximal phase shift when D is short, over what range of durations does K produce the same phase shift?]. For visual responses, reciprocity holds for durations that are measured in milliseconds; at longer durations, adaptation to the stimulus occurs and drastically modifies the response. For phase-shifting of the circadian system of the hamster, reciprocity holds for durations of at least 45 min. (attempts to use longer pulses were often thwarted because the animal went to sleep in the light-exposure chamber). It seems unlikely that a system with this property could provide a reasonable representation of a series of rapidly changing images. However, since this photoreceptive system appears to count photons, it might provide a very accurate representation of the photic environment during the hours around dawn or dusk.

It appears that the mammalian circadian system receives its light information from a sensory system that is neuroanatomically and physiologically non-visual. Is this information transduced by the normal visual photoreceptors (in the case of nocturnal rodents, the rods) or might there be a subset of retinal photoreceptors specialized for clock photoreception in mammals, as there are extraretinal photoreceptors in the other vertebrates?

Mice, homozygous for the mutation *rd,* lose first their rods and then their cones during post-natal life (Carter-Dawson et al. 1978). By 108 days, the age at which a single light pulse was delivered to the *rd* mouse whose record is shown in Fig. 5, *all* the rod outer and inner segments have degenerated and 90 % of the cones are also gone. Although we have not yet compared the phase shift produced in *rd* mice with that produced in their normal litter-mates by the same light pulse, the size of the phase shift that we have obtained from the *rd* animals (of which Fig. 5 is an example) is within the range reported by Daan and Pittendrigh (1976) for normal mice of this strain (C57 blacks).[4] In normal mice, about 3 % of the retinal photoreceptors are cones. Phase shifts within the normal range can be produced in mutant mice that have lost all rods and retain only 10 % of their cones (thus 0.3 % of the normal photoreceptor population).

[4] Our results with C57BL/6J *le rd/le rd* mice are consistent with the results reported by Ebihara and Tsuji (1980) in the only paper in the literature dealing with circadian rhythmicity in this mutant, although our interpretation is at variance with theirs. They used a different strain of mouse, the C3H bearing the same *rd* mutation. They showed that C3H mice with degenerate retinae entrained to light cycles but emphasized that these mice were less sensitive to light than were the normal controls which, however, were of a different strain (C57 blacks).

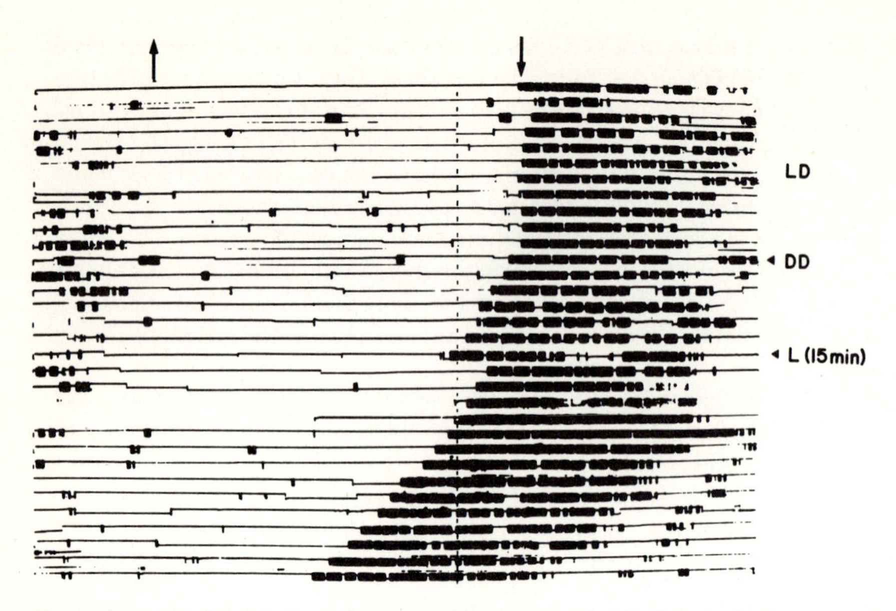

Fig. 5. Phase shift (2-h delay), produced by a 15-min pulse of white light, in the locomotor rhythm of a mouse with a degenerate retina. The mouse, a C57BL/6J *le rd* animal, was 108 days of age on the day of the pulse. *Arrows* at the top of the figure indicate the beginning (↑) and end (↓) of the light portion of the LD cycle. *First arrow on the right* indicates the day of transition from LD to DD; *second arrow* indicates the day on which the light pulse was administered 4 h after activity onset. (Unpublished data of Hudson and Menaker)

These very preliminary results suggest strongly that there is a subset of retinal photoreceptors (either cones or an as yet unidentified photoreceptive cell type) that are specialized for clock photoreception.

The unusual physiological properties of the retinal photoreceptors that supply photic information to the circadian system may reflect their evolutionary history as well as the immediate requirements of the system they serve. The state of a process is often best monitored by a mechanism not itself subject to that process. If the clock-related subset of retinal receptors originated to provide light information to the melatonin-synthesizing machinery in the retina and if, as I have suggested above, that machinery initially controlled light and dark adaption in other retinal photoreceptors, then one might expect the clock-related receptors to be resistant to adaptation. Our reciprocity measurements indicate that that is the case, at least in the hamster retina.

Acknowledgement The experimental work reported here was partially supported by NIH grants HD 13162 and AM 26972 and by a grant from the Medical Research Foundation of Oregon.

References

Aschoff J (ed) (1981) Handbook of behavioral neurobiology, 4. Biological rhythms. Plenum Press, New York London

Blanks JC, Adinolfi AM, Lolley RN (1974) Photoreceptor degeneration and synaptogenesis in retinal-degenerative *(rd)* mice. J Comp Neurol 156:95-106

Banerjee S, Margulis L (1973) Mitotic arrest by melatonin. Exp Cell Res 78:314-318

Cardinali DP, Freire F (1975) Melatonin effects on brain. Interactions with microtubule protein, inhibition of fast axoplasmic flow and induction of crystalloid and tubular formations in the hypothalamus. Molec and Cell Endocrinol 2:317-330

Carter-Dawson LD, LaVail MM, Sidman RL (1978) Differential effect of the *rd* mutation on rods and cones in the mouse retina. Invest Ophthalmol Vis Sci 17:489-498

Daan S, Pittendrigh CS (1976) A functional analysis of circadian pacemakers in nocturnal rodents. II. The variability of phase response curves. J Comp Physiol 106:253-266

Ebihara S, Tsuji K (1980) Entrainment of the circadian activity rhythm to the light cycle: Effective light intensity for a Zeitgeber in the retinal degenerate C3H mouse and the normal C57BL mouse. Physiol Behav 24:523-527

Gwinner E (1978) Effects of pinealectomy on circadian locomotor activity rhythms in European starlings, *Sturnus vulgaris.* J Comp Physiol 126:123-129

Levinson G, Burnside B (1981) Circadian rhythms in teleost retinomotor movements. Invest Ophthalmol Vis Sci 20:294-302

McMillan JP, Keatts HC, Menaker M (1975) On the role of eyes and brain phtoreceptors in the sparrow: Arrhythmicity in constant light. J Comp Physiol 102:263-268

Menaker M (1968) Extraretinal light perception in the sparrow. I. Entrainment of the biological clock. Proc Nat Acad Sci USA 59:414-421

Moore RY (1979) The retinohypothalamic tract, suprachiasmatic hypothalamic nucleus and central neural mechanisms of circadian rhythm regulation. In: Suda M, Hayaishi O, Nakagawa, H (eds) Biological rhythms and their central mechanism: A Naito Foundation symposium. Elsevier/North-Holland Biomedical Press, Amsterdam

Morin LP, Fitzgerald KM, Zucker I (1977) Estradiol shortens the period of hamster circadian rhythms. Science 196:305-307

Pang SF, Yew DT (1979) Pigment aggregation by melatonin in the retinal pigment epithelium and choroid of guinea pig. Experientia 35:231-233

Pittendrigh CS (1981) Circadian organization and the photoperiodic phenomena. In: Follett BK, Follett DE (eds) Biological clocks in seasonal reproductive cycles. John Wright and Sons Ltd., Bristol, pp 1-35

Reiter RJ (ed) (1981) The pineal gland. I. Anatomy and biochemistry. CRC Press Inc., Boca Raton, Florida

Reiter RJ (1982) Chronobiological aspects of the mammalian pineal gland. In: Mayersbach HV, Scheving LE, Pauly JE (eds) Biological rhythms in structure and function. Alan R Liss, New York

Reppert FM, Perlow MJ, Ungerleider LG, Mishkin M, Tamarkin L, Orloff DG, Hoffman HJ, Klein DC (1981) Effects of damage to the suprachiasmatic area of the anterior hypothalamus on the daily melatonin and cortisol rhythms in the rhesus moneay. Neuroscience 1:1414-1425

Simpson SM, Follett BK (1981) Pineal and hypothalamic pacemakers: Their role in regulating circadian rhythmicity in Japanese quail. J Comp Physiol 144:381-389

Suda M. Hayaishi O, Nakagawa H (eds) (1979) Biological rhythms and their central mechanism: A Naito Foundation symposium. Elsevier/North-Holland Biomedical Press, Amsterdam

Takahashi JS (1981) Neural and endocrine regulation of avian circadian systems. Ph.D. dissertation, Department of Biology and Institute of Neuroscience, University of Oregon

Takahashi JS, Menaker M (1979) Brain mechanism in avian circadian systems. In: Suda M, Hayaishi O, Nakagawa, H (eds) Biological rhythms and their central mechanism: A Naito Foundation symposium. Elsevier/North-Holland Biomedical Press, Amsterdam, pp 95-109

Takahashi JS, Hamm H, Menaker M (1980) Circadian rhythms of melatonin release from individual superfused chicken pineal glands in vitro. Proc Nat Acad Sci USA 77:2319-2322

Underwood H (1977) Circadian organization in lizards: The role of the pineal organ. Science 195: 587-589

Underwood H (1981) Circadian clocks in lizards: Photoreception, physiology and photoperiodic time measurement. In: Follett BK, Follett, DE (eds) Biological clocks in seasonal reproductive cycles. John Wright and Sons Ltd., Bristol, pp 137-152

Underwood H, Menaker M (1976) Extraretinal photoreception in lizards. Photochem and Photobiol 23:227-243

Nelson RJ, Zucker I (1981) Absence of extraocular photoreception in diurnal and nocturnal rodents exposed to direct sunlight. Comp Biochem Physiol 69A:145-148

1.2 Zeitgebers, Entrainment, and Masking: Some Unsettled Questions

J. Aschoff[1], S. Daan[2], and K.-I. Honma[3]

1 Introduction

To be of full functional significance for the organism, circadian rhythms have to be synchronized to the 24-h day. This is done by response to periodic signals from the environment, the zeitgebers (Pittendrigh 1981). The entrained state is characterized by a stable phase angle difference ψ between rhythm and zeitgeber; sign and amount of ψ depend on the period τ of the rhythm (as measured in constant conditions) and on the period T of the zeitgeber. Furthermore, circadian systems can be entrained to periods deviating from 24 h only within certain limits (Aschoff and Pohl 1978).

It has been emphasized that entrainment is a concept specifically indicating the synchronization of self-sustaining oscillations and applicable to a biological periodicity only if a free-running rhythm can be demonstrated in that system. Such a restrictive definition has its drawbacks. The range of external conditions within which a free-running circadian rhythm can be expressed is often limited. Two conditions outside this range, e.g., two intensities of illumination, each of which causes arrhythmia of the circadian system if given constantly, can result in „entrainment" if given in periodic alternation (Wever 1980). The question then arises to what extent a circadian rhythm continues to have the properties of a self-sustaining oscillation when entrained by a zeitgeber. The criterion that ψ depends on T is of little use because this rule applies to damped as well as to self-sustaining oscillations. In pinealectomized sparrows self-sustainment of circadian rhythms is lost in constant conditions, yet their activity pattern in LD cycles has been called „entrained" (Gaston and Menaker 1968).

It is thus a point of debate where and when the terms „zeitgebers" and „entrainment" become inappropriate. We should further be aware that periodic environmental signals may affect circadian systems in a variety of ways, and may in fact influence parts of these systems differentially. In this essay, we attempt to summarize such effects and to outline some of the problems facing their physiological analysis.

2 Zeitgeber Modalities

There are many ways in which environmental periodicity can keep circadian systems in synchrony with the earth's rotation. For nearly all organisms, the light-dark cycle (LD)

[1] Max-Planck-Institut für Verhaltensphysiologie, D-8138 Andechs/FRG
[2] Zoological Laboratory, Groningen University, Haren, Netherlands
[3] Department of Physiology, Hokkaido University School of Medicine, Sapporo 060, Japan

Vertebrate Circadian Systems
Ed. by J. Aschoff, S. Daan, and G. Groos

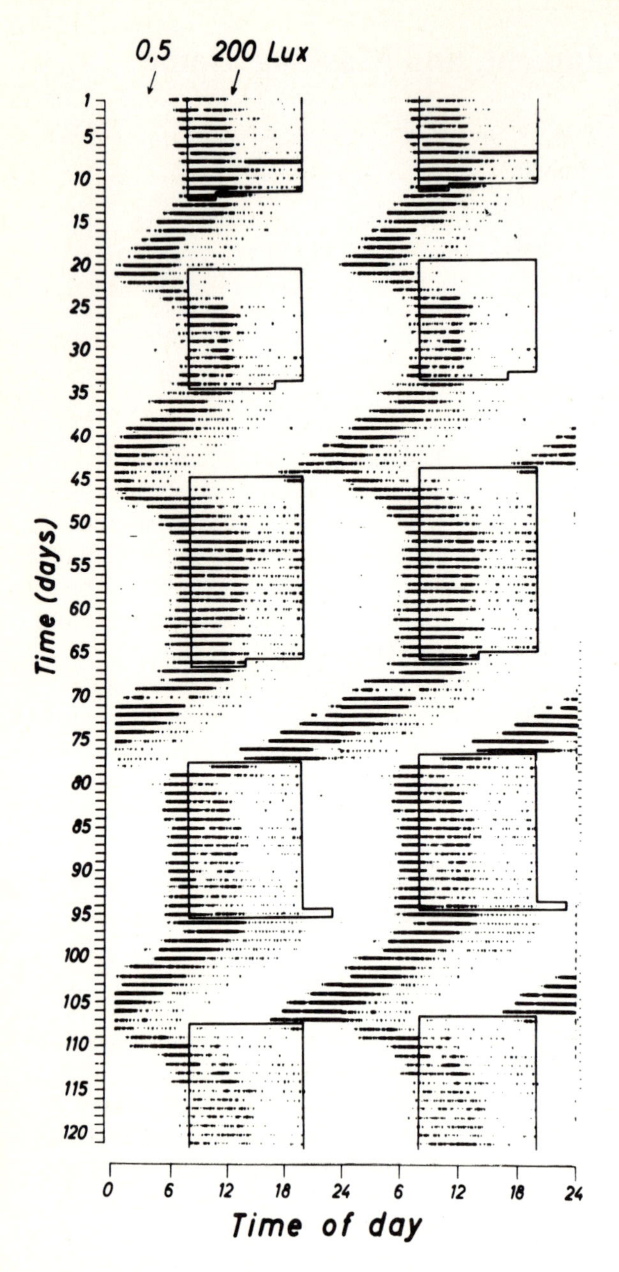

Fig. 1. Circadian activity rhythm of a chaffinch, *Fringilla coelebs*, alternatively kept in a light-dark cycle and in constant dim illumination. Original 24-h record plotted twice

is the most powerful zeitgeber. As an example, Fig. 1 shows the locomotor activity rhythm of a chaffinch with repeated transitions from a free-running state to entrainment by LD. Each introduction of the zeitgeber results in entrainment, either via delay phase shifts (days 20, 45 and 108) or by advance shifts (day 78).

A cycle of low and high temperature easily entrains the rhythm of poikilothermic animals (Hoffmann 1969a), but can also be effective in birds and mammals. Figure 2

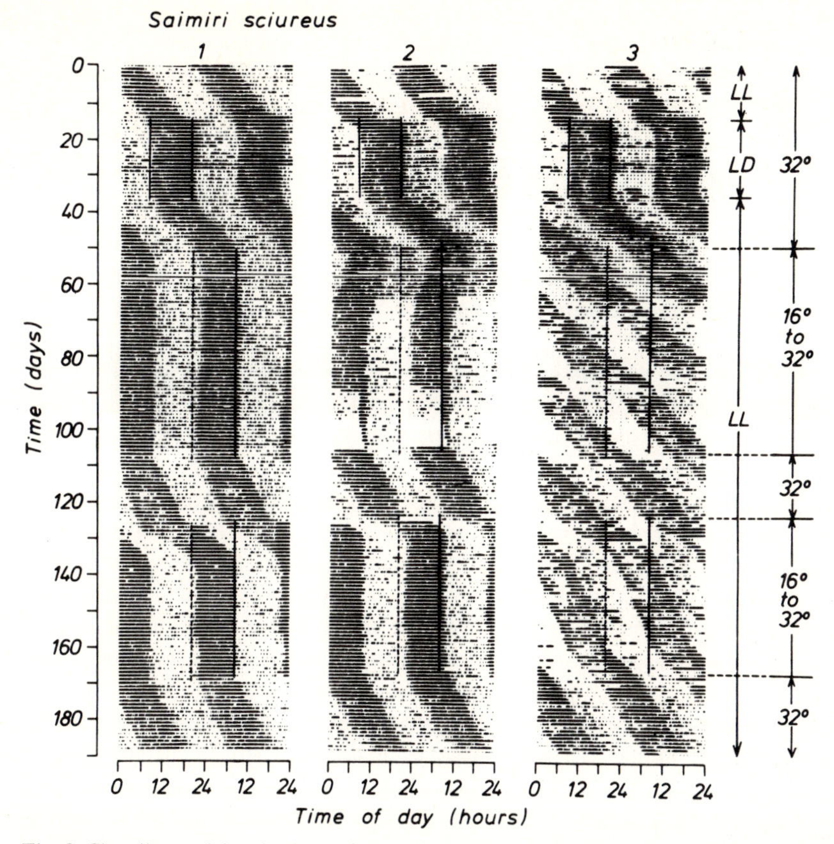

Fig. 2. Circadian activity rhythms of three squirrel monkeys, kept in conditions of constant dim light *LL* and temperature and alternatively exposed to cycles of light and dark *LD)* or of ambient temperature (16° to 32° C) (*dashed line* onset of decreasing temperatures). Original 24-h records plotted twice. (Tokura and Aschoff 1981)

presents activity records of three squirrel monkeys which were alternatively kept in constant light and temperature or exposed to either LD or a temperature cycle. Animals 1 and 2 were equally well entrained by temperature as by light; in animal 3, the temperature cycle failed to entrain the rhythm but still exerted an influence on it as shown by the occurrence of „relative coordination", i.e., slight period modulation of the free-running rhythm. In other species, temperature cycles failed to entrain the rhythm but only „masked" activity, e.g., in the flying squirrel (DeCoursey 1960) and in the wood mouse (cf. Fig. 3B).

Less well documented, but not necessarily of less relevance, is entrainment by social cues. Some avian species have been shown to be entrainable by species-specific song cycles (Gwinner 1966). Social entrainment is also indicated by the synchrony in free-running rhythms between the members of a beaver family living under snow-covered ice (Bovet and Oertly 1974). Recently Marimuthu and coworkers (1981) have shown that the free-running activity rhythm of a singly caged bat, recorded in the darkness of an otherwise uninhabited cave, becomes entrained to 24 h when the animal is surrounded by free-moving conspecifics.

For the effectiveness of other environmental factors as potential zeitgebers, the evidence is still scarce. To be mentioned are cycles in air pressure (Hayden and Lindberg 1969), in an electric a.c. field (Wever 1979), and in food availability (cf. Sect. 4).

3 The Problem of Masking

A zeitgeber often affects a circadian rhythm in at least two different modes: it entrains the rhythm by controlling the phase of the pacemaker's oscillation, and it may influence the variable measured (the overt rhythm) in a more direct way with or without a relationship to the process of entrainment. In Fig. 1 the records of day 45 and 108 provide examples of „negative masking": the activity is suppressed for several hours after the light has been turned off. In another record (Fig. 3A) positive masking is indicated by the daily bursts of activity elicited by daily light pulses of 15-min duration. The light pulses in this case are not sufficient for entrainment, but they reach the pacemaker as indicated by the relative coordination between rhythm and pulse. It cannot be concluded that the signal responsible for relative coordination and for masking follow the same pathway. It is conceivable that the light signals reach the pacemaker via the retino-hypothalamic tract (cf. Part 2, this Vol.) but at the same time bypasses it and produces masking by reaching directly a centre controlling locomotor activity. It is also possible that the pacemaker is influenced indirectly by feedback from locomotor activity.

Figure 3B demonstrates negative masking of activity in a wood mouse by temperature. Exposure of the animal to a temperature cycle did not result in entrainment; however, after a phase shift of the temperature cycle by 9 h, activity was suppressed as long as the activity time coincided with decreasing and low temperatures, and activity was resumed as soon as the ambient temperature reached a certain threshold. The sharp onsets of activity which in the record run parallel to the line indicating the daily temperature maxima could easily be mistaken as representing entrainment.

There are then good reasons not to confuse entrainment with masking, and to separate two kinds of masking. Masking by signals which bypass the pacemaker and hence do not result in relative coordination of a free-running rhythm is illustrated for blindes squirrel monkeys by their reaction to the daily attendances of the caretaker (cf. Figs. 6 and 21 in Richter 1968), as well as by the effects of restricted daily feeding on the activity rhythm of intact monkeys, free-running in dim LL (cf. Fig. 4). On the other hand, masking effects which are transmitted by the pacemaker can be expected not only to produce relative coordination (in a free-running system) but also to vary in intensity with circadian time (also in an entrained system). For masking by light pulses, such a dependence on phase has been demonstrated in brain temperature increases of chicken (Aschoff and Saint Paul 1976; cf. their Figs. 2, 5 and 6), as well as in colonic temperature of monkeys (cf. Fig. 3A in Ghap. 5.6).

4 Entrainment by Periodic Feeding

It has been claimed that a schedule in which food is offered to animals once per day for a few hours only can act as a zeitgeber. In squirrel monkeys which were kept in LL

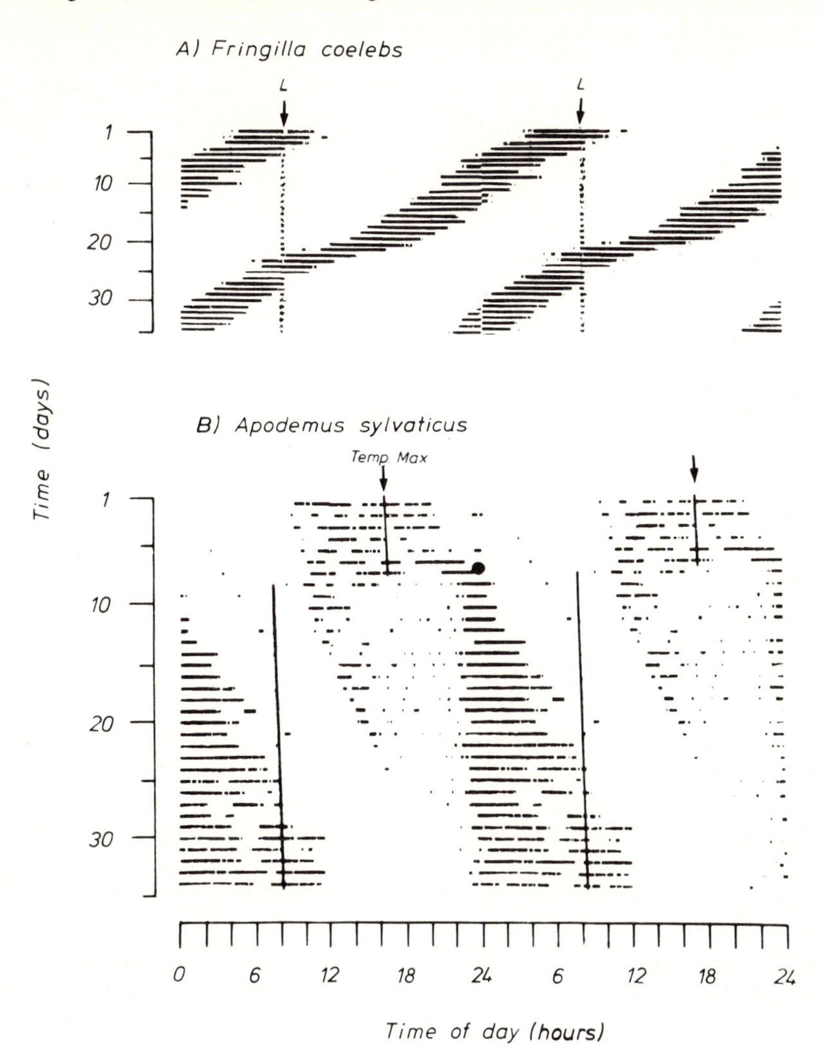

Fig. 3. Circadian activity rhythms of a chaffinch **(A)** and a wood mouse **(B)**, free-running in conditions of constant dim illumination. The bird was exposed daily to a 15-min pulse of bright light *(L)*, the mouse to a temperature cycle of 10° to 36° C (*vertical lines* maximum of temperature). Original 24-h records plotted twice. (**A** Enright 1965, **B** Hoffmann 1969b)

and had access to food daily from 8:00 to 11:00, the rhythms of water intake, body temperature and urinary excretion were synchronized to 24 h; they followed an 8-h shift of the meal time immediately, i.e., nearly without transients (Sulzman et al. 1978a). Although suggestive of entrainment, some of the data do not exclude the possibility that free-running rhythms had only been masked by the imposed feeding regimen (or were only partially entrained), an interpretation which seems to be supported by recent findings in the same species (Fig. 4; cf. the comments at the end of the following paragraph). On the basis of an extensive study on activity rhythms in rats, Gibbs (1979) concluded that „fixed interval feeding only masks the activity rhythm caused by the

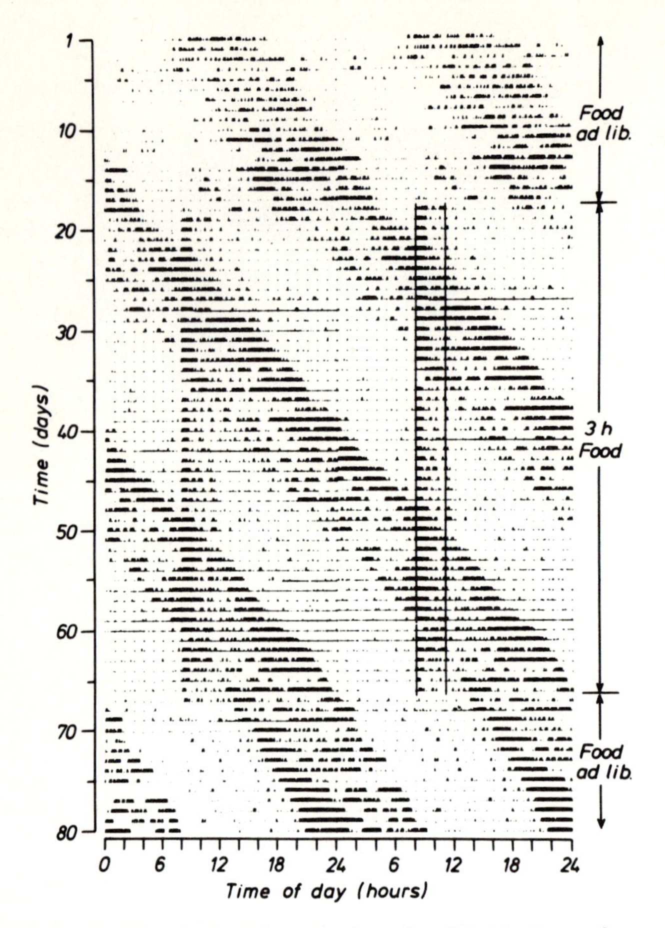

Fig. 4. Activity rhythm of a squirrel monkey *Saimiri sciureus*, free-running in constant dim illumination either with food available all the time or for three hours per day only. Double plot of the original record; restricted feeding between the two vertical lines (indicated only in one record). (Aschoff and v. Goetz, unpublished)

pacemaker". However, in the case of the rat this explanation does not seem to fully account for the facts. In Fig. 5, activity records are presented from rats which were kept in dim LL and were either fed ad lib ord had access to food daily for a few hours only. The interval between the beginning of subsequent meal times (T) was 24 h in one series of experiments (left diagram) and 23.5 h in another series (right diagram). Several phenomena can be noticed: (a) In the free-running activity rhythm neither period nor phase was altered by periodic feeding; (b) The food schedules not only produced anticipatory activity before the meals but also after a certain latency were followed by a block of activity which was in synchrony with the meal period T; (c) after the termination of the feeding schedule, the induced (anticipatory) activity component continued for several cycles within the activity band of the free-running rhythm (left diagram) or merged into it via a few delay shifts (right diagram); (d) The duration of anticipatory

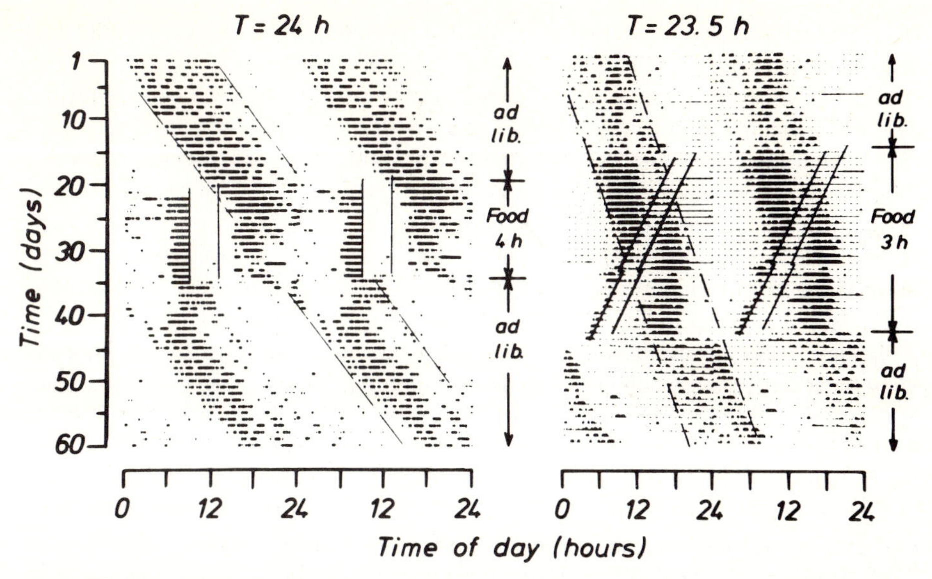

Fig. 5. Circadian activity rhythms of two rats, kept in constant dim illumination and either fed ad lib or for a few hours per day only. Meal times indicated by *two parallel solid lines. Dashed lines* through onset and end of activity drawn by eye. *T* interval between meal times. Original 24-h records plotted twice. (Honma and Aschoff 1981)

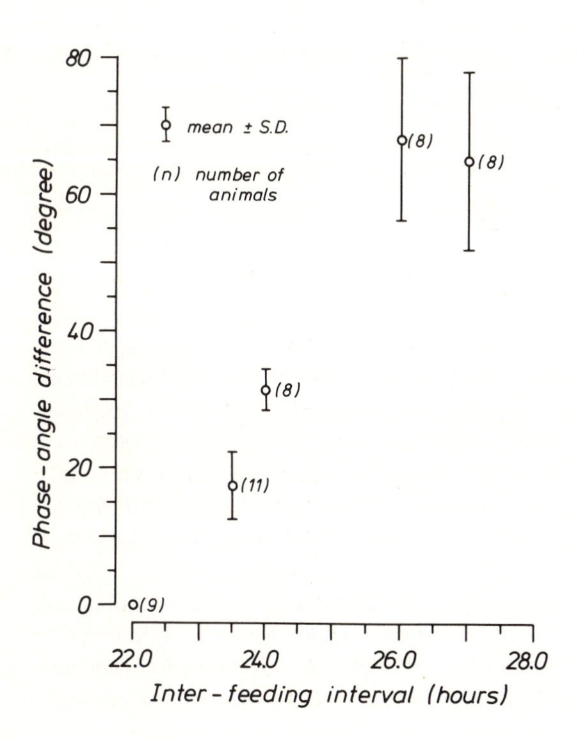

Fig. 1. Phase-angle difference between onset of anticipatory activity and onset of feeding, measured on rats during periodic feeding with varying intervals. (Aschoff, v. Goetz, Honma, unpubl.).

activity was positively correlated with T as documented in Fig. 6, which includes data from three more experiments with T = 22, 26 and 27 h, respectively.

The findings illustrated in Figs. 5 and 6 suggest a process similar to entrainment. The meal-induced component of activity cannot be due to masking because it anticipates the signal, it can outlast the feeding schedule for a few cycles, and it seems to have a limited range of entrainment (cf. Fig. 6: no entrainment with T = 22 h). Data published by other authors support the view that a circadian-like time structure is involved: in free-running rhythms, anticipatory activity could be induced by meal times with T values from 23 to 25 h, but not with T = 18 h or T = 30 h (Boulos et al. 1980), and meal-induced rhythms in plasma corticosterone and in intestinal enzymes persisted during starvation for several cycles (Kato et al. 1980, Suda and Saito 1979). Since the duration of anticipatory activity depends on T (Fig. 6), caution is needed in the interpretation of the monkey data shown in Fig. 4. It could be that the absence of anticipatory activity in this experiment represents a zero phase-angle difference and that in experiments with greater T values a positive phase-angle difference (anticipation) would occur.

All these phenomena can also be observed in animals which have been made arrhythmic by lesions of the suprachiasmatic nucleus (SCN) (Krieger et al. 1977, Stephan 1981, Boulos et al. 1980). There must be systems outside the SCN which are capable of at least damped oscillations and which under the influence of periodic meal signals show properties of an entrained circadian system. There are similarities to the entrainment of activity by light in a pinealectomized bird. On the basis of his studies in starlings, Gwinner (1978) has postulated the existence of oscillating systems which, in the arrhythmic bird, are out of phase with each other and which become synchronized by the light signals. In the intact rat, a feeding schedule may dissociate such systems from the SCN-driven rhythm. It is tempting to speculate on a relationship of these systems to the „bout-oscillators", postulated by Davis and Menaker (1980), and on the involvement of a pacemaker-like structure (cf. Chap. 3.7).

5 Complexitiy of the System

It is now well established that in (adult) mammals the entraining light signals are perceived by the eyes only, and that the retino-hypothalamic tract (RHT) is an essential but not exclusive route for entrainment; the primary optic tract (POT) contributes to entrainment, and seems in this regard to be redundant with the RHT (cf. Chaps. 3.1, 3.4). Birds have retinal as well as extraretinal photoreceptors, both cooperating in entrainment, in affecting the period of the free-running rhythm, and in producting arrhythmicity by bright light (Takahashi and Menaker 1979). There is further good evidence that the SCN plays a major role in generating circadian rhythms in mammals, with only minor influences from the pineal organ; in birds, both the SCN and the pineal seem to be involved in the generation of circadian rhythms (cf. Parts 3 and 4, this Vol.).

In contrast to light, nearly nothing is known about the perception and transmission of other zeitgeber signals. With regard to temperature cycles, changes in deep body temperature could be involved in entrainment, although there is no positive evidence. In homoiothermic animals, such effects cannot be ruled out a priori, but the participation of peripheral temperature receptors is more likely. To where are those signals transmitted? Do they have to reach the SCN or the pineal organ? Similar questions can be asked for acoustical or olfactory signals such as are possibly involved in social entrainment. The entraining effects of periodic feeding also follow unknown pathways. Only one study sheds some light on where the signals enter the organism. As Suda and Saito (1979) have shown, the meal-induced rhythms in plasma corticosterone and intestinal enzymes are abolished when food is given parenterally instead of per os.

There is further evidence that various zeitgebers affect circadian rhythms differentially, opening the possibility of „selective entrainment" (Aschoff 1981). In animals which are entrained by LD, some rhythms such as plasma corticosterone and urinary excretion can be shifted by periodic feeding while others, e.g., the rhythms in body temperature and in corneal mitosis, keep an unchanged ψ to LD (Krieger and Hauser 1978, Philippens et al. 1977, Sulzman et al. 1978b). Such differential effects are of special interest in the case of „partial entrainment" (Aschoff 1978), i.e., when in presence of a zeitgeber some rhythms become entrained while others are free-running. Figure 7 presents free-running rhythms of two human subjects who were living in an isolation unit and who both developed „internal desynchronization" by a sudden lenghtening of their sleep-wake cycles. After desynchronization had occurred, the rhythm of rectal temperature continued to free-run with a period of 25.1 h in the one subject who was exposed to constant illumination (upper diagram), but became entrained in the other subject who was exposed to a 24-h LD cycle.

The multitude of circadian rhythms observed in the physiology and behaviour of an organism usually behaves as though controlled by a central oscillator or pacemaker through which zeitgebers exert their synchronizing influence. However, a steadily increasing body of observations demands modification of such a simplified concept. Environmental changes may bypass the central pacemaker not only in directly affecting overt rhythms by masking, but also by entrainment-like effects on circadian subsystems. Apparently, such subsystems, like the rat's meal anticipation, have oscillatory capacity independent of a central, at least the SCN pacemaker. Support for such a concept comes from squirrel monkey studies, showing that after SCN-lesions a self-sustaining circadian oscillation in body temperature persists and can be entrained by LD cycles (Albers et al. 1980). Whether in the absence of such a pacemaker they are damped or capable of self-sustained oscillations, remains unanswered. It is further unknown if such subsystems are controlled by pacemakers of their own. The mutual coupling and feedback influences between different elements in the circadian system, by which each is timed appropriately with respect to the ensemble and to the outside world, will present a major research challenge in the years to come. The use of conflicting zeitgebers of different modalities may become a powerful tool for this purpose, expecially since direct physiological assays of pacemaker activity (Schwartz et al. 1979 and Chap. 3.5) have become available. The following chapters adress several of these questions more specifically.

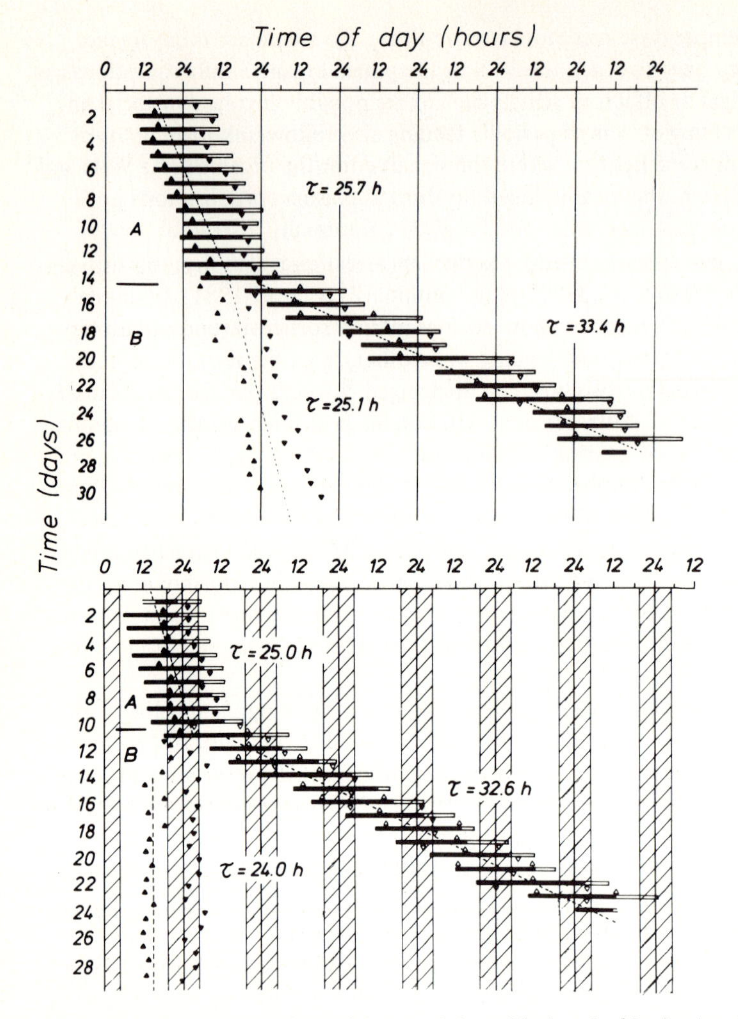

Fig. 7. Circadian rhythms of wakefulness and sleep (*black* and *white bars*) and of rectal temperature (*triangles above and below bars* for maxima and minima, respectively) in two subjects, each living singly in an isolation unit. *Upper diagram* subject exposed to constant illumination throughout the experiment; *Lower diagram* subject exposed to a light-dark cycle (*shaded area* darkness) and provided with a small reading lamp. In each diagram, synchronized rhythms in part *A* and internal desynchronization in part *B* τ = circadian period. (Wever 1979)

References

Albers HE, Lydic R, Moore-Ede MC (1980) Light-dark cycle entrainment of the persisting circadian rhythm of core body temperature on SCN-lesioned primates. Néurosci Abstr 6:708

Aschoff J (1978) Features of circadian rhythms relevant for the design of shift schedules. Ergonomics 21:739-754

Aschoff J (1981) Circadian rhythms: interference with an dependence on work-rest schedules. In: Johnson LC, Tepas DI, Colquhoun W P, Colligan MJ (eds) Biological rhythms and shift work. Advances in sleep research, vol VII. Spectrum Publ, New York

Aschoff J, Pohl H (1978) Phase relations between a circadian rhythm and its zeitgeber within the range of entrainment. Naturwissenschaften 65:80-84
Aschoff J, Saint Paul U von (1976) Brain temperature in the unanaesthetized chicken: its circadian rhythm of responsiveness to light. Brain Res 101:1-9
Boulos Z, Rosenwasser AM, Terman M (1980) Feeding schedules and the circadian organization of behavior in the rat. Behav Brain Res 1:39-65
Bovet J, Oertli EF (1974) Freerunning circadian activity rhythms in freeliving beaver (Castor canadensis). J Comp Physiol 92:1-10
Davis FS, Menaker M (1980) Hamsters through time's window: temporal structure of hamster locomotor rhythmicity. Am J Physiol 239: R149-R155
DeCoursey P (1960) Phase control of activity in a rodent. Cold Spring Harbor Symp Quant Biol 25:49-54
Enright JT (1965) Synchronization and ranges of entrainment. In: Aschoff J (ed) Circadian Clocks. North Holland Publ Comp, Amsterdam pp 112-124
Gaston S, Menaker M (1968) Pineal function: The biological clock in the sparrow? Sciense 160: 1125-1127
Gibbs FP (1979) Fixed interval feeding does not entrain the circadian pacemaker in blind rate. Am J Physiol 236:R249-253
Gwinner E (1966) Entrainment of a circadian rhythm in birds by species-specific song cycles *(Aves, Fringillidae: Carduelis spinus, Serinus serinus).* Experientia 22:765
Gwinner E (1978) Effects of pinealectomy on circadian locomotor activity rhythms in European starlings, *Sturnus vulgaris.* J Comp Physiol 126:123-129
Hayden P, Lindberg RG (1969) Circadian rhythm in mammalian body temperature entrained by cyclic pressure changes. Science 164:1288-1289
Hoffmann K (1969a) Zum Einfluß der Zeitgeberstärke auf die Phasenlage der synchronisierten circadianen Periodik. Z Vergl Physiol 62:93-110
Hoffmann K (1969b) Die relative Wirksamkeit von Zeitgebern. Oecologia 3:184-206
Honma KI, Aschoff J (1981) Effects of periodic feeding on circadian rhythms in the rat (in prep.)
Kato H, Saito M, Suda M (1980) Effect of starvation on the circadian adrenocortical rhythm in rats. Endocrinology 106:918-921
Krieger DT, Hauser H (1978) Comparison of synchronization of circadian corticosteroid rhythms by photoperiod and food. Proc Natl Acad Sci 75:1577-1581
Krieger D, Hauser H, Krey LC (1977) Suprachiasmatic nuclear lesions do not abolish food-shifted circadian adrenal and temperature rhythmicity. Science 197:398-399
Marimuthu G, Rajan S, Chandrashekaran MK (1981) Social entrainment of the circadian rhythm in the flight activity of the microchiropteran bat, *Hipposideros speoris.* Behav Ecol Sociol 8:147-150
Philippens KMH, Mayersbach H von, Scheving LE (1977) Effects of the scheduling of meal feeding at different phases of the circadian system in rats. J Nutr 107:176-193
Pittendrigh CS (1981) Circadian systems: entrainment. In: Aschoff J (ed) Biological Rhythms. Handbook of Behavioral Neurobiology, vol IV. Plenum Press, New York, pp 95-124
Richter CP (1968) Inherent twenty-four hour and lunar clocks of a primate – the squirrel monkey. Comm Behav Biol Part A 1:305-332
Schwartz WJ, Smith CB, Davidsen LC (1979) In vivo glucose utilization of the suprachiasmatic nucleus. In: Suda M, Hayaishi O, Nakagawa H (eds) Biological Rhythms and their Central Mechanism. Elsevier/North Holland Biomedical Press, Amsterdam, pp 355-367
Stephan FK (1981) Limits of entrainment to periodic feeding in rats with suprachiasmatic lesions. J Comp Physiol A 143:401-410
Suda M, Saito M (1979) Coordinative regulation of feeding behavior and metabolism by a circadian timing system. In: Suda M, Hayaishi O, Nakagawa (eds) Biological Rhythms and their Central Mechanism. Elsevier/North-Holland Biomedical Press, Amsterdam, pp 263-271
Sulzman FM, Fuller ChA, Moore-Ede MC (1977) Feeding time synchronizes primate circadian rhythms. Physiol Behav 18:775-779
Sulzman FM, Fuller ChA, Hiles G, Moore-Ede MC (1978a) Circadian rhythm dissociation in an environment with conflicIting temporal information. Am J Physiol 235:R175-R180

Sulzman FM, Fuller ChA, Moore-Ede MC (1978b) Comparison of synchronization of primate circadian rhythms by light and food. Am J Physiol 234:R130-R135
Takahashi JS, Menaker M (1979) Brain mechanisms in avian circadian systems. In: Suda M, Hayaishi O, Nakagawa H (eds) Biological Rhythms and their Central Mechanism. Elsevier/North-Holland Biomedical Press, Amsterdam, pp 95-109
Tokura H, Aschoff J (1981) Entrainment of circadian activity rhythms by temperature cycles in primates (in prep)
Wever R (1979) The circadian system of man. Springer, Berlin Heidelberg New York
Wever R (1980) Circadian rhythms of finches under bright light: is self-sustainment a precondition for circadian rhythmicity? J Comp Physiol 139:49-58

2.1 Comparative Aspects of Retinal and Extraretinal Photosensory Input Channels Entraining Endogenous Rhythms

H.-G. Hartwig[1]

With some minor exceptions „all of the energy for life on earth is contained within a narrow spectral range of electromagnetic radiation received from the sun" (Seliger and McElroy 1965). The total amount of incident electromagnetic radiation exhibits considerable daily and seasonal variations. Light, by virtue of its daily and seasonal variations acts as a zeitgeber (cf. Aschoff 1965) synchronizing circadian and circannual rhythms by which living organisms adjust their autonomic functions to periodic changes in their environment (Scharrer 1964). A response to light has been observed in large numbers of different biological systems (for review, see Wurtman 1975). However, in vertebrates, entrainment of endogenous rhythms by natural or artificial photoperiods is exclusively mediated via photoreceptive systems located in a circumscribed region of the central nervous system, the diencephalon (for review, see Rusak and Zucker 1979). In the following brief survey attention will be focussed predominantly on extraretinal photosensory systems.

Vertebrate photoreceptive systems engaged in the control of autonomic rhythmic events evolved along three different lines: (1) epithalamic pineal sense organs, (2) deep encephalic photoreceptors, and (3) retinohypothalamic projections (Fig. 1). In poikilo-

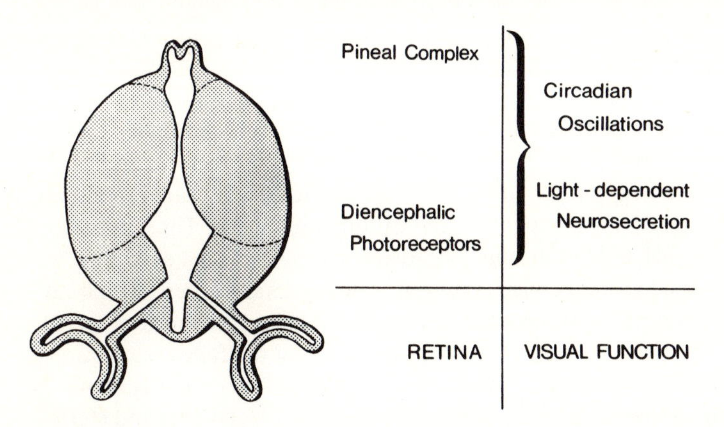

Fig. 1. Diagrammatic presentation of a cross-sectioned embryonic diencephalon showing location and most probable function of different evolutionary lines of photosensory systems

[1] Department of Anatomy, University of Kiel, FRG

Vertebrate Circadian Systems
Ed. by J. Aschoff, S. Daan, and G. Groos

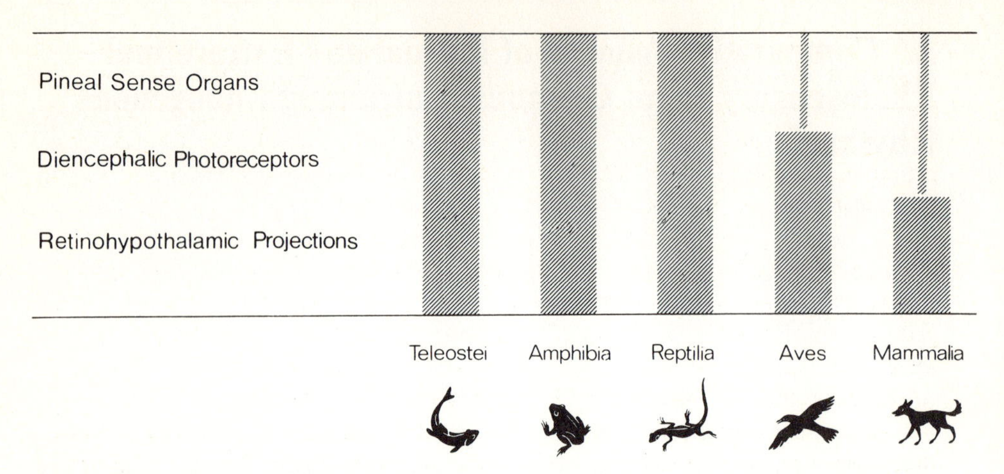

Fig. 2. Comparative survey indicating the presence of (1) pineal sense organs, (2) diencephalic photoreceptors, and (3) retinohypothalamic projections in the vertebrate phylum. *Hatched columns* presence of photosensory system indicated on the left side of the diagram. *Hatched line* photosensory capacity not shown in threshold experiments. Only limited numbers of species have been investigated

thermic vertebrates there is a considerable body of morphological and experimental evidence for the presence of all three different photosensory input channels. Pineal sense organs seem to be absent in homeothermic vertebrates, and in mammals retinohypothalamic projections apparently are the exclusive mediators capable of entraining endogenous rhythmic events with an external light regimen (Fig. 2; for review of comparative aspects, see Hartwig 1975; Hartwig and Oksche 1982). The questions arises how the different photosensory input channels present in poikilothermic vertebrates and birds may interact in the control of light-dependent autonomic events. In this respect only limited numbers of species have been thoroughly investigated. Menaker and his coworkers presented evidence that in the house sparrow the eyes do not participate in the photoperiodic photoreception (McMillan et al 1975). Van Veen (1981) showed in threshold experiments that entrainment of circadian locomotor behavior requires identical amounts of periodically applied light in intact and in blinded and pinealectomized European eels (LD 12:12; white light with a color temperature of 2,700 K and an incident energy of $10^{-2}\mu W/cm^2$ at the surface of the aquaria; this energy value is approximately 2 log units below that of full moon light). Thus, in vertebrates endowed with retinal and extraretinal photoreceptive systems the eyes may be devoted primarily to visual functions, whereas pineal sense organs and encephalic photoreceptors may serve to control photoneuroendocrine events (Fig. 1; for review, see Oksche and Hartwig 1979). This view can be further supported by the following observations characterizing pineal sense organs. Pineal sense organs contain photopigments resembling those found in the retina (Hartwig and Baumann 1974; Oksche and Hartwig 1979). A light-dependent change of spontaneous discharge activity has been recorded electrophysiologically in the pineal tract running from the epiphysis cerebri toward the posterior commissure (chromatic and achromatic types of responses; for review, see Dodt 1973). However, results from the laboratories of Oksche (see Oksche and Hartwig 1979) and Collin (see Collin 1976, Falcon 1978) strongly support the hypothesis that pineal sense organs, in addition to the nervous message, might influence other brain areas via light-dependent synthesis and release of neurohumoral mediators. The pineal sensory cell probably represents an ideal photoneuroendocrine element for the investigation of the transduction of photic information into neuroendocrine signals. Cells of the receptor line seem to be capable of indoleamine metabolism (for details and review, see Falcon 1978). Using a more sensitive modification of the formaldehyde-induced fluorescence technique we have shown in *Rana temporaria* that in animals not pretreated pharmacologically serotonin fluorescence is predominantly found within apical poles of pinealocytes in close association with the pineal lumen. This observation is in agreement with regional differences in the uptake of radioactively labeled precursors of the indoleamine biosynthesis in the pineal organ of *Xenopus laevis* (Charlton 1966).

Collin (1976) emphasized the specific morphological character of two different pineal cell lines, the „photoreceptor cell" and the „secretory rudimentary receptor cell". Thus, it remains to be elucidated whether all sensory elements of the pineal complex are capable of metabolizing indoleamines. In this respect it should be noted that only the intracranial component of the pineal complex, the epiphysis cerebri, has been proven to contain indoleamines. Extracranial derivatives of the pineal complex existent in some amphibians (frontal organ) and reptiles (parietal eye) apparantly are not endowed with the capacity of indoleamine metabolism. Indoleamines of the epiphysis cerebri may be engaged in the control of daily changes in the threshold sensitivity of the parietal eye in the collared lizard (*Crotaphytus collaris,* Engbretson and Lent 1976).

Recently, Hartwig and Calas (1981) observed that in continuous darkness ^{3}H-deoxyglucose is specifically taken up by individual pineal receptor cells bearing well-developed outer segments and by individual pineal neurons (incubation for 15 min). Stimulation with white light (15 min; 10 lx; 2,000 K) abolished this distinct pattern in the cellular distribution of silver grains. Electron microscopic investigations on this material are in progress to reveal possible correlations between the light-dependent uptake of ^{3}H-deoxyglucose and ultrastructural parameters indicating a secretory activity.

To date rhythmic aspects of sensory and neuroendocrine activities of pineal sense organs are still poorly understodd (cf. Hartwig and Oksche 1981). In this context it is remarkable that the light-dependent shedding and renewal of photopigment-containing membranes has not been investigated in pineal sense organs (for review of this phenomenon in retinal photoreceptor cells, see Young 1978).

In comparison with pineal sense organs the knowledge dealing with the morphology of encephalic photoreceptive elements is extremely poor. The existence of functional extraretinal and extrapineal photosensory systems mediating color change and light-dependent conditioned reflexes has been shown in European minnows by von Frisch (1911) and Scharrer (1928). Van Veen et al. (1976) obtained similar results in European eels (Fig. 3). In 1935, Benoit (for review, see Benoit 1964) showed that domestic mallards possess a light-sensitive diencephalon mediating the photoperiodically induced gonadal growth (for results obtained in House sparrows, see Menaker and Underwood 1976). Yokoyama et al. (1978) stimulated various sites of the diencephalon in American white-crowned sparrows, *Zonotrichia leucophrys gambelii,* via stereotaxically implanted small-diameter optic fibers with long photoperiods. According to their results the photosensitive region mediating the photoperiodic induction of gonadal growth is located

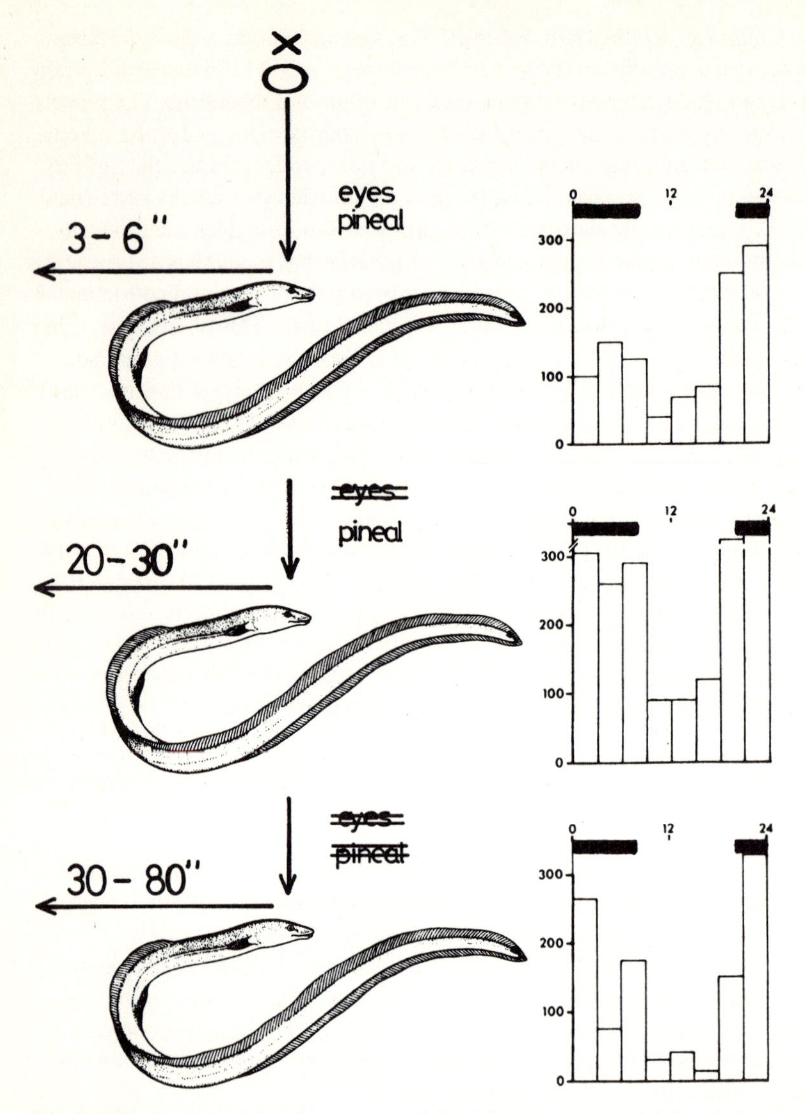

Fig. 3. Light-dependent circadian motor activity (diagrams on the *right*) and photonegative behavior (diagrams on the *left*) in intact (*top*), blinded (*middle*), and blinded and pinealectomized European eels (after van Veen et al. 1976). *Left* photonegative behavior observed with a latency indicated in seconds on the arrow showing direction of characteristic light-induced backward swimming following local illumination of the head region. *Right* circadian motor activity recorded as interruptions of an IR light beam (y-axis) in an artificial photoperiod. On top of each diagram a *thick bar* represents the dark period

within the mediobasal hypothalamus. However, microspectrophotometric recordings failed to demonstrate a photolabile compound in the corresponding brain region of house sparrows (Hartwig 1975). In small individuals of *Salmo gairdneri* and *Phoxinus phoxinus* a photolabile compound has been demonstrated microspectrophotometrically in an ependymal area covering the anterodorsal hypothalamus (see Oksche and Hartwig 1975, 1979, Hartwig and Oksche 1982).

Considering threshold values for the extraretinally mediated light-dependent gonadal growth in the house sparrow (cf. McMillan et al. 1975) and spectral transmission values of the tissues covering the hypothalamus (for details and review, see Hartwig and van Veen 1979) it can be computed that the threshold sensitivity of encephalic photoreceptor systems is in the range of the threshold of the electrophysiological response of the dark adapted retina in a mammal (for details, see Hartwig and Oksche 1982). Consequently, one could expect to find a hypothalamic system of photosensory elements and integrating neuronal circuitry resembling at least in some respect that of the retina and that of pineal sense organs. Since various attempts failed to identify encephalic photoreceptive systems by morphological techniques additional mechanisms of photosensitivity have to be considered, e.g., photosensitive enzymes or modification of enzyme activity mediated by photochromic compounds (for references and discussion, see Hartwig and van Veen 1979, Hartwig and Oksche 1982). The unique concentration of photosensory and neuroendocrine properties in a single cell might have offered specific functional advantages in the phylogeny of vertebrates realized in pineal sense organs and in hypothalamic secretory active neurons. In this respect it should be noted that primitive neurons of invertebrates may be capable of sensory, conductile and secretory functions (Lentz 1968).

Acknowledgement The investigations of the author cited in this contribution were supported by grants of the Deutsche Forschungsgemeinschaft (Ha 726/5 and Ha 726/6) and have been predominantly performed at the Department of Anatomy and Cytobiology, Justus Liebig-University Giessen.

References

Aschoff J (1965) Circadian clocks. Elsevier/Noth Holland, Amsterdam

Benoit J (1964) The structural component of the hypothalamo-hypophyseal pathway, with particular reference to photostimulation of the gonads in bird. Ann NY Acad Sci 117: 23-34

Charlton HM (1966) The uptake of C^{14}-5-hydroxytryptamine creatinine sulphate and C^{14} c-methyl-methionine by the epiphysis of *Xenopus laevis* Daudin. Comp Biochem Physiol 17: 777-784

Collin JP (1976) La rudimentation des photorécepteurs dans l'organe pinéal des vertébrés. In: Mécanismes de la rudimentation des organes chez les embryos de vertébrés. Coll Int CNRS No 266. CNRS, Paris, pp 393-408

Dodt E (1973) The parietal eye (pineal and parapineal organs) of lower vertebrates. In: Autrum H, Jung R, Loewenstein WR, MacKay DM, Teuber HL (eds) Handbook of sensory physiology, vol VII/3 B. Springer, Berlin Heidelberg New York, pp 113-140

Engbretson GA, Lent CA (1976) Parietal eye of the lizard: neuronal photoresponses and feedback from the pineal gland. Proc Natl Acad Sci USA 73: 654-657

Falcon J (1978) Pluralité et sites d'élaboration des messages de l'organe pinéal. Etude chez une vertébrés inférieur: Le brochet (*Esox lucius*, L). Thesis, Univ Poitiers, pp 1-91

Frisch K von (1911) Beiträge zur Physiologie der Pigmentzellen in der Fischhaut. Pfluegers Arch 138: 319-397

Hartwig HG (1975) Neurobiologische Studien an photoneuroendokrinen Systemen. Habilitationsschrift, Bereich Humanmed, Justus Liebig-Univ Giessen

Hartwig HG, Baumann Ch (1974) Evidence for photosensitive pigments in the pineal complex of the frog. Vision Res 14: 597-598

Hartwig HG, Calas A (1981) Light-dependent uptake of triated deoxyglucose by retinal and pineal photosensory system in *Rana temporaria*. Acta Anatomica 111:58

Hartwig HG, Oksche A (1982) Neurobiological aspects of extraretinal and extrapineal photoreceptive systems. Experientia (in press)

Hartwig HG, Veen van T (1979) Spectral characteristics of visible radiation penetrating into the brain and stimulating extraretinal photoreceptors. J Comp Physiol 130: 277-282

Lentz TL (1968) Primitive nervous systems. Yale Univ Press, New Haven, Conn

McMillan JP, Underwood HA, Elliot JA, Stetson MH, Menaker M (1975) Extraretinal light perception in the sparrow. 4. Further evidence that the eyes do not participate in photoperiodic photoreception. J Comp Physiol 97: 205-214

Menaker M, Underwood H (1976) Extraretinal photoreception in birds. Photochem Photobiol 23: 299-306

Oksche A, Hartwig HG (1975) Photoneuroendocrine systems and the third ventricle. In: Knigge KM et al. (eds). Brain-endocrine interaction II. The ventricular system. 2nd Int Symp Shizuoka 1974, Karger, Basel, pp 40-53

Oksche A, Hartwig HG (1979) Pineal sense organs-components of photoneuroendocrine systems. Prog Brain Res 52: 113-130

Rusak B, Zucker I (1979) Neural regulation of circadian rhythms. Physiol Rev 59 (3): 449-513

Scharrer E (1928) Die Lichtempfindlichkeit blinder Elritzen. (Untersuchungen über das Zwischenhirn der Fische). Z Vergl Physiol 7: 1-38

Scharrer E (1964) Photo-neuro-endocrine systems: general concepts. Ann NY Acad Sci 117: 13-22

Seliger HH, McElroy WD (1965) Light: Physical and biological action. Academic Press, London New York

Veen van TH (1981) A study on the basis for zeitgeber entrainment. With special reference to extraretinal photoreception in the eel. Thesis, Dep Zool, Univ of Lund, Sweden

Veen van TH, Hartwig HG, Müller K (1976) Light-dependent motor activity and photonegative behavior in the eel (*Anguilla anguilla* L.). Evidence for extraretinal and extrapineal photoreception. J Comp Physiol 111: 209-219

Wurtman RJ (1975) The effects of light on man and other mammals. Annu Rev Physiol 37: 467-483

Yokoyama K, Oksche A, Darden THR, Farner DS (1978) The sites of encephalic photoreception in photoperiodic induction of the growth of the testes in the white-crowned sparrow. *Zonotrichia leucophrys gabelii.* Cell Tissue Res 189: 441-467

Young RW (1978) Visual cells, daily rhythms, and vision research. Vision Res 18: 5730578

Note added in proof:

For recent results dealing with light-dependent uptake of tritiated deoxyglucose by pineal and encephalic photosensory systems and the distribution pattern of pineal indoleamines in *Rana temporaria* and *Rana esculenta* see the following references:

Hartwig HG (1982) Retinale und extraretinale Photorezeptoren: Meßfühler und Integratoren der ‚Biologischen Uhr'. Verh. Anat. Ges., 77. Vers. (Hannover) (im Druck)

Hartwig HG, Oksche A (1981) Photoneuroendocrine cells and systems: a concept revisited. In: Oksche A, Pevét P (eds.). The Pineal Organ – Photobiology, Biochronometry, Endocrinology, Developments in Endocrinology 14: 49-59

Hartwig HG, Reinhold C (1981) Microspectrofluorometry of biogenic monoamines in pineal systems. In: Oksche A, Pevét P (eds.). The Pineal Organ – Photobiology, Biochronometry, Endocrinology, Developments in Endocrinology 14: 237-246

2.2 Neuroanatomical Pattern of Endocrine and Oscillatory Systems of the Brain: Retrospect and Prospect

A. Oksche[1]

1 Introduction

The aim of the present contribution is to elucidate the anatomical basis of neural mechanisms in avian and mammalian circadian systems. Although biorhythmicity can be considered as a fundamental property of the single cell, this analysis will be focussed on complex structures of the brain, especially the diencephalon. In this connection two diencephalic structures, the suprachiasmatic nucleus and the pineal organ, deserve particular attention. The former is located in the hypothalamic, the latter in the epithalamic region of the diencephalon. Both regions belong to the phylogenetically ancient portion of the brain. The hypothalamus is exceedingly rich in neuroendocrine effectors, which apparently have evolved from primitive pluripotent senso-neuroendocrine cells (cf. Oksche 1978b, 1981). In phylogeny these elements were successively re-arranged to form increasingly complex aggregates encompassing a variety of afferents and interneurons. On the other hand, in the pineal organ photoreceptor cells capable of indoleamine metabolism were transformed into glandular elements that receive different kinds of central nervous inputs, including an input from the lateral eyes (cf. Oksche and Hartwig 1979).

These statements may serve as a rather simplified introduction to a more detailed analysis of neuroanatomical patterns. For the input channels to the circadian oscillators of the brain, see Hartwig (Chap. 2.1), for the physiological organization of the avian and mammalian circadian system, see Parts 3 and 4 of this volume.

2 General Considerations

At the present state of our knowledge, it is impossible to give a clear-cut anatomical model of a circadian oscillator of the vertebrate brain. The two best-known structures involved in generation and control of circadian locomotor and neuroendocrine rhythms, the suprachiasmatic nucleus and the pineal organ, are rather different in their basic structural design. Another important problem is the principle and degree of coupling among the single circadian clocks located within the central nervous system. The classical neuroanatomy of the hypothalamus is largely based on the existence of cytoarchitec-

[1] Department of Anatomy and Cytobiology, Justus Liebig University, Giessen, Federal Republic of Germany

Vertebrate Circadian Systems
Ed. by J. Aschoff, S. Daan, and G. Groos

turally defined, circumscribed nuclear areas. In mammals, these nuclear areas are rather distinct, and they have been mapped in great detail (cf. Diepen 1962). However, in addition to conspicuous nuclear entities numerous scattered neurons were shown to occur in the hypothalamus. The evidence obtained from experimental work with hypothalamic lesions indicates that hypothalamic functions can only partly be localized in or correlated to certain nuclear areas. Such discrepancies were explained by (1) damage extending to adjacent neural tracts, and/or (2) the existence of heterogeneous cell populations in the nuclear areas.

Recent immunocytochemical studies have revealed that most of the peptidergic neuronal systems exhibit a scattered pattern, complemented by local clustered aggregations of immunoreacitve neurons of a certain type. Such aggregations may become manifest in multiple, even spatially remote, regions of the brainstem. This pattern is, e.g., characteristic of the LHRH- or somatostatin-immunoreactive neurons; it resembles a neuronal network extending over a wide area of the hypothalamus. Circumscribed clusters of neuronal perikarya displaying common histochemical features may then be regarded as local nodular aggregations within this network. A further excessive proliferation of the neuronal units may finally lead to anatomical entities corresponding to (or identical with) classical nuclear areas. There is no doubt that most of the well-defined nuclear areas of the mammalian hypothalamus are a product of proliferative processes during the course of phylogenetic development. In lower vertebrates only the magnocellular nuclei, composed primarily of vasotocin- and mesotocin-immunoreactive cells, are cytoarchitecturally conspicuous and well delimited. In phylogeny the concentric periventricular layers of hypothalamic neurons seggregate into patterned groups of neuronal clusters. The latter are encompassed by regionally highly specific neuropil areas rich in synaptic contacts. This synaptic apparatus is essential for isolation of or interaction among adjacent hypothalamic formations. In comparative terms, the hypothalamus of reptiles offers the key to the interpretation of the pattern of the avian and mammalian hypothalami.

This concept of the hypothalamic architecture may offer a better basis for understanding hypothalamic functions. Time-consuming, systematic neuroanatomical investigations will be required to prove or to disprove details of the concept. However, the classical idea of hypothalamic nuclei as structural and functional entities appears to be too crude to serve as a model in the analysis of neuroendocrine systems and circadian oscillators. In this respect, the suprachiasmatic nucleus is one of the most fascinating hypothalamic areas.

3 Suprachiasmatic Nucleus

For functional aspects of the suprachiasmatic nucleus of birds and mammals, especially the control of circadian neuroendocrine and locomotor rhythms, Part 3 of this volume should be consulted.

Neuroanatomically, the suprachiasmatic nucleus of the rat has been examined thoroughly by means of silver-impregnation methods, immunocytochemistry and electron microscopy. In addition to its retinal input (Hendrickson et al. 1972, Moore 1973), the suprachiasmatic nucleus receives projections from different other areas of the brain.

A detailed quantitative inventory of the synapses in the suprachiasmatic nucleus of the rat was published by Güldner (1976) (cf. Chap. 2.3). The homologous area of the avian brain is much more discrete; the retinal input to the avian suprachiasmatic nucleus was demonstrated by Hartwig (1974; house sparrow) and Bons (1974, 1976; domestic mallard). Since the circadian system and the extraretinal photoreception of the house sparrow *(Passer domesticus)* have been subjects of systematic investigations (Menaker and Underwood 1976), my further comments will be focused on the suprachiasmatic nucleus of this avian species.

In our laboratory systematic electron-microscopic studies of the suprachiasmatic nucleus of *P. domesticus* have revealed a number of structural peculiarities. The neurons of this nucleus and the adjacent periventricular formation tend to aggregate in clusters. Within these clusters neuronal perikarya were found in direct apposition, being completely devoid of interposed glial lamellae (Fig. 1 A). At certain sites the plasmalemma of such neurons displayed distinct densities (Fig. 1 B). Direct apposition of neuronal perikarya may occur also in other formations of the avian brain; however, in the area of the suprachiasmatic nucleus, this structural arrangement is very conspicuous. In addition, in some cases an afferent established contacts with two adjacent neuronal perikarya; due to their membrane densities and accumulations of vesicular inclusions these axon terminals appeared to be presynaptic to both neuronal somata (Fig. 2). Although this peculiar anatomical feature needs further investigation, one may assume that the above-described contact areas serve the interaction (coupling) between two or among several neuronal units. The suprachiasmatic area of *P. domesticus* is rich in patterned monoaminergic afferents from the lower brainstem. In contrast to mammals, these fibers belong exclusively to the noradrenergic system; there is no microspectrofluorimetric evidence for the presence of 5-HT-containing terminals in the suprachiasmatic nucleus of *P. domesticus* (Hartwig 1975).

Beyond the level of the nuclear area proper, the suprachiasmatic nucleus of mammals is reciprocally linked to the mediobasal tuberal area; it is also suggested to possess a projection to the median eminence (Swanson and Cowan 1975; cf. Oksche 1978 b). Comparable experimental data for avian species are as yet not available. However, the anterior median eminence of passerine birds receives a heavy input from different areas of the rostral hypothalamus (Oksche and Farner 1974). In the suprachiasmatic nucleus of mammals vasopressin-immunoreactive perikarya have been shown with certainty (cf. Krisch 1980). Sporadic claims of the presence of other types of immunoreactive neurons (e.g., LHRH-perikarya) in the suprachiasmatic nucleus must be regarded with great caution; the staining of these elements may depend on some sort of cross-reaction. To date, there are no reports on immunoreactive peptidergic perikarya in the suprachiasmatic nucleus proper of birds.

4 Pineal Organ

In birds and mammals, the pineal organ is another major component of the system of biological clocks; for fundamental functional differences (circadian versus seasonal control) in these two classes of vertebrates, see section 4. The glandular activity of the avian and mammalian pineal organs must be viewed in close context with the evolu-

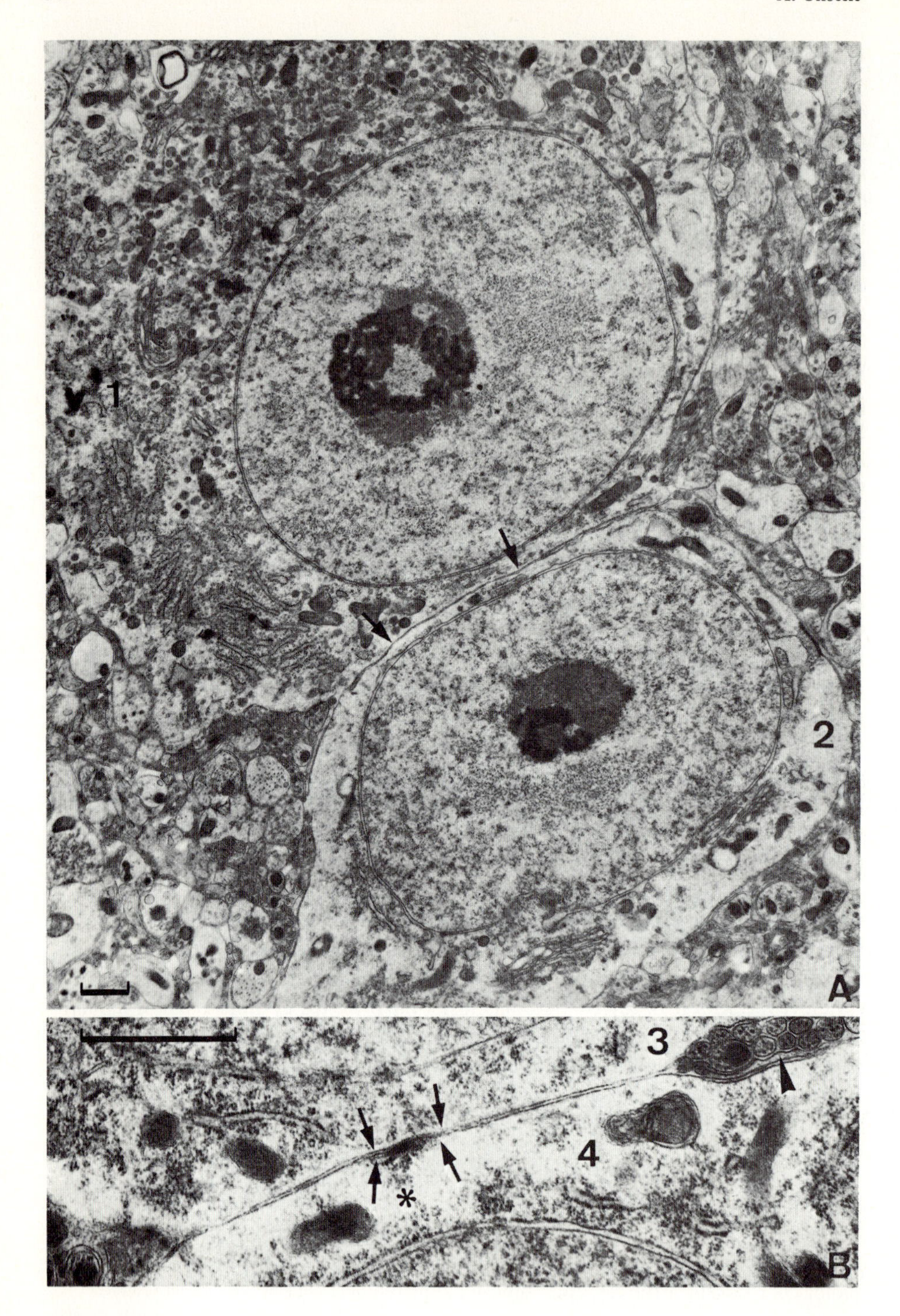
1
2
A
3
4
*
B

tionary history of this organ (Fig. 3; cf. Vollrath 1981, Oksche and Pévet 1981). In phylogeny, the pineal organ is transformed from a sense organ responding directly to light (Dodt 1973) into an endocrine gland controlled by the lateral eyes via the sympathetic nervous system (for references, see Collin 1971, Oksche 1971, Vollrath 1981).

There is now a convincing body of evidence that the pinealocytes of mammals have evolved from pineal photoreceptor cells of lower vertebrates (Collin 1971, Oksche 1971, for detailed references, see Vollrath 1981). In phylogenetic terms, a certain type of pinealocyte endowed with sensory structures also appears to be capable of neuroendocrine functions mediated by release of indoleamines (Oksche 1971). At least a certain line of pineal photoreceptor cells is capable of indoleamine synthesis and metabolism. The controversial ideas concerning the existence of two different sensory cell lines (photoreceptors and photoneuroendocrine cells) are open to discussion (Meiniel 1981, versus Collin 1981, cf. Collin 1976); although of great phylogenetic and cytobiological interest (cf. Oksche and Pévet 1981), this problem is of lesser importance for considerations concerning circadian mechanisms. However, it should not be overlooked that the pinealocytes of mammals display synaptic ribbons (vesicle-crowned rodlets); these peculiar structures speak in favor of the neuronal origin of the pinealocytes and may serve the functional coupling of individual pinealocyte cells (cf. Vollrath 1981). Finally, it is of importance to note that the irregular lamellar whorls of avian pinealocytes (homologues of the regular outer segments of pineal sensory cells) display a positive immunoreaction to an antibody against opsin, thus indicating some sort of modified neurosensory capacity (Vigh and Vigh-Teichmann 1981). These findings may help to understand some of the recent functional observations (see Chaps. 4.4 and 4.5). Previous electrophysiological studies (cf. Dodt 1973) were not able to show a direct response of the avian pineal organ to light (see, however, Chap. 4.3). Further, it should not be overlooked that the pineal organ of some avian families (e.g., passerine birds) is endowed with a conspicuous instrinsic neuronal apparatus.

This brief review emphasizes principal differences in the cytological composition and neuroanatomical organization of the suprachiasmatic nucleus and the pineal organ. Nevertheless, the pineal organ possesses reciprocal anatomical pathways to and from the brain, reflecting phylogenetically ancient neural connections of the pineal complex. In *Passer domesticus,* the biorhythmic message of the pineal organ to the brain appears to be exclusively humoral (Headrick Zimmermann 1976); however, the abundant neural projections of this organ to the brain may possess another yet undiscovered function. In this connection the target areas of the afferent pineal pathways are of particular interest; they have been studied in detail in our laboratory by use of neurohistological degeneration methods and tracer techniques.

Paul et al. (1971) were the first to show that, in the frog, the pineal tract projects to the reticular system of the mesencephalic portion of the brainstem and to the pretectal area. Korf and colleagues (cf. Korf and Wagner 1980, 1981) studied these connec-

Fig. 1. *Passer domesticus.* **A.** Two secretory neurons *(1, 2)* in the area of the suprachiasmatic nucleus and the adjacent periventricular formation. *Arrows* direct contact between the perikarya of these neurons; there is no evidence for an interposed glial lamella. The perikarya are embedded in a neuropil rich in axodendritic and axosomatic synapses. x 6,900. Bar 1 μm **B** Direct contact of two other hypothalamic neurons *(3, 4). Arrow* contact site containing electron-dense material with adjacent structures in neuron 4 *(asterisk). Arrowhead* neuropil. x 23,000. Bar 1 μm
Note: Neuronal complexes of this type show intimate vascular contacts

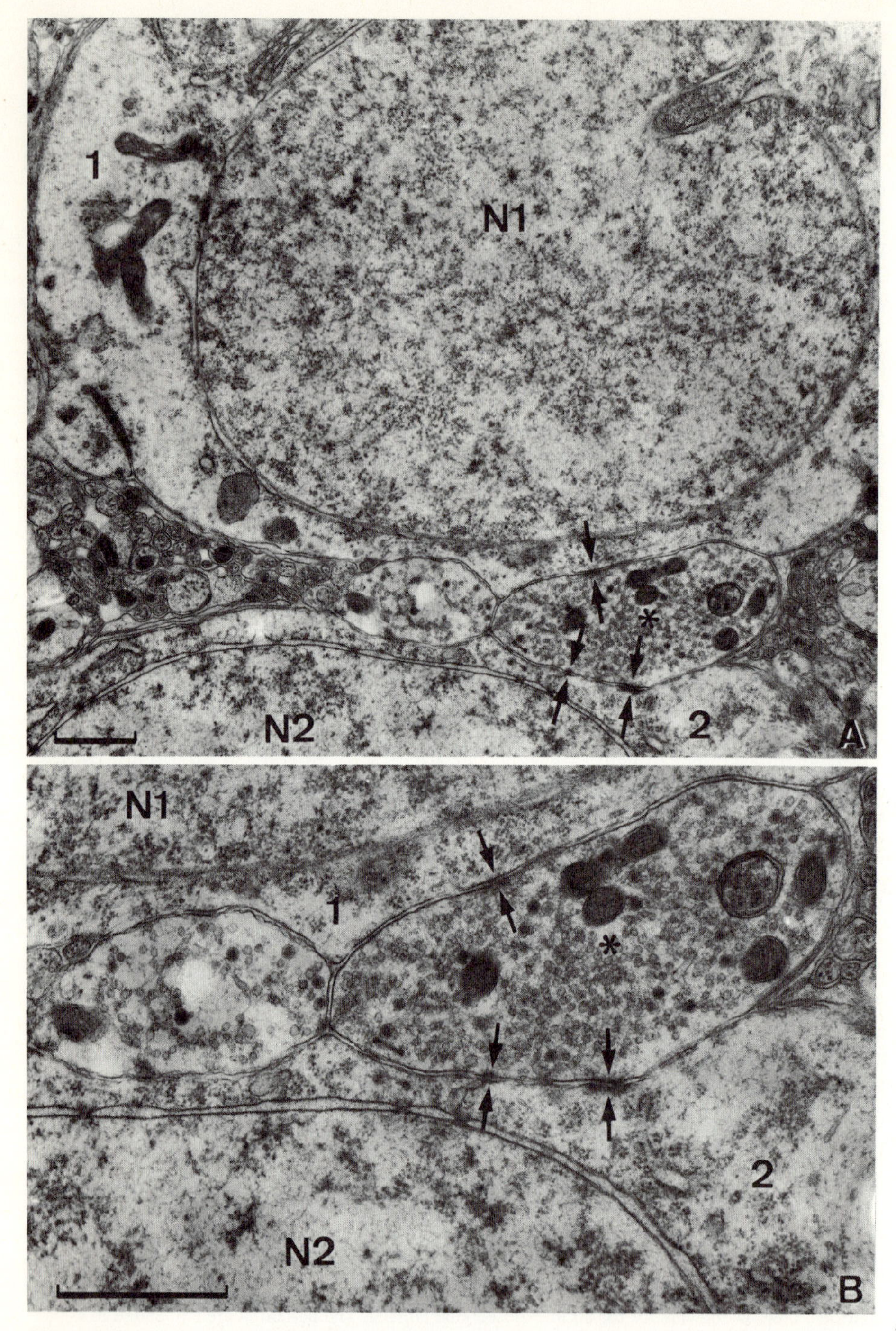

Fig. 2. *Passer domesticus.* Area of the suprachiasmatic nucleus and the adjacent periventricular formation. **A** Nerve terminal (*) displaying membrane thickenings *(arrows)* at the contact sites with two secretory perikarya *(1, 2)*. *N1, N2* cell nuclei. **B** Higher magnification of the contact sites (see **A** for abbreviations). **A** x 12,000; **B** x 24,000. Bar 1 μm (cf. Oksche et al. 1974)

tions in several lacertilian *(Lacerta sicula),* avian *(Passer domesticus)* and mammalian *(Cavia cobaya)* species by use of microiontophoretically administered horseradish peroxidase. In *L. sicula,* labeled axonal fibers and terminations of the parietal nerve were found in the left habenular ganglion, dorsolateral thalamic nucleus, hypothalamic periventricular gray, preoptic region, and pretectal area. On the other hand, scattered perikarya located in the area of the paraventricular nucleus establish an efferent connection to the parietal eye. In *Passer domesticus,* pinealopetal nerve fibers originate from the periventricular portion of the paraventricular nucleus and the medial habenular ganglion[2]. In *Cavia cobaya,* discrete pinealopetal axons were traced back to a circumscribed area of the paraventricular nucleus. The pineal organ of *Erinaceus europaeus* receives an input of vasopressin and oxytocin-immunoreactive nerve fibers, apparently of paraventricular origin (Nürnberger and Korf 1981).

In functional terms it appears noteworthy that the target areas of certain pinealofugal (afferent) projections (to the preoptic hypothalamus and the pretectal area, i.e., neuroendocrine and motor centers) show an overlap with optic projections from the lateral eyes. Even more important, these results indicate a nervous coupling of the pineal organ to the hypothalamus. Further work is in progress in this line of neuroanatomical work. It may help to understand the differences in the degree of coupling of the pineal organ to the suprachiasmatic nucleus and other biological clocks located in the hypothalamus and in adjacent regions of the brainstem (cf. Chap. 7.2). To date, it has not been possible to show a direct pinealofugal or pinealopetal connection between the pineal organ and the suprachiasmatic nucleus (Korf, pers. commun.); this leaves the problem of a multisynaptic type of connectivity open to discussion.

5 Conclusions

The classical nuclear areas of the hypothalamus and the lower brainstem provide only to a very limited extent an appropriate basis for understanding the physiological organization of the vertebrate circadian system. The clearly outlined character of the suprachiasmatic nucleus as a cytoarchitectural entity is not in principal contradiction to this statement; the structural design of the other circadian oscillators of the brainstem appears to be much more discrete. This holds true also for the single components of the neuroendocrine apparatus, which is more or less closely linked to the circadian system. For the sites of deep encephalic photoreception, see Hartwig (Chap. 2.1) and Yokoyama et al. (1978).

In general, endocrine and pacemaking functions follow only partly the anatomical boundaries of the classical hypothalamic nuclei (cf. Oksche 1978a, b). These functions can be more readily correlated to networks formed by patterned complexes of neurons displaying stronger or weaker local aggregations of neuronal perikarya. Within this framework an occasional exessive accumulation of nerve-cell somata may assume the appearance of a nuclear area. The above-mentioned networks encompass highly specialized neuropil zones rich in different types of synaptic connections. Local peptidergic neurons and extrahypothalamic peptidergic projections may contribute to this synaptic apparatus; the role of neuropeptides as transmitters or neuromodulators is a matter of actual discussion.

---

[2] Added in proof: cf. Korf et al. (1982)

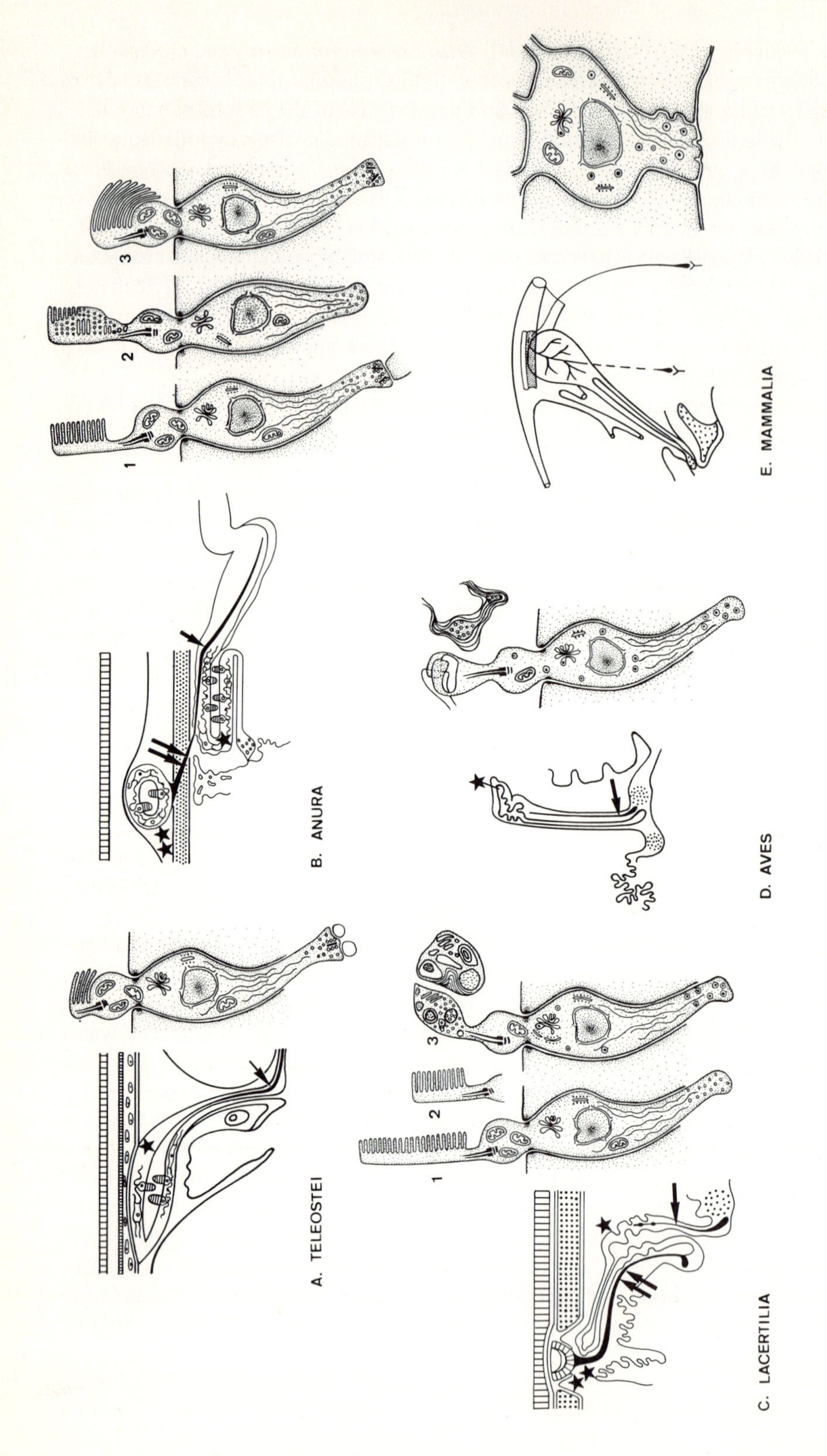
A. TELEOSTEI
B. ANURA
C. LACERTILIA
D. AVES
E. MAMMALIA
1
2
3
1
2
3

In addition to definite channels serving a progressive propagation of excitation, return circuits and self-reexciting chains are established; the latter may be an essential pre-requisite of integrative centers and biological oscillators. Interneurons may play a crucial role in the establishment of subunits in extended nuclear areas known to encompass a variety of different functions. The patterned synaptic apparatus of the neuronal networks is responsible for convergence or divergence as well as for isolation of or interaction among adjacent functional systems. Finally, the mode of this interaction depends on the pattern of the synaptically defined and polarized channels developed and modified during the course of phylogeny. This evolutionary process is characterized by a consecutive rearrangement of the structural relationships of basic nervous and neuroendocrine units (cf. Oksche 1978 b).

Acknowledgements. The reported investigations were supported by the following grants from the Deutsche Forschungsgemeinschaft to the author: Ok 1/19, 1/22 (SPP Biologie der Zeitmessung) and Ok 1/24, 1/25 (SPP Mechanismen biologischer Uhren)

References

Bons N (1974) Mise en évidence au microscope électronique, de terminaisons neuveuse d'origine rétinienne dans l'hypothalamus anterieur du Canard. C R Acad Sci 288: 319-321

Bons N (1976) Retinohypothalamic pathway in the duck *(Anas platyrhynchos)*. Cell Tissue Res 168: 343-360

Collin J P (1971) Differentiation and regression of cells of the sensory line in the epiphysis cerebri. in: Wolstenholme GEW, Knight J (eds) The pineal gland. (A Ciba Foundation Symposium). Churchill, London, pp 79-125

Collin J P (1976) La rudimentation des photorécepteurs dans l'orange pinéal des Vertébrés. In: Mécanismes de la Rudimentation des Organes chez les Embryons des Vertébrés. Coll Int CNRS No 266. Paris, CNRS, pp 393-408

Collin J P (1981) New data and vistas on the mechanisms of secretion of proteins and indoles in the mammalian pinealocyte and its phylogenetic precursors; the pinealin hypothesis and preliminary comments on membrane traffic. In: Oksche, A, Pévet P (eds) The Pineal Organ. Photobiology – Biochronometry – Endocrinology. Developments in Endocrinology Vol. 14, Amsterdam, Elsevier/North Holland Biomedical Press, pp. 187-210

Fig. 3. Comparative anatomy of pineal complexes (derivatives of pineal and parapineal primordia). Diagrammatic midsagittal sections in relation to the characteristic cells of the sensory line. **A** Teleostei: *star,* pineal organ; *arrow,* pineal tract. The outer segment of the receptor cell is relatively short and mostly overlaps the inner segment. **B** Anura: *double star,* frontal organ; *double arrow,* frontal-organ nerve; *star,* epiphysis cerebri; *arrow,* pineal tract. The outer segments may be *(1)* regular (conelike), *(2)* irregular (vesiculated), or *(3)* domelike. Occasionally rudimentary (bulbous) outer segments have been described. **C** Lacertilia: *double star,* parietal eye (apparently derivative of the parapineal primordium); *double arrow,* parietal nerve; *star,* epiphysis cerebri; *arrow,* pineal tract. Long, regular outer segments *(1)* predominate in the parietal eye. In the epiphysis cerebri, in addition to short regular outer segments *(2)*, irregular forms *(3)* and bulbous cilia characteristic of rudimentary photoreceptor cells have been observed; the latter contain numerous dense-cored vesicles. **D** Aves: *arrow,* pineal tract. The avian pinealocytes belong to the type of rudimentary photoreceptor cells. **E** Mammalia. Pinealocytes of adult mammals are secretory cells devoid of outer segment structures and endowed with vesicle-crowned rodlets („synaptic ribbons"). (According to Oksche; from M. Menaker and A. Oksche 1974, Courtesy Academic Press)

Diepen R (1962) Der Hypothalamus. In: Bargmann W (ed) Handbuch der mikroskopischen Anatomie des Menschen, Bd IV/7. Springer, Berlin Göttingen Heidelberg

Dodt E (1973) The parietal eye (pineal and parapineal organs) of lower vertebrates. In: Autrum H, Jung R, Loewenstein WR, Mac Kay D M, Teuber H L (eds) Handbook of sensory physiology, vol VII/3 B. Springer, Berlin Heidelberg New York, pp 113-140

Güldner F-H (1976) Synaptology of the rat suprachiasmatic nucleus. Cell Tissue Res 165;509-544

Hartwig H G (1974) Electron microscopic evidence for a retinohypothalamic projection to the suprachiasmatic nucleus of *Passer domesticus.* Cell Tissue Res 153: 89-99

Hartwig H G (1975) Neurobiologische Studien an photoneuroendokrinen Systemen. Habilitationsschrift, Bereich Humanmed, Justus Liebig-Univ, Giessen

Headrick Zimmerman N (1976) Organization within the circadian system of the house sparrow: Hormonal coupling and the location of a circadian oscillator. Dissertation, Univ of Texas, Austin

Hendrickson A E, Wagoner N, Cowan W M (1972) An autoradiographic and electron microscopic study of retinohypothalamic connections. Z Zellforsch 135: 1-26

Korf H W., Wagner U (1980) Evidence for a nervous connection between the brain and the pineal organ in the guinea pig. Cell Tissue Res 209: 505-510

Korf H W, Wagner U (1981) Nervous connections of the parietal eye in adult *Lacerta s. sicula* Rafinesque as demonstrated by anterograde and retrograde transport of horseradish peroxidase. Cell Tissue Res (in press)

Krisch B (1980) Immunocytochemistry of neuroendocrine systems (vasopressin, somatostatin, luliberin). Prog Histochem Cytochem 13/2: 1-166

Meiniel A (1981) New aspects of the phylogenetic evolution of sensory cell lines in the vertebrate pineal complex. In: Oksche A, Pévet P (eds) The pineal organ. Photobiology – biochronometry – endocrinology. Developments in Endocrinology vol 14, pp. 27-42, Elsevier/North-Holland Biomedical Press, Amsterdam

Menaker M, Oksche A (1974) The avian pineal organ. In: Farner DS, King JR (eds) Avian biology, vol IV. New York, Academic Press, London New York, pp 79-118

Menaker M, Underwood H (1976) Extraretinal photoreception in birds. Photochem Photobiol 23: 299-306

Moore R Y (1973) Retinohypothalamic projection in mammals: a comparative study. Brain Res 49: 403-409

Nürnberger F, Korf H W (1981) Oxytocin- and vasopressin- immunoreactive nerve fibers in the pineal organ of the hedgehog, *Erinaceus europaeus* L. Cell Tissue Res 220: 87-97

Oksche A (1971) Sensory and glandular elements of the pineal organ. In: Wolstenholme GEW, Knight J (eds) The pineal gland (A Ciba Foundation Symposium). Churchill, London, pp. 127-146

Oksche A (1978) Evolution, differentiation and organization of hypothalamic systems controlling reproduction. Neurobiological concepts. In: Scott DE, Kozlowski SP, Weindl A (eds) Brain-endocrine interaction III. Neural hormones and reproduction. 3rd Int Symp Würzburg 1977. Karger, Basel, pp 1-15

Oksche A (1978b) The neurosecretory cell in the organization of the central nervous system: phylogenetic aspects. In: Cell biology of hypothalamic neurosecretion. Coll Int CNRS No 280. Paris, CNRS, pp 27-41

Oksche A (1981) Evolution of neurosecretory cells and systems. In: Gersch M, Karlson P (eds) The evolution of hormonal systems (Leopoldina-Symposium Reinhardsbrunn, 1981). Nova Acta Leopoldina N.F. (in press)

Oksche A, Farner D S (1974) Neurohistological studies of the hypothalamo-hypophysial system of *Zonotrichia leucophrys gambelii* (Aves, Passeriformes). With special attention to its role in the control of reproduction. Adv Anat Embryol Cell Biol 48/4: 1-136

Oksche A, Hartwig H G (1979) Pineal sense organs – components of photoneuroendocrine systems. In: Ariëns Kappers J, Pévet P (eds) The pineal gland of vertebrates including man. (Prog Brain Res 52). Elsevier/North-Holland Biomedical Press, Amsterdam, pp 113-130

Oksche A, Pévet P (eds) (1981) The pineal organ: Photobiology – biochronometry – endocrinology. Developments in Endocrinology vol 14 pp 366. Elsevier/North-Holland Biomedical Press, Amsterdam

Oksche A, Kirschstein H, Hartwig H G, Oehmke H J, Farner D S (1974) Secretory parvocellular neurons in the rostral hypothalamus and in the tuberal complex of *Passer domesticus*. Cell Tissue Res 149: 363-370

Paul E, Hartwig H G, Oksche A (1971) Neurone und zentralnervöse Verbindungen des Pinealorgans der Anuren. Z Zellforsch 112: 466-493

Swanson L W, Cowan V M (1975) The efferent connections of the suprachiasmatic nucleus of the hypothalamus. J Comp Neurol 160: 1-14

Vigh B, Vigh-Teichmann I (1981) Immunocytochemical demonstration of rhodopsin in the pinealocytes of various vertebrates. In: Pévet P, Tapp E (eds) EPSG Newsletter, Suppl 3, Abstr 2nd Colloquium of the EPSG. Amsterdam, EPSG, p 56

Vollrath L (1981) The pineal organ. In: Oksche A, Vollrath L (eds) Handbuch der mikroskopischen Anatomie des Menschen, vol VI/7, Springer, Berlin Heidelberg New York

Yokoyama K, Oksche A, Darden T R, Farner D S (1978) The sites of encephalic photoreception in photoperiodic induction of the growth of the testes in the White-crowned Sparrow, *Zonotrichia leucophyrs gambelii*. Cell Tissue Res 189: 441-467

Added in proof:

Korf H-W, Zimmerman NH, Oksche A (1982) Intrinsic neurons and neural connections of the pineal organ of the house sparrow, *Passer domesticus*, as revealed by anterograde and retrograde transport of horseradish peroxidase. Cell Tissue Res 222: 243-260

Vigh B, Vigh-Teichmann I (1981) Light- and electron-microscopic demonstration of immunoreactive opsin in the pinealocytes of various vertebrates. Cell Tissue Res 221: 451-463

2.3 Unspecific Optic Fibres and Their Terminal Fields

J.K. Mai[1] and L. Teckhaus[2]

1 Introduction

Besides the principal connections of the retina with the main subcortical centres, additional pathways derived from retinal ganglion cells have been distinguished. These are formed by fibres, which separate from the optic tract along its way to the superior colliculus to innervate structures within hypothalamus, meso-diencephalic junction area (synencephalon) and subthalamic region. Since more than the retino-hypothalamic (RHS) and accessory optic fibres (AOS) (for review see Mai 1978a) are comprised, the designation „unspecific optic fibre system" (UOS) is used to collectively describe these visual pathways. The intrinsic architecture of the UOS, its relationship with the principal retinal afferents, and information concerning the probable role in feedforward and feedback control mechanisms of circadian rhythmicity is largely lacking. However, data are available on the organization and architecture of the UOS terminal fields and these studies are reviewed in the present paper.

2 Terminations of the Unspecific Optic fibres

2.1 Hypothalamus

At the level of the suprachiasmatic area several characteristic subdivisions of the so-called retino-hypothalamic fibres can be separately distinguished by using autoradiographic methods (Fig. 1). Some of these fibres establish contacts within localized areas of the ipsi- and contralateral suprachiasmatic nucleus (SCN) as has been demonstrated by various electron microscopical techniques (Moore and Lenn 1972, Mai and Junger 1977, Güldner 1978). Apparently, these fibres do not impigne on the central parvicellular part of the nucleus (Fig. 2).

Corresponding to the circumscribed topography of the retinal afferents is the intrinsic diversity of the morphological and chemical architecture of the SCN (cf. Mai 1978b). Atrophy following enucleation of the rat does not occur within the typical central parvicellular compartment of the SCN but is confined to the external and auto-

[1] C.u.O. Vogt-Institut für Hirnforschung der Universität Düsseldorf,
[2] Max-Planck-Institut für Systemphysiologie, Dortmund (FRG)

Vertebrate Circadian Systems
Ed. by J. Aschoff, S. Daan, and G. Groos

radiographically labelled ventrolateral part, indicating that in the rat this region of the SCN is predominantly innervated by these fibres (Mai et al. 1982).

The area between the SCN (4 in Fig. 1), which encompasses also neurosecretory, peptidergic, corticohypothalamic, commissural and amygdalofugal fibres, contains accumulations or clusters of magnocellular neurons. These probably correspond to the median hypothalamic nucleus. The functional significance of these neurons is presently obscure.

Studies by axonal or terminal degeneration methods have provided conflicting data regarding the termination of retinal fibres within other hypothalamic nuclei (cf. Mai 1978a). More refined experimental techniques have suggested as terminal areas the preoptic (Silver and Brand 1979), the anterior hypothalamic (Conrad and Stumpf 1975, Mai 1976) and the arcuate nuclei (Sousa-Pinto 1970, Mason and Lincoln 1976) of the periventricular and medial hypothalamus. Within the lateral hypothalamic area, a morphologically heterogeneous labelled field was found by light microscopical autoradiography just dorsal to the contralateral supraoptic nucleus (Mai 1979). Moreover, Golgi studies in the rat, reported by Riley et al. (1981) indicate a diffuse neural connection along the principal and accessory optic tract with dendrites of lateral hypothalamic neurons.

2.2 Synencephalic Region

2.2.1 The Classic Accessory Optic System (AOS)

According to the classic schema (Hayhow et al. 1959-1966), the AOS of mammals consists of a meshwork of fibres which can be roughly separated into inferior and superior fasciculi. The inferior fasciculus recruits mainly from the fibre component which crosses the midline between the SCN (see above). At the lateral hypothalamic area the fibres run alongside the medial part of the cerebral peduncle (Fig. 3a). More caudally, they run in an essentially horizontal direction towards the emerging oculomotor nerve fibres.

The superior fasciculus is built up of fibres which segregate from the optic tract throughout its course along the cerebral peduncle. The peduncular surface becomes thereby covered by a very thin sheath of fibres, directed transversly to those forming the cerebral peduncle (Fig. 3b-e); therefore, the macroscopically visible component of these fibres was termed the transpeduncular tract (von Gudden 1870). By evaluation of autoradiographic pictures, a distinction between superior and inferior fibres, or between those which run superficially or lie deep inside the brain appears questionable, since peri- and transpeduncular components of the UOS surround and penetrate the cerebral peduncle (Fig. 3b, d; 4). In addition to the original descriptions by Hayhow, an ipsilateral representation of the AOS also exists.

Termination of the accessory optic fibres occurs within the synencephalon of the brain. Hayhow and co-workers have described the three fields where they observed terminal degeneration as the dorsal, lateral and medial terminal nuclei of the accessory optic system.

The *dorsal terminal nucleus* belongs topographically to the pretectal area. The neuropil contains few fusiform cells; most of them are scattered between the fibres forming the transpeduncular tract, resembling an interstitial nucleus of this component.

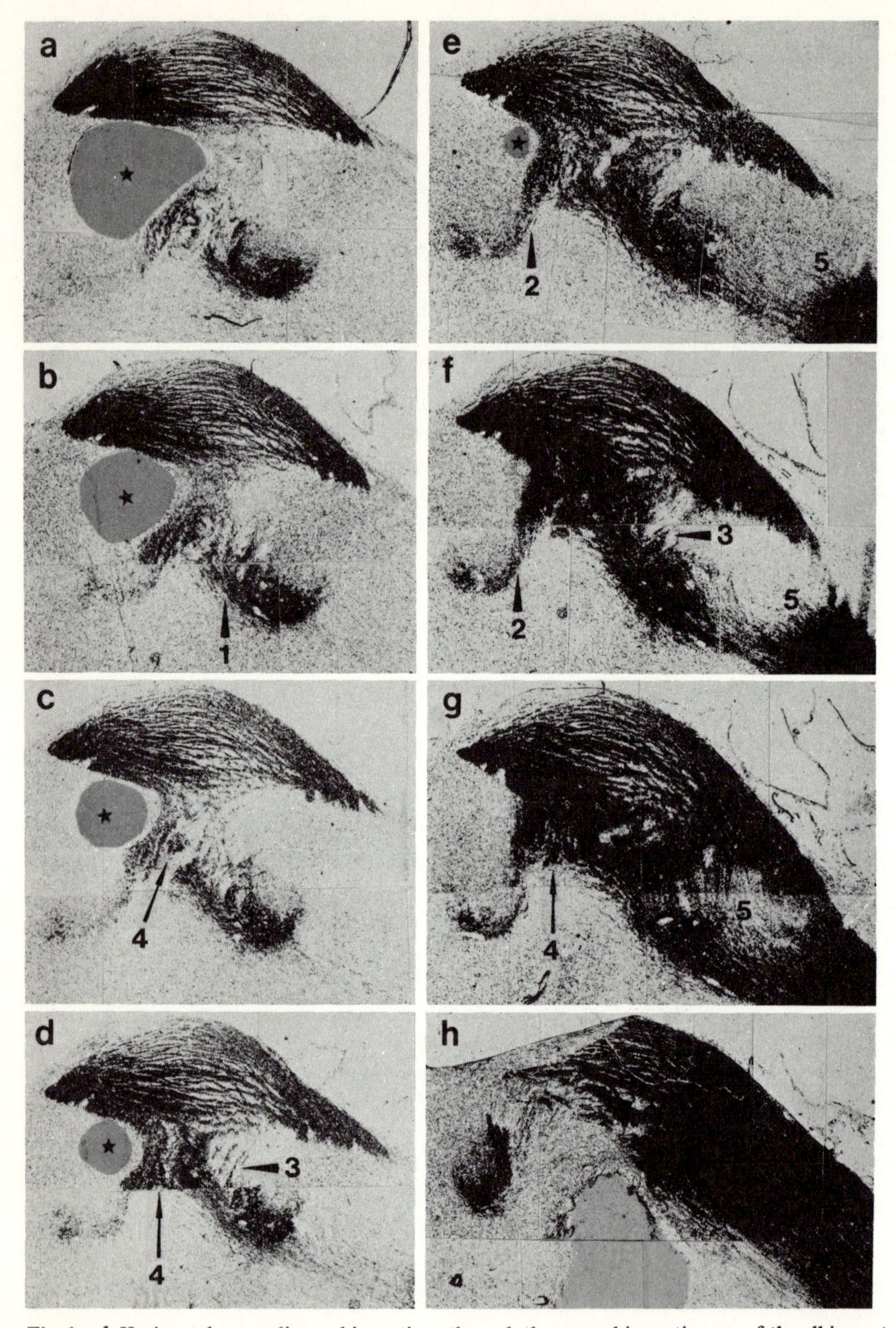

Fig. 1. a-h Horizontal autoradiographic sections through the suprachiasmatic area of the albino rat. Survival time after the injection of tritiated tyrosine into one eye was 24 h. Before embedding, the brain was slightly pressed between two object holders in order to obtain sections in parallel to the optic tract. The autoradiographs, therefore, do not give an authentic image of the spatial distribution

Fig. 2. a, b Three-dimensional reconstruction of frontal sections at actual distances (either 50 or 100 μm), seen from the right and left side, respectively. The contour lines around the central areas *(c)* as well as of the outer boundaries of the SCN *(e)* are derived from densitometry of cresyl violet stained sections. The contour lines selected for representation of silver grain distribution, determined by quantitative reflectance photometry, enclose same ranges for all sections. Note the distance between central nuclear areas and labelled areas (marked by *grey colour*) of the optic chiasma *(CO)* and the ventrolateral part of the SCN at all planes of section. (*V* third ventricle)

The *lateral terminal nucleus* (LTN) is part, distinguished by its optic afferents, of the peripeduncular (Papez and Aronson 1934, Stern 1936) or suprapeduncular (Keyser 1979) area which extends between the cerebral peduncle and dorsolateral part of the substantia nigra ventrally and the ventral margin of the medial geniculate body dorsally until the lateral lemniscal and the parabigeminal nuclei. Anteriorly, the LTN is continuous with the zona incerta, forming two cellular extensions: one that originates at the medial part of the ventral (division of the lateral) geniculate body (Fig. 3e), reaching the dorsal tip of the medial terminal nucleus; another which commences more dorsally at the basal medullary layer of the dorsal geniculate body, and is directed medially towards the lateral part of the ncl. ruber. The caudal continuation of the peripeduncular area as well as both extensions exhibit consistently high silver grain densities.

The *medial terminal nucleus* (MTN) represents an elongated structure oriented along the medial border of the substantia nigra. In the rat, this nucleus shows an extremely high level of incorporated radioactivity, if compared to other structures of the UOS.

←

of the labelled optic fibres, but they allow tracing of single fibres for long distances and distinction of various separate clearly marked components (1-5), as evidenced by the reduced silver grains: Some give rise to well localized projection fields in the posterior part of the contra- *(1)* and ipsilateral *(2)* nucleus. Others form strands of silver grains deviating from the posterior part of the optic chiasma *(3)* or voluminous twisted bundles *(4)* crossing between both SCN. The bulk of these fibres passes laterally, where some fibres become part of the AOS, others rejoin the main optic tract. The most prominent portion of silver grains, however, is delivered by retinal fibres which pass through the ventrolateral part of the contralateral nucleus *(5)* in order to reach other destinations. (*: third ventricle)

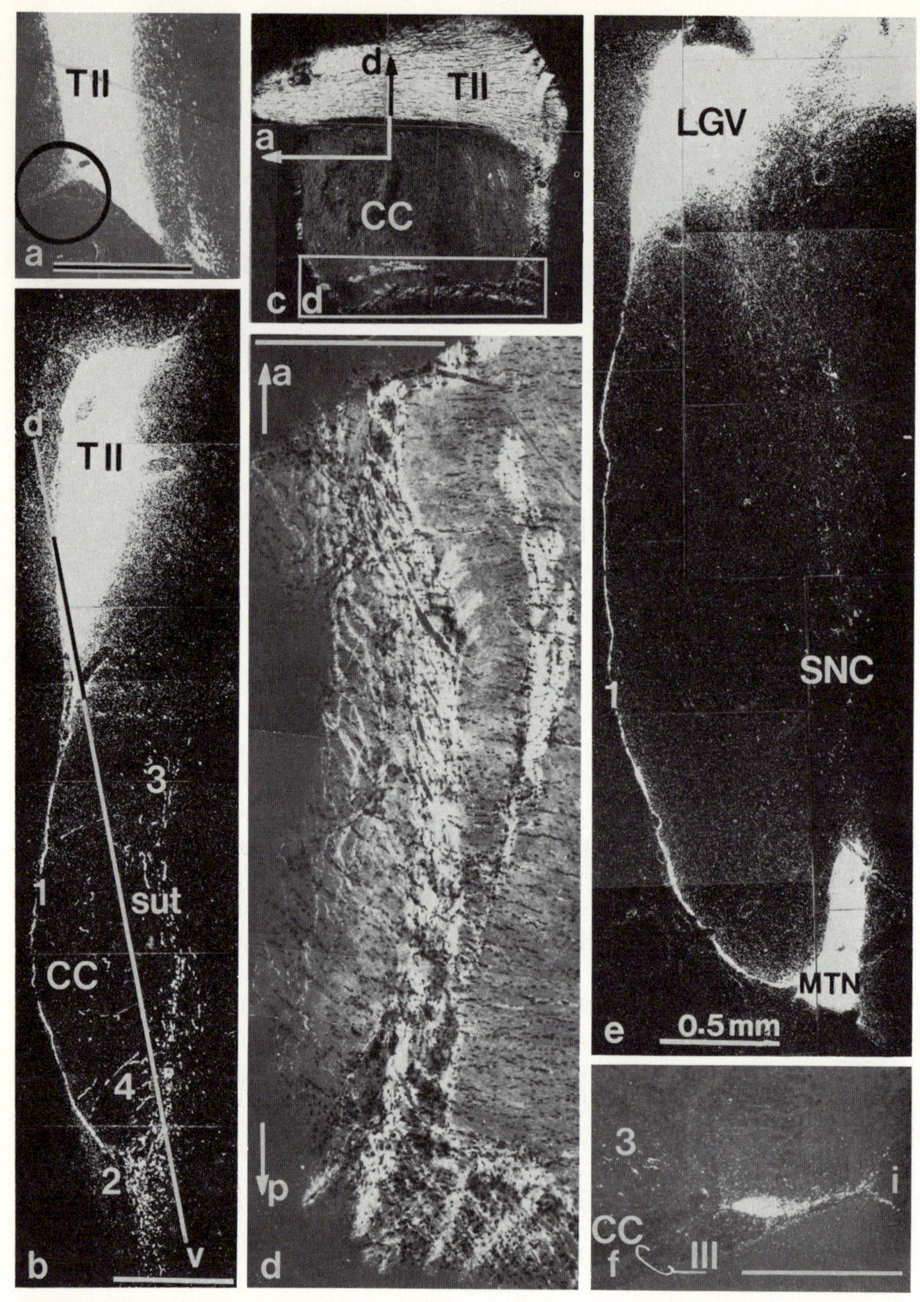

Fig. 3. Darkfield-illuminated autoradiographic sections after monocular injection of tritiated amino acids. **a** Optic tract *(TII)* and unspecific optic fibres *(right)*. The existence of labelling from the laterobasal margin of the optic tract to the temporal lobe *(circle)* suggests relationship with the amygdalo-piriform complex. **b** Frontal section, approximately 1.2 mm more caudally, showing peripeduncular fibres [covering the lateral *(1)*, ventral *(2)* and medial *(3)* surfaces of the cerebral peduncle *(CC)*] and transpeduncular fibres *(4)* of the UOS. The line, indicated by *d* (dorsal) and *v* (ventral), marks the section plane of figures **c** and **d.** (*sut* subthalamic ncl.). **c, d** Unspecific optic fibres, sec-

2.2.2 Substantia Nigra

Autoradiographs show a diffuse pattern of labelling within the pars compacta at its most anterior levels. A band of label stretches from the peripeduncular area along the ventral extension described above towards the apex of the medial terminal nucleus (Fig. 3e). Since the labelling is found in the vicinity of that nucleus only, it is not yet clear whether the labelled terminals impinge on neurons of the substantia nigra or on dendritic arborizations which extend outside the medial terminal nuclei.

3 Subthalamic Area

The subthalamic nucleus displays a diffuse increase in silver grain density, besides a very localized silver grain pattern above single nerve fribres, indicating that this nucleus is not only penetrated by accessory optic fibres, but actually provides synaptic contact with UOS-fibres. This supports earlier findings based on degeneration methods (cf. Mai 1978a). By degeneration techniques also the entopeduncular nucleus was claimed to exhibit signs of terminal degeneration (Holbrook and Shapiro 1974). The zona incerta proper, as recipient for retinal afferents, has been discussed above.

4 Other Areas

Monosynaptic retinal pathways have also been described to the mediodorsal nucleus of the thalamus by Conrad and Stumpf (1975) and to the raphe nuclei (B8 according to Dahlström and Fuxe 1965) by Foote et al. (1978). Moreover, autoradiography provides evidence for additional connections (e.g., by fibres which distribute among the ansa peduncularis towards the amygdalo-piriform complex and also by fibres among the supramamillary decussation and below the interpeduncular nucleus (Fig. 3f) with unknown destinations), whose monosynaptic nature, specifity and significance have yet to be established.

5 Comparative Anatomy

The components of the unspecific optic fibres show pronounced variation in number, size, arrangement and distribution in different vertebrates. Most striking is the replacement of a single set of accessory optic fibres in submammals (the basal optic root) into

←

tioned tangentially, which run in anterior *(a)* – posterior *(p)* direction along the basal part of the cerebral peduncle. (*d* dorsal direction). **e** Frontal section, approximately 1.4 mm more caudal than **b.** External or lateral unspecific optic fibres (1) in relation to the medial terminal ncl. *(MTN).* Labelling above the inferior extension from the ventral part of the lateral geniculate body *(LGV)* towards the pars compacta of the substantia nigra *(SNC)* and from the internuclear fiber plexus becomes discernable. Labelling becomes more pronounced if seen in relation to the lateral terminal ncl. **f** Frontal section through the posterior part of the MTN. Silver grains can be followed towards the interpeduncular ncl. *(i).* (*III* emerging oculomotor nerve fibres)

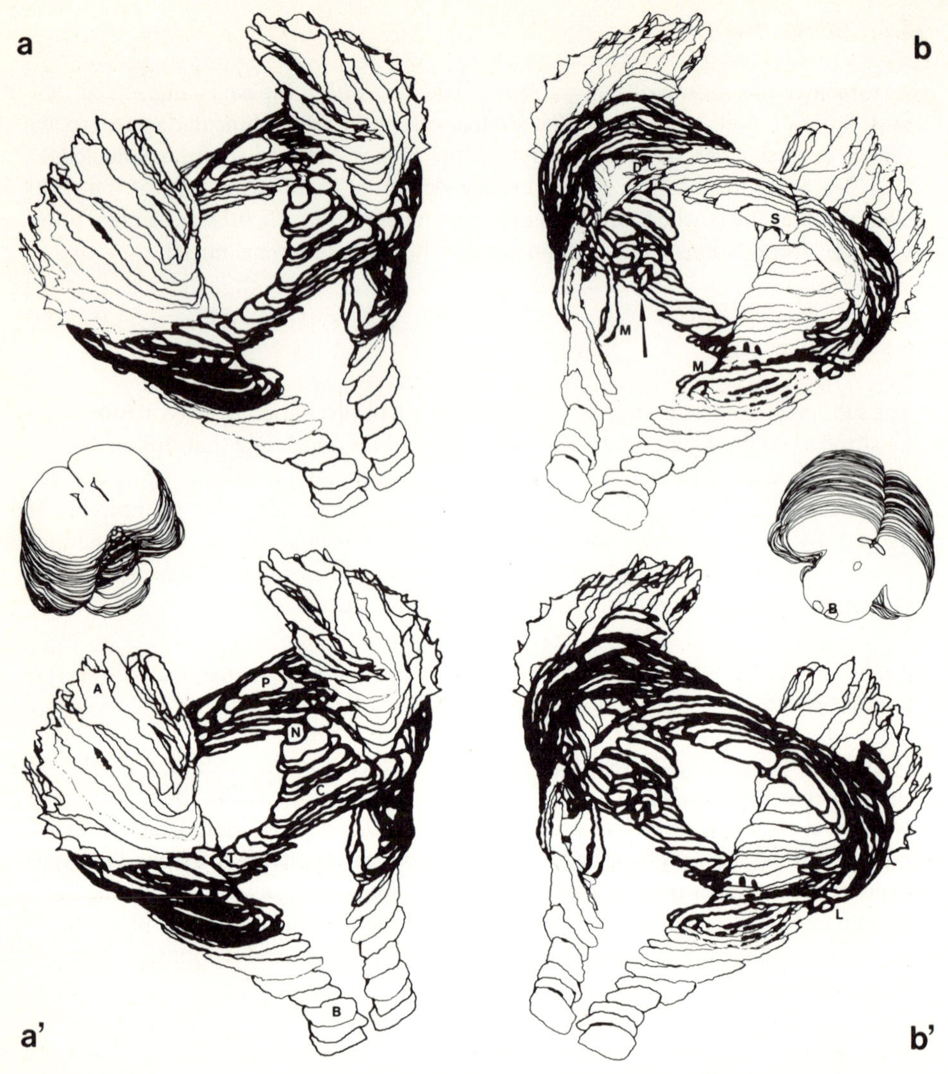

Fig. 4. Reconstruction of the distribution of the main and unspecific optic fibre systems (heavy lines in **a′** and **b′**) in the albino rat and their relationship with the internal capsule *(A)* and the cerebral peduncle *(B)*. This view does not include the unspecific optic fibres situated along the inner surface of the cerebral peduncle. With the aid of a stereoviewer *, the corresponding figures **a, a′** and **b, b′**, respectively, can be viewed stereoscopically. The angle of view in relation to the pial surface is given by the small graphs. (*C* optic chiasma, *D* dorsal-, *L* lateral-, *M* medial terminal ncl. of the accessory optic system (according to Hayhow and coworkers); *P* pretectal area; *S* superior colliculus; *T* optic tract; in **a** and **b** pretectal area and superior colliculi are drawn by *thin lines*)

* Stereo-Vertrieb Nesh, G. Neubauer POB 5726, D-44 Münster

a dual or even more extended organization in mammals. Even among mammals great differences are evident: in carnivora, insectivora, ungulata and primates, including man, the ventral fibres appear either lacking or only scarcely developed (see Lin and Giolli 1979). A generally accepted interpretation of this variation is missing, although various explanations have been given (cf. Mai 1978a). It appears that the shift in position of the trans- and peripeduncular components of the UOS (constituing the superior fasciculus of Hayhow) and of their terminal fields occurs concommitantly with the evolution of the cerebral peduncles and the transformation of the lateral forebrain bundle (cf. Schnitzlein et al. 1973).

6 Function

While considerable progress has been made in the elucidation of the function of the SCN, the function of the remaining divisions of the UOS, especially to its synencephalic termini, remains unsettled. Various interpretations advanced in the past were discussed by Marg (1964). Most of these implicit functional relations expressed by the topographic connections with proprioceptive, somatic and special sensory signals. Although none of these concepts can be regarded as established, the involvement of the UOS in processing self-induced visual motion has been substantiated also in mammalian species (Simpson et al. 1979).

Different considerations, however, arise from the following observations: first, that retinal fibres mediate light effects to several hypothalamic areas which might provide rhythmic signals in the absence of suprachiasmatic neuropil (cf. Rusak 1977) and second, that unspecific optic fibres are related to various neurotransmitter and peptidergic systems at the synencephalic level. These connections might be important for the modulation of the neural and endocrine environment of the hypothalamic structures thus influencing the expression of rhythmic signals.

Both the LTN and MTN represent subdivisions (distinguished by their optic afferents) of histochemically specialized regions. They are not only traversed by various aminergic and peptidergic pathways, but LTN and MTN provide neurotransmitter-producing cells. Dopaminergic axons from these regions as well as serotoninergic neurons, possibly influenced directly by visual stimuli (Foote et al. 1978) exert antagonistic effects on suprachiasmatic neurons (Johansson 1976). Involvement of neuropeptides in circadian regulation is suggested by the fact that the concentration of opiate receptors associated with the accessory optic structures and the content of substance P and enkephalin becomes greatly diminished following visual deprivation (Snyder and Simantov 1977, Pradelles et al. 1979).

It is evident from this review that existing morphological and histochemical data cannot yet provide a functional description of circadian control at the neurotransmitter and neuropeptidergic level. Evidence is, however, increasing for a regulation of suprachiasmatic function by at least some of the unspecific optic terminal fields.

Acknowledgements. Encouragement and support by Prof. Dr. A. Hopf and Prof. Dr. D.W. Lübbers is greatly appreciated. The authors are grateful to a number of collaborators, particularly I. Krauthausen, R. Schmidt-Kastner and H.-B. Tefett for their contributions and also to Dr. R. Lydic for critically reading the manuscript.

References

Conrad CD, Stumpf WE (1975) Direct visual input to the limbic system: Crossed retinal projections to the nucleus anterodorsalis thalami in the tree shrew. Exp Brain Res 23: 141-149

Dahlström A, Fuxe K (1965) Evidence for the existence of monoamine-containing neurons in the central nervous system. I. Demonstration of monoamines in the cell bodies of brain stem neurons. Acta Physiol Scand 232: (62 Suppl) 1-55

Foote WE, Taber-Pierce E, Edwards L (1978) Evidence for a retinal projection to the midbrain raphe of the cat. Brain Res 156: 135-140

Gudden von B (1870) Über einen bisher nicht beschriebenen Nervenfaserstrang im Gehirn der Säugetiere und des Menschen. Arch Psychiat 2: 364-366

Güldner FH (1978) Synapses of optic nerve afferents in the rat suprachiasmatic nucleus. I. Identification, qualitative description, development and distribution. Cell Tissue Res 194: 17-35

Hayhow WR (1959) An experimental study of the accessory optic fiber system in the cat. J Comp Neurol 113: 281-314

Hayhow WR (1966) The accessory optic system in the marsupial phalanger Trichosurus vulpecula. An experimental degeneration study. J Comp Neurol 126: 653-672

Hayhow WR, Webb C, Jervie A (1960) The accessory optic fiber system in the rat. J Comp Neurol 115: 187-216

Holbrook JR, Schapiro H (1974) The accessory optic tract in the dog; a retino-entopeduncular pathway. J Hirnforsch 15: 365-377

Johansson NG (1976) Biosynthesis and secretion of hypothalamic hormones. In: Hafez ESE, Reel JR (eds) Hypothalamic Hormones. Ann Arbor Sci Publ Inc. Michigan, pp 17-51

Keyser A (1979) Development of the hypothalamus in mammals. In: Morgane PJ, Panksepp J (eds) Handbook of the hypothalamus, vol I. Anatomy of the hypothalamus. Marcel Dekker, New York Basel, pp 65-136

Lin H, Giolli RA (1979) Accessory optic system of rhesus monkey. Exp Neurol 63: 163-176

Mai JK (1976) Quantitative autoradiographische Untersuchungen am subcorticalen optischen System der Albinoratte. Dissertation, Univ Düsseldorf

Mai JK (1978a) The accessory optic system and retino-hypothalamic system. A review. J Hirnforsch 19: 213-288

Mai JK (1978b) Morphologische Untersuchungen zum photo-neuro-endokrinen System der Albinoratte. Habilitationsschrift, Düsseldorf

Mai JK (1979) Distribution of retinal axons within the lateral hypothalamic area. Exp Brain Res 34: 373-377

Mai JK, Junger E (1977) Quantitative autoradiographic light and electron microscopic studies on the retino-hypothalamic connections in the rat. Cell Tissue Res 183: 221-237

Mai JK, Krauthausen I, Teckhaus L, Weindl A, Schrell V, Sofroniew MV (1982) Stereological evaluation of histological and histochemical properties within the suprachiasmatic area of the albino rat. (submitted for publication)

Marg E (1964) The accessory optic system. Ann N Y Acad Sci 117: 35-51

Mason CA, Lincoln DW (1976) Visualization of the retino-hypothalamic projection by cobalt precipitation. Cell Tissue Res 168: 117-131

Moore RY, Lenn NJ (1972) A retinohypothalamic projection in the rat. J Comp Neurol 146: 1-9

Papez JW, Aronson LR (1934) Thalamic nuclei of pithecus (macacus) rhesus. I. Ventral thalamus. Arch Neur Psychiatry 32: 1-26

Pradelles P, Cros C, Humbert J, Dray F, Ben-Ari Y (1979) Visual deprivation decreases met-enkephalin and substance P content of various forebrain structures. Brain Res 166: 191-193

Riley JN, Card JP, Moore RY (1981) A retinal projection to the lateral hypothalamus in the rat. Cell Tissue Res 214: 257-269

Rusak B (1977) The role of the suprachiasmatic nuclei in the generation of circadian rhythms in the golden hamster, Mesocricetus auratus. J Comp Psychol 118: 145-164

Schnitzlein HN, Hamel EG Jr, Carey JH, Brown JW, Hoffman HH, Faucette JR, Showers MJC (1973) The interrelations of the striatum with subcortical areas through the lateral forebrain bundle. J Hirnforsch 13: 409-455

Silver J, Brand S (1979) A route for direct retinal input to the preoptic hypothalamus: Dendritic projections into the optic chiasm. Am J Anat 155: 391-402

Simpson J, Soodak RE, Hess R (1979) The accessory optic system and its relation to the vestibulocerebellum. In: Granit R, Pompeinao O (eds) Reflex control of posture and movement. Prog Brain Res 50: 715-724

Snyder SH, Simantov R (1977) The opiate receptor and opioid peptides. J Neurochem 28: 13-20

Sousa-Pinto (1970) Electron microscopic observations on the possible retinohypothalamic projection in the rat. Exp Brain Res 11: 528-538

Stern K (1936) Der Zellaufbau des menschlichen Mittelhirns. Z Gesamte Neurol Psychiatr 154: 521-598

2.4 Characteristics of the Interaction Between the Central Circadian Mechanism and the Retina in Rabbits

A.C. Bobbert and J. Brandenburg[1]

1 Introduction; Materials and Methods

During an investigation into the influences of non-visual factors on the rabbit's cortical response to flashes (Bobbert et al. 1977), it was observed that this Visually Evoked Potential (V.E.P.) shows marked circadian changes. Because in mammals the visual pathway functions as the input channel for the main zeitgeber, it was considered worth while studying the phenomenon in detail.

Rabbits were provided with chronically implanted electrodes and exposed to the natural alternations of daylight and darkness (nLD) or to an artificial 24-h light-dark schedule (aLD). After at least 3 weeks the animal was fixed within an isolated and usually darkened experimental compartment in which it was subjected to 3 μs flashes, repeated at intervals of 1503 ms. The responses to these stimuli were simultaneously recorded from several – up to 4 – locations within the visual pathways and nearly always averaged over 100 consecutive flashes. Such samples, usually of both electroretinograms (ERG's) and cortical V.E.P.'s, were automatically taken at 3 or 5 min intervals and stored for off-line processing. Their analysis was routinely performed over periods of 250 or 500 ms and started with the last 10 or 20 ms before the flash. The experimental set up was usually arranged in such a way that the effective illuminance during each flash was about 2.10^7 lx at the location of the left eye – the fibres of which project for some 95 % to the right hemisphere (Shkol'nik-Yarros 1971), whereas at the same time the effective illuminance at the location of the right eye – from which the cortical responses on the left side are mainly derived – was some 4.10^4 times lower. This means that simultaneous recordings were made of the responses to strong and to very weak flashes.

Records of the cortical and subcortical V.E.P.'s were obtained from implanted electrodes while ERG's were recorded by way of contact lens electrodes that were provided with cannulas for perfusion of the conjunctival sacs with fluid containing drugs whose action on the retina, or its rhythm, is to be tested (for further details cf. Bobbert et al. 1978b; Brandenburg et al. 1981).

[1] Department of Physiology and Physiological Physics, University of Leiden, Wassenaarseweg 62, 2333 AL Leiden, The Netherlands

Vertebrate Circadian Systems
Ed. by J. Aschoff, S. Daan, and G. Groos

2 Results

2.1 The Circadian Rhythm in Visual Evoked Potentials

2.1.1 Exposure to nLD

Figure 1 shows a few of the V.E.P.'s in response to strong (left) and weak flashes (right), recorded during an experiment of 50 h duration and performed on a rabbit that had

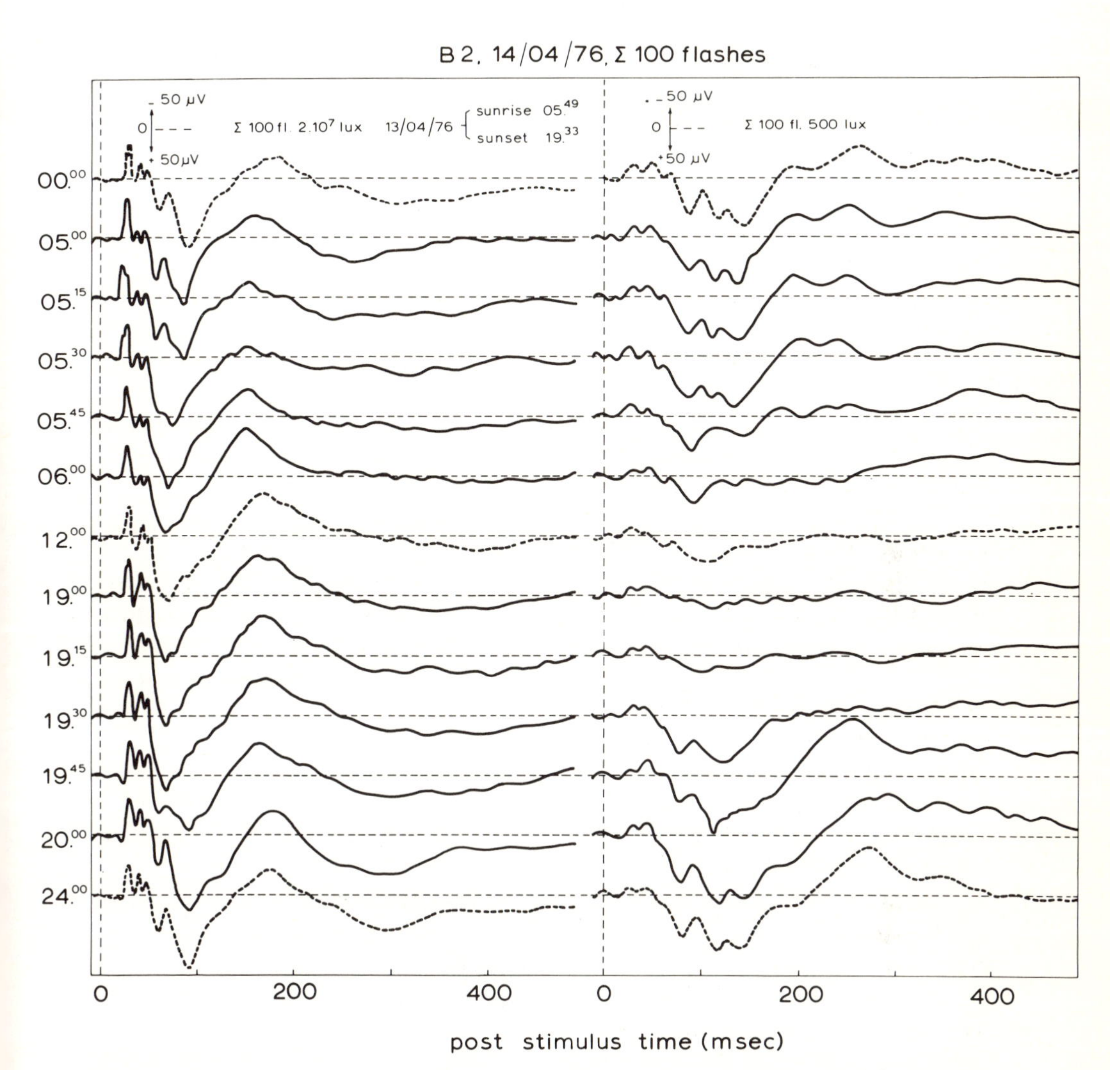

Fig. 1. Samples of the cortical Visually Evoked Potentials in response to very strong *(left)* and weak flashes *(right)*, recorded in darkness simultaneously from a rabbit that had previously been exposed to the natural LD alternations, with sunrise at 0549 and sunset at 1933 on the last day before the experiment. Note the simultaneous changes in the V.E.P.'s from Day Time Potentials towards Night Time Potentials and vice versa

previously been exposed to the nLD prevailing at 52° N during April. It is clear that there are actually *two types of V.E.P.* of which one occurs during a large part of the 24-h cycle and the other type during the rest of it, with fairly rapid transitions between these *two stable phases.* This is better seen when the amplitudes of some of the waves and the duration of one of their latencies are read off at suitable post-stimulus times (pst's) and then plotted vs. time for the whole experiment. This has been done in Fig. 2, from which it appears that in complete darkness there are spontaneous changes in shape of the V.E.P.'s which recur after 24 h, and that this circadian rhythm runs a peculiar time course with 2 phases in which the cortical photic responses are markedly different. There is a stable phase during which the animal responds with V.E.P.'s referred to as *„Day Time Potentials"* (D.T.P.'s), coinciding with the period of daylight on the last day before the experiment was started, and a phase with *„Night Time Potentials"* (N.T.P.'s) corresponding with the last night before the experiment. Since the transitions from

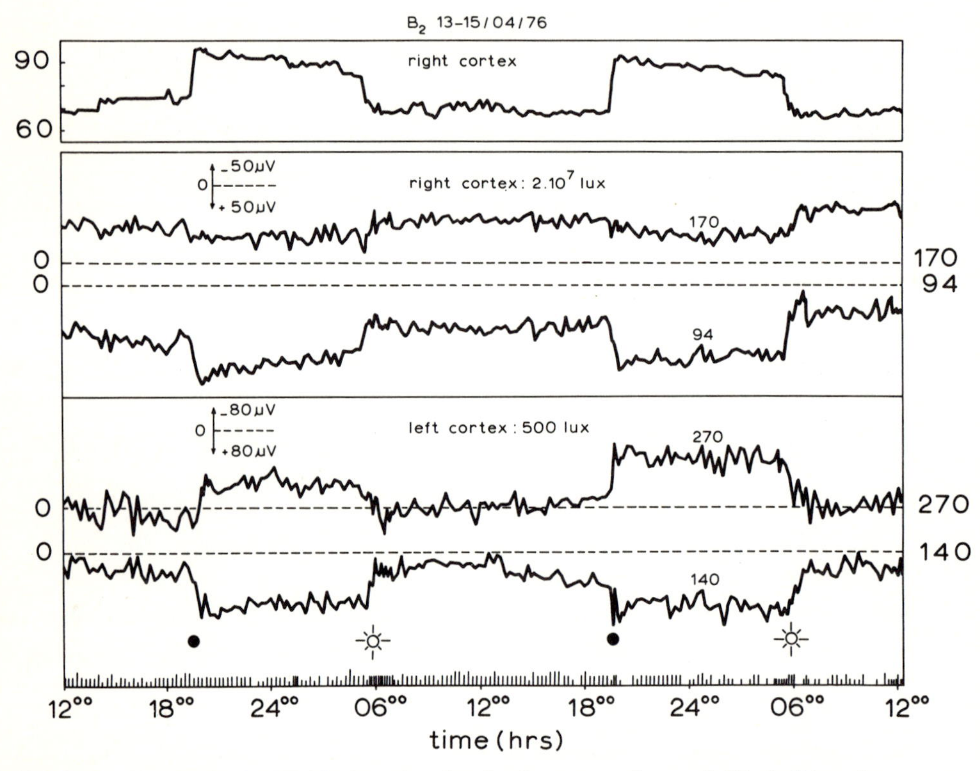

Fig. 2. Circadian changes in V.E.P.'s plotted vs. time for the same specimen as in Fig. 1. Curves show, from top to bottom, the time courses for the latency of the peak of one wave in the V.E.P. to strong flashes in ms and for amplitude and polarity of this response read off at post stimulus times 170 and 94 ms, and for the amplitude and polarity of the response to weak flashes, measured at the pst's 270 and 140 ms. Note the close correspondence between the onsets of transitions between the two phases of the rhythm and the moments of sunrise and sunset (marked by the symbols) on the day before each experiment was started. Reproduced, with permission from (Bobbert et al. 1978a)

one phase to the other are completed within 20-25 min and start very close to the times of sunrise and sunset during the previous day, it would seem that the durations of the two phases of this rhythm are predeterminated by the length of „day" and „night".

2.1.2 Programming of the Rhythm

The possible effect of the seasonal fluctuation in the nLD cycles was investigated in experiments performed at intervals of 8-12 weeks on two rabbits that were exposed for 15 months to nLD at 52° N. The results indicate that the transitions from the N.T.P.- to the D.T.P.-phase, and in the reverse direction, closely followed the moments of sunrise and sunset over the major part of the year (cf. Fig. 3 in Bobbert et al. 1978a). This phenomenon, henceforth denoted as „programming", is illustrated in the right half of Fig. 3 which shows for these, and two additional rabbits, the close relationship between the duration of the D.T.P. phases and the duration of daylight.

Programming of the rhythm by aLD was investigated in eight rabbits from which the V.E.P.'s were recorded after they had been exposed for at least 4 weeks to an aLD

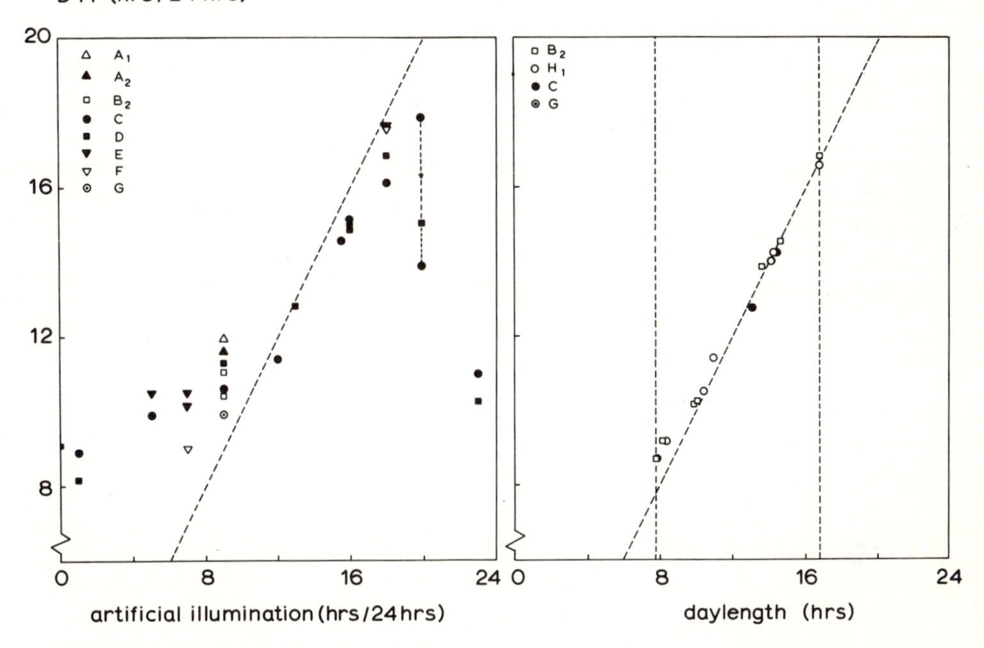

Fig. 3. The *right side* shows, for four rabbits programmed by the natural LD alternations, the way in which the duration of the D.T.P. phase of the rhythm changes with daylength between the durations of the (locally) shortest and longest days, of 7 h 45 min and 16 h 45 min respectively (indicated by the *interrupted vertical lines*). The *oblique line* represents ideal programming. The *left half* shows a similar plot for the results of experiments on eight animals which had previously been exposed to several types of artificial 24 h LD alternation. Note the similarity of the curves for both types of photoperiodic programming. (Bobbert et al. 1978a)

with a fixed photoperiod, resulting in a similar duration of the D.T.P. phases. Thereafter, the animal was exposed to another photoperiod for 4 weeks or more, whereupon the V.E.P. rhythm was recorded anew. This was done several times for each specimen while the photoperiod was varied in a non-systematical way. The results are summarized in the left half of Fig. 3, from which it can be derived that the duration of the D.T.P. phase corresponds to the photoperiod within a range from about 10 up to 16 h (Bobbert et al. 1978c). From a comparison between the data obtained after aLD (left diagram) with those after nLD (right diagram), it follows that the programming ranges differ only slightly. Outside these ranges the close relation between the time courses of the V.E.P. rhythm and of the LD schedule is lost, although the D.T.P. phase can still be shortened or lengthened a bit further, down to some 8.5 h and up to about 18.5 h.

Since an aLD 14:10 regimen is suitable for programming the rhythm, this regimen was used as pretreatment preceding nearly all of the succeeding experiments.

2.1.3 Accuracy of the Programming

Many rabbits were exposed to aLD 14:10 for some time with lights-OFF at 14^{00} h. Subsequently, simultaneous records were taken in darkness of the ERG and V.E.P. from both sides, at intervals of 3 min. Using the average onset of the changes in these four recordings, the moment at which the transition started from D.T.P.- to the N.T.P.-phase could be established with fair precision. This was done in 52 experiments on 16 rabbits, from which it appeared that this transition started, on the average, at 13 h 59 min 23 s, with a standard deviation of 6 min 23 s. This means that a suitable LD program can be copied with high accuracy by the circadian mechanism which gives rise to the rhythm in photic responses.

2.1.4 Exposure to Continuous Illumination (LL) or Darkness (DD)

Seven rabbits were exposed, first, to the standard-LD 14:10 for at least 3 weeks, thereupon for 4 up to 14 weeks to LL (75-125 lx), and, finally, to DD for 8 up to 20 weeks. After the first and the second section of these experiments the rhythm was recorded once for 50 h, and several times during the third section when the animals were exposed to DD. After aLD, there occurred typically programmed rhythms, with periods of 24 h and an average duration for the D.T.P.-period of 13 h 55 min (range: 13 h 47 min – 14 h 22 min). After exposure to LL for 4 weeks or more, no rhythm in photic responses could be detected while the V.E.P.'s were steadily of the type of N.T.P.'s (four animals) or of a type in between N.T.P.'s and D.T.P.'s. The circadian V.E.P. rhythm reappeared within 5 weeks of exposure to DD, but was now free-running with an average period of 24 h 25 min (range: 24 h 04 min - 24 h 43 min) and an average duration of the D.T.P. phase of only 8 h 38 min (range: 7 h 31 min - 9 h 29 min). Under these conditions this rhythm in photic responses was closely coupled to the circadian rhythm in eating activity, the transition from N.T.P.- to D.T.P.-phase coinciding with a marked drop in the amount of food intake per unit of time.

2.1.5 Simultaneous Changes in Photic Responses of Subcortical Structures and in the Retina

In experiments on four rabbits in which electrodes had been implanted both epidurally and subcortically (within the optic tract, the dorsal nucleus of the lateral geniculate body, and in the optic radiation) it was observed that, concurrently with the V.E.P. rhythm, circadian changes occur at all subcortical levels of the visual pathway, including the retina (ERG), with a time course that is programmed by the preceding LD schedule. In as much as the occurrence of simultaneous changes in the response of the visual cortex to *electrical stimulation at subcortical levels* – and changes in threshold for this type of stimulation – were completely absent, it may be concluded that the circadian changes in V.E.P.'s and in subcortical photic responses are merely the result of a circadian rhythm in photic responses of the *retina*, with an identical time course.

2.2 Evidence for the Existence of a Retino-Hypothalamo-Retinal Loop

2.2.1 Effects of Optic Nerve Sectioning

After *unilateral* sectioning of the optic nerve in rabbits exposed to LD 14:10, there are still the usual and synchronously running programmed rhythms in the ERG's of both sides and in the V.E.P.'s and subcortical photic responses on the ipsilateral side. This proves that the circadian changes in retinal photic responses are not brought about by means of the centrifugal fibres present within the optic nerves. On the other hand, after *bilateral* sectioning of the optic nerve in four rabbits there were, also, the usual and simultaneously occurring circadian changes in both ERG's but the rhythms were not programmed any more. They were free-running, with an average period of 24 h 20 min (range: 24 h 8 min - 24 h 34 min) and with, on the average, a short D.T.P. phase of 8 h 27 min (range 7 h 58 min - 8 h 54 min). These values are to such a degree similar to the data obtained from intact rabbits which have been exposed to DD (cf. Chap. 3.4), that they may be considered to represent some „basal" rhythm. Its period closely corresponds to that of the free-running rhythm in food intake of rabbits (to be published). In mammals, it has been shown that such circadian rhythms almost certainly originate in the suprachiasmatic nuclei of the hypothalamus (see Rusak, Chap. 5.5, this Vol).

2.2.2 Effects of Sectioning and Stimulation of the Cervical Sympathetic Nerves, and of Pinealectomy

After *unilateral* sectioning of the cervical sympathetic nerve (CSN), the time course of the rhythm is unchanged although it may be reduced in amplitude for the ipsilateral ERG and for the contralateral V.E.P. However, after bilateral transsection of the CSN no rhythm could be found in the photic responses recorded from anywhere within the visual pathway. Afterwards, the responses were constantly of the D.T.P.-type which suggests that bilateral blocking of the CSN leaves the retinae in a permanent state of

D.T.P.'s. This would imply that the rhythm in photic responses results from a raised CSN discharge during the N.T.P.-phase as compared with the other phase, and that the central circadian mechanism induces the retinal rhythm by way of the CSN's. This is supported by the effects of electrical stimulation of one CSN (Brandenburg et al. 1981) and of the application of noradrenalin and α-sympathomimetics to one of the eyes (Brandenburg et al. 1981, 1982), which always resulted in marked rises in ERG amplitude and in changes in shape of the V.E.P.'s, from the type of D.T.P.'s to that of N.T.P.'s.

Pinealectomy, performed on another four rabbits, had no influence at all on the rhythm; the usual programmed changes in photic responses occurred in each animal although its pineal gland had completely been removed.

2.2.3 Programming in the Presence of a Retino-Hypothalamo-Retinal Loop

In summary, it may be stated that there is ample evidence for the existence of a loop in which the basal rhythm, generated within the hypothalamus, is programmed by the external LD program via the retina and subsequently fed back to the retina. Looking at the phenomenon of „programming" in this way, it may be considered to consist of both an entrainment of the basal rhythm to 24 h, and a lengthening of the basal D.T.P. phase to the duration of the external photoperiod. The question whether this occurs in a closed loop remains to be solved. It would imply that the CSN fibres involved influence the ganglion cells giving rise to the retinohypothalamic tract (Rusak and Zucker 1979). Several observations argue in favour of a closed loop, among them the large amplitude of the retinal rhythm (Fig. 2), and the fact that the rhythm can accurately be re-programmed even by a *single exposure* to a slightly different LD regimen, such as apparently occurs under natural conditions from one day to another (see Sect. 2.1.1).

2.3 Nature of the Circadian Changes in Retinal Function

2.3.1 Sensitivity to Flashes is Higher During the N.T.P. Phase

In five rabbits, which had been subjected to the usual LD 14:10, the stimulus-response relationship was measured over a wide range of flash intensities during both phases of the rhythm. Taking the threshold for evoking the ERG's and V.E.P.'s as criteria, it was established that retinal photosensitivity is, on the average, 8 times higher during the N.T.P.-phase than in the other phase (with extreme values of 4 and 16 times). Furthermore, such a difference exists over a wide range of flash intensities, from the threshold intensity for evoking an ERG during the D.T.P. phase up to an intensity more than 4.10^7 times higher.

2.3.2 Functional Organization of the Retina in the Two Phases and the Process of Dark Adaption

In the aforementioned experiments (Sect. 2.3.1) while the sensitivity to flashes of any intensity was always lower during the D.T.P. phase, the shapes of the ERG and V.E.P.

evoked during the D.T.P. phase by flashes of whatever intensity could *never* be reproduced in the N.T.P. phase using flash intensities lower than those used in the other phase. This, together with the observed difference in photosensitivity (Sect. 2.3.1) indicates that, if the process of dark adaption is considered to reflect an increase of the signal to noise ratio during which the level of biological noise is reduced by removal of the metabolites accumulated during illumination, then the change from the D.T.P.- to the N.T.P.-phase of the circadian rhythm may be regarded as reflecting a *change in the functional organization* of the neuronal network within the retina. This idea is strongly supported by the fact that the ERG's recorded during both phases of the rhythm differ mainly in the amplitude and latency of their b-waves, which originate from the bipolar cells of the retinal layer also containing the horizontal and amacrine cells, the very cells capable of changing the structure and extent of the receptive fields of optic nerve fibres. This would imply that fibres of the CSN terminate within the retina close to or upon these cells.

The conception, that the transition from the D.T.P.- to the N.T.P.-phase of the rhythm differs basically from the change in photosensitivity during adaption to darkness, is supported by observations made in five other rabbits. In these, the course of dark adaption was followed for the ERG's and V.E.P.'s over a 130-min period both during the D.T.P.- and the N.T.P.-phase of a programmed rhythm. It appeared that the relative velocity of adaption to darkness – with the final and strongly different values for the amplitude of the b-wave taken as the 100 % criterion – was exactly the same in both phases (Brandenburg et al. 1982). This further suggests that the mechanisms responsible for dark adaption are not influenced by the CSN's, in contrast with those involved in circadian changes in the functional organization of the retina.

2.3.3 Transitions from One Phase of the Rhythm to the Other During Illumination

Eight rabbits were programmed by the standard LD 14:10, with lights-OFF at 1400. After more than 4 weeks, experiments were made in which the ERG's and V.E.P.'s were recorded from both sides for 9 h in such a way that the D.T.P.-N.T.P.-transition had to occur about half-way through this period. The effective illuminances during each flash were the usual ones (see Sect. 1) but now the left eye was provided with a contact lens (electrode) with an artificial pupil 2 mm in width, whereas the right eye was uncovered and provided with a completely translucent contact lens. Two types of experiment were performed in these circumstances.

In one type of experiment the animals were subjected to steady diffuse illumination from 1200 to 1600, with an effective illuminance, at the location of the eyes, of 1200 lx for four animals and of 2500 lx for the others. During the rest of each experiment, from about 0930 till 1200 and between 1600 and 1830, they were kept in complete darkness, apart from the 3 μs flashes delivered every 1503 ms. Therefore, during these experiments the D.T.P.-N.T.P.-transition had to take place during illumination. In the other experiments the same specimens were exposed to steady illumination from 0900 till 1130 and from 1600 until 1800; hence the transition could now occur in complete darkness between 1130 and 1600. It is clear that in both types of experiment, as long as the background illumination is on, the signal to noise ratio (S/N) for the flashes is fairly good for the left eye which looks directly into the flashes, though through a small pupil of 2 mm width. The S/N is very low for the right eye which receives only reflected light from the flashes through a pupil that is at least 4 times larger in diameter than on the left side, so that the background illumination creating noise must be more than 16 times higher for the right eye. The ERG's were recorded at 3-min intervals and only the mean amplitude of the b-waves of the last 10 ERG's of each period was used for making comparisons. During these experiments it was observed that a D.T.P.-N.T.P.-transition occurred also during illumination, and that the mean amplitude of the ERG b-wave increased by, on the average, 15.4 % for the left eye and 12.6 % on the right eye. When the transition occurred in darkness, the corresponding values for these rabbits were 25.6 % and 18.1 %, respectively (to be published). These transitions therefore definitely occur also in the presence of bright background illumination while, if the ERG amplitude is taken as the criterion, it can be stated that a transition is clearly diminished in amplitude by steady illumination.

The experiments described in the preceding paragraph further allowed a comparison between the amplitudes of the ERG b-waves during illumination vs. those during darkness in both the D.T.P.- and N.T.P.-phase of the rhythm. The ratio ERG during illumination/ERG in darkness was, on the average, 0.515 for the left eye and 0.172 for the right eye during the D.T.P. phase, whereas the corresponding values were 0.475 and 0.162, respectively, in the N.T.P. phase. From this it is clear not only that in the presence of a continuous background illumination the response to flashes is diminished, as was to be expected, but also that this steadily suppressive influence is stronger during the N.T.P. phase than during the other phase. These observations suggest that the afore-described (Sect. 2.3.1, 2.3.2) differences in photic responses between the two phases are applicable to flashes of *any* duration longer than 3 μs and, moreover, that they are also relevant for the rabbit's daily life (Bobbert et al. 1978a; Brandenburg et al. 1982).

3 Concluding Remarks

The results obtained until now can be looked at in the light of the two-oscillator model (Pittendrigh and Daan 1976) developed to explain the link between the phenomena of circadian rhythmicity and animal photoperiodism (Elliott 1976). In this model the induction of photoperiodic events is coupled with the occurrence of distinct differences in the phase-angle between two oscillators one of which, the „morning oscillator", responds to lights-ON whereas the other one, the „evening oscillator", reacts to lights-OFF. The phenomenon of „programming" of the time course for the rhythm in photosensitivity by natural or artificial LD cycles (Sect. 2.1.2) over a wide range of photoperiods (Fig. 3), and with a high accuracy with respect to the timing of its course (Sect. 2.1.3) is in agreement with this model.

The existence of a rhythm in photic responses, based on the presence of a retino-hypothalamo-retinal loop (Sect. 2.2), during which a „programmed" rise in photic sensitivity (Sect. 2.3.1) occurs very close to the time of sunset and a lowering at dawn (Sect. 2.1.2) while photosensitivity is continuously higher (Sect. 2.3.3) during the N.T.P. phase – i.e., the very phase corresponding with the period of nocturnal dark – would seem to be of great advantage for the organism. Apart from this, such a rhythm in retinal sensitivity to light may be useful for daily re-programming of the central circadian mechanism (Sect. 2.1.2), provided that the retino-hypothalamo-retinal loop represents a closed circuit by the existence of intraretinal connections between CSN fibres and the retinal structures projecting towards the hypothalamus. This remains to be solved, together with the problem whether two distinctive circadian oscillators are, indeed, situated within the tiny suprachiasmatic nuclei which are said to function as „the" pacemaker of the central circadian system in rodents (see Chap. 3, this Vol.).

References

Bobbert AC, Krul WH, Brandenburg J (1977) Nonvisual influences on the rabbit's visual system. Act Nerv Super 19: 186-187

Bobbert AC, Brandenburg J, Krul WH (1978a) Seasonal fluctuations of the circadian changes in rabbit Visual Evoked Potentials. Int J Chronobiol 5: 519-532

Bobbert AC, Krul WH, Brandenburg J (1978b) Diurnal changes in the rabbit's Visual Evoked Potential. Int J Chronobiol 5: 307-325

Bobbert AC, Krul WH, Brandenburg J (1978c) Photoperiodic programming of diurnal changes in rabbit Visual Evoked Potentials. Int J Chronobiol 5: 327-344

Brandenburg J, Bobbert AC, Eggelmeyer F (1981) Evidence for the existence of a retino-hypothalamo-retinal loop in rabbits. Int J Chronobiol 8: 13-29

Brandenburg J, Bobbert AC, Eggelmeyer F (1982) Circadian changes in the response of the rabbit's retina to flashes, mediated by sympathetic fibres. Behav Brain Res (in press)

Elliott JA (1976) Circadian rhythms and photoperiodic time measurement in mammals. Fed Proc 35: 2339-2346

Pittendrigh CS, Daan S (1976) A functional analysis of circadian pacemakers in nocturnal rodents. V. Pacemaker structure: a clock for all seasons. J Comp Physiol 106: 333-355

Rusak B, Zucker I (1979) Neural regulation of circadian rhythms. Physiol Rev 59: 449-526

Shkol'nik-Yarros EG (1971) Neurons and interneural connections of the central visual system. Plenum Press, New York

3.1 Physiological Models of the Rodent Circadian System

B. Rusak[1]

1 Background

C.P. Richter tried for a number of years to identify the physiological substrate underlying the endogenous generation of circadian rhythms in rats. After attempting a variety of neural and endocrine ablations, he reported the successful abolition of freerunning rhythms of activity, eating and drinking by lesions in the hypothalamus. He did not identify the locus of the effective lesions except to indicate that they were near the "ventral median nucleus" of the hypothalamus (Richter 1967).

The specification of the critical hypothalamic site at which lesions eliminated rhythmicity resulted from another line of research. This approach sought to identify the visual projections responsible for photic entrainment of daily rhythms of physiological and behavioral functions (Chase et al. 1969, Stephan and Zucker 1972a). Despite evidence that retinal efferents are essential for entrainment in mammals (Browman 1943, Richter 1967), interruption of the identified targets of these efferents caudal to the chiasm failed to eliminate entrainment. The paradox presented by these observations was resolved when two laboratories used autoradiographic techniques to describe a retinal projection that terminated in the suprachiasmatic nuclei (SCN) immediately dorsal to the chiasm (Moore and Lenn 1972, Hendrickson et al. 1972, Mai 1978).

Ablation of the SCN in order to interrupt the newly identified retinohypothalamic tract (RHT) prevented photic entrainment of adrenal corticosterone, activity and drinking rhythms in rats (Moore and Eichler 1972, Stephan and Zucker 1972b). Unlike peripherally blinded rats, animals with SCN lesions did not generate free-running circadian rhythms. Instead, the distribution of the endpoints measured was apparently random with respect to time of day. Small lesions centered on the SCN eliminated the capacity for both photic synchronization and endogenous generation of circadian rhythms in rodents (Fig. 1).

These findings have been repeated for a variety of behavioral and physiological parameters (see Rusak and Zucker 1979), and have led to the hypothesis that the SCN and RHT are essential to the generation and entrainment of mammalian circadian rhythms. This basic hypothesis is represented by the model illustrated in Fig. 2. This chapter summarizes the experimental results obtained during the last decade that are relevant to the assessment of this hypothesis and develops a revised model to incorporate these findings.

1 Department of Psychology Dalhousie University Halifax, Nova Scotia Canada B3H 4J1

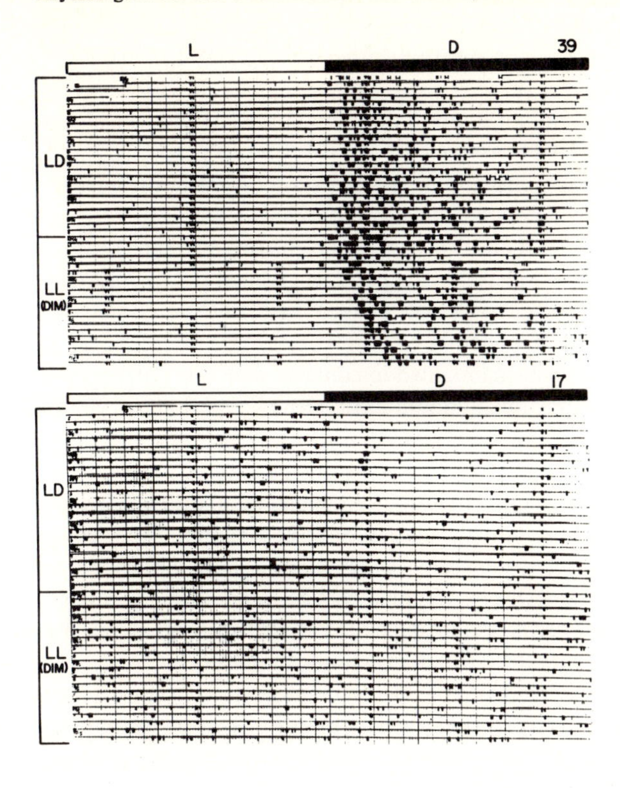

Fig. 1. 24-h records of drinking-tube contacts in an intact **(A)** and a SCN-ablated **(B)** hamster under a light-dark cycle *(LD)* and in constant dim illumination *(LL)*. The intact hamster entrained and free-ran with a circadian period. Both entrainment and the free-run are abolished by SCN lesions. (Rusak 1977a)

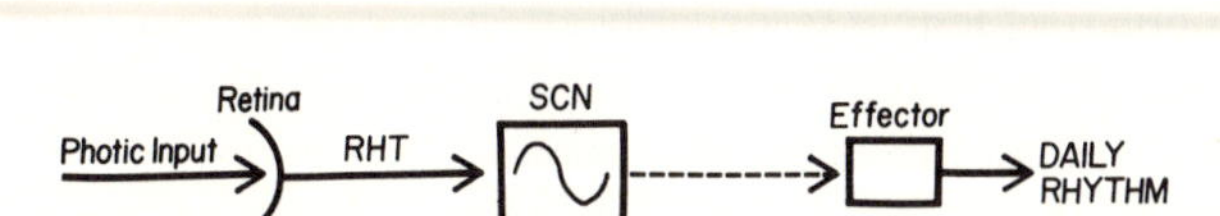

Fig. 2. A model of circadian rhythm generation and entrainment in mammals. In this and subsequent figures, the following code applies: an endogeneously self-sustained oscillator is represented by (∿); an anatomically identified pathway is represented by a *solid line;* a structure or pathway that is unidentified but must exist on the basis of functional evidence is represented by a *dashed line;* a pathway that may exist but for which there is no compelling functional evidence is represented by a *dotted line*

2 Entrainment

The observation that loss of the RHT prevents entrainment while damage to other optic projections does not is rendered ambiguous by the fact that RHT interruptions are accomplished in rodents by ablation of the SCN. Since RHT interruptions involve destruction of the putative oscillator in the SCN while other visual pathway lesions do not, one cannot attribute differential effects on entrainment solely to the different visual projections damaged. Destruction of the rhythmic substrate on which visual input can act may prevent entrainment via projections other than the RHT, even if these projections are competent to produce entrainment.

The studies that indicated that visual projections other than the RHT are not involved in entrainment were limited in scope. They generally asked only whether normal nocturnality under a light-dark (LD) cycle survived interruption of these projections. More detailed studies conducted under several lighting conditions have revealed effects of visual system damage that indicate a role for visual projections other than the RHT in mediating photic effects on rhythms.

Lesions of the optic tracts that did not impinge directly on the RHT in hamsters altered several aspects of circadian activity rhythms. In one study lesions centered on the lateral geniculate nuclei (LGN) did not change steady-state entrainment but severely retarded the rate of re-entrainment to a 12-h shift in the LD cycle (Zucker et al. 1976). In a second study, lesions of the optic tracts at the level of the chiasm lengthened the active phase (α) under a LD cycle, slowed phase-shifting and lengthened the freerunning period (τ) in constant dim illumination (LL) (Rusak 1977b). In a later study some of these changes were found to occur independently of others after optic tract lesions. Period lengthening was observed in animals maintained in constant darkness (DD), as well as in LL, although τ lengthened proportionally more in LL (Rusak and Boulos 1981; Fig. 3).

These results imply that the lesions affected both the organization of rhythms in DD and their photic responsiveness. Changes that do not depend on photic input might still reflect visual system damage. Many retinal ganglion cells show spontaneous activity in DD ("dark discharge"); lesion-induced damage to ganglion cells could modify the function of their target structures even in DD. The complex changes produced may reflect altered coupling among component oscillators of the circadian system as well as changes in the influence of light on these oscillators (Rusak and Boulos 1981).

Several possible mechanisms may mediate the observed lesion effects. The primary and accessory optic tracts may reach targets that mediate photic effects on rhythms independently of the RHT-SCN system. Although entrainment survives optic tract interruption, one report described large reductions in nocturnality in rats bearing striate cortex or superior colliculus lesions (Altman 1960).

Another possibility is that optic tract lesions secondarily affect rhythms by modifying visual input to the SCN. A projection from the ventral nucleus of the LGN (LGN_V) reaches the SCN (Swanson et al. 1974); its function might be altered by the loss of afference to the LGN. Reduced photic responsiveness of SCN neurons has been reported after LGN lesions in rats (Groos and Mason 1978). In addition, at least some of the axons forming the RHT are collaterals of larger fibers that run in the optic tracts (Mason et al. 1977, Millhouse 1977). Destruction of the optic tracts may cause degeneration of these collateral fibers.

The loss of RHT fibers, the loss of a LGN_V-SCN projection or consequent sprouting or pruning responses of surviving RHT fibers (Schneider 1973) may contribute to functional changes in visual input reaching the SCN. The structure of the RHT after optic tract lesions has not been studied, but a limited developmental plasticity of the RHT has been described (Stanfield and Cowan 1976). It is uncertain, however, whether any of these mechanisms would be rapid enough to mediate the changes in rhythm parameters which occur soon after surgery (Fig. 3).

It is also possible that some of the photic or nonphotic effects of the lesions are attributable to damage to lateral hypothalamic structures adjacent to the optic tracts. Recent studies have described direct routes for visual input to the lateral and anterior hypothalamus (Silver and Brand 1979, Riley et al. 1981, Mai 1979 and Chap. 2.4, this Vol.); whether these projections mediate photic influences on rhythms has not been established, but an earlier report has described effects of lateral hypothalamic lesions on rhythms in rats (Rowland 1976).

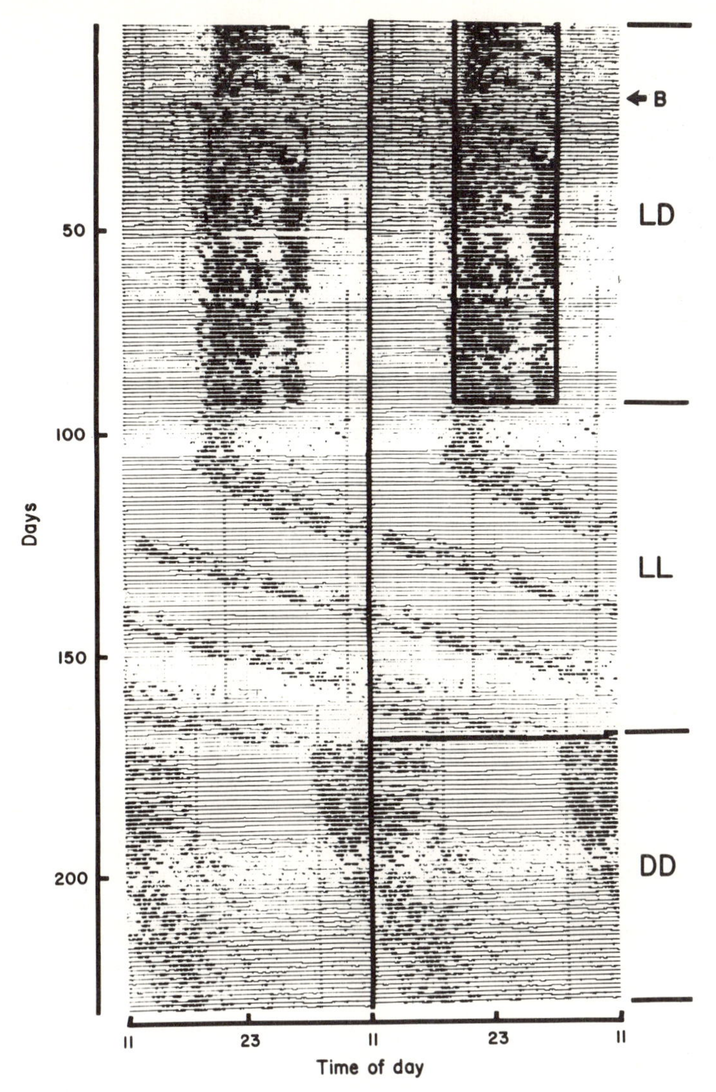

Fig. 3. Double-plotted wheel-running activity record of a male hamster that received bilateral lesions of the primary and accessory optic tracts *(arrow at B)* while in LD 14 : 10. Under entrained conditions there is an increase in the duration of activity and an apparent change in the entrained phase angle. Periods expressed under constant conditions were unusually long (25.33 h in LL, 24.20 h in DD). (Rusak and Boulos 1981)

The results of SCN lesion studies also indicate that visual projections other than the RHT can mediate photic effects on rhythms. Lesions of the SCN/RHT do not uniformly prevent photic synchronization of rhythms. Some hamsters bearing complete SCN lesions synchronized wheel-running activity to the D phase of LD cycles (Fig. 4); others showed intermittent or unstable synchronization (Rusak 1975, 1977a). Synchronization demonstrates the existence of an effective visual input, but these hamsters did not meet the usu-

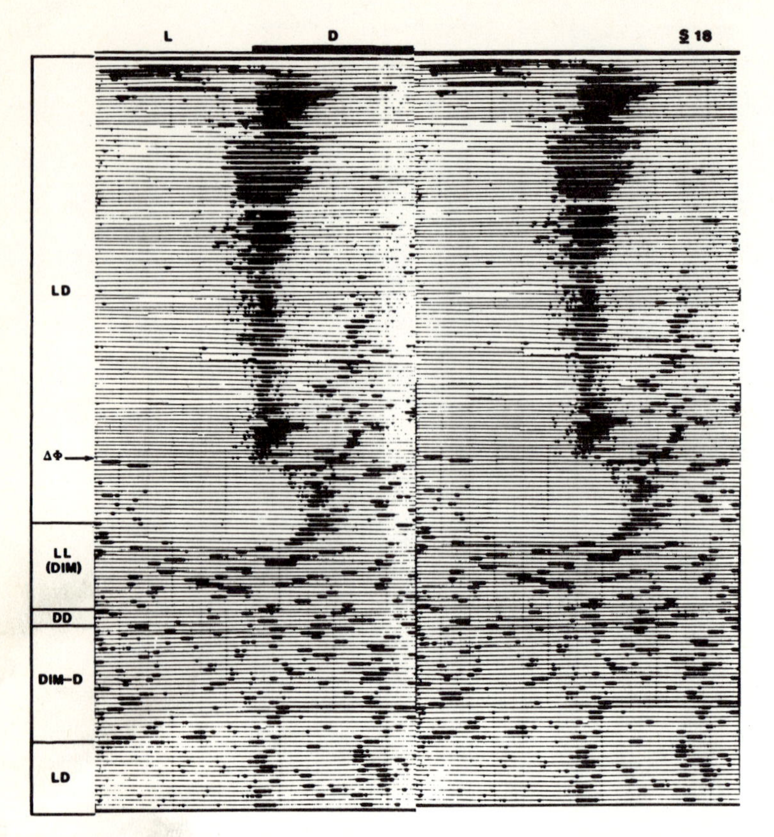

Fig. 4. Double-plotted wheel-running activity record of a SCN-ablated hamster. The hamster entrained activity to the LD cycle, followed a phase shift ($\Delta\Phi$) of the cycle, but failed to generate a circadian rhythm in constant conditions (LL). The noisy pattern generated in LL follows an apparent dissociation between short- and long-period oscillations that appear when LL is imposed. (Rusak 1977a)

al criteria for entrainment (Menaker and Eskin 1966) since they failed to generate intact circadian rhythms under constant conditions.

However, several features of the activity records indicated that circadian oscillators were being entrained. Activity onset sometimes anticipated the beginning of the D phase and the rhythm phase shifted with several transient cycles after a shift of the LD cycle. The synchronized rhythm sometimes persisted for a few cycles after being released into constant conditions, although the integrity of the rhythm was then rapidly lost (Fig. 4). All of these features suggest that circadian oscillators outside the SCN were being entrained via visual projections other than the RHT.

Reports of continued photic synchronization of some rhythms in rats bearing SCN lesions are consistent with this conclusion (e.g., Abe et al. 1979). The visual projections that mediate photic influences on rhythms after SCN/RHT ablations might include the classical primary and accessory optic tracts and the recently described routes for visual input to the preoptic area, anterior and lateral hypothalamus (Silver and Brand 1979, Mai 1979 and Chap. 2.4, this Vol., Riley et al. 1981).

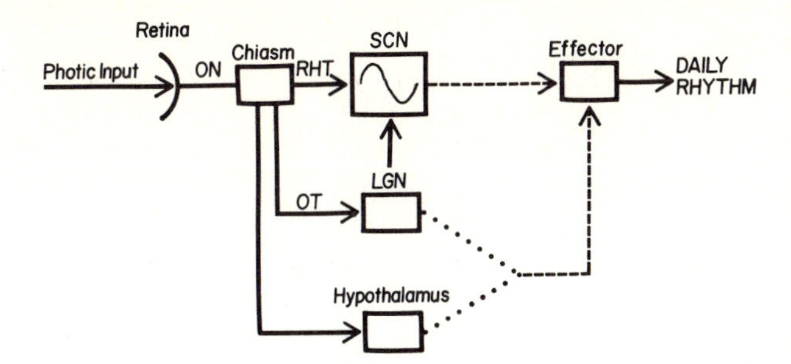

Fig. 5. A model of circadian rhythm generation and entrainment that incorporates a role for visual projections other than the RHT in entrainment. See text and Fig. 2 for details

The persistence of photic synchronization of rhythms after SCN/RHT ablations and the multiple changes in rhythms after optic tract ablations that spare the SCN provide functional evidence for the complexity of the visual afference to the mammalian circadian system. This complexity is also reflected in the emerging anatomical evidence for multiple direct and indirect routes by which photic information can reach parts of the hypothalamus, including the SCN. These conclusions require modification of the hypothesis concerning entrainment mechanisms. The model presented in Fig. 5 summarizes the known relations among visual projections that may play a role in photic entrainment (see Groos, Chap. 3.4, this Vol.).

3 Generation

The loss of normal circadian organization after SCN ablations does not necessarily imply that the oscillators responsible for rhythm generation have been destroyed. The evidence that rhythms can be entrained (rather than merely photically driven) after SCN lesions, and the evidence for transient persistence of rhythms in constant conditions imply the survival of circadian oscillators (Fig. 4).

This suggestion is reinforced by the pattern of activity generated by lesioned hamsters maintained in constant conditions. The records of many hamsters include activity bouts that express circadian periods; several bouts can sometimes be identified and followed over a number of cycles as they run with independent circadian rhythms. These bouts appear to be the products of circadian oscillators which may remain uncoupled or may become stably coupled and generate circa 8 or 12 h activity rhythms (Fig. 6). The loss of circadian organization appears to result from the dissociation of a number of these component oscillators (Rusak 1975, 1977a).

These findings may be explained in a number of ways; for example, the circadian generating mechanism may consist of a large number of (circadian?) oscillators that depend on mutual interactions to remain coupled (Winfree 1967, Pavlidis 1969). The size of the oscillator population may be crucial to the coupling process and SCN lesions may eliminate part of this population. As a result, the remaining oscillators may be incapable of maintaining sufficient coupling strength to express an integrated circadian rhythm. An external input such as a circadian illumination cycle may exert phase control over many of these oscillators and thereby reintegrate the overt rhythm (Fig. 4). The oscillators pre-

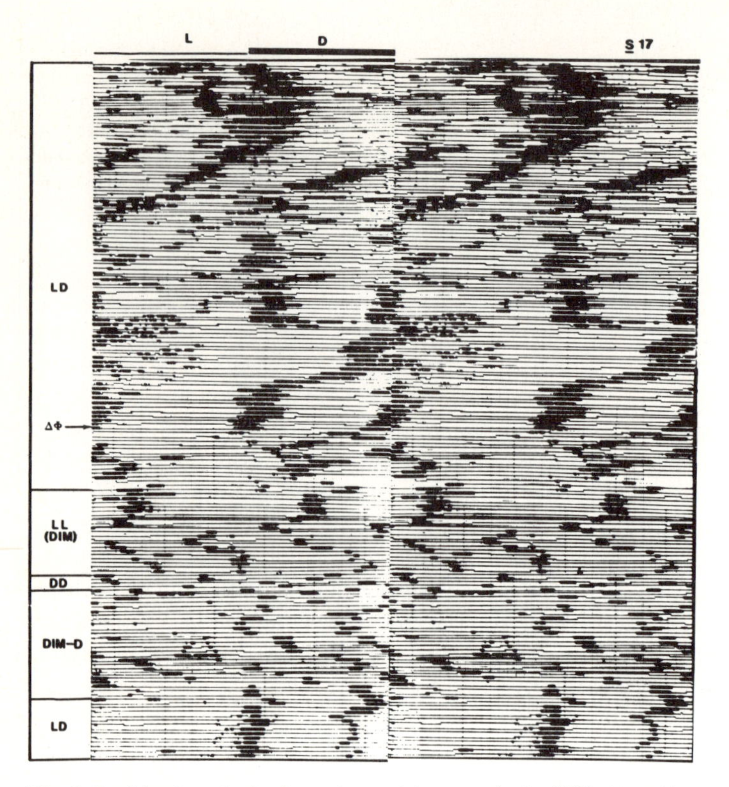

Fig. 6. Double-plotted wheel running activity record of a SCN-ablated hamster. This record illustrates the apparent existence of multiple circadian oscillators with different periods after SCN ablation. These oscillators may be synchronized by the LD cycle and may couple to each other, as well as freerunning with independent periods. See Fig. 4 for details. (Rusak 1977a)

sumed to be in the SCN may be functionally similar to those outside the SCN or they may be specialized to hierarchically regulate the rhythms of other oscillators.

An alternative model derives from evidence that the output of component (or "bout") oscillators may be observed even in intact rodents. Davis and Menaker (1980) hypothesized that the output of a bout oscillator is normally expressed in overt behavior only when it coincides with the daily active phase (α), the limits of which are set by a dominant oscillator. They suggested that SCN ablations may eliminate this superordinate oscillator and permit behavioral expression of the free-running bout oscillators at all times.

These models have the following features in common: that circadian oscillators exist outside the SCN, that they continue to function after SCN destruction, and that they are subject to regulation by the SCN. Coupling of these oscillators into stable ultradian patterns indicates a capacity for mutual interaction, and their photic synchronization demonstrates access to light information that is independent of the SCN. The SCN may influence effector systems both through control of the component oscillators and through non-oscillatory mechanisms. These features are summarized in Fig. 7.

The location of the component oscillators is unknown but it is likely that some of them are in the hypothalamus. In one study, large lesions that destroyed both the SCN

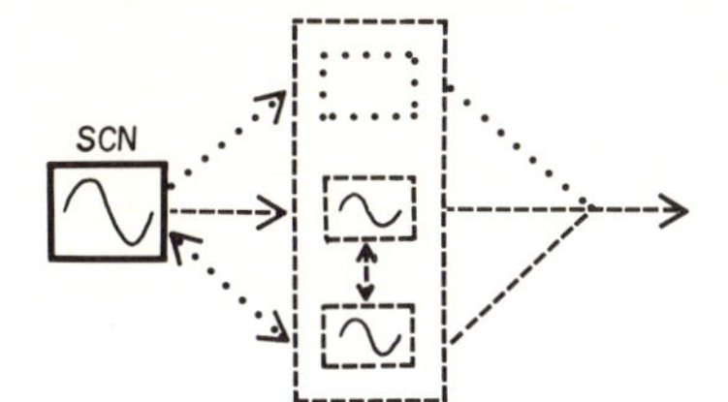

Fig. 7. A portion of a model for circadian rhythm generation, incorporating a population of circadian oscillators that are dominated by the SCN and that show mutual interactions as well. The box that does not contain a symbol for an oscillator represents the possibility that the SCN can influence effector systems directly as well as through intervening oscillators. See Fig. 2 for details.

and considerable surrounding tissue produced the most severe disruptions of rhythmicity (Rusak 1975, 1977a). Since extra-SCN oscillators can be synchronized by LD cycles, the evidence that structures near the SCN receive visual input is consistent with the suggestion that some of these oscillators are in the hypothalamus.

4 Nonphotic Entrainment

Environmental cycles other than that of illumination can influence the circadian system of mammals. For example, rats increase their activity levels in anticipation of a limited daily period of food availability. The involvement of a circadian mechanism in timing the anticipatory activity was demonstrated by Bolles and Stokes (1965) who showed that rats failed to anticipate noncircadian schedules of food availability. However, a free-running circadian activity rhythm can coexist with a restriction-synchronized activity peak (Boulos et al. 1980), and the free-running rhythm's phase may be unaffected by the feeding schedule (Boulos et al. 1980, Gibbs 1979; see Boulos and Terman 1979 and Chap. 1.2, this Vol. for reviews). These observations indicate that the feeding schedule affects a portion of the circadian system, but does not have access to the dominant pacemaker of the system.

Activity, drinking and corticosterone secretion can also be synchronized by feeding schedules in SCN-ablated rats, and only circadian cycles are effective (Krieger et al. 1977, Boulos et al. 1980; see Stephan, Chap. 3.7, this Vol.). This indication that extra-SCN oscillators are involved was reinforced by evidence that the schedule-induced rhythm persisted with a circadian period for a few cycles after the restriction schedule was ended (Boulos et al. 1980). The physiological consequences of rhythmic feeding cycles may synchronize a population of dissociated oscillators outside the SCN just as photic input does.

Whether the SCN also receive input from nonphotic cycles is unknown, but there is indirect evidence that they do not. The integrated endogeneous activity rhythm of an intact rat is apparently a product of the SCN, and this rhythm is not entrained by food restriction. Even when this rhythm is not expressed overtly during the feeding schedule, there is evidence that it continues its free-run (Boulos et al. 1980). A target of the unspecified synchronizing stimuli resulting from food restriction may be the ventromedial hypothalamus since lesions in this area abolished the schedule-induced corticosterone rhythm in rats (Krieger 1980). Whether these lesions also abolish other restriction-induced rhythms remains to be established. These findings are consistent with the suggestion that at least some extra-SCN oscillators are in adjacent hypothalamic areas.

5 Nonlesion Approaches

The technique of ablating a structure in order to determine its function is valuable but leaves many interpretational problems that the application of other techniques may help to resolve (see Schoenfeld and Hamilton 1977, Rusak and Zucker 1979). A few different approaches to studying SCN function that have been attempted are introduced below (see also Chap. 3.4, 3.5, and 5.1, this Vol.).

The technique of 2-deoxy-D-glucose autoradiography has been developed as a means of assessing levels of metabolic activity in brain structures (Kennedy et al. 1975). Schwartz and Gainer (1977) described a rhythm of metabolic activity in the SCN that peaked in the L phase; the rhythm persisted in DD and metabolic activity was affected by light exposure differentially depending on circadian phase (Schwartz et al. 1980). These findings suggest that SCN neurons undergo a daily rhythm of activity that is modulated by light exposure. These are the features that should characterize a central oscillator mechanism that can be entrained by LD cycles.

Neurophysiological studies have confirmed these two features: a rhythm of neural activity and its modulation by photic input. Multiple unit activity of SCN neurons varies with circadian phase and this rhythm persists after the hypothalamus has been surgically isolated (Kawamura and Inouye, 1979, Chap. 3.5, this Vol.). Several laboratories have studied the photic responsiveness of SCN neurons: most cells respond to light with increased firing, but some are inhibited by light (Nishino et al. 1976, Sawaki 1977, Groos and Mason 1978, see Groos, Chap. 3.4, this Vol.).

These findings are consistent with the hypothesis that the SCN generate a rhythm of neural activity that is modulated by photic input. The rhythmic output of the SCN may in turn synchronize the rhythms of other oscillators or influence effector mechanisms directly. Modification of SCN neural activity should therefore have significant effects on circadian rhythms. A recent study examined the effects of electrical stimulation in the SCN on rodent circadian rhythms. Brief electrical stimulation resulted in phase shifts and changes in freerunning periods of activity and eating rhythms in hamsters and rats (Fig. 8, Rusak and Groos 1982). The phase-shifting effects of electrical stimulation generally resembled the effects that light stimulation would have at equivalent circadian phaes. These findings support the idea that visual inputs to the circadian system converge on the suprachiasmatic region and that neural activity in this region can regulate the output of the circadian system.

Another feature that should be included in a model of the rodent circadian system is the influence of hormones on rhythms. Testosterone and estrogen have been shown to influence the period and the integrity of circadian rhythms in several rodent species (Daan et al. 1975, Morin et al. 1977, Morin 1980, Albers 1981, Eskes and Zucker 1978, see Zucker 1979, and Chap. 5.1, this Vol.). These hormones are probably produced rhythmically under the control of the central circadian system regulating pituitary function. Thus, organs that can be seen as effectors of the circadian system can modify its function. The target of these hormonal influences has not been established, but gonadal hormones are known to be taken up in the preoptic-suprachiasmatic region (e.g., Pfaff and Keiner 1973).

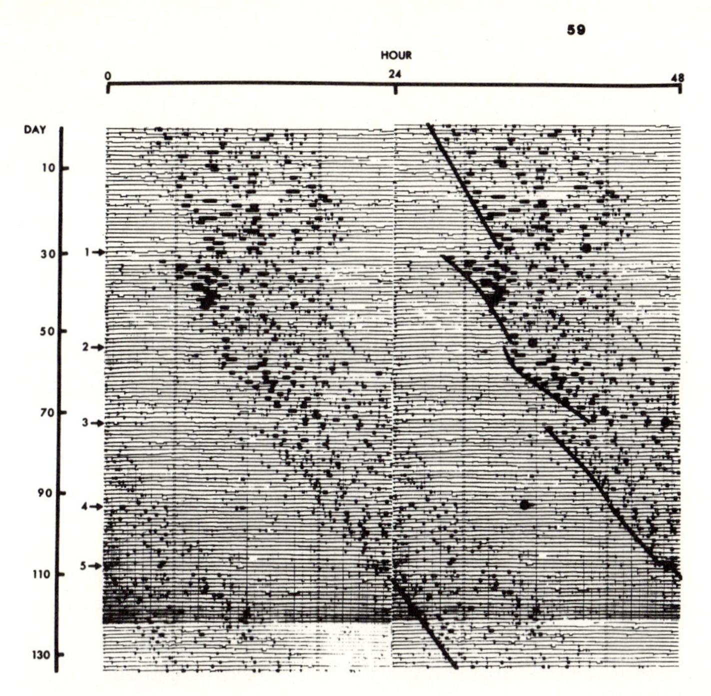

Fig. 8. Double-plotted record of wheel-running activity in a blinded hamster with stimulating electrodes implanted in the suprachiasmatic area. Electrical stimulation (indicated by *arrows*) caused no phase shift when administered for 10 min during the subjective day (#4), phase advances late in the subjective night (#'s 1 and 3), and period lengthening (#2) or a small delay (#5) early in the subjective night. (Rusak and Groos 1982)

6 Conclusions

The mammalian circadian system is comprised of a number of circadian oscillators whose integration depends on the SCN. Neural activity in the SCN may synchronize activity in extra-SCN oscillators. The RHT is important for photic entrainment, but other projections can synchronize components of the circadian system after SCN ablation and may modify features of rhythms in intact animals.

The hypothalamus may contain many endogenously rhythmic circadian oscillators, some of which are in the SCN. Since circadian systems evolved in the presence of daily illumination cycles, the domination of the circadian system by the SCN may have resulted from their receiving much of the visual input that entrains the entire system. The SCN may represent a population of hypothalamic oscillators that were "captured" by retinal projections and therefore came to exert phase control over other oscillators that have only limited access to critical photic information.

Some nonvisual inputs to the circadian system can affect oscillators outside the SCN, and there is indirect evidence that these inputs may not reach the SCN. Hormones can affect circadian mechanisms but the targets of these inputs are not known.

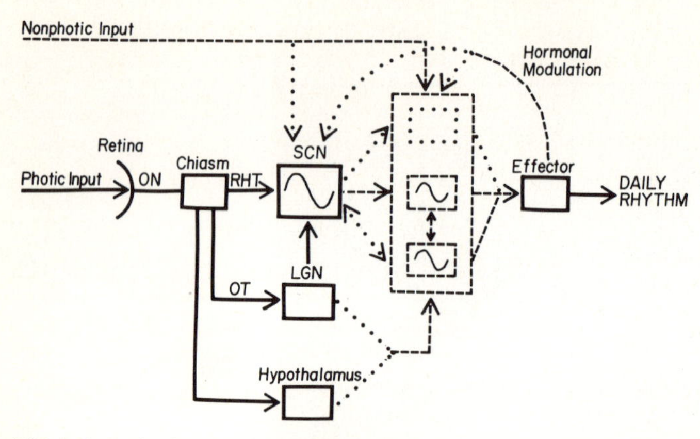

Fig. 9. A model for circadian rhythm generation and entrainment in mammals. This model incorporates those illustrated in Figs. 5 and 7, and adds mechanisms for nonphotic entrainment and for endocrine effects on the circadian system. See text and Fig. 2 for details

These features of the rodent circadian system are illustrated in Fig. 9. This model is a product of attempts to test the propositions represented in the simple model shown in Fig. 2. The studies generated in response to the original model have changed our understanding of the mechanism underlying mammalian circadian rhythmicity, although the SCN continue to occupy a central position in that mechanism. The model in Fig. 9 attempts to summarize both our current knowledge and our ignorance of the system. As with the earlier model, its usefulness depends not on its accuracy but on its capacity to indicate research problems that will help refine our understanding of the physiology of circadian rhythms.

Acknowledgments. Preparation of this manuscript and performance of some of the research described were supported by grant A-0305 from NSERC of Canada. I am grateful to Marilyn and Sarah Klein for preparing the figures and to Priti Jana for typing the manuscript.

References

Abe K, Kroning J, Greer MA, Critchlow V (1979) Effects of destruction of the suprachiasmatic nuclei on the circadian rhythms in plasma corticosterone, body temperature, feeding and thyrotropin. Neuroendocrinology 29: 119-131

Albers HE (1981) Gonadal hormones organize and modulate the circadian system of the rat. Am J Physiol 241: R62-R66

Altman J (1960) Diurnal activity rhythm of rats with lesions of superior colliculus and visual cortex. Am J Physiol 202: 1205-1207

Bolles RC, Stokes LW (1965) Rat's anticipation of diurnal and a-diurnal feeding. J Comp Physiol Psychol 60: 290-294

Boulos Z, Terman M (1979) Food availability and daily biological rhythms. Neurosci Biobehav Rev 4: 119-131

Boulos Z, Rosenwasser A, Terman M (1980) Feeding schedules and the circadian organization of behavior in the rat. Behav Brain Res 1: 39-65

Browman LG (1943) The effect of controlled temperatures upon the spontaneous activity rhythms of the albino rat. J Exp Zool 94: 477-489

Chase PA, Seiden LS, Moore RY (1969) Behavioral and neuroendocrine responses to light mediated by separate visual pathways in the rat. Physiol Behav 4: 949-952

Daan S, Damassa D, Pittendrigh CS, Smith ER (1975) An effect of castration and testosterone replacement on a circadian pacemaker in mice *(Mus musculus)*. Proc Natl Acad Sci USA 72: 3744-3747

Davis FC, Menaker M (1980) Hamsters through time's window: temporal structure of hamster locomotor rhythmicity. Am J Physiol 239: R149-R155

Eskes GA, Zucker I (1978) Photoperiodic regulation of the hamster testis: dependence on circadian rhythms. Proc Natl Acad Sci USA 75: 1034-1038

Gibbs FP (1979) Fixed interval feeding does not entrain the circadian pacemaker in blind rats. Am J Physiol 236: R249-R253

Groos G, Mason R (1978) Maintained discharge of rat suprachiasmatic neurons at different adaptation levels. Neurosci Lett 8: 59-64

Hendrickson AE, Wagoner N, Cowan WM (1972) An autoradiographic and electron microscopic study of retino-hypothalamic connections. Z Zellforsch 135: 1-26

Kawamura H, Inouye ST (1979) Circadian rhythm in an island containing the suprachiasmatic nucleus. In: Suda M, Hayaishi O, Nakagawa H (eds) Biological rhythms and their central mechanism. Elsevier/North-Holland, Amsterdam, pp 335-341

Kennedy C, DesRosiers MH, Jehle JW, Reivich M, Sharp F, Sokoloff L (1975) Mapping of functional neural pathways by autoradiographic survey of local metabolic rate with [^{14}C] deoxyglucose. Science 187: 850-853

Krieger DT (1980) Ventromedial hypothalamic lesions abolish food-shifted circadian adrenal and temperature rhythmicity. Endocrinology 106: 649-654

Krieger DT, Hauser H, Krey LC (1977) Suprachiasmatic nuclear lesions do not abolish food-shifted circadian adrenal and temperature rhythmicity. Science 197: 398-399

Mai JK (1978) The accessory optic system and the retino-hypothalamic system. A review. J Hirnforsch 19: 213-288

Mai JK (1979) Distribution of retinal axons within the lateral hypothalamic area. Exp Brain Res 34: 373-377

Mason CA, Sparrow N, Lincoln DW (1977) Structural features of the retinohypothalamic projection in the rat during normal development. Brain Res 132: 141-148

Menaker M, Eskin A (1966) Entrainment of circadian rhythms by sound in *Passer domesticus*. Science 154: 1579-1581

Millhouse OE (1977) Optic chiasm collaterals afferent to the suprachiasmatic nucleus. Brain Res 137: 351-355

Moore RY, Eichler VB (1972) Loss of a circadian adrenal corticosterone rhythm following suprachiasmatic lesions in the rat. Brain Res 42: 201-206

Moore RY, Lenn NJ (1972) A retinohypothalamic projection in the rat. J Comp Neurol 146: 1-14

Morin LP (1980) Effect of ovarian hormones on synchrony of hamster circadian rhythms. Physiol Behav 24: 741-749

Morin LP, Fitzgerald KM, Zucker I (1977) Estradiol shortens the period of hamster circadian rhythms. Science 196: 305-307

Nishino H, Koizumi K, Brooks CM (1976) The role of suprachiasmatic nuclei of the hypothalamus in the production of circadian rhythm. Brain Res 112: 45-59

Pavlidis T (1969) Populations of interacting oscillators and circadian rhythms. J Theor Biol 22: 418-436

Pfaff D, Keiner M (1973) Atlas of estradiol-concentrating cells in the central nervous system of the female rat. J Comp Neurol 151: 121-158

Richter CP (1967) Sleep and activity: their relation to the 24-hour clock. Res Publ Assoc Nerv Ment Dis 45: 8-27

Riley JN, Card JP, Moore RY (1981) A retinal projection to the lateral hypothalamus in the rat. Cell Tissue Res 214: 257-269

Rowland N (1976) Circadian rhythms and partial recovery of regulatory drinking in rats after lateral hypothalamic lesions. J Comp Physiol Psychol 90: 383-393

Rusak B (1975) Neural control of circadian rhythms in behavior of the golden hamster. *Mesocricetus auratus.* Dissertation, Univ California, Berkeley

Rusak B (1977a) The role of the suprachiasmatic nuclei in the generation of circadian rhythms in the golden hamster. *Mesocricetus auratus.* J Comp Physiol 118: 145-164

Rusak B (1977b) Involvement of the primary optic tracts in mediation of light effects on hamster circadian rhythms. J Comp Physiol 118: 165-172

Rusak B, Boulos Z (1981) Pathways for photic entrainment of mammalian circadian rhythms. Photochem Photobiol 34: 267-273

Rusak B, Groos G (1982) Suprachiasmatic stimulation phase shifts rodent circadian rhythms. Science 215: 1407-1409

Rusak B, Zucker I (1979) Neural regulation of circadian rhythms. Physiol Rev 59: 449-526

Sawaki Y (1977) Retinohypothalamic projection: electrophysiological evidence for the existence in female rats. Brain Res 120: 336-341

Schneider GE (1973) Early lesions of superior colliculus: factors affecting the formation of abnormal retinal projections. Brain Behav Evol 8: 73-109

Schoenfeld TA, Hamilton LW (1977) Secondary brain changes following lesions: a new paradigm for lesion experimentation. Physiol Behav 18: 951-967

Schwartz WJ, Gainer H (1977) Suprachiasmatic nucleus: use of ^{14}C-level deoxyglucose uptake as a functional marker. Science 197: 1089-1091

Schwartz WJ, Davidsen LC, Smith CB (1980) *In vivo* metabolic activity of a putative circadian oscillator, the rat suprachiasmatic nucleus. J Comp Neurol 189: 157-167

Silver J Brand S (1979) A route for direct retinal input to the preoptic hypothalamus: dendritic projections into the optic chiasm. Am J Anat 155: 391-401

Stanfield B, Cowan WM (1976) Evidence for a change in the retinohypothalamic projection in the rat following early removal of one eye. Brain Res 104: 129-136

Stephan FK, Zucker I (1972a) Rat drinking rhythms: central visual pathways and endocrine factors mediating responsiveness to environmental illumination. Physiol Behav 8: 315-326

Stephan FK, Zucker I (1972b) Circadian rhythms in drinking behavior and locomotor activity of rats are eliminated by hypothalamic lesions. Proc Natl Acad Sci USA 69: 1583-1586

Swanson LW, Cowan WM, Jones EG (1974) An autoradiographic study of the connections of the ventral lateral geniculate nucleus in the albino rat and the cat. J Comp Neurol 156: 143-164

Winfree AT (1967) Biological rhythms and the behavior of populations of coupled oscillators. J Theor Biol 16: 15-42

Zucker I (1979) Hormones and hamster circadian organization. In: Suda M, Hayaishi O, Nakagawa H (eds) Biological rhythms and their central mechanism. Elsevier/North-Holland, Amsterdam, pp 369-381

Zucker I, Rusak B, King RG Jr (1976) Neural bases for circadian rhythms in rodent behavior. In: Riesen AH, Thompson RF (eds) Advances in psychobiology, vol. III. Wiley, New York, pp 35-74

3.2 Neuroanatomical Organization and Connections of the Suprachiasmatic Nucleus

M.V. Sofroniew[1] and A. Weindl[2]

1 Introduction

The mammalian suprachiasmatic nucleus (SCN) consists of a cluster of small round perikarya lying bilaterally adjacent to the third ventricle just dorsal to the optic chiasm (Christ 1969, Diepen 1962). The SCN appears to be one of the basic structures involved in generating biological rhythms; it receives a direct input from the retina (Hendrickson et al. 1972, Moore and Lenn 1972), and destruction of the SCN disrupts a variety of circadian rhythms (Moore and Eichler 1972, Stephan and Zucker 1972, van den Pol and Powley 1979). In this chapter, the neuroanatomical organization and afferent and efferent connections of the SCN will be reviewed, giving particular attention to aspects possibly related to SCN control of circadian rhythms. Most of the information currently available is derived from studies conducted in rodents; however, where possible, comparative studies will be considered.

2 Organization of the Neural Elements Within the SCN

In early studies on the anatomy of the hypothalamus, the SCN has been described as an ovoid nucleus consisting of a homogeneous population of small (6-12 μm) round or fusiform very densely packed neurons (see Christ 1969). Information obtained more recently using a variety of neuroanatomical techniques indicates that the SCN is a complex nucleus with recognizable subdivisions and a heterogeneous population of different types of neurons. Two major dorsomedial and ventrolateral subdivisions are recognizable not only on the basis of neuromorphology and neurocytology (van den Pol 1980), but also in the relative distributions of different types of peptidergic neurons (Fig. 1), as well as in the innervation of the nucleus (see Sect. 4). In a recent detailed analysis of the intrinsic anatomy of the rat SCN using classical neuroanatomical techniques, van den Pol (1980) found that SCN neurons generally have relatively simple dendritic arbors of which different types could be identified. These included simple bipolar, curly bipolar, radial, monopolar and spinous neurons. He also found that neurons tend to be smaller and packed more densely in the dorsomedial as opposed to the ventrolateral portion of

1 Present address: University Department of Human Anatomy, South Parks Road, Oxford 0X1 3QX, U.K.

2 Department of Anatomy, Ludwig-Maximilians University, Munich and Department of Neurology, Technical University, Munich, FRG

Vertebrate Circadian Systems
Ed. by J. Aschoff, S. Daan, and G. Groos

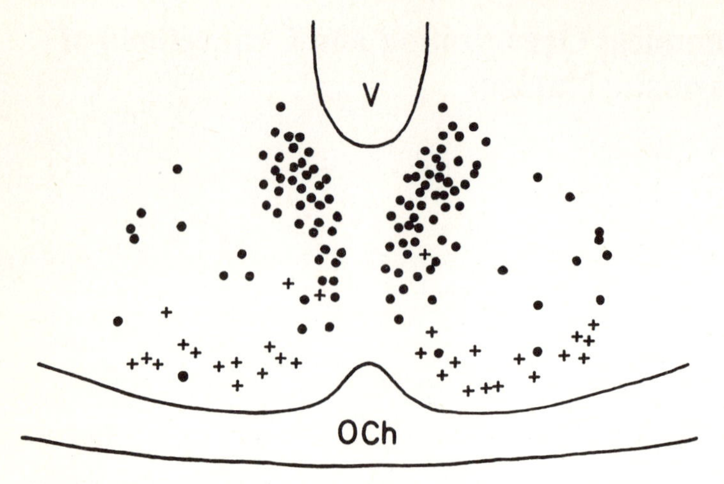

Fig. 1. Drawing of a frontal section through the rodent suprachiasmatic nucleus depicting the location within the nucleus of the vasopressin *(filled circles)* or VIP *(crosses)* neurons, prepared according to the findings of Lorén et al. (1979), Sims et al. (1980) and Sofroniew and Weindl (1978). *OCh* optic chiasm; *V* third ventricle

the nucleus, and that significantly more somatic appositions occur in the dorsomedial part of the nucleus. Furthermore, within any given area of the SCN, ultrastructural differences between neighboring neurons were found suggesting a heterogeneous population of neurons (van den Pol 1980). At the ultrastructural level, Güldner (1976) has analyzed in detail the complex synaptology within the SCN, finding two types of Gray type I synapses and three types of Gray type II synapses.

To date, two specific substances have been identified as being produced by SCN neurons, the peptides vasopressin and vasoactive intestinal polypeptide (VIP). The neurohypophyseal hormone vasopressin and its associated neurophysin were first identified within SCN neurons of the rat during the course of immunohistochemical studies on the hypothalamus (Vandesande et al. 1974, 1975). Parvocellular vasopressin neurons have now been identified in a wide variety of mammals including the human and other primates (Table 1) (Sofroniew and Weindl 1980). Vasopressin neurons comprise approximately 17% of SCN neurons in the rat and approximately 31% in the human (Sofroniew and

Table 1. Mammalian orders and species in which parvocellular vasopressin/neurophysin neurons have been found in the SCN (Sofroniew and Weindl 1980).

Marsupials	Rodents	Lagomorphs	Artiodactyls	Carnivores	Primates
Opossum	Hamster	Rabbit	Cow	Cat	Tree shrew[a]
	Mouse		Pig		Squirrel monkey
	Rat				Rhesus monkey
	Guinea pig				Human

[a] The tree shrew is listed as a primate according to the classification of Walker (1968)

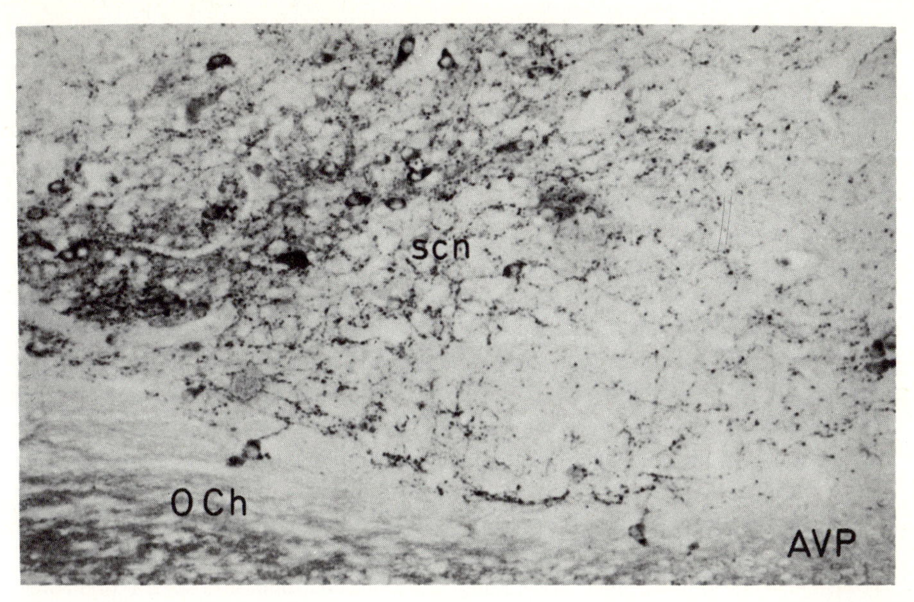

Fig. 2. Sagittal section through the suprachiasmatic nucleus *(scn)* of a rat, immunohistochemically stained for arginine-vasopressin *(AVP)*. Rostral is to the *left.* Vasopressin neurons are more concentrated in the rostral-dorsal part of the nucleus. Nevertheless, many vasopressin fibers are present in the ventral-caudal portion of the nucleus. *OCh* optic chiasm. x 180

Weindl 1980). At the ultrastructural level, the vasopressin within SCN neurons is contained within granules (Van Leeuwen et al. 1978). VIP is a peptide originally isolated from the duodenum (Said and Mutt 1970) and is present in peripheral as well as central neurons (Fuxe et al. 1977). Neurons within the SCN containing VIP have been reported for the mouse and rat using immunohistochemistry (Lorén et al. 1979, Sims et al. 1980). There appear to be somewhat fewer VIP than vasopressin neurons in the SCN, although cell counts have not yet been performed. Taken together, the vasopressin and VIP neurons in the SCN still probably constitute less than 50% of the total population of SCN neurons, indicating that the SCN consists of a fairly heterogeneous population of different types of neurons. None of the other currently known neuropeptides, including oxytocin, LRH, substance P, enkephalin, neurotensin, somatostatin, cholecystokinin, β-endorphin, ACTH or α-MSH, have to date been reported present in SCN neurons. Comparing the distribution within the SCN of vasopressin or VIP neurons, it appears that the peptides are present in separate neurons, with vasopressin neurons being more concentrated in the rostral dorsomedial portion and with VIP neurons being more concentrated in the ventral portion of the nucleus (Figs. 1 and 2).

Although the SCN has a generally ovoid appearance in rodents, such a shape is not readily distinguishable in the SCN of various primates including man (Lydic and Moore-Ede 1980, Lydic et al. 1980, Sofroniew and Weindl 1980). Nevertheless, the presence of subdivisions within the SCN appears not to be confined to rodents: a three-dimensional analysis of the squirrel monkey SCN has revealed a complex, heterogeneous structure (Lydic and Moore-Ede 1980) and the distribution of vasopressin neurons within the primate SCN is such that they are generally more concentrated in the dorsomedial part of the nucleus (Sofroniew and Weindl 1980).

3 Projections from SCN Neurons

3.1 Projections from Peptidergic Neurons

We have previously described projections of parvocellular vasopressin neurons of the SCN using immunohistochemistry in various rodents and primates (Sofroniew 1980, Sofroniew and Weindl 1978, 1981, Sofroniew et al. 1981). The findings are summarized in Fig. 5. In contrast to magnocellular vasopressin neurons in hypothalamic supraoptic and paraventricular nuclei, the parvocellular vasopressin neurons in the SCN do not project to the neurohypophysis. No fibers arising from these neurons could be traced to the median eminence or neural lobe (Sofroniew and Weindl 1978), nor were any SCN neurons retrogradely labelled following injection of horseradish peroxidase into the neurohypophysis (Kelly and Swanson 1980). The major projections of SCN vasopressin neurons appear to be to neural targets in the lateral septum, mediodorsal thalamus, lateral habenula, rostral mesencephalic central grey and periventricular posterior hypothalamus. In these areas axosomatic as well as axodendritic contacts are made. Less prominent projections appear to include the nucleus of the diagonal band of Broca, interpeduncular nucleus, medial amygdala, ventral hippocampus and possibly the locus coeruleus and several brain stem structures. These findings, summarized in Fig. 5, are based on tracing immunohistochemically stained fibers in serial sections, and the major pathways are prominent and easily followed. The smaller projections are not completely certain, since there are also vasopressin projections to extrahypothalamic sites from neurons in the paraventricular nucleus and in some areas pathways overlap (Sofroniew 1980, Sofroniew and Weindl 1981). For this reason, studies which combine experimental neuroanatomical tracing techniques with immunohistochemical identification of peptides will be necessary to firmly establish these projections. Some support for these projections can be derived from the autoradiographic or horseradish peroxidase tracing studies described in Section 3.2.

Although VIP neurons have been identified within the SCN, no reports have yet appeared on the specific projections of these neurons. Prominent terminal fields of VIP fibers are present in various forebrain areas (Lorén et al. 1979, Sims et al. 1980), however, there are also numerous VIP neurons outside the SCN in several forebrain areas (Loren et al. 1979, Sims et al. 1980) and it is not yet certain which neurons project where. It is interesting that the mediodorsal thalamus receives VIP fibers, and that vasopressin neurons in the SCN project to this area.

3.2 SCN Projections Identified by Neuroanatomical Tracing Techniques

There are relatively few reports on projections of SCN neurons determined by neuroanatomical tracing techniques. Using the autoradiographic method for anterograde demonstration of projections, Swanson and Cowan (1975) describe findings obtained from a single rat in which the injection of ^{3}H-proline was confined to the SCN, primarily in the ventrolateral portion. The projection in this rat was strictly ipsilateral and consisted only of caudally moving fibers which remained within the hypothalamus. The label was heaviest in the periventricular posterior hypothalamus and in the ventromedial/arcuate area. More recently, Berk and Finkelstein (1982) and Stephan et al. (1982) , also using the autoradiographic method report more extensive projections from SCN to the mediodorsal thalamus, lateral habenula (Stephan et al. 1982 only), hypothalamic paraventricular, dorsomedial, ventromedial and arcuate nuclei, median eminence, posterior periventricular hypothalamus, mesencephalic central grey, septum and anterior amygdala. Kucera and Favrod (1979) using both anterograde and retrograde tracing procedures, report a projection from SCN neurons to the mesencephalic central grey in the mouse. In an analysis of the afferent projections to the cat locus coeruleus using retrograde horseradish peroxidase tracing, Sakai et al. (1977) found that in some cases SCN neurons were labelled. The results of these studies, summarized in Fig. 5, confirm many of the projections we have described using immunohistochemistry (see Sect. 3.1.).

3.3 Local Circuit Axons and Interaction Between SCN Neurons

Based on analysis of Golgi impregnated material, van den Pol (1980) has reported that short axons or axon collaterals arising and terminating within the SCN are a common feature of this nucleus. He found locally terminating axons or collaterals arising from the majority of SCN neurons. Some of these locally terminating axons may be collaterals of long projection axons, indicating that the organization of the efferent connections of SCN neurons is extremely complex. In this context it is interesting to note that although most parvocellular vasopressin neurons are located mediodorsally within the SCN and the major projection pathways from these neurons leave the SCN dorsally, a large number of vasopressin fibers are present in the ventral and lateral portions of the nucleus (Fig. 2). The possibility that these represent local collaterals should be considered. There is also a substantial number of VIP fibers located within the SCN.

Other probable forms of interaction between SCN neurons are through dendro-dendritic synapses (Güldner and Wolff 1974) or through the large number of soma-somatic appositions (van den Pol 1980). Taking into consideration the numerous local axons as well as the dendro-dendritic, dendro-somatic, and soma-somatic contacts, the potential for communication within the SCN is extremely high. In addition, van den Pol (1980) noted that axons crossing the midline to enter the controlateral SCN could be identified with both Golgi impregnation and horseradish peroxidase tracing, providing evidence that the left and right SCN also communicate with one another.

4 Afferent Input into the SCN

4.1 Neuroanatomically Identified Afferent Connections

At present, the most well-defined input into the SCN derives from the retina as demonstrated by the autoradiographic method (Hendrickson et al. 1972, Moore and Lenn 1972). Retinohypothalamic projections terminating in the SCN have been identified in a variety of mammals including primates (Hendrickson et al. 1972, Moore 1973). In all species examined, the retinal afferents appear to be bilateral, and in general terminate most heavily in the ventral and caudal SCN (Hendrickson et al. 1972, Mai and Junger 1977, Moore 1973). In addition, the SCN receives a projection from the ventral lateral geniculate, al-

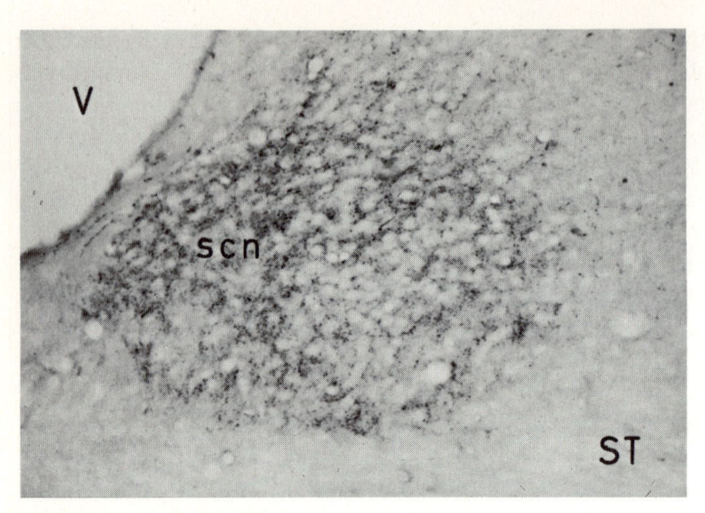

Fig. 3. Frontal section through the suprachiasmatic nucleus *(scn)* of a rat, immunohistochemically stained for somatostatin *(ST)*. Somatostatin fibers and terminals are present throughout the nucleus. *V* third ventricle. x 90

so terminating most heavily in the ventral SCN (Ribak and Peters 1975, Swanson et al. 1974).

4.2 Histochemically Identified Innervation

Serotonin (5-hydroxytryptamine) visualized by fluorescence histochemistry was the first substance identified within fibers innervating the SCN (Fuxe 1965). These fibers probably derive from the dorsal raphe nucleus (Aghajanian et al. 1969), and are most concentrated in the ventrolateral part of the nucleus (Fuxe 1965, Steinbusch 1981). The enzyme glutamate decarboxylase has been found by immunohistochemistry within fibers of the SCN (Perez de la Mora et al. 1981, W. Oertel, pers. commun.), providing evidence for the presence of the transmitter GABA within these fibers. These fibers also appear most concentrated in the ventral SCN. Of the various peptides present in the central nervous system there is a substantial innervation of the SCN by somatostatin fibers which are distributed throughout the nucleus (Fig. 3) (see also Hökfelt et al. 1978). There are also many vasopressin or VIP fibers within the SCN which may represent local circuit axons (see Sect. 3.3.). There appear to be some, but not many, substance P fibers (Cuello and Kanazawa 1978, Ljungdahl et al. 1978) and no neurotensin (Kahn et al. 1980) or enkephalin (Sar et al. 1978) fibers in the SCN. It is also interesting that while fibers deriving from the β-endorphin/ACTH producing neurons of the arcuate nucleus (Bloom et al. 1978, Sofroniew 1979, Watson et al. 1978) innervate the surrounding anterior hypothalamus fairly heavily, these fibers conspicuously spare the SCN (Fig. 4). In this context it should be noted that relatively few dendrites of SCN neurons appear to leave the boundaries of the nucleus (van den Pol 1980), thus, most afferent projections to SCN neurons

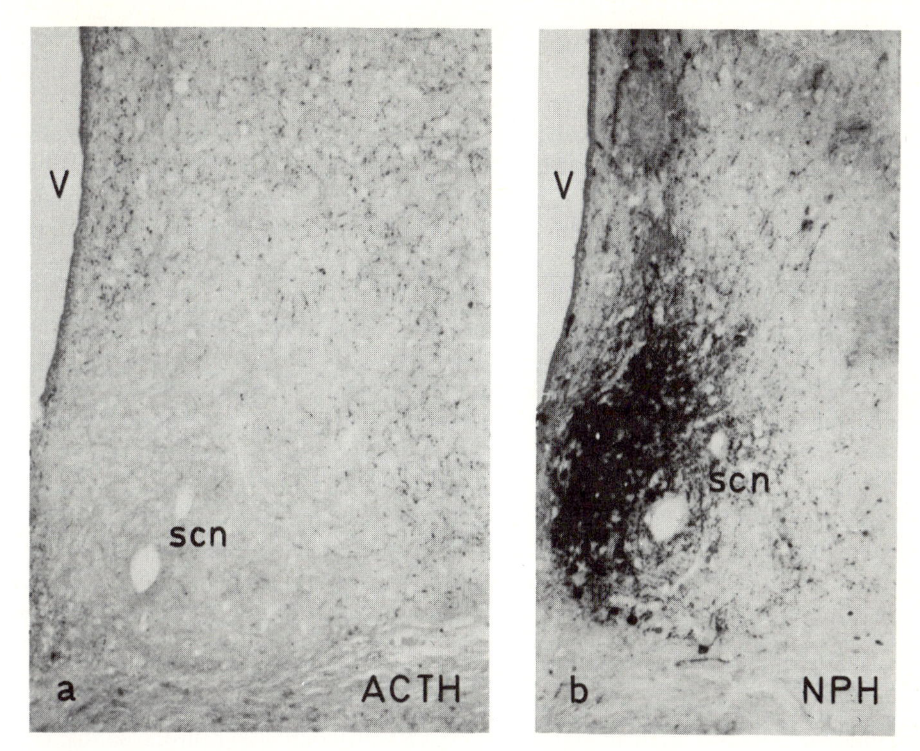

Fig. 4 a, b. Closely neighboring frontal sections through the suprachiasmatic nucleus *(scn)* of a rat, immunohistochemically stained for ACTH **(a)** or neurophysin (NPH, **b**). Note that while finely beaded ACTH fibers densely innervate the surrounding anterior hypothalamus, the fibers spare the SCN. *V* third ventricle. x 90

probably enter the nucleus. At present it appears that most chemically, as well as most neuroanatomically, identified afferents to the SCN terminate in the more ventral, lateral and caudal portions of the nucleus. These findings are summarized in Fig. 5.

5 Relationships Between Neuroanatomical and Physiological Findings

Evidence is now quite strong that the SCN plays a critical role in the generation and entrainment of circadian rhythms. Destruction of the SCN abolishes a variety of circadian rhythms (Moore and Eichler 1972, Stephan and Zucker 1972, van den Pol and Powley 1979), and the SCN shows a circadian variation in its uptake of 2-deoxyglucose (Schwartz and Gainer 1977) and the amino acid lysine (van den Pol 1981). Furthermore, it appears that the rhythmicity of the electrical activity of the SCN is independent of all afferent inputs since it persists after complete isolation of the SCN by knife cuts (Inouye and Kawamura 1979). Thus, rhythmicity, at least of multiple unit activity, appears to be intrinsic to the SCN. The role of the various afferent inputs into the SCN, from the visual system and other areas in the brain, may therefore be more of a modulatory nature. In line with this, serotonin terminals in the SCN may be involved in the phasic release of luteinizing hormone (Coen and MacKinnon 1980). The extent to which circadian rhythmicity originates in the SCN itself, and the roles of the various afferent inputs currently known, are questions likely to receive much attention in the future.

A good deal of the available neuroanatomical data suggest that the SCN is a heterogeneous nucleus with two major subdivisions and several subpopulations of neurons. This raises the possibility that the SCN may also be involved in functions distinct from the generation of rhythms, and there is indeed some evidence to support this. In various mammalian species about 15%-30% of SCN neurons produce vasopressin (Sofroniew and Weindl 1980), and these neurons have substantial projections to various areas in the forebrain (see Fig. 5). Nevertheless, rats with a genetic inability to produce vasopressin (Long Evans, Brattleboro strain) show no alterations or deficits in their circadian rhythmicity (Peterson et al. 1980, Stephan and Zucker 1974). It therefore seems possible that these vasopressin neurons, which constitute about 1/5 of the SCN neurons in the rat, are involved in another function.

In this context it is also interesting to consider the evidence in favor of two major subdivisions of the SCN. Most vasopressin neurons, which at present seem not to be involved in SCN control of rhythms, are found in the rostral and dorsomedial portion of the SCN. In contrast, most of the afferent inputs from the visual system and various other areas appear to terminate most heavily in the ventral, lateral and caudal SCN. VIP neurons are also located primarily in the ventral SCN, in further support of dorsomedial/ventrolateral subdivisions of the nucleus. Nevertheless, these segregations are not exclusive, and a good deal of overlap exists. Furthermore, analysis of local circuitry indicates the potential for extensive intercommunication between different SCN neurons, so that the different subdivisions probably interact extensively with one another. At present little information is available regarding the possible segregation of the SCN into physiologically active subunits. In most reported lesion studies, not only the SCN, but the surrounding area as well, was destroyed. Van den Pol and Powley (1979) and van den Pol (1980), using small rediofrequency lesions, report that destruction of the medial dorsal posterior SCN and area just dorsal to this appear effective in disrupting rhythms. Thus, either neurons, or probably axons passing through this area, are critical for circadian rhythms. At present, little is known about whether different neurons in the SCN have different projections, and may therefore subserve different functions. In general, it seems likely that caudally directed projections from the SCN are essential for maintenance of rhythms (Moore and Eichler 1972, van den Pol and Powley 1979), although possible relay connections formed by SCN neurons involved in control of rhythms are currently unknown. The projection of SCN neurons to the mesencephalic central grey (see Sect. 3.2) may represent an SCN connection with a poly-synaptic relay to sympathetic neurons innervating the pineal (Moore et al. 1968).

6 Concluding Remarks and Summary

On the basis of a number of different types of studies, it seems clear that the mammalian SCN is, neuroanatomically speaking, a heterogeneous nucleus comprised of subdivisions and several subpopulations of neurons. This evidence is derived from the neuromorphological and neurocytological analysis of SCN neurons, from the distribution with-

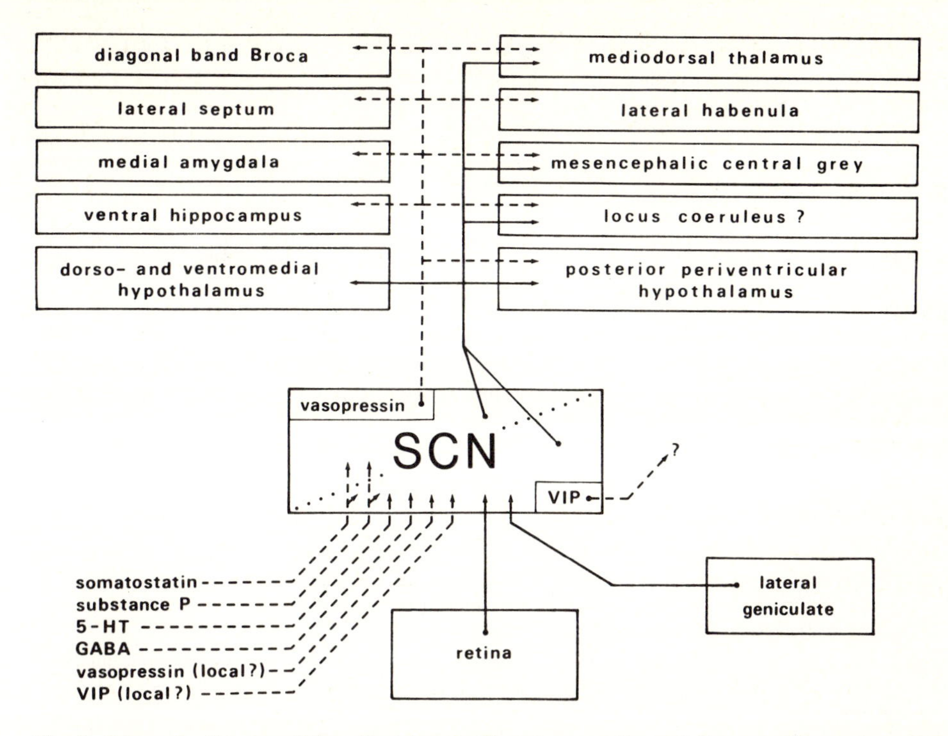

Fig. 5. Schematic diagram of the afferent and efferent connections of the suprachiasmatic nucleus *(scn).* The *dotted line* symbolizes the two major subdivisions of the SCN; *dorsomedial,* containing most of the vasopressin neurons; *ventrolateral,* containing most of the VIP neurons. Afferent or efferent connections detected by immuno- or fluorescenthistochemistry are depicted by *dashed lines.* Connections established by experimental neuroanatomical tracing techniques are depicted by *solid lines.* The diagram is compiled from numerous reports, for references see text

in the SCN of different types of peptidergic neurons, and from the distribution within the SCN of various afferent inputs. Figure 5 summarizes schematically the two major subdivisions of the SCN, the afferent input into the SCN and the efferent projections from the SCN. Two types of peptidergic neurons have been identified within the SCN, vasopressin neurons and VIP neurons, which taken together still constitute less than 50% of the total number of SCN neurons. Most vasopressin neurons appear to lie in the rostral dorsomedial portion of the nucleus, while most VIP neurons are in the ventral portion. The retinal afferents and other inputs from the visual system appear to innervate the ventral, lateral and caudal portions of the SCN most heavily, as does the serotonin input. Somatostatin and a few substance P fibers are also present throughout the SCN. A difference in the projections from neurons in the different subdivisions is not certain. Analysis of local circuitry shows the potential for extensive intercommunication between SCN neurons, indicating that interaction between these subdivisions as well as between the left and right SCN is likely. Certain aspects of the complex and heterogeneous neuroanatomical organization of the SCN suggest it may be involved in activities in addition to its central role in the generation of circadian rhythms, although at present there is no physiological evidence to support this.

Acknowledgments. The authors thank R. Köpp-Eckmann, I. Wild and J. Werner for technical, and P. Campbell for editorial assistance. This work was supported by the Deutsche Forschungsgemeinschaft, grant We 608/5.

References

Aghajanian GK, Bloom FE, Sheard MH (1969) Electron microscopy of degeneration within the serotonin pathway of rat brain. Brain Res 13: 266-273

Berk ML, Finkelstein JA (1982) An autoradiographic determination of the efferent projections of the suprachiasmatic nucleus of the hypothalamus. Brain Res 226: 1-14

Bloom F, Battenberg E, Rossier J, Ling N, Guillemin R (1978) Neurons containing β-endorphin in rat brain exist separately from those containing enkephalin: Immunocytochemical studies. Proc Natl Acad Sci USA 75: 1591-1595

Christ JF (1969) Derivation and boundaries of the hypothalamus, with atlas of hypothalamic grisea. In: Haymaker W, Anderson E, Nauta WJH (eds) The hypothalamus. CC Thomas, Springfield, pp 13-60

Coen CW, MacKinnon PCB (1980) Lesions of the suprachiasmatic nuclei and the serotonin-dependent phasic release of luteinizing hormone in the rat: Effects on drinking rhythmicity and on the consequences of preoptic area stimulation. J Endocrinol 84: 231-236

Cuello AC, Kanazawa I (1978) The distribution of substance P immunoreactive fibers in the rat central nervous system. J Comp Neurol 178: 129-156

Diepen R (1962) Der Hypothalamus. In: v. Möllendorf W, Bargmann W (eds) Handbuch der mikroskopischen Anatomie des Menschen, Bd 4, Teil 7. Springer, Berlin Heidelberg New York, pp 37-63

Fuxe K (1965) Evidence for the existence of monoamine neurons in the central nervous system. IV. Distribution of monoamine nerve terminals in the central nervous system. Acta Physiol Scand 64: (Suppl 247) 37-85

Fuxe K, Hökfelt T, Said S, Mutt V (1977) Vasoactive intestinal polypeptide and the nervous system: Immunohistochemical evidence for localization in central and peripheral neurons, particularly intracortical neurons of the cerebral cortex. Neurosci Lett 5: 241-246

Güldner FH (1976) Synaptology of the rat suprachiasmatic nucleus. Cell Tissue Res 165: 509-544

Güldner FH, Wolff JR (1974) Dendro-dendritic synapses in the suprachiasmatic nucleus of the rat hypothalamus. J Neurocytol 3: 245-250

Hendrickson AE, Wagoner N, Cowan WM (1972) An autoradiographic and electron microscopic study of retino-hypothalamic connections. Z Zellforsch 135: 1-26

Hökfelt T, Elde R, Fuxe K, Johannson O, Ljungdahl A, Goldstein M, Luft R, Efendic S, Nilsson G, Terenius L, Ganten D, Jeffcoate SL, Rehfeld J, Said S, Perez de la Mora M, Possani L, Tapia R, Teran L, Palacios R (1978) Aminergic and peptidergic pathways in the nervous system with special reference to the hypothalamus. In: Reichlin S, Baldessarini RJ, Martin JB (eds) The hypothalamus. Raven Press, New York, pp 69-135

Inouye ST, Kawamura H (1979) Persistence of circadian rhythmicity in a mammalian hypothalamic "island" containing the suprachiasmatic nucleus. Proc Natl Acad Sci USA 76: 5962-5966

Kahn D, Abrams GM, Zimmerman EA, Carraway R, Leeman SE (1980) Neurotensin neurons in the rat hypothalamus: An immunocytochemical study. Endocrinology 107: 47-54

Kelly J. Swanson LW (1980) Additional forebrain regions projecting to the posterior pituitary: Preoptic region, bed nucleus of the stria terminalis, and zona incerta. Brain Res 197: 1-9

Kucera P, Favrod P (1979) Suprachiasmatic nucleus projection to mesencephalic central grey in the woodmouse *(Apodemus sylvaticus L.)*. Neuroscience 4: 1705-1716

Ljungdahl A, Hökfelt T, Nilsson G (1978) Distribution of substance P-like immunoreactivity in the central nervous system of the rat-I. Cell bodies and nerve terminals. Neuroscience 3: 861-943

Lorén I, Emson PC, Fahrenkrug J, Björklund A, Alumets J, Hakanson R, Sundler F (1979) Distribution of vasoactive intestinal polypeptide in the rat and mouse brain. Neuroscience 4: 1953-1976

Lydic R, Moore-Ede MC (1980) Three dimensional structure of the suprachiasmatic nuclei in the diurnal squirrel monkey *(Saimiri sciureus).* Neurosci Lett 17: 295-299

Lydic R, Schoene WC, Czeisler CA, Moore-Ede MC (1980) Suprachiasmatic region of the human hypothalamus: Homolog to the primate circadian pacemaker? Sleep 2: 355-361

Mai JK, Junger E (1977) Quantitative autoradiographic light- and electron microscopic studies on the retinohypothalamic connections in the rat. Cell Tissue Res 183: 221-237

Moore RY (1973) Retinohypothalamic projection in mammals: a comparative study. Brain Res 49: 403-409

Moore RY, Eichler VB (1972) Loss of a circadian adrenal corticosterone rhythm following suprachiasmatic lesions in the rat. Brain Res 42: 201-206

Moore RY, Lenn NJ (1972) A retinohypothalamic projection in the rat. J Comp Neurol 146: 1-14

Moore RY, Heller A, Bhatnager RK, Wurtman RJ, Axelrod J (1968) Central control of the pineal gland: Visual pathways. Arch Neurol 18: 208-218

Pérez de la Mora M, Possani LD, Tapia R, Teran L, Palacios R, Fuxe K, Hökfelt T, Ljungdahl A (1981) Demonstration of central γ-aminobutryate-containing nerve terminals by means of antibodies against glutamate decarboxylase. Neuroscience 6: 875-895

Peterson GM, Watkins WB, Moore RY (1980) The suprachiasmatic hypothalamic nuclei of the rat. VI. Vasopressin neurons and circadian rhythmicity. Behav. Neural Biol 29: 236-245

Ribak CE, Peters A (1975) An autoradiographic study of the projections from the lateral geniculate body of the rat. Brain Res 92: 341-368

Said SI, Mutt V (1970) Polypeptide with broad biological activity: Isolation from small intestine. Science 169: 1217-1218

Sakai K, Touret M, Salvert D, Leger L, Jouvet M (1977) Afferent projections to the cat locus coeruleus as visualized by the horseradish peroxidase technique. Brain Res 119: 21-41

Sar M, Stumpf WE, Miller RJ, Chang KJ, Cuatrecasas P (1978) Immunohistochemical localization of enkephalin in rat brain and spinal cord. J Comp Neurol 182: 17-38

Schwartz WJ, Gainer H (1977) Suprachiasmatic nucleus: Use of ^{14}C-labeled deoxyglucose uptake as a functional marker. Science 197: 1089-1091

Sims KB, Hoffman DL, Said SI, Zimmerman EA (1980) Vasoactive intestinal polypeptide (VIP) in mouse and rat brain: An immunocytochemical study. Brain Res 186: 165-183

Sofroniew MV (1979) Immunoreactive β-endorphin and ACTH in the same neurons of the hypothalamic arcuate nucleus in the rat. Am J Anat 154: 283-289

Sofroniew MV (1980) Projections from vasopressin, oxytocin, and neurophysin neurons to neural targets in the rat and human. J Histochem Cytochem 28: 475-478

Sofroniew MV, Weindl A (1978) Projections from the parvocellular vasopressin- and neurophysin-containing neurons of the suprachiasmatic nucleus. Am J Anat 153: 391-430

Sofroniew MV, Weindl A (1980) Identification of parvocellular vasopressin and neurophysin neurons in the suprachiasmatic nucleus of a variety of mammals including primates. J Comp Neurol 193: 659-675

Sofroniew MV, Weindl A (1981) Central nervous system distribution of vasopressin, oxytocin and neurophysin. In: Martinez JL, Jensen RA, Messing RB, Rigter H, McGaugh JL (eds) Endogenous peptides and learning and memory processes. Academic Press, London New York pp 327-369

Sofroniew MV, Weindl A, Schrell U, Wetzstein R (1981) Immunohistochemistry of vasopressin, oxytocin and neurophysin in the hypothalamus and extrahypothalamic regions of the human and primate brain. Acta Histochem (Suppl) 24: 79-95

Steinbusch HWM (1981) Distribution of serotonin-immunoreactivity in the central nervous system of the rat – cell bodies and terminals. Neuroscience 6: 557-618

Stephan FK, Zucker I (1972) Circadian rhythms in drinking behavior and locomotor activity of rats are eliminated by hypothalamic lesions. Proc Natl Acad Sci USA 69: 1583-1686

Stephan FK, Zucker I (1974) Endocrine and neural mediation of the effects of constant light on water intake of rats. Neuroendocrinology 14: 44-60

Stephan FK, Berkeley KJ, Moss RL (1982) Efferent connections of the rat suprachiasmatic nucleus. Neuroscience 6: 2625-2641

Swanson LW, Cowan WM (1975) The efferent connections of the suprachiasmatic nucleus of the hypothalamus. J Comp Neurol 160: 1-12

Swanson LW, Cowan WM, Jones EG (1974) An autoradiographic study of the efferent connections of the ventral lateral geniculate nucleus in the albino rat and the cat. J Comp Neurol 156: 143-163

Vandesande F, De Mey J, Dierickx K (1974) Identification of neurophysin producing cells. I. The origin of the neurophysin-like substance-containing nerve fibres of the external region of the median eminence of the rat. Cell Tissue Res 151: 187-200

Vandesande F, Dierickx K, De Mey J (1975) Identification of the vasopressin-neurophysin producing neurons of the rat suprachiasmatic nuclei. Cell Tissue Res 156: 377-380

van den Pol AN (1980) The hypothalamic suprachiasmatic nucleus of rat: Intrinsic anatomy. J Comp Neurol 191: 661-702

van den Pol AN (1981) Amino acid incorporation in medial hypothalamic nuclei: circadian, perikarya, and neuropil variations. Am J Physiol 240: R16-R22

van den Pol AN, Powley T (1979) A fine-grained anatomical analysis of the role of the rat suprachiasmatic nucleus in circadian rhythms of feeding and drinking. Brain Res 160: 307-326

van Leeuwen FW, Swaab DF, Raay De C (1978) Immuno-electron-microscopic localization of vasopressin in the rat suprachiasmatic nucleus. Cell Tissue Res 193: 1-10

Walker EP (1968) Mammals of the world, vol. I. Johns Hopkins Press, Baltimore

Watson SJ, Richards CW, Barchas JD (1978) Adrenocorticotropin in rat brain: Immunocytochemical localization of cells and axons. Science 220: 1180-1182

3.3 CNS Structures Controlling Circadian Neuroendocrine and Activity Rhythms in Rats

I. Assenmacher[1]

1 Introduction

Although most pituitary hormones are produced in pulsatile sequences of varying frequencies, a fairly stable circadian secretory pattern usually emerges from this episodic background. This raises the first basic question applying to all circadian rhythms, as to whether neuroendocrine circadian rhythms are primarily driven by an endogenous, environmentally entrained oscillator or circadian pacemaker. So far, the existence of such a pacemaker generating an overt neuroendocrine rhythm has only been clearly demonstrated in the pineal (Moore and Klein 1974) and in the adrenocorticotropic system. For the latter, the period of freerunning cortisol rhythm was thus estimated at 24.8 h in blind men (Wever 1979), and for freerunning ACTH and corticosterone rhythms, it ranged from 24.2 h to 24.4 h in blind or dark-housed (DD) rats (Szafarczyk et al. 1980a, 1981b).

2 Functional Characteristics of the Circadian Pacemaker

Interestingly, close phase relationships have been observed between SCN-controlled endocrine rhythms and the sleep/wake cycle, so that most endocrine rhythms start their daily phase of increased release during sleep. In man, this pattern holds for GH, PRL, ACTH, TSH and, before puberty, for LH (see Krieger 1979), and in rats, for ACTH, TSH, PRL and, in estradiol-supplemented castrates, for LH (see Szafarczyk et al. 1980b). However, no clear indication has yet been found that the particular phase relationships between circadian neuroendocrine and sleep-wake rhythms have any adaptive significance, except in the case of the adrenocorticotropic system, and perhaps of other metabolic hormones whose daily sleep-related activation might increase fuel production in anticipation of enhanced energy requirements starting at arousal. Nevertheless, these phase relationships raise the problem of whether there is a master-clock driving basic CNS-controlled behavioral and endocrine circadian rhythms.

To answer this question, a series of chronobiological studies in man (Aschoff 1978, 1979, Wever 1979) and in rats (Szafarczyk et al. 1978) recently collected clear evidence

1 Laboratoire de Neuroendocrinologie, ERA 85 du CNRS, Université de Montpellier II, 34060-Montpellier, France

Vertebrate Circadian Systems
Ed. by J. Aschoff, S. Daan, and G. Groos

in favour of a dual circadian pacemaker system driving at least the sleep-wake and adrenocorticotropic rhythms, since both were shown to display internal desynchronization either after exposure to unusual ahemeral zeitgebers, i.e., T ≠ 24 h, or during re-entrainment after external phase-shifts. No data are available on the comparative responses of other neuroendocrine rhythms under similar conditions, except that in man, the nocturnal GH peak is clearly more firmly sleep-locked than any other endocrine rhythm.

Another basic characteristic of circadian pacemakers also emerged from these chronobiological studies, i.e., the complex structure of the circadian systems controlling both general activity and adrenocortical rhythms. Thus, ahemeral zeitgebers occasionally "split" the circadian system, and combined periodicities (e.g., an ahemeral and a freerunning period) coexisted for the same function (Aschoff 1978, Szafarczyk et al. 1978). For extreme ahemeral environments, i.e., T <23 h or >27 h, a variety of ultradian components was usually detectable, together with an endogenous circadian rhythm, as well as an environmental period and the corresponding harmonics. In a typical experiment on rats exposed for 6 weeks to T = 21 h photoperiod [10L · (10 lx) : 11D], spectral analysis of motor activity records revealed a predominant circadian period of 24.6 h, in conjunction with a one-third powerful environmental period (21 h), and various related harmonics. But it also showed the existence of a further series of ultradian periods (15.9 h, 11.5 h and 6.7 h) with 5% to 7% of the circadian power (Nouguier-Soulé et al. 1982). For unknown reasons, a split was also observed in the endogenous oscillator driving ACTH, corticosterone and activity rhythms of rats exposed for 8 weeks to constant light [LL, 10 lx, (Fig. 1)]. These rats displayed an intriguing motor activity rhythm with a predominant circadian period (25.7 h with various harmonics), combined with several ultradian periods, for instance of 13.3 h, 12.2 h and 8.8 h, which exhibited 16% to 18% of the circadian power (Nouguier-Soulé et al. 1981, see also Honma and Hiroshige, 1978). Clearly, these observations are in line with the theory of circadian pacemakers consisting of complex multi-ocsillator systems (Wever 1979).

3 Structure of Circadian Pacemakers

Whatever the insight into the functional structure of circadian systems provided by conventional chronobiological studies, closer dissection of the pacemakers which up till now were considered as "black boxes", obviously required additional investigations based on neurophysiological and neuropharmacological methods. Clearly, the preeminent part played in overall behavioral and neuroendocrine regulations by the hypothalamus in conjunction with several limbic structures and with the brain stem reticular system indicates that the circadian pacemakers driving behavioural and neuroendocrine rhythms a priori might be located at that level. In this connection, Moore and Eichler (1972) and Stephan and Zucker (1972) succeeded in opening up a most fruitful area of research when they showed for the first time that bilateral lesions of the suprachiasmatic nuclei (SCN) badly disrupted a number of circadian rhythms (see Moore 1979).

In a series of recent experiments on rats performed in light-controlled bunkers (12L-12D when not otherwise specified), we studied the effects of several exclusion procedures on the SCN on the circadian rhythms of the corticotropic system (plasma ACTH and corticosterone levels were estimated in sequential samples from individual rats), and on

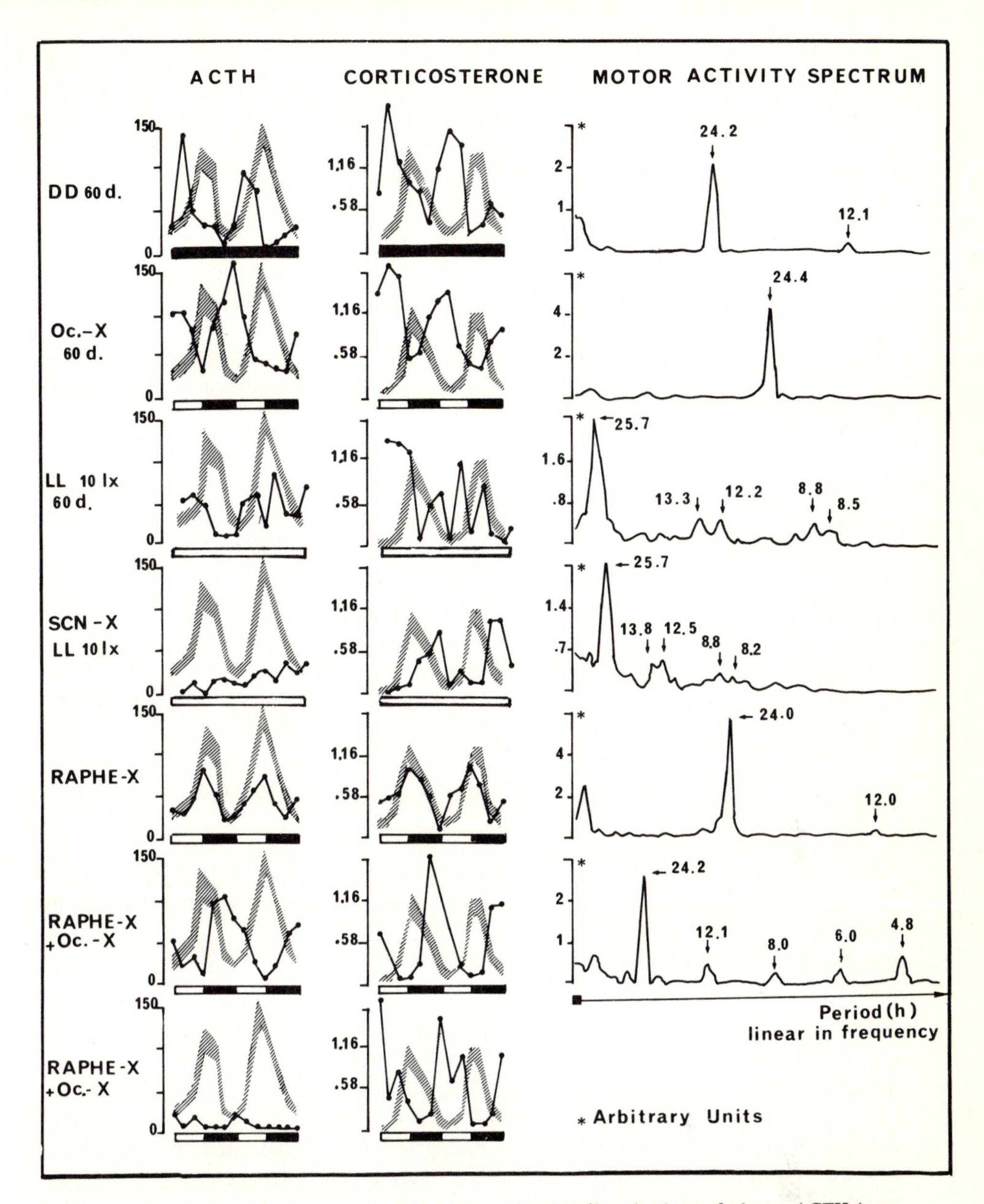

Fig. 1. Effects of various experimental treatments on the circadian rhythms of plasma ACTH (ng. 1^{-1}) and corticosterone (μM.1^{-1}) levels, and of general activity (power spectra). For each treatment, representative specimens are shown of female rats exposed to a photoperiod of 12L (10 lx) –12D, or to LL (10 lx), or to DD (0 lx) as indicated on the abscissae. *Shaded areas* represent mean values ± SE of controls (n = 11) exposed to 12L-12D. *Oc.-X* ocular enucleation; *SCN-X* bilateral lesions of SCN; *Raphe-X* 5-HT denervation of SCN by midbrain lesion. The two specimens of blind raphe-X rats illustrate two different patterns in ACTH responses. The abscissae of the power spectra refer to linear frequencies, but different scales were used for each spectrum. The prominent periods are indicated on each graph. Ref. cited in text.

overall motor activity measured by stabilimeters with piezoelectric transducers. The results were as follows:

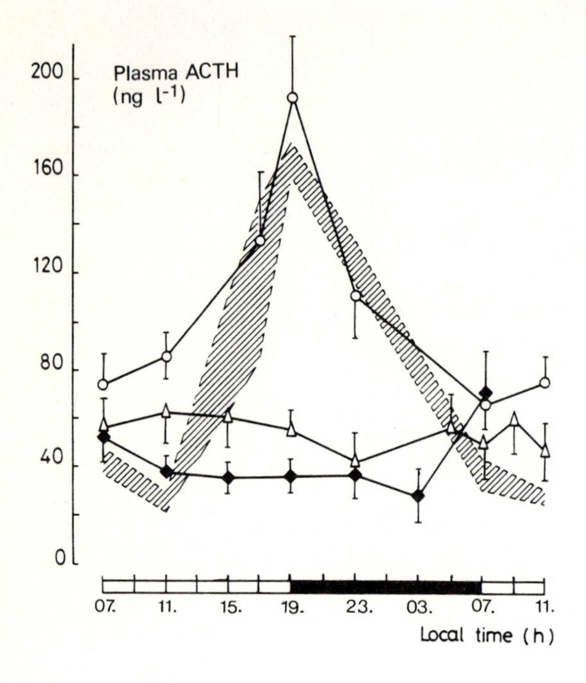

Fig. 2. Plasma concentrations in ACTH ($ng.1^{-1}$) from female rats exposed to 12L (50 lx) : 12D after blockade of the 5-HT system by pCPA. Means ± SE (n = 8). *Shaded area*: confidence limits for controls. –◆–◆–: pCPA injected i.p. 2 x 300 mg/kg on days 3 and 2 before first sampling. *Open symbols*: pCPA treatment with additional 5-HTP administration (60 mg/kg) (i.p.) for 3 days, either at 1100 h (–○–○–), or at 2300 h (–△–△–). Note that 5-HTP given at 1100 h restored a normal ACTH cycle. (Safarczyk et al. 1980d)

3.1 Lesions of the SCN

Bilateral SCN lesions led to a very characteristic syndrome of internal desynchronization, first observed in rats kept on a 12L:12D lighting regimen (Szafarczyk et al. 1979). This syndrome was marked by total obliteration of the ACTH rhythm, strangely associated with a persisting corticosterone rhythm of circadian or ultradian periodicity and normal mean level and amplitude, as well as with the maintenance of a rhythm of activity whose mean level and amplitude were depressed by two-thirds; nevertheless, a circadian periodicity was still detectable (Fig. 3), although it was generally combined with various ultradian components indicating a dissociation of oscillators. It is extremely interesting to note that the same syndrome was recently observed in SCN-lesioned rats exposed before and after lesion to 10 lx constant light, a finding which clearly excluded the interference of any environmental "masking effect" (Figs. 1 and 3). Three weeks after SCN lesion under LL, power spectrum analysis of a typical activity recording, thus showed a circadian period of 25.7 h to coexist with several ultradian periods (13.7 h, 12.5 h and 8.7 h). The power ratios of the ultradian/circadian periods were 15%, 21% and 25%, respectively (Fig. 1) (Szafarczyk et al. 1981b).

The elimination of any detectable ACTH rhythm after SCN lesion is in keeping with the similar inhibition observed for other neuroendocrine rhythms like LH (Héry 1977), TSH (Jordan et al. 1979) and PRL (Dunn et al. 1980). On the other hand, the maintenance of corticosterone rhythms in the absence of ACTH rhythms certainly recalls the

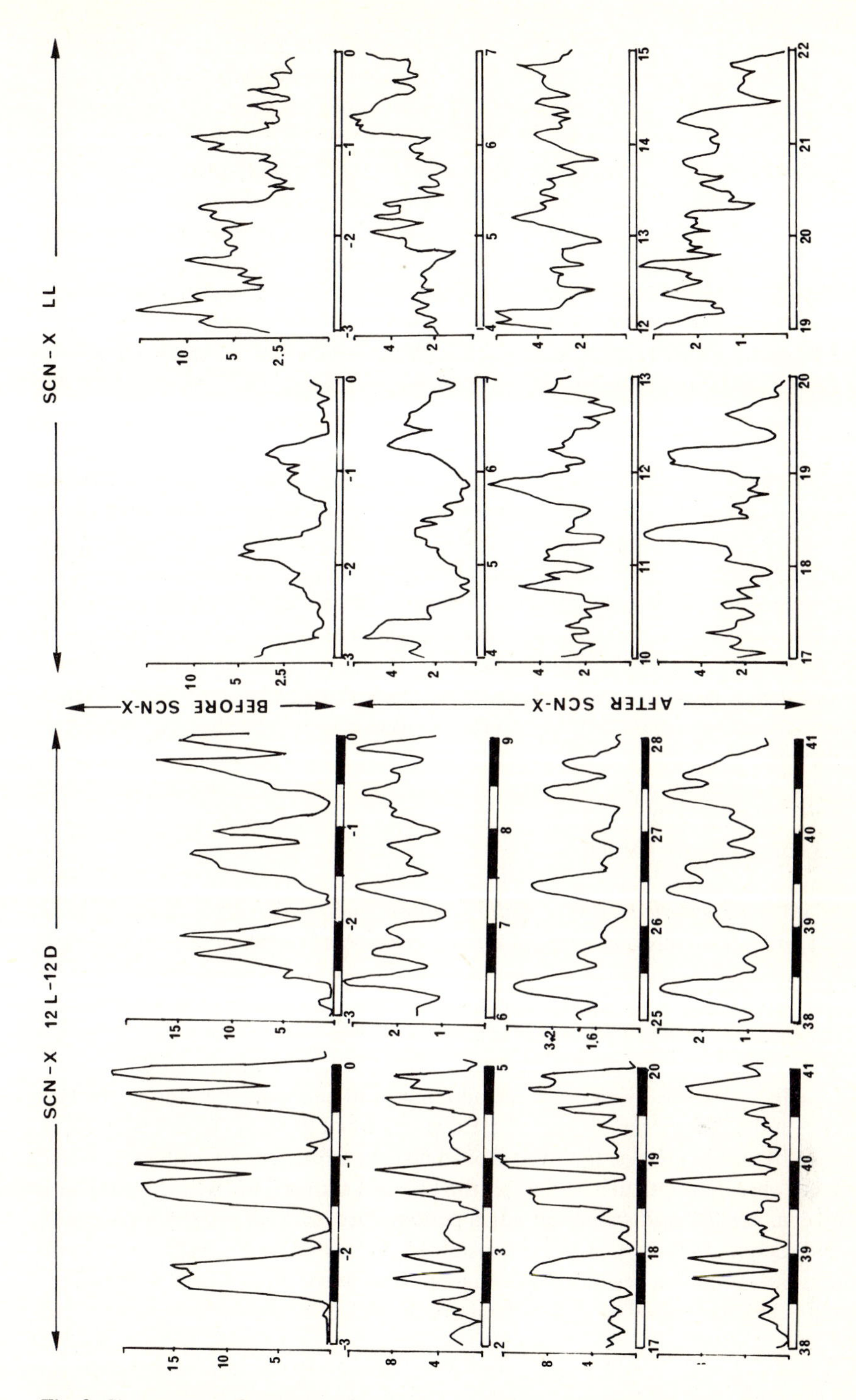

Fig. 3. Chronograms of general activity cycles (stabilimeter recordings) from representative specimens of female rats bearing bilateral SCN lesions, and exposed to 12L (10 lx) : 12D, or to LL (10 lx). *Abscissae* indicate days before and after SCN lesion. *Figures on ordinates* are arbitrary units. (Szafarczyk et al. 1979, 1981b).

persistence of corticosterone rhythms reported in organ-cultured hamster adrenals (Andrews 1971), as well as the maintenance of circadian rhythms for adrenal cyclic nucleotide activity observed in hypophysectomized rats (Guillemant et al. 1980). Whether these adrenal rhythms are driven by oscillators located within or outside the adrenals themselves requires further elucidation. However, the results just quoted certainly disagree with the generally accepted notion that the corticosterone rhythm is obliterated after SCN lesions (see rev. in Krieger 1979). Finally, the persistence of circadian rhythmicity in general activity, albeit with a drastically decreased amplitude and frequently emerging ultradian periodicities denoting a dissociation of circadian pacemakers, obviously contrasts with most reports that SCN lesions are followed by a complete loss of rhythmicity in locomotor activity, sleep-wake rhythm, feeding behaviour and body temperature (see reviews in Ibuka and Kawamura 1975, Stephan and Nunez 1977, Moore 1979, Abe et al. 1979).

In the present experiments, the very sensitive actographic device used, which, incidentally, measured overall activity patterns rather than the locomotor activity estimated by revolving wheel devices, together with computerized analysis of activity records, may account for the discrepancy between the results reported and those of other investigators. Nevertheless, a few other findings have been published for SCN-lesioned rats showing the maintenence of circadian rhythms in sleep-wake behaviour (Coindet et al. 1975), feeding (Abe et al. 1979), and body temperature (Dunn et al. 1977, Powell et al. 1977), even though the phases of these rhythms sometimes shifted consistently compared to the phases recorded in intact animals (Coindet et al. 1975, Powell et al. 1977). However, since these experiments were conducted·in LD a masking effect by the photoperiod can not be excluded.

3.2 Lesions of the Raphe Nuclei

In sighted rats, *electrolytic lesions to midbrain dorsal and median raphe nuclei,* which adequately suppressed serotoninergic (5-HT) innervation of the SCN, or alternatively, local destruction of such innervation by microinjections of the neurotoxin 5,7-di-hydrotryptamine into the SCN, halved the amplitude and mean level of otherwise entrained ACTH rhythms, but left corticosterone and motor activity rhythms unaltered (Szafarczyk et al. 1980c) (Fig. 1). Under similar conditions, raphe lesions in sighted rats were also shown to reduce only the mean levels and amplitudes of circadian rhythms in LH (Héry et al. 1978) and TSH (Jordan et al. 1979). However, when the same lesions were made in rats blinded by ocular enucleation, whose circadian rhythms were originally freerunning, their effects were more perturbing. In half the lesioned animals, the ACTH rhythm either displayed coexistent circadian and ultradian periodicities, or, as after SCN-lesions, was obliterated. Concomitantly, plasma corticosterone fluctuated with circadian and sometimes additional ultradian periodicities, while motor activity rhythm persisted with a freerunning circadian period of 24.2 h (Szafarczyk et al. 1981 a).

3.3 Blockade of the Serotininergic System

Strangely enough, the pharmacological blockade of the entire serotonergic system by systemic administration of p-chlorophenylalanine (pCPA) had more striking effects than simply suppression 5-HT innervation of the SCN. The former treatment always eliminated the ACTH rhythm, but a corticosterone rhythm persisted, with either circadian or ultradian rhythmicity (Szafarczyk et al. 1979). Similarily, systemic pCPA treatment obliterated the circadian rhythm in LH (Héry et al. 1976) and in TSH (Jordan et al. 1979). To explore further the role of the 5-HT system on circadian neuroendocrine rhythms, an attempt was made to bypass its pCPA induced blockade by daily injections of the 5-HT precursor 5-hydroxytryptophan (5-HTP). In view of the daily parallel rise observed about 4 h after the onset of the light phase in 5-HT concentrations of SCN punches (Bessone 1979), and in the level of plasma ACTH, 5-HTP was injected either after dawn (1100 h) or 12 h later at 2300 h. Clearly, 5-HTP administered at 1100 h restored a daily ACTH rhythm similar to that of the intact controls while the same dose given at 2300 h was found to be totally ineffective (Fig. 2) (Szafarczyk et al. 1980d).

4 Conclusions

Taken together, the chronobiological and neurophysiological studies discussed here lead to several main conclusions:

a) Under experimental conditions involving extreme manipulation of environmental synchronizers like ahemeral photoperiods, or environmental phase-shifts or, again the suppression of discrete brain areas (SCN, midbrain raphe, 5-HT system, etc.), the respective periodicities of the adrenocorticotropic and motor activity rhythms may respond quite differently. This obviously argues against the theory that a master clock drives both the neuroendocrine and basic behavioural rhythms and suggests, instead, that these rhythms might depend on separate circadian pacemakers, observed to be essentially phase-locked under basal conditions.

b) Furthermore, the circadian pacemarkes respectively generating the adrenocorticotropic and motor activity rhythms both appear to consist of complex multi-oscillator systems, as inferred from the splitting effect in the form of multiple period rhythms produced in both adrenocorticotropic and motor activity, again by unusual environments or minute CNS lesions.

c) The SCN definitely emerges as a basic component of the circadian pacemaker driving neuroendocrine rhythms, and to lesser extent motor activity rhythm. It may tentatively be assumed that this neuronal complex integrates a variety of periodic inputs impinging on the SCN in fairly precise phase-relationships (see the 5-HTP experiment). In that case, such inputs would come, in particular, from the photoperiodic synchronizer via the retino-hypothalamic pathway, from the 5-HT midbrain raphe and possibly other CNS areas like the limbic system, which receive a supply of 5-HT (see the pCPA experiment).

d) The emergence of ultradian periods together with freerunning circadian periodicity, detected when experimental conditions led to pacemaker splitting, is strongly reminiscent of the ontogenetic development of circadian systems. In the newborn, ultradian periodicities for most biological rhythms have in fact been reported to precede and later to accomplany the circadian periodicities that eventually persist alone (Hellbrügge 1960).

Acknowledgments. I am grateful to my colleagues Dr. A. Szafarczyk, G. Ixart, G. Alonso, F. Malaval and J. Nouguier-Soulé, with whom these experiments were undertaken.

References

Abe K, Greer MA, Critchlow V (1979) Effects of destruction of the suprachiasmatic nuclei on the circadian rhythms in plasma corticosterone, body temperature, feeding and plasma thyrotropin. Neuroendocrinology 29: 119-131

Andrews RV (1971) Circadian rhythms in adrenal organ cultures. Gegenbaurs Morphol Jahrb 177: 89-99

Aschoff J (1978) Circadian rhythms within and outside their ranges of entrainment. In: Assenmacher I, Farner DS (eds) Environmental endocrinology. Springer, Berlin Heidelberg New York, pp 172-181

Aschoff J (1979) Circadian rhythms: general features and endocrinological aspects. In: Krieger DT (ed) Endocrine rhythms, Raven Press New York, pp 1-61

Bessone R (1979) Etude morphologique et fonctionnelle de l'innervation sérotoninergique du noyau suprachiasmatique chez le rat. Thèse endocrinol, Univ Montpellier II, pp 61

Coindet J, Chouvet G, Mouret J (1975) Effects of lesions of the suprachiasmatic nuclei on paradoxical and slow-wave sleep circadian rhythms in the rat. Neurosc Lett 1: 243-247

Dunn JD, Castro AJ, MacNulty JA (1977) Effect of suprachiasmatic ablation on the daily temperature rhythm. Neurosci Lett 6: 345-348

Dunn JD, Johnson DC, Castro AJ, Swenson R (1980) Twenty-four hour pattern of prolactin levels in female rats subjected to transection of the mesencephalic raphe or ablation of the suprachiasmatic nuclei. Neuroendocrinology 31: 85-91

Guillemant J, Guillemant S, Reinberg A (1980) Circadian variations of adrenocortical cyclic nucleotides (cAMP and cGMP) in hypophysectomized rats. Experientia 36: 367-368

Hellbrügge T (1960) The development of circadian rhythms in infants. Cold Spring Harbor Symp Quant Biol 25: 311-323

Héry ME (1977) Rôle de la sérotonine dans le contrôle de la sécrétion cyclique de LH chez la ratte. Thèse Doct Sci, Univ Paris-VI, pp 300

Héry ME, Laplante E, Kordon C (1976) Participation of serotonin in the phasic release of LH. I Evidence from pharmacological experiments. Endocrinology 79: 496-503

Héry ME, Laplante E, Kordon C (1978) Participation of serotonin in the phasic release of LH. II Effects of lesions of serotonin-containing pathways in the CNS. Endocrinology 102: 1019-1025

Honma K, Hiroshige T (1978) Endogenous ultradian rhythms in rats exposed to prolonged continuous light. Am J Physiol 235: R250-R256

Ibuka N, Kawamura H (1975) Loss of circadian rhythm in sleep-wakefulness cycle in the rat by suprachiasmatic nucleus lesion. Brain Res 94: 76-81

Jordan D, Pigeon P, MacRae-Degueurce A, Pujol FJ, Mornex R (1979) Participation of serotonin in thyrotropin release. II Evidence for the action of serotonin on the phasic release of thyrotropin. Endocrinology 105: 975-979

Krieger DT (1979) Rhythms in CRF, ACTH and corticosteroids. In: Krieger DT (ed) Endocrine rhythms. Raven Press, New York, pp 123-142

Moore RY (1979) The anatomy of central neural mechanisms regulating endocrine rhythms. In: Krieger DT (ed) Endocrine rhythms. Raven Press, New York, pp 63-87

Moore RY, Eichler VB (1972) Loss of circadian adrenal corticosterone rhythm following suprachiasmatic lesions in the rat. Brain Res 42: 201-206

Nouguier-Soulé J, Szafarczyk A, Ixart G, Alonso G, Malaval F, Assenmacher I (1982) Dissociation du pacemaker circadian du rythme de l'activité générale chez des rattes soumises à des environnements anhéméraux ou à des lésions nerveuses centrales. CR Acad Sci (in prep)

Powell EW, Pasley JN, Brockway B, Scheving L, Lubanovic W, Halberg F (1977) Suprachiasmatic dinuclear lesions alter temperature rhythm's amplitude and timing in light-dark synchronized rats. Chronobiologia 4: (Suppl 1) 146-147

Stephan FK, Nunez A (1977) Elimination of circadian rhythm in drinking activity, sleep and temperature by isolation of the suprachiasmatic nucleus. Behav Biol 20: 1-16

Stephan FK, Zucker I (1972) Circadian rhythms in drinking behavior and locomotor activity are eliminated by suprachiasmatic lesions. Proc Natl Acad Sci USA 69: 1583-1586

Szafarczyk A, Boissin J, Nouguier-Soulé J, Assenmacher I (1978) Effect of ahemeral environmental periodicities on the rhythms of adrenocortical and locomotor functions in rats and japanese quail. In: Assenmacher I, Farner DS (eds) Environmental endocrinology. Springer, Berlin Heidelberg New York, pp 182-184

Szafarczyk A, Ixart G, Malaval F, Nouguier-Soulé J, Assenmacher I (1979) Effects of lesions of the the suprachiasmatic nuclei and of p-CPA on the circadian rhythms of ACTH and corticosterone in the plasma and on locomotor activity in rats. J Endocrinol 83: 1-16

Szafarczyk A, Ixart G, Malaval F, Nouguier-Soulé J, Assenmacher I (1980a) Corrélations entre les rythmes circadiens de l'ACTH et de la corticostérone plasmatique, et de l'activité motrice, évoluant en "libre cours" aprés énucléation oculaire chez le rat. C R Acad Sci 290: 587-599

Szafarczyk A, Héry M, Laplante E, Ixart G, Assenmacher I, Kordon C (1980b) Temporal relationships between the circadian rhythmicity in plasma levels of pituitary hormones and in hypothalamic concentrations of releasing factors. Neuroendocrinology 30: 369-376

Szafarczyk A, Alonso G, Ixart G, Malaval F, Nouguier-Soulé J, Assenmacher I (1980c) The serotoninergic system and the circadian rhythms in ACTH and corticosterone in rats. Am J Physiol 239: 482-489

Szafarczyk A, Ixart G, Malaval F, Nouguier-Soulé J, Assenmacher I (1980d) Influence de l'heure d' administration de 5-hydroxytryptophanne sur la restauration de la stimulation circadienne de la sécrétion de l'ACTH chez les rats traités à la pCPA. C R Soc Biol 174: 170-175

Szafarczyk A, Ixart G, Alonso G, Malaval F, Nouguier-Soulé J, Assenmacher I (1981a) Effects of raphe lesions on circadian ACTH corticosterone and motor activity rhythms in freerunning blinded rats. Neurosci Lett 23; 87-92

Szafarczyk A, Ixart G, Alonso G, Malaval F, Nouguier-Soulé J, Assenmacher I (1981b) Effects de la destruction des noyaux suprachiasmatiques sur les rythmes circadiens de l'ACTH, de la corticostérone et de l'activité générale chez des rattes soumises à un environment apériodique. C R Soc Biol 175: 801-810

Wever RA (1979) The circadian system of man. Springer, Berlin Heidelberg New York, pp 276

3.4 The Neurophysiology of the Mammalian Suprachiasmatic Nucleus and Its Visual Afferents

G.A. Groos[1]

1 Introduction

In 1967 Richter concluded that the circadian clock controlling the rest-activity cycle of the rat is probably located "somewhere in the hypothalamus" (Richter 1967). He added: "...we assume that (the clock) is constituted of many cells, each one of which is programmed to function at a rate of 24 or nearly 24 h, that under ordinary conditions these cells all function together but that under certain conditions they may become desynchronized at least to some extent". These statements represent perhaps the earliest speculation about the localization and neurophysiological organization of a circadian pacemaker in the mammalian hypothalamus. More recently Enright developed a more sophisticated neural model which includes specific mechanisms both for the generation of circadian oscillations and their photic entrainment. This model for the circadian pacemaker is primarily theoretical in nature. In order to construct a physiological model for the circadian pacemaker, a study of the functional properties of a "real" biological clock is essential. With the identification of the suprachiasmatic nuclei (SCN) as a putative pacemaker in the mammalian hypothalamus such an experimental program has become possible.

The study of the SCN is naturally guided by at least two basic questions. Firstly it needs to be established if and in what way the neural network of the SCN is capable of generating self-sustaining circadian oscillations. In the second place the role of the SCN and its retinal afferents in the process of photic entrainment will have to be investigated. The present chapter will deal with the latter of these two problems. In particular the neuroanatomy of the visual afferents to the SCN and their visual physiology is discussed in relation to the mechanism of entrainment of circadian rhythms by the daily illumination cycle.

2 The Visual Afferents of the SCN

While in submammalian vertebrates the entraining photic signals from the environment reach the circadian system through multiple pathways, including the retina, pineal gland

1 National Institute of Mental Health, Bethesda, Maryland 20205, USA

Vertebrate Circadian Systems
Ed. by J. Aschoff, S. Daan, and G. Groos

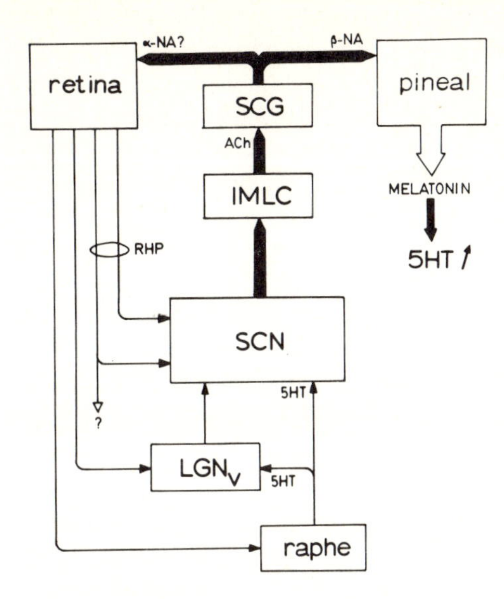

Fig. 1. The afferent and re-afferent visual inputs to the SCN. The RHP constitutes the only direct pathway from the retina to the SCN. It comprises both collaterals of primary optic fibres and axons which have their principal target in the SCN. A part of the optic tract terminates in the ventral portion of the lateral geniculate nucleus *(LGN_V)* or the raphe complex. Both latter structures project to the SCN. The raphe complex sends a serotonergic projection to the LGN_V. The retino-geniculo-SCN and the retino-raphe-SCN pathways constitute the putative indirect visual projections to the SCN. The re-afferent pathway originates in the SCN and projects via an unknown route to the upper thoracic intermediolateral cell column of the spinal cord *(IMLC)*. From the IMLC pre-ganglionic fibres reach the superior cervical ganglion *(SCG)* which projects both to the retina and the pineal gland. In so far as the transmitters involved in these pathways have been identified they are indicated in the diagram *(ACh* acethylcholine; *5HT* serotonin; α–, *β–NA* α–, and β-noradrenalin)

and deep brain photoreceptors, the situation in adult mammals is apparently less complicated (Underwood and Groos 1982). Their pineal gland is not directly sensitive to light, while there is strong evidence against a role for brain photoreceptors in the entrainment of their circadian rhythms (Groos 1979, Groos and Vanderkooy 1981).

In rodents the retino-hypothalamic projection (RHP) to the SCN is sufficient for entrainment (Rusak and Boulos 1981). It would be an oversimplification, however, to conclude that the RHP is the only pathway involved in the photic control of rodent circadian rhythms. In a recent paper Rusak and Boulos (1981) conclude that a variety of retino-fugal pathways "interact in a complex fashion to produce the features of normal photic entrainment". The intricate organization of the visual inputs to the SCN implied by this statement is summarized in Fig. 1. From this diagram it is evident that, besides the direct retinal projection via the RHP, a number of putative, indirect visual pathways terminate in the SCN.

The direct bilateral RHP to the SCN is a common feature of the mammalian visual system (Moore 1973, Mai, Chap. 2.3, this Vol.). The RHP comprises fine collaterals of retinal ganglion cell axons terminating in the lateral portion of the SCN, and larger diameter fibers, the majority of which terminate in the ventromedial SCN (Mason et al. 1977). In autoradiographic material the RHP is seen to project principally to the ventrocaudal region of the contra-lateral SCN (Moore 1973). Although this spatial distribution of RHP terminals is largely substantiated by electrophysiological recordings (Groos and Mason 1980), the topographical organization of the RHP may vary among different species (Pickard 1980). The majority of RHP fibres form Gray type I synapses on SCN cells, while a smaller proportion of optic terminals make Gray type II contacts (Gueldner and Ingham 1979, 1980).

There is extensive evidence for the existence of a projection from the ventral lateral geniculate nucleus (LGN_V), itself a terminal nucleus for retinal fibres, to the SCN (Swanson et al. 1974). Similarly the midbrain raphe complex receives a retinal input (Foote et al. 1978) and, in turn, gives rise to a prominent serotonergic (5HT) projection to the SCN as well as to the LGN_V (Azmitia and Segal 1978). Thus, the two indirect retinal projections to the SCN may not be independent.

The anatomical evidence suggests that, in addition to a direct visual input, the SCN receives indirect retinal projections via the LGN_V and the raphe nuclei. However, it needs to be demonstrated that those raphe and LGN_V cells that project to the SCN are connected to retino-fugal fibres and consequently exhibit visual responsiveness. This has not yet been shown for the raphe projection to the SCN. Evidence that electrical stimulation of the raphe complex can alter the discharge of visual SCN cells (Nishino 1976) is not conclusive in this respect. This effect may be mediated by non-visual raphe cells projecting either directly or indirectly (via the LGN_V) to the SCN.

On the other hand there is recent evidence that visually responsive LGN_V neurones in the rat can be antidromically activated by electrical stimulation of their fibres in the SCN (Groos and Rusak in prep.). From this observation it can be tentatively concluded that the retino-geniculo-suprachiasmatic pathway represents an indirect visual projection to the SCN.

If multiple visual inputs converge upon the SCN it is important to establish which effects of light on this central circadian pacemaker is mediated by each. Lesion studies involving the primary optic system, the LGN_V and the raphe complex show that the RHP alone is sufficient for entrainment as well as for the effects of constant light on the freerunning period (Rusak and Boulos 1981). These investigations also suggest that the indirect retinal projections to the SCN may be involved in the photic control of circadian rhythms. Primary optic tract or LGN lesions alter the re-entrainment rate to a phase-shifted LD cycle and affect the freerunning period (Rusak and Boulos 1981). It is conceivable that these effects are the result of retrograde degeneration of optic fibres, including those that give rise to the RHP collaterals. It seems plausible, however, that at least part of these effects are produced by the interruption of the retino-geniculo-suprachiasmatic pathway. In contrast, it is unlikely that the raphe nuclei play an important role in transmitting photic information to the SCN, since raphe lesions neither affect steady-state entrainment nor result in an abnormal freerunning period in constant illumination (Rusak and Zucker 1979).

3 The Feedback Control of the SCN

From Fig. 1 it can be seen that the SCN may exert feedback control on its visual input. The pathways mediating such feedback have a common origin in the superior cervical division of the sympathetic nervous system (Fig. 1).

Since stimulation of the SCN inhibits the discharge of the pre-ganglionic fibres projecting to the superior cervical ganglion (Nishino et al. 1976), a pathway must exist from the SCN to this ganglion, probably via the intermediolateral cell column of the spinal cord. The post-ganglionic fibres innervate the eyes and the pineal gland. In the eyes they can affect the retinal response to light in two ways: indirectly by dilating the pupil, thereby increasing the photic energy reaching the photoreceptors, and directly by imposing a circadian rhythm in sensitivity to light on the retina itself (Bobbert, Chap. 2.4, this Vol.). The sympathetic control of pineal activity has been intensively studied in the past decades. Electrical activity in the post-ganglionic fibres stimulates pineal enzyme activity resulting in the nocturnal production of melatonin (Klein 1978, Nishino et al. 1976). An increased concentration of circulating melatonin leads to an increase of 5HT levels in the diencephalon while pinealectomy results in decreased 5HT levels (Anton-Tay et al. 1968, Sugden and Morris 1979). Therefore, it is conceivable that the SCN regulates the activity of its putative visual afferents from the raphe and the LGN_V via its rhythmic control of pineal melatonin production. This idea is supported by the observation that 5HT may be involved as a transmitter in these pathways and that a large proportion of visual SCN cells are sensitive to 5HT and raphe stimulation (Nishino 1976).

To establish if the feedback pathways are involved in the photic control of circadian rhythms, the most straightforward experiment would be to interrupt the pre-ganglionic sympathetic nerves. Bilateral sympathectomy, which abolishes both the retinal sensitivity rhythm (Bobbert, Chap. 2.4, this Vol.) and the rhythm of pineal melatonin synthesis (Klein 1978), does not have a marked effect on steady-state entrainment (for a review, see Rusak 1982). Similarily, sympathectomy does not prevent normal phase delays and advances of the freerunning hamster activity rhythm induced by short light pulses presented during the early and late subjective night, respectively (Groos and Aschoff unpub.). Pinealectomy, which eliminates the effect of pineal melatonin on 5HT in the LGN_V, the raphe and the SCN, fails to affect normal entrainment of the locomotor activity rhythm in several rodent species (Rusak 1982). Also, pinealectomized hamsters seem to have a normal phase response curve (Groos and Aschoff unpub.). Although there have been a number of reports describing an increased re-entrainment rate in pinealelectomized rodents after a phase-shift in the LD cycle these obersvations can be interpreted in terms of an increased sensitivity to the masking effects of light after pinealectomy (Rusak 1982).

In summary there is little evidence that the feedback pathways of the SCN are involved in the control of circadian rhythms by light; however, whether they mediate different or more subtle influences on the circadian pacemaker remains an open question. In particular, the effects of melatonin on 5HT terminals (Yates and Herbert 1979) and the presence of both substances in the SCN (Azmitia and Segal 1978, Bubenik et al. 1976) should make us cautious in ignoring the interaction between SCN and pineal as a factor in the central regulation of circadian rhythms and, perhaps more likely, photoperiodic phenomena.

4 The Visual Neurophysiology of the SCN

Given direct and indirect retinal projections converging on the SCN, it is not surprising that a considerable proportion of SCN cells respond to visual stimulation of the retina or electrical stimulation of the optic nerves (Groos and Mason 1978, 1980, Nishino et al. 1976, Sawaki 1979). In striking contrast to the large majority of neurones in other parts of the mammalian visual system, they do so by either a sustained increase or a sustained decrease of their mean discharge rate (Fig. 2). It is therefore useful to classify vis-

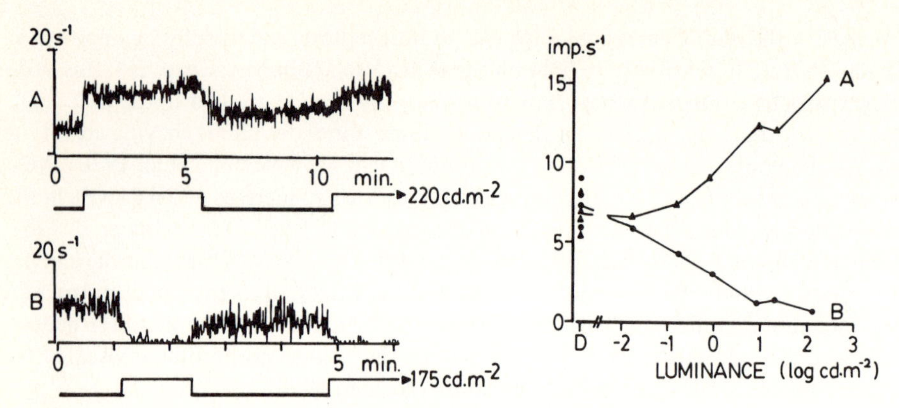

Fig. 2. Light-activated **(A)** and suppressed **(B)** neurons in the rat SCN. The mean discharge rate of both types of cell is monotonically related to the luminance level as illustrated in the *graph on the right* for the cells in **A** and **B**, respectively. In this graph *D* indicates darkness

ually responsive SCN neurones as activated or suppressed cells, depending on whether an increase in retinal illumination increases or decreases their electrical activity, respectively (Groos and Mason 1980). The activated cell type, which is the more common one, exhibits a discharge rate which is an increasing function of environmental luminance within a wide range of light intensities. Within the same range the suppressed cells decrease their firing with increasing luminance. Thus it is evident that visual SCN cells behave like luminance detectors. They do so by integrating photic energy from large areas of the retina. This is demonstrated by the large size and homogeneous organization of their receptive fields, a feature which is rarely encountered in other parts of the mammalian visual system.

A functional organization as described above for the SCN can be obtained in two distinct ways. On one hand it is possible that tonic W-cells, a rare class of retinal luminance detectors with small receptive fields, converge on individual visual SCN cells. This convergence would extend the size of the receptive fields of SCN neurones but also make them functionally homogeneous. On the other hand, there could be a specialized class of retinal ganglion cells with appropriate functional properties which projects predominantly to the SCN. A choice for the former of these two possibilities would be most parsimonious at present but it should be noted that there is anatomical evidence supporting the latter suggestion (Pickard 1980).

Recording visual responses in the SCN does not answer the question whether the activated and suppressed cells are innervated by the RHP, by axons from the LGN, or by a combination of these anatomical alternatives. From the LGN_V, itself part of the primary optic system, neurons project to the SCN. Interestingly, a sizeable subpopulation of LGN_V cells can be classified as luminance detectors with functional properties which are very similar to those of the SCN (Fig. 3A, B). These LGN_V cells are either tonically activated or suppressed by a sustained increase in overall retinal illumination (Hale and Sefton 1978). The amplitude of their response increases with increasing illumination levels and the receptive fields of these cells are frequently uniform and extend over an extremely large area of the retina (Richard et al. 1977).

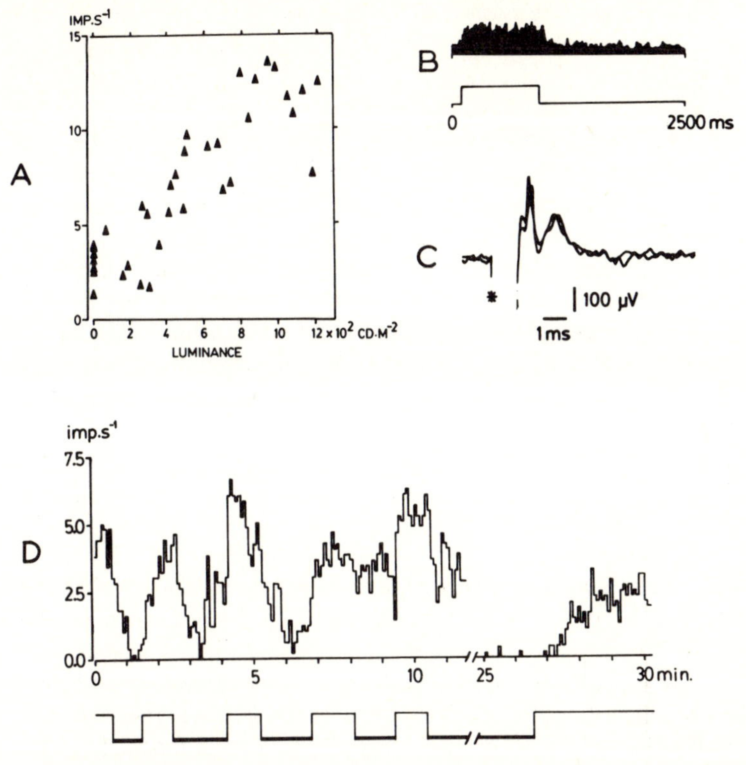

Fig. 3. This figure presents an example of a cell in the LGN_V of the rat which is tonically activated by light (**B** averaged response to flash stimulation). The cell exhibits a mean discharge rate in constant light which is an increasing function of ambient luminance **(A)**. In these respects this behaviour is very similar to that of a visually activated SCN cell. Interestingly, this LGN_V cell could be antidromically activated by stimulation of the SCN (**C** the *asterisk* marks the electrical stimulus). It is therefore likely that this cell projects to the SCN. The record in **D** illustrates the persisting visual responsiveness of a cell in the SCN in a rat sustaining complete bilateral LGN_V lesions

Considering the striking similarities between this class of LGN_V cells and the visual cells of the SCN, the question arises if these particular LGN_V cells project to the SCN. Recent studies have demonstrated that at least some of the tonically responding LGN_V cells terminate in the SCN. Electrical stimulation of their terminals in the rat SCN results in antidromic activation of their cell bodies in the LGN_V (Groos and Rusak in prep.). This finding is illustrated in Fig. 3C. Antidromic activation is also occasionally observed in LGN_V cells which do not respond to diffuse retinal illumination. It can therefore only be concluded that a part of the LGN_V input to the SCN is comprised of cells which are functionally similar to visual SCN neurones.

As was mentioned earlier, the integrity of the retino-geniculo-suprachiasmatic projection is not necessary for the photic control of circadian rhythms, while the RHP alone is sufficient to mediate such photic effects. At the electrophysiological level, these observations are parallelled by the finding that SCN cells are still capable of responding to light at various times (1 h-4 months) after bilateral lesions of the primary optic pathway or the LGN_V (Groos and Mason 1978, Sawaki 1979). Moreover, at present there is no indication that the visual behavior of the SCN is drastically changed after such lesions (Fig. 3D).

From the observations summarized above, the following picture of the visual physiology of the SCN emerges. There are at least two projections from the retina to the SCN, the RHP and the retino-geniculo-suprachiasmatic pathway. Both pathways transmit information about the overall environmental luminance to the SCN. This nucleus contains two subpopulations of neurons, the activated and the suppressed cells, the discharge of which is either positively or negatively related to the ambient light intensity. Intuitively, this functional organization seems appropriate in view of the circadian pacemaker function of the SCN. The visual behavior of SCN cells allows for a redundant coding of dawn and dusk, as well as photoperiod. It seems worthwhile, however, to have a closer look at how well this behavior can be related to some of the formal properties of the photic entrainment process.

5 The Effects of Light on Circadian Rhythms and the SCN

One of the most striking and consistent effects of light on the mammalian circadian pacemaker(s) is the differential sensitivity of the pacemaker to light presented at different phases of its cycle. This property is reflected in the phase-response curve. In addition, a variety of effects of constant light are known, including the dependence of τ on light intensity and such phenomena as splitting and arrhythmicity in LL. There is only little evidence that these effects are paralleled by physiological processes in the SCN or its visual afferents. For example, no extensive studies have been undertaken to establish if the functional properties of visual SCN cells remain constant throughout the circadian cycle. Variations in visual responsiveness could be at the basis of the differential effects of light at different circadian phases which, in turn, underlie the non-parametric mode of entrainment. From our studies it has become evident that the occurrence of activation or suppression by light in SCN cells is not restricted to particular phases of the circadian cycle (Groos and Mason 1980) but long-term recordings are needed to establish whether or not quantitative or qualitative changes occur in the visual behaviour of the SCN as a function of phase.

If it is assumed that the visual responsiveness of SCN cells is invariant with phase, it is very likely that the pacemaker of the SCN shows a phase-dependent response to light. Even in blinded rats and hamsters, in which all visual afferents to the SCN have degenerated, the SCN exhibits a phase-dependent response to electrical stimulation (Rusak and Groos in prep.). The phase-shifts induced by local stimulation can be summarized in a phase-response curve which is very similar to that for short light pulses. Intraventricular injections of carbachol in blinded mice similarly cause phase delays and advances which mimic the action of light pulses at the phase of injection (Zatz and Herkenham 1981). It is very likely that carbachol in these experiments causes phase shifts by a direct action on the SCN rather than on the other brain structures (Zatz and Brownstein 1981). Evidence that the SCN as a whole is differentially sensitive to light is derived from studies which show that the change in SCN glucose consumption in response to a 46-min light pulse is phase-dependent. Light at dawn decreases, while light at dusk increases the glucose utilization of the rat SCN (Schwartz et al. 1980). These findings illustrate that regardless of whether the visual cells change their properties with time of day, the neural network of the SCN presumably responsible for the generation of circadian rhythms is itself differentially sensitive to photic, electrical and chemical stimulation at different phases.

When animals are kept in constant conditions, their circadian rhythms are affected according to the light intensity to which they are exposed. At the physiological level, a synaptic re-organization of the SCN is observed under these conditions (Gueldner and Ingham 1979, 1980). In constant darkness or after binocular enucleation the relative number of optic and non-optic synapses in the SCN with thick postsynaptic density material increases while in constant light this number gradually decreases. Ultrastructurally synapses can be classified as Gray type I (with thick post-synaptic density) and Gray type II synapses (which virtually lack postsynaptic density material). In most cases the Gray type I synapse is excitatory and the Gray type II synapse inhibitory in nature (Gueldner and Ingham 1979, 1980). If this correlation between morphology and physiology holds for the SCN, it can be concluded that in constant light a higher proportion of synapses SCN becomes inhibitory. Although this suggestion is perhaps too speculative, the morphological changes observed in the SCN after exposure to constant light may underlie at least some of the effects and after-effects of light on the circadian pacemaker.

Constant light not only alters the synaptology of the SCN but also causes degeneration of the rod photoreceptors in the retina of rodents (Williams and Baker 1980). In spite of the massive generation of rods, such animals still entrain their circadian rhythms to LD cycles (Rusak and Zucker 1979). Apparently the few cones in the rodent retina are sufficient to mediate photic entrainment. This conclusion is supported by the finding that rats entrain their activity rhythm to red light pulses of a wavelength far beyond the action spectrum of the rods (McCormack and Sontag 1980). If rods and cones both mediate entrainment, it is plausible to assume that when rodents are exposed to constant light of photopic intensities an increasing number of cones are recruited and may eventually dominate the inflow of visual information to the SCN. Perhaps the effects of constant photopic illumination on rodent circadian rhythms are partially a result of the dominant activity of cones under these conditions. In this case, it can be expected that constant red light facilitates splitting and arrhythmicity. There is some evidence to support this suggestion. Splitting is frequently observed in hamsters exposed to constant red light (Wirz-Justice and Wehr unpubl.), while continuous red light produces persistent estrous in female rats (Lambert 1975). Since persistent estrous in constant light generally occurs together with circadian arrhythmicity (Lambert 1975, Rusak and Zucker 1979), it is conceivable that this level of constant red light produces arrhythmicity in rats. If the ratio of rod-versus cone-mediated activity in the visual projection to the SCN is important in the nocturnal animals with rod-dominated retinae, a similar but reverse situation may exist in diurnal species, where cones predominate. Perhaps Hoffman's observation (Hoffmann 1971) that the activity rhythm of the diurnal tree shrew splits at low rather than high intensities illustrates this different organization in diurnal species.

In contrast to the prevalent concept of the mid-1970's that the mammalian SCN is a dominant circadian pacemaker which is entrained by LD cycles via the RHP, the actual physiological and anatomical organization of the entrainment pathways is more complex. In this chapter evidence was reviewed that multiple pathways, notably the RHP and the

LGN_V projection to the SCN, mediate entrainment. Their visual behaviour is very similar and reflects their specialized function of coding the changes in environmental illumination between day and night. However, many problems remain unsolved. Particularly, there is no indication at present what function, if any, can be ascribed to the feedback pathways of the SCN. Also the nature of the interaction between the pacemaker of the SCN and its visual input is largely unknown, Finally, at the level of the retina, the initial stage of the entrainment process, we have just begun to speculate on the contribution of the rods and cones to the entrainment process.

References

Anton-Tay F, Chou C, Anton S, Wurtman RJ (1968) Brain serotonin concentration: elevation following intraperitoneal administration of melatonin. Science 162: 277-278

Azmitia EC, Segal M (1978) An autoradiographic analysis of the differential ascending projections of the dorsal and medial raphe nuclei in the rat. J Comp Neurol 179: 641-668

Bubenik GA, Brown GM, Grota LJ (1976) Differential localization of N-acetylated indolealkylamines in CNS and Harderian gland using immunohistology. Brain Res 118: 417-423

Enright JT (1980) The timing of sleep and wakefulness. Springer, Berlin Heidelberg New York

Foote WE, Taber-Pierce E, Edwards L (1978) Evidence for a retinal projection to the midbrain of the cat. Brain Res 156: 135-140

Groos GA (1979) Electrophysiological evidence for the absence of photosensitive neurons in the rat suprachiasmatic nucleus. IRCS Med Sci 7: 342

Groos GA, Mason R (1978) Maintained discharge of rat suprachiasmatic neurons at different adaptation levels. Neurosc Lett 8: 59-64

Groos GA, Mason R (1980) The visual properties of rat and cat suprachiasmatic neurones. J Comp Physiol 135: 349-356

Groos GA, Vanderkooy D (1981) Functional absence of brain photoreceptors mediating entrainment of circadian rhythms in the adult rat. Experientia 37: 71-72

Gueldner FH, Ingham CA (1979) Plasticity in synaptic appositions of optic nerve afferents under different lighting conditions. Neurosci Lett 14: 235-240

Gueldner FH, Ingham CA (1980) Increase in post-synaptic density material in optic target neurons of the rat suprachiasmatic nucleus after bilateral enucleation. Neurosci Lett 17: 27-31

Hale PT, Sefton AJ (1978) A comparison of the visual and electrical response properties of cells in the dorsal and ventral lateral geniculate nuclei. Brain Res 153: 591-595

Hoffmann K (1971) Splitting of the circadian rhythm as a function of light intensity. In: Menaker M (ed) Biochronometry. Natl Acad Sci, Washington DC, pp 134-151

Klein DC (1978) The pineal gland: a model of neuroendocrine regulation. In: Reichlin S, Baldessarini RJ, Martin JB (eds) The hypothalamus. Raven Press, New York, pp 303-327

Lambert HH (1975) Continuous red light induces persistent estrus without retinal degeneration in the albino rat. Endocrinology 97: 208-210

Mason CA, Sparrow N, Lincoln DW (1977) Structural features of the retinohypothalamic projection in the rat during normal development. Brain Res 132: 141-148

McCormack CE, Sontag CR (1880) Entrainment by red light of running activity and ovulation rhythms of rats. Am J Physiol 239: R450-R4543

Moore RY (1973) Retino-hypothalamic projections in mammals: a comparative study. Brain Res 49: 403-409

Nishino H (1976) Suprachiasmatic nuclei and circadian rhythms. Folia Pharmacol 72: 941-954

Nishino H, Koizumi K, McBrooks C (1976) The role of suprachiasmatic nuclei of the hypothalamus in the production of circadian rhythm. Brain Res 112: 49-59

Pickard GE (1980) Morphological characteristics of retinal ganglion cells projecting to the suprachiasmatic nucleus: a horseradish perioxidase study. Brain Res 183: 458-465

Richard D, Koszul MF, Buser P (1977) Size and characteristics of visual receptive fields in nucleus ventralis lateralis in cat under chloralose anesthesia. Brain Res 138: 175-179

Richter CP (1967) Sleep and activity: their relation to the 24-hour clock. Proc Assoc Res Nerv Dis 45: 8-27

Rusak B (1982) Circadian organization in mammals and birds: role of the pineal gland. In: Reiter RJ (ed) The pineal fland: extra-reproductive effects. CRM Press, Boca Raton, Florida (in press)

Rusak B, Boulos Z (1981) Pathways for photic entrainment of mammalian circadian rhythms. Photochem Photobiol 34: 267-273

Sawaki Y (1979) Suprachiasmatic nucleus neurones: excitation and inhibition mediated by the direct retino-hypothalamic projection in female rats. Exp Brain Res 37: 127-138

Schwartz WJ, Davidsen LC, Smith CB (1980) In vivo metabolic activity of a putative circadian oscillator, the rat suprachiasmatic nucleus. J Comp Neurol 189: 157-167

Sugden D, Morris RD (1979) Changes in regional brain levels of tryptophan, 5-hydroxytryptamine,5-hydroxyindoleacetic acid, dopamize and nor-adrenaline after pinealectomy in the rat. J Neurochem 32: 1593-1596

Swanson LW, Cowan WM, Jones EG (1974) An autoradiographic study of the efferent connections of the ventral lateral geniculate nucleus in the albino rat and the cat. J Comp Neurol 156: 143-164

Underwood H, Groos GA (1982) Vertebrate circadian rhythms: retinal and extraretinal photoreceptors. Experientia (in press)

Williams TP, Baker BN (1980) The effects of constant light on visual processes. Plenum Press, New York

Yates CA, Herbert J (1979) The effects of different photo-periods on circadian 5HT rhythms in regional brain areas and their modulation by pinealectomy, melatonin and oestradiol. Brain Res 176: 311-326

Zatz M, Brownstein MJ (1981) Injection of alpha-bungarotoxin near the suprachiasmatic nucleus blocks the effects of light on nocturnal pineal enzyme activity. Brain Res 213: 438-442

Zatz M, Herkenham MA (1981) Intraventricular carbachol mimics the phase-shifting effect of light on the circadian rhythm of wheel-running activity. Brain Res 212: 234-238

3.5 Neurophysiological Studies of the SCN in the Rat and in the Java Sparrow

Hiroshi Kawamura, Shin-ichi T. Inouye, Shizufumi Ebihara, and Setsuko Noguchi[1]

Lesion experiments of the suprachiasmatic nucleus (SCN), a tiny nucleus in the hypothalamus, revealed abolishment of circadian rhythms of various physiological and behavioral activities in the rat. Stephan and Zucker (1972) indicated elimination of circadian rhythms of drinking and eating behavior as well as wheel-running activity in the rat after bilateral lesions of the SCN. Moore and Eichler (1972) reported the abolishment of the adrenal corticosterone rhythms after lesions of the SCN. Later on, we also observed that lesions of the rat SCN abolished the circadian rhythm of sleep and wakefulness in this nocturnal animal without changing the total amount of slow wave sleep and paradoxical sleep in a day (Ibuka and Kawamura 1975). Distribution of sleep during daytime and nighttime became even after lesions of the SCN. These studies suggested that the nucleus contains an endogenous oscillator generating circadian rhythms. Alternatively, this nucleus could be a coupler of many circadian rhythm oscillators within the brain. From such lesion experiments alone, it is difficult to conclude that the SCN really generates a circadian rhythm. For such a conclusion, we should obtain a tissue culture preparation of the SCN demonstrating that cellular metabolic or physicochemical activity of the SCN in vitro shows a circadian rhythm. However, presently, the in vitro culture of adult mammalian brain tissue is technically difficult. Therefore we developed a method for the in vivo isolation of a part of the hypothalamus containing the SCN as an island using a modified Halasz technique. We succeeded in making a relatively intact island in the anterior hypothalamus (Inouye and Kawamura 1979). In addition, we inserted one or two bipolar electrodes into the island which were aimed at the SCN.

Multiple unit activity (MUA) from several tens of neurons was recorded by a bipolar electrode (Fig. 1) which was used to analyze the rhythmic activity. We chose this method because continuous recording of single unit activity for more than 2-3 h is difficult, whereas MUA recording can be done more easily for longer periods. MUA exceeding the noise level of recording system was counted using an electronic slicer and counter, and the number of detected spikes was printed out automatically every 5 min. The MUA was simultaneously recorded on a polygraph using a cumulative recorder. In intact rats, simultaneous recording of MUA from two locations in the brain showed a circadian rhythm with increase of activity during nighttime when the rat is highly active and decrease of activity during daytime when the animal is frequently asleep. These changes in intact rats brain were usually parallel regardless of electrode tip location. One exception to this

1 Mitsubishi-Kasei Institute of Life Sciences, Machida-shi, Tokyo 194, Japan

Vertebrate Circadian Systems
Ed. by J. Aschoff, S. Daan, and G. Groos

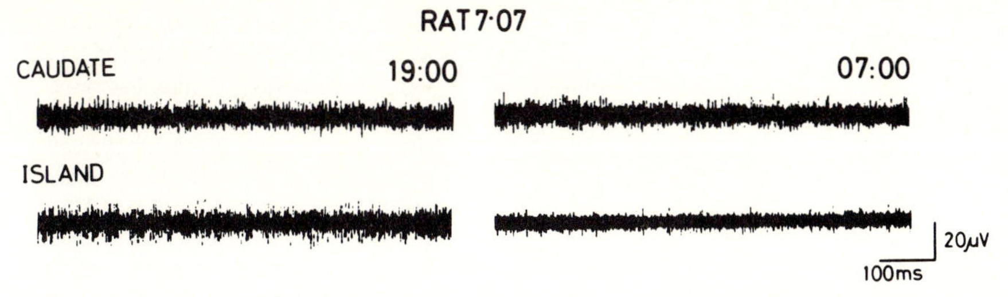

Fig. 1. Multiple unit activity simultaneously recorded from caudate nucleus *(upper trace)* and inside the hypothalamic island *(lower trace)*. Marked difference within the island depending on the time of the day (Inouye)

observation is that when the electrode tip is within the SCN or very close to the SCN the MUA shows a 180° reversed phase compared to the MUA from any other location in the brain. Thus, when MUA outside the SCN shows high activity during subjective nighttime or dark phase of the LD cycle, MUA within the SCN shows low activity. However, after making a hypothalamic island containing the bilateral SCN, no circadian rhythm could be seen in MUA recorded outside the island. On the other hand, within the hypothalamic island, we could record a clear circadian rhythm of MUA (Fig. 2). When the hypothalamic island was completely isolated from the rest of the brain, behavioral circadian rhythms of sleep-waking, drinking or eating (Fig. 3) were completely abolished. This finding implies that the circadian rhythm oscillator is clearly located inside the hypothalamic island and not elsewhere in the brain. Outside the island, the reticular activating system which controls sleep-waking is intact. Therefore, if a circadian rhythm oscillator or oscillators existed elsewhere, a spectral analysis of sleep and wakefulness should indicate the persistence of a circadian rhythmicity. However, Ibuka et al. (1977) demonstrated a complete loss of rhythms with a period of about 24 h after bilateral SCN lesions. Of course, within the island, not only the SCN but also various adjacent hypothalamic nuclei are included. Schwarz and Gainer (1977), using the radioactive deoxyglucose uptake method, found an increase of metabolic activity in the SCN during daytime which matches the increase of MUA within the SCN during daytime (including subjective daytime during free-running). Our MUA data demonstrate that MUA showed reversed activity only within the SCN compared to the activity of the surrounding hypothalamic areas. This means that the SCN has a particular characteristic among the hypothalamic nuclei.

SCN activity in the hypothalamic island has been successfully recorded by Inouye for periods up to 35 days after surgery. In these cases, a very beautiful sinusoidal circadian rhythm shows damped oscillations. So far, we do not know whether such damping is due to a characteristic of the oscillator or it is only due to development of gliosis detrimental to the recording of neuronal activity of the SCN. In the hypothalamic island with a portion of optic nerves uncut, thus leaving a neural connection between eyes and the SCN, when a free-running circadian rhythm of MUA in the SCN in constant darkness is observed, a light pulse given at a certain phase of the circadian rhythm can sometimes induce an instantaneous shift of the rhythm. For instance, when a light pulse with a duration of 3 h was presented during the second half of the subjective night, the phase of the circadian rhythm advanced up to 3 h, whereas when the pulse fell during the first half of the subjective night, the phase delayed also in the range of up to 3 h. But

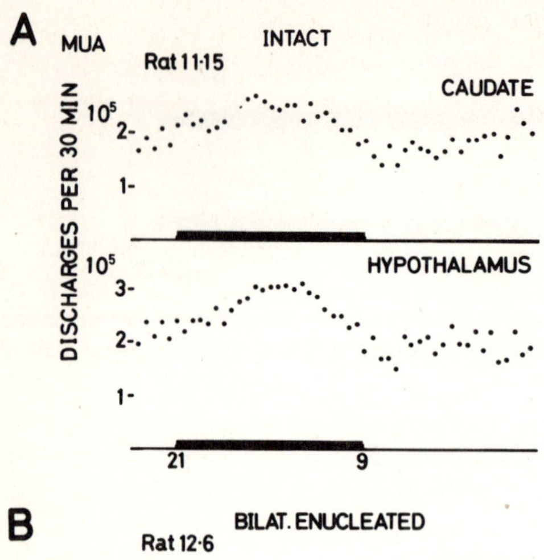

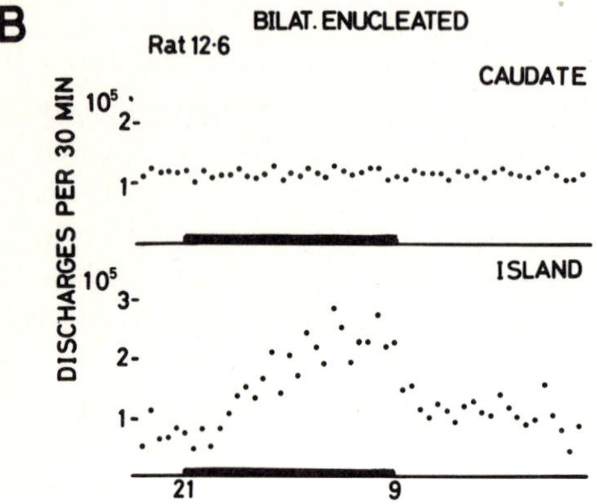

Fig. 2. MUA in an intact rat **(A)** and in a rat with hypothalamic island **(B)**. MUA's were plotted every 30 min. *Ordinates* Number of discharge per 30 min. *Abscissae* Time of day, *black bar* indicates dark period of LD cycle. In the intact rat the simultaneously recorded MUA's (from caudate nucleus and hypothalamus outside the SCN) show parallel changes. In a blinded rat with hypothalamic island **(B)**, the circadian rhythm persisted, but no circadian rhythm was found outside the island. (Inouye)

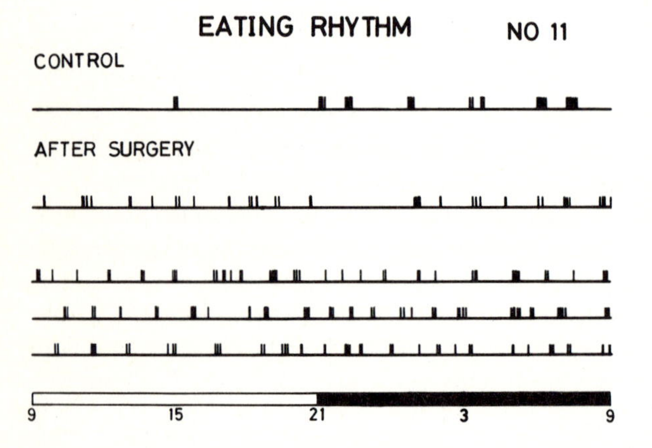

Fig. 3. Abolishment of circadian eating rhythm after surgical isolation of the SCN. Control record shows frequent eating activity during dark period *(black bar)*, whereas after making hypothalamic island even distribution of eating activity can be seen. A food pellet was provided to the rat each time when the rat pressed a bar, which was then recorded on a recorder (Noguchi)

when such a light pulse was given during the subjective day of the animal, no phase shift of the circadian rhythm was observed. Such instantaneous phase shifts also represent a characteristic of the circadian rhythm oscillator in the rodent. However, when the rat with a hypothalamic island which is still connected with the eyes is exposed to a LD 12 : 12 cycle, and the phase of the environmental LD cycle is shifted abruptly, for instance for 6 h, then entrainment of the MUA circadian rhythm to the new environmental LD cycle takes 4-5 days. (Inouye and Kawamura 1982). This is similar to our experience during jet lag after transmeridian flight.

In the case of birds, the pineal organ is considered to be a circadian rhythm oscillator. This suggestion is supported by organ culture experiments by several authors and also by cell culture experiments (cf. Chaps. 4.4 and 4.5, this Vol.). However, Gwinner's (1978) results showing persistence of a circadian rhythm after removal of the pineal organ suggest the existence of another oscillator in the avian brain. To study this problem, we used the Java sparrow, *Padda oryzivora*, a passerine bird. This bird, which is the cheapest one we can obtain from pet shops in Japan, is suitable for mounting on a stereotaxic instrument (Ebihara and Kawamura 1981). Since location of the SCN had not yet been established in this bird, radioactive leucine was injected into one eye of the bird and the distribution of the silver grain was examined autoradiographically to identify the location of the SCN (Fig. 4). In the Java sparrow, after lesions of the SCN, free-running perch-hopping activity in constant dim light was completely abolished. If we exposed the bird to an environmental LD cycle, the activity rhythm was synchronized to the LD cycle, perch-hopping activity being more frequent during light phase of the LD cycle and the bird being quiet during the dark (dim light) phase. When the environmental

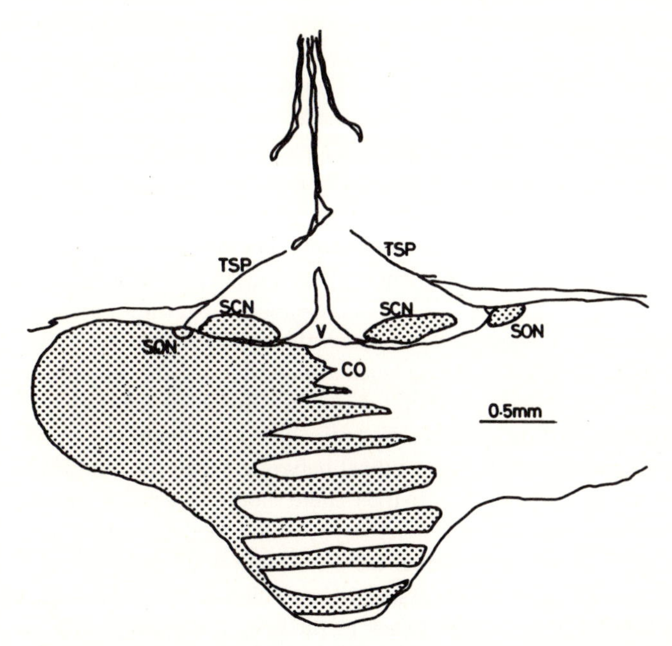

Fig. 4. Localization of SCN in the Java sparrow as revealed by autoradiography. *SCN* suprachiasmatic nucleus; *SON* supraoptic nucleus; *CO* optic chiasma; *TSP* septo-preoptic tract (Ebihara)

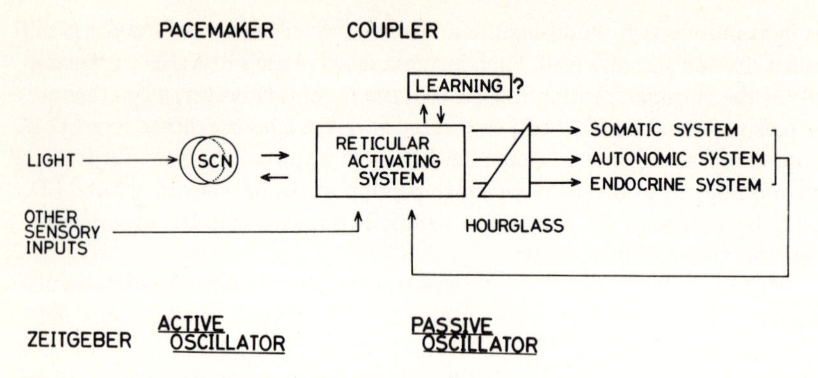

Fig. 5. Tentative scheme of a circadian clock organization in the rat

LD cycle was phase-shifted by several hours, the activity rhythm phase-shifted immediately. Lesions of the SCN completely eliminate the free-running circadian rhythm. On the other hand, after removal of the pineal organ, free-running rhythm is eliminated, but after exposure to the LD cycle a residual rhythm could be seen in constant dim light. Also when the pinealectomized bird was kept under a LD cycle in which total darkness was used for the dark phase, the anticipatory perch-hopping activity before the L phase was not seen, but when dim light was used during dark phase the anticipatory activity before the L phase was present. Here again the role of SCN as a circadian rhythm oscillator is clearly indicated. From these data it is apparent that the SCN is functioning as an important part of the circadian system in this bird.

Figure 5 presents a tentative scheme of the circadian clock system in the rat. Admittedly, this is rather a rough and ready idea which must be revised as our knowledge of the circadian oscillator system develops. In this scheme, the SCN is considered as an active oscillator producing a circadian rhythm. This nucleus may generate a rhythm under normal conditions as a pacemaker. This does not necessarily deny the possibility that left and right SCN's function to some extent as independent oscillators. The SCN is anatomically closely and directly connected with the visual input. The effectiveness of photic stimuli as a zeitgeber for entrainment of the circadian rhythm is clear from this anatomical organization. Other sensory inputs may affect the circadian oscillator (SCN) indirectly, presumably via the reticular activating system (of which the hypothalamus is a part), the effectiveness of these inputs as zeitgebers being weaker than that of light. The reticular formation may play a role as an amplifier which, for instance, amplifies relatively weak oscillators of the SCN, passing it on to the entire brain. This again induces a circadian sleep-waking rhythm which affects the whole animal behavior. When the oscillator activity is relayed to induce various functional activities, there should be a characteristic time lag depending on the individual physiological activity. For instance, the motor system will not have much time lag. Autonomic system will have more delay and probably in general the endocrine system will have the longest time lag. This can be a physiological basis of an hourglass. Individual activity may feed back to the brain and some may be stored as memory through a specific memory channel which will induce a residual rhythm for several cycles whenever the circumstances allow.

References

Ebihara S, Kawamura H (1981) The role of the pineal organ and the suprachiasmatic nucleus in the control of circadian locomotor rhythms in the Java sparrow, *Padda oryzivora.* J Comp Physiol 141: 207-214

Gwinner E (1978) Effects of pinealectomy on circadian locomotor activity rhythms in European starlings, *Sturnus vulgaris.* J Comp Physiol 126: 123-129

Ibuka N, Kawamura H (1975) Loss of circadian rhythm in sleep-wakefulness cycle in the rat by suprachiasmatic lesion. Brain Res 96: 76-81

Ibuka N, Inouye ST, Kawamura H (1977) Analysis of sleep-wakefulness rhythms in male rats after suprachiasmatic nucleus lesions and ocular enucleation. Brain Res 122: 33-47

Inouye ST, Kawamura H (1979) Persistence of circadian rhythmicity in a mammalian hypothalamic "island" containing the suprachiasmatic nucleus. Proc Natl Acad Sci USA: 76: 5962-5966

Inouye ST, Kawamura H (1982) Characteristics of a circadian pacemaker in the suprachiasmatic nucleus. J Comp Physiol 146: 153-160

Moore RY, Eichler VB (1972) Loss of a circadian adrenal corticosterone rhythm following suprachiasmatic lesions in the rat. Brain Res 42: 201-206

Schwarz WJ, Gainer H (1977) Suprachiasmatic nucleus: use of ^{14}C-labeled deoxyglucose uptake as a functional marker. Science 197: 1089-1091

Stephan FK, Zucker I (1972) Circadian rhythm in drinking behavior and locomotor activity of rats are eliminated by hypothalamic lesions. Proc Natl Acad Sci USA 69: 1583-1586

3.6 Neural Mechanisms in Avian Circadian Systems: Hypothalamic Pacemaking Systems

J.S. Takahashi[1]

1 Introduction

Since the initial discovery that pinealectomy abolishes circadian locomotor rhythms in house sparrows (Gaston and Menaker 1968), the precise role of the pineal gland within the avian circadian system has been studied extensively (Menaker et al. 1978, 1981, Takahashi and Menaker 1979a). At present, there exists sufficient comparative data among avian species to suggest that the pineal plays a major role in the regulation of circadian rhythms in passerine birds, but only a minor role in gallinaceous birds (Takahashi and Menaker 1979a). The interspecific differences in the effects of pinealectomy among birds demonstrate that structures other than the pineal play an important role in regulating avian circadian rhythms.

In mammals, the suprachiasmatic nuclei (SCN) of the hypothalamus are necessary for the maintenance of circadian rhythmicity. Lesions of these nuclei abolish the circadian component in a number of physiological and behavioral rhythms, and disrupt entrainment to light-dark cycles (Moore 1978, Rusak and Zucker 1979). These results and others (Inouye and Kawamura 1979, Schwartz et al. 1980) demonstrate that the SCN are crucial elements within the mammalian circadian system.

In the house sparrow, the suprachiasmatic nuclei are described (Crosby and Showers 1969) and a direct retinal projection to the contralateral nucleus is present (Hartwig 1974). Because of the anatomical similarities of the SCN of birds and mammals, we asked whether functional similarities might also exist by assessing the effects of SCN lesions on the locomotor rhythms of house sparrows.

2 Effects of SCN Lesions on Circadian Activity Rhythms

Lesions that destroyed at least 90% of both SCN severly disrupted free-running locomotor rhythms of sparrows held in constant darkness (Fig. 1A). No circadian and few ultradian components were detected in the power spectra from SCN-lesioned sparrows (Fig. 2). Partial or unilateral SCN lesions did not abolish circadian rhythmicity. In some cases the precision of the rhythm was reduced, or the free-running period length was changed (Fig. 1B). In some birds with large but incomplete SCN lesions, circadian rhythmicity

1 Institute of Neuroscience, Department of Biology, University of Oregon, Eugene, Oregon 97403

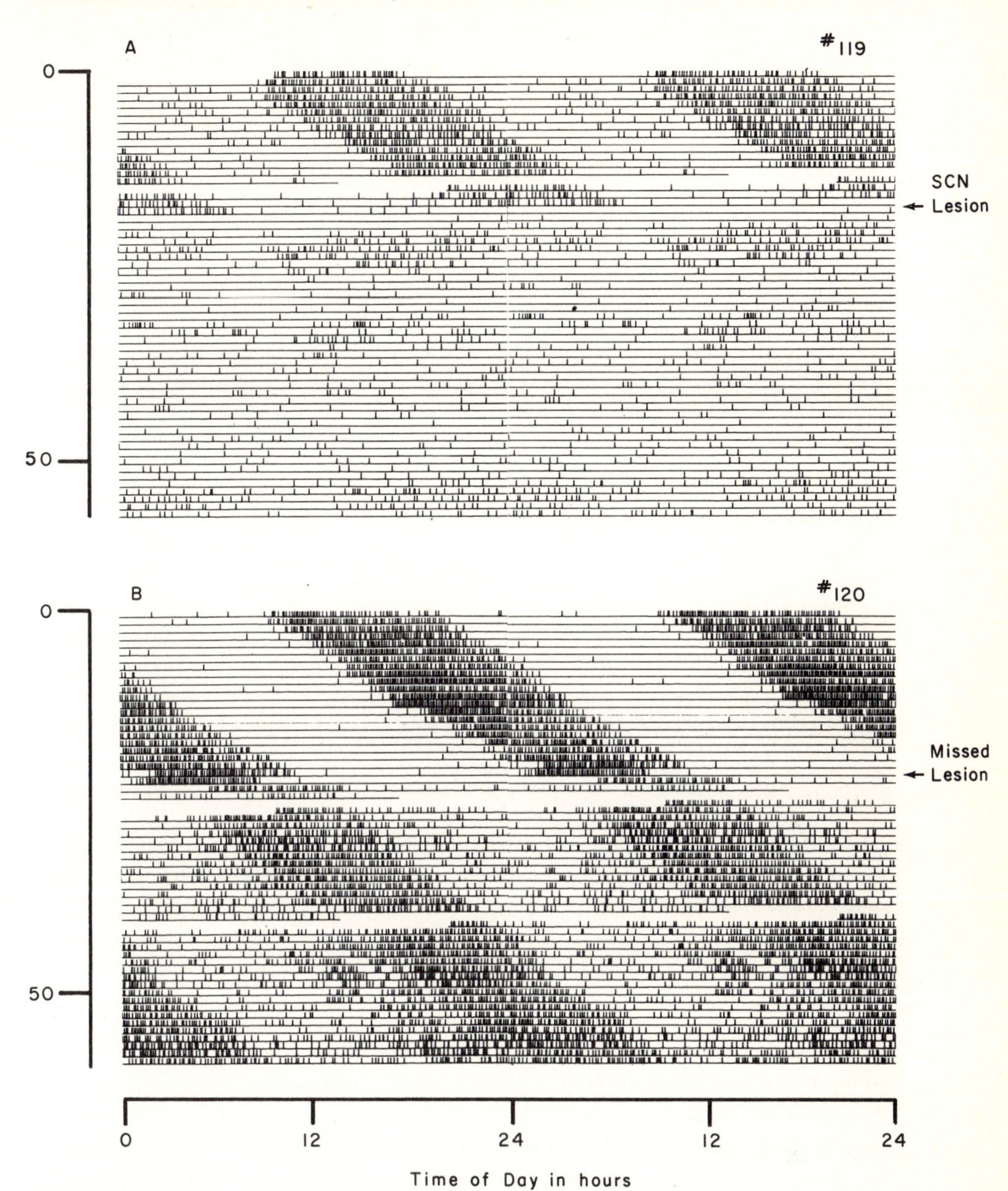

Fig. 1. Effects of hypothalamic lesions on circadian activity rhythms in sparrows. The activity records are continuous. Time progresses from left to right and days from top to bottom. Each record has been photographically duplicated and double plotted on a 48-h time scale to aid in visual inspection. The bird is active where the record is dense. **A** Effect of a lesion that destroyed 95%-100% of the SCN. The bird is in constant darkness for the duration of the record. The lesion disrupts the free-running circadian rhythm. **B** Effect of a lesion that destroyed about 30% of the SCN unilaterally. Same experimental conditions as in **A**. The free-running rhythm persists with a slight increase in "rest time" activity. Sham lesions were completely without effect

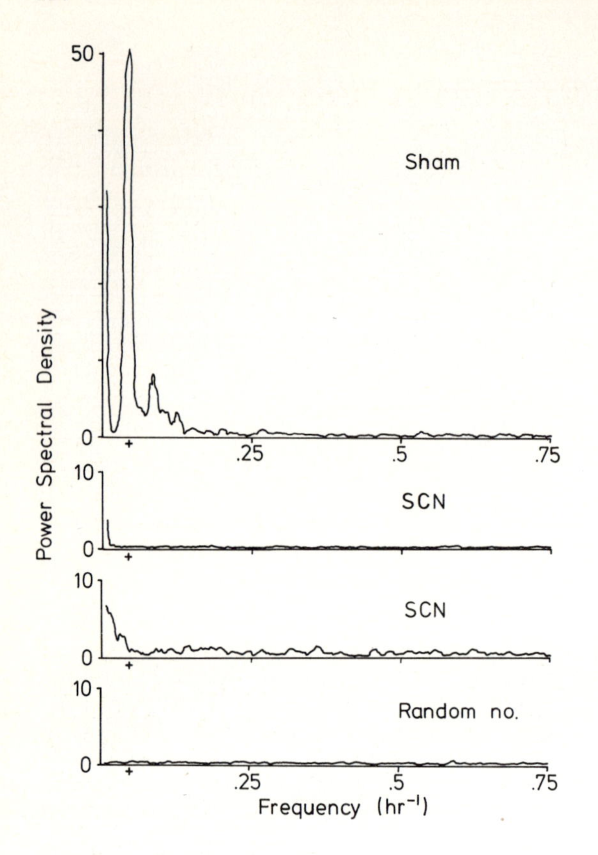

Fig. 2. Spectral analysis of sparrow locomotor activity. Continuous time series of quantitative locomotor data were analyzed for rhythmicity by direct computation of the power spectral density function using the fast Fourier transform method (Cooley and Tukey 1965). Details of the spectral analysis procedure are described in Takahashi (1981). Power spectra from a sham-operated sparrow *(Sham)*, two SCN-lesioned sparrows *(SCN)*, and a series of computer generated pseudorandom numbers *(Random no.)* are shown. The frequency corresponding to a 24-h period length is indicated by a cross (0.04167 cycles per hour). There is a major peak in the power spectrum in the circadian range in the record from the sham-operated bird; this circadian component is abolished in the records from SCN-lesioned birds

alternated with transient intervals of apparent arrhythmicity. In some of these records, there was weak evidence for components of the locomotor activity which appeared to dissociate and free-run.

3 Effects of SCN Lesions on Entrainment

When sparrows that were arrhythmic as a result of SCN lesions were exposed to light-dark cycles, they expressed rhythmic activity patterns which were similar to the entrainment patterns of unoperated birds (Fig. 3). All lesioned sparrows synchronized normally on LD 12 : 12. Since activity coincided with the light portion in LD 12 : 12, it was not clear whether the rhythmic pattern was a result of activity directly forced by the light (masking), or entrainment of an oscillator (cf. Chap. 1.2, this Vol.). To examine this further, sparrows were exposed to an LD 1 : 24 light cycle chosen because intact sparrows reliably entrain to it with a stable phase lead of the activity rhythm. Since the light portion is short, this light cycle is probably a weaker entraining stimulus than LD 12 : 12.

Sparrows that were completely arrhythmic in constant darkness did not entrain normally to the LD 1 : 24 cycle (Fig. 4B). Sparrows that showed weak circadian components in constant darkness or that were only transiently arrhythmic entrained normally (Fig. 4C). The disruption of entrainment in lesioned sparrows on LD 1 : 24 is consistent with

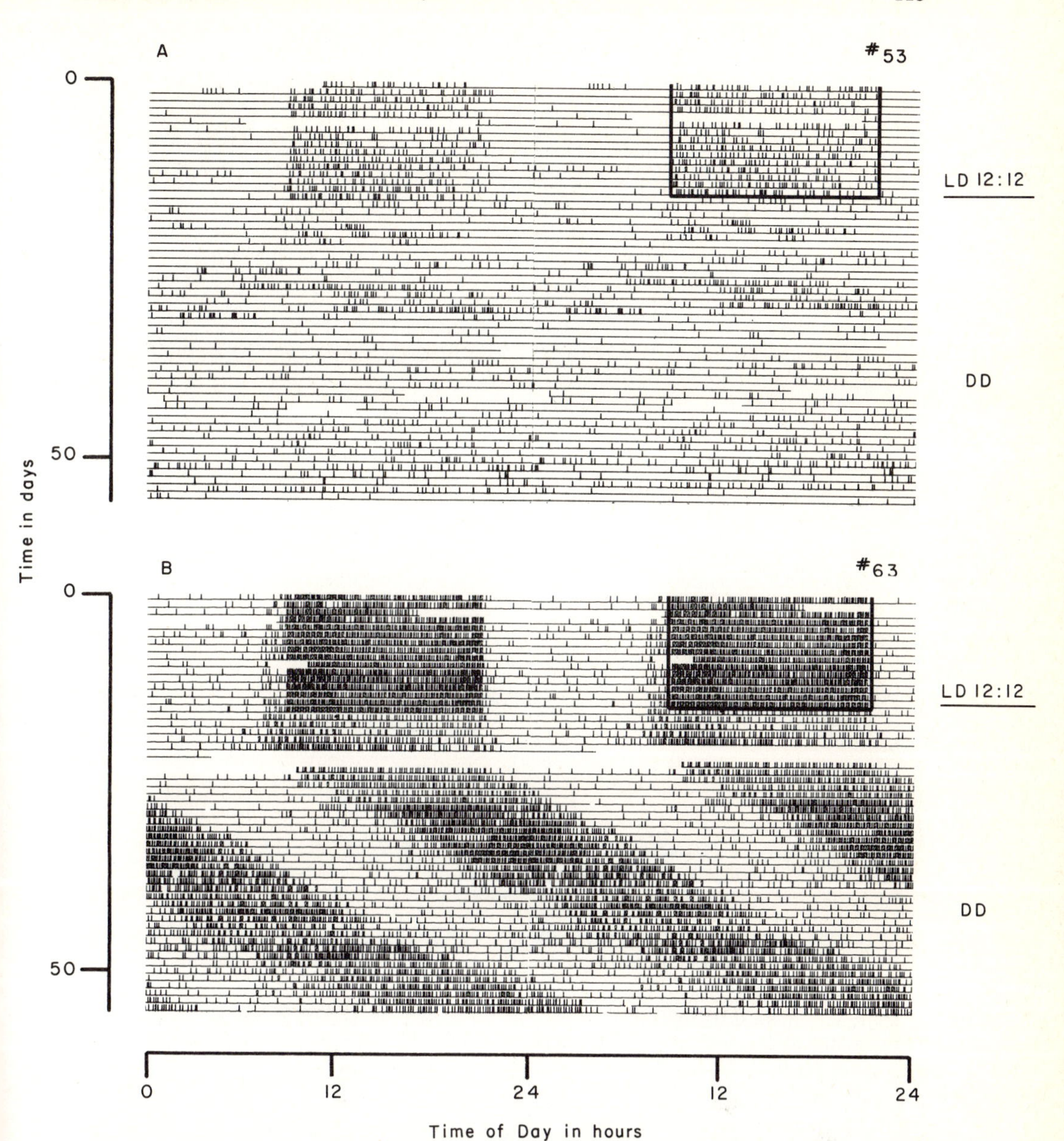

Fig. 3. Rhythmic activity of sparrows in light-dark cycles. **A** Activity record of a sparrow bearing SCN lesions. The bird's activity is rhythmic in *LD 12 : 12,* but arrhythmic in constant darkness *(DD).* **B** Activity record of a sparrow with a lesion that missed the SCN. The bird is entrained in *LD 12 : 12* and free runs in constant darkness. Its behavior is similar to unoperated sparrows

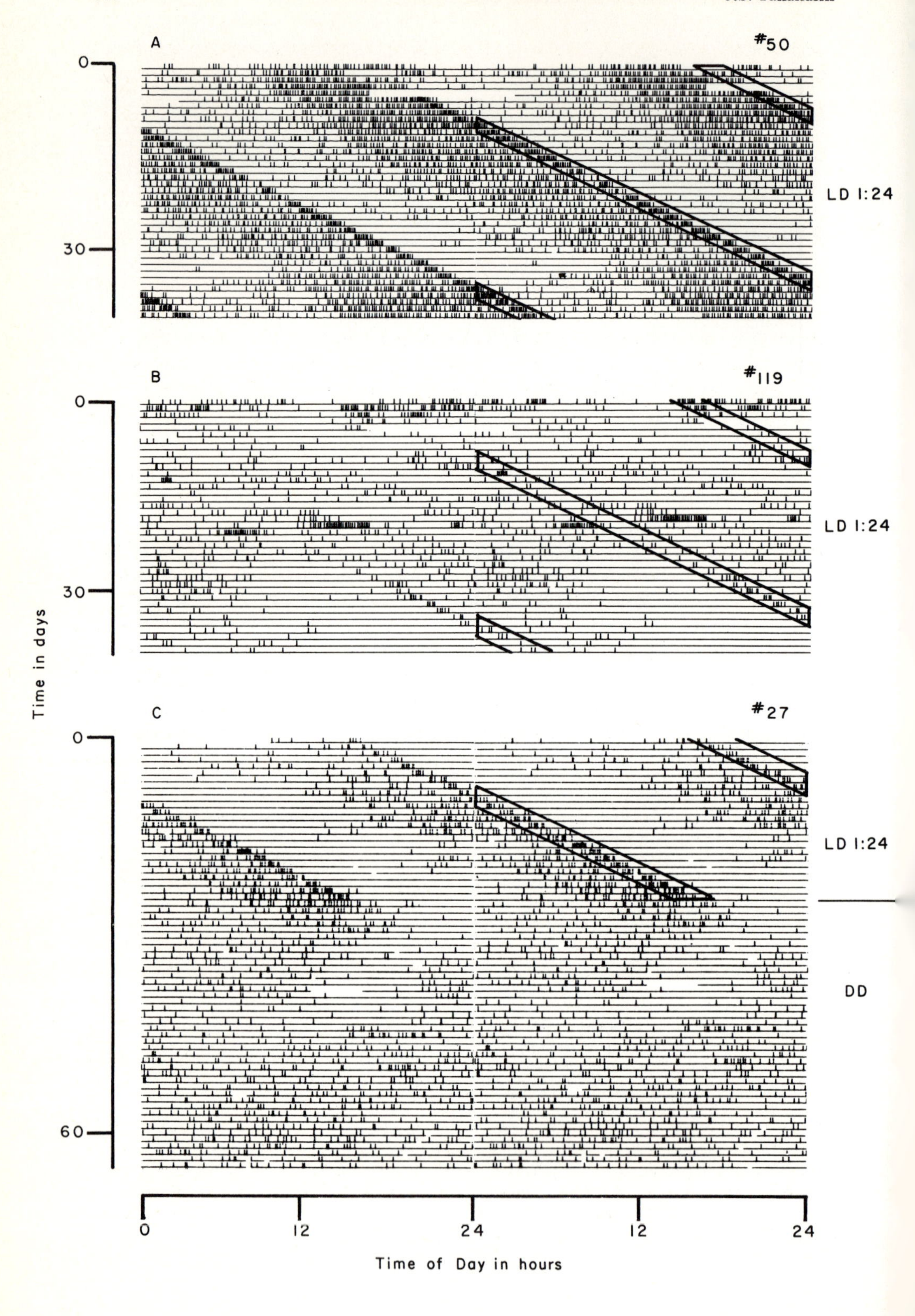
A
#50
0
30
LD 1:24
B
#119
0
30
LD 1:24
C
#27
0
60
LD 1:24
DD
Time in days
0
12
24
12
24
Time of Day in hours

what is known concerning the involvement of both retinal and extraretinal photoreceptors in mediating entrainment information in this species (Menaker and Underwood 1976). Because retinal and extraretinal photoreceptors are additive in their contribution to entrainment, SCN lesions would be expected to reduce the total photic input for entrainment by reducing retinal input mediated by the retinohypothalamic tract.

4 Discussion

The SCN of house sparrows play a crucial role in the regulation of circadian rhythmicity at the organismal level. The sparrow SCN are necessary for the persistence of locomotor rhythms in constant darkness and contribute to the entrainment of activity rhythms to light cycles. Lesions of the SCN severely disrupted free-running circadian locomotor rhythms of sparrows maintained in constant darkness. Spectral analysis of the locomotor records revealed that circadian periodicities were absent and that few ultradian periodicities were present. Sparrows bearing SCN lesions synchronized normally to LD 12 : 12 light cycles, but did not entrain to a relatively weak zeitgeber (entraining agent).

To a first approximation, the effects of SCN lesions are similar to those of pinealectomy in the sparrow. Both surgical procedures abolish free-running rhythmicity in constant darkness, but do not abolish synchronization to light-dark cycles. In the case of SCN lesions, however, we did not observe a phase lead of the activity rhythm to the light cycle; and in LD 1 : 24, the entraining cycle may have been too weak to synchronize the system after lesioning. In sparrows, there are multiple photoreceptive inputs to the circadian system (Menaker and Underwood 1976). The avian pineal is both photoreceptive and contains circadian oscillators (Deguchi 1979, Takahashi et al. 1980). Similarly, the sparrow SCN receive direct retinal input through the retinohypothalamic tract (Hartwig 1974) and could also contain circadian oscillators. In addition, there is, at minimum, a third photic input from brain photoreceptors that are capable of entraining sparrows whose retinae and pineal have been removed (Menaker 1971). Whether the disruption of entrainment to weak zeitgebers is due to a reduction in photic input, or to removal of an oscillator, or both, remains to be determined.

The relationship between the avian pineal and the SCN remains to be more precisely determined. At the present time, there are several testable hypotheses that could account for the behavior of sparrows bearing SCN lesions:

1. The suprachiasmatic nuclei could be components of the output pathway between a circadian pacemaker located elsewhere and the overt locomotor rhythm that we meas-

←

Fig. 4. Entrainment of sparrows on a short light portion cycle, *LD 1 : 24*. The period of the light cycle is 25 h and therefore is delayed by 1 h each day so that it appears to drift toward the right. The light portion is drawn only on the right half of the activity record. **A** Activity record of a bird with a lesion that destroyed about 35% of the SCN. The behavior of this bird is typical of intact sparrows on LD 1 : 24; it is entrained with the activity onset leading the light by about 12 h. **B** Same lesioned bird as in Fig. 1A showing disrupted entrainment in this light cycle. **C** Another SCN-lesioned sparrow (90% SCN destroyed) which shows normal entrainment in LD 1 : 24. When transferred to constant darkness, the rhythm free runs with a short period length before the amplitude of the rhythm decays some 20 cycles later

ure. On this hypothesis, overt arrhythmicity in SCN-lesioned birds would be the result of uncoupling the pacemaker from its output by interrupting this pathway.

2. The SCN could play a permissive role and be necessary for sustaining pineal rhythmicity in vivo in constant conditions. Although neural connections of the pineal do not appear to be necessary for sustaining circadian locomotor rhythmicity in sparrows (Zimmerman and Menaker 1975), hormonal pathways to the pineal involving the SCN cannot be excluded.
3. The SCN and the pineal gland may interact and function together as a complex pacemaker. In this case neither structure alone would be able to sustain the normal overt rhythmicity of the bird if the other was absent.

While it is well established that the mammalian SCN are important in the regulation of circadian phenomena (Moore 1978, Rusak and Zucker 1979), there are few studies of the role of this region in birds. Lesions or cuts in the anterior hypothalamus have been shown to abolish several physiological rhythms in avian species. Ralph (1959) has shown that such lesions in the hen prevent ovulation, which occurs at approximately 25-h intervals. Davies (1980) has clearly demonstrated that lesions of the supraoptic region of the Japanese quail interfere with ovulation if the lesions destroy the SCN as well as the supraoptic nuclei. Bouillé et al. (1975) have shown that hypothalamic deafferentation abolishes the diurnal plasma corticosterone rhythm in pigeons. Recently, Simpson and Follet (1980) have shown that anterior hypothalamic lesions abolish circadian locomotor activity rhythms in Japanese quail. Using a passerine species, the Java sparrow *(Padda oryzivora)*, Ebihara and Kawamura (1981) have now confirmed our initial report (Takahashi and Menaker 1979b) that SCN lesions abolish circadian rhythmicity. In all three species that have been examined (house sparrow, Java sparrow, and Japanese quail), the suprachiasmatic region appears critical for the maintenance of normal circadian rhythmicity. Clearly there are similarities in the neural control of circadian rhythms in birds and mammals. There are also striking differences. Photoreceptive inputs to the circadian system of birds are numerous and varied (Menaker and Underwood 1976). Both retinal and extraretinal photoreceptors influence the avian circadian system whereas in mammals only retinal photoreceptors are involved (Rusak and Zucker 1979). The pineal plays a central role in the control of circadian rhythms in passerine birds (Takahashi and Menaker 1979a); while the pineal of mammals plays only a minor role, if any (Rusak and Zucker 1979). The similarities and differences between birds and mammals suggest that although homologies exist there is a rich variety, yet to be discovered, in the pattern of organization of circadian systems among the vertebrates.

Acknowledgments. I thank Michael Menaker for his enthusiastic support throughout all phases of this work. Special thanks are due to R. Van Buskirk, G. Vaaler, G. Wyche, E. Kluth and D. Schaffer. Research was supported by NIH grants HD-03803 and HD-07727 to M. Menaker, and a National Science Foundation graduate fellowship and National Institute of General Medical Sciences National Research Award 5T32 GM 07257 to J. Takahashi.

References

Bouillé C, Herbute S, Baylé JD (1975) Effects of hypothalamic deafferentation on basal and stress-induced adrenocortical activity in the pigeon. J Endocrinol 66: 413-419

Cooley JW, Tukey JW (1965) An algorithm for machine calculation of complex Fourier series. Math Comput 19: 297-301

Crosby EC, Showers MJC (1969) Comparative anatomy of the preoptic and hypothalamic areas. In: Haymaker W, Anderson E, Nauta WJH (eds) The hypothalamus. CC Thomas, Springfield, pp 61-135

Davies DT (1980) The neuroendocrine control of gonadotrophin release in the Japanese quail. III. The role of the tuberal and anterior hypothalamus in the control of ovarian development and ovulation. Proc R Soc London Ser B 206: 421-437

Deguchi T (1969) Circadian rhythm of serotonin N-acetyltransferase activity in organ culture of chicken pineal gland. Science 203: 1245-1247

Ebihara S, Kawamura H (1981) The role of the pineal organ and the suprachiasmatic nucleus in the control of circadian locomotor rhythms in the Java sparrow, *Padda oryzivora.* J Comp Physiol 141: 207-214

Gaston S, Menaker M (1968) Pineal function: The biological clock in the sparrow? Science 160: 1125-1127

Hartwig HG (1974) Electron microscopic evidence for a retino-hypothalamic projection to the suprachiasmatic nuclei of *Passer domesticus.* Cell Tissue Res 153: 89-99

Inouye ST, Kawamura H (1979) Persistence of circadian rhythmicity in a mammalian hypothalamic "island" containing the suprachiasmatic nucleus. Proc Natl Acad Sci USA 76: 5962-5966

Menaker M (1971) Rhythms, reproduction and phtoreception. Biol Reprod 4: 295-308

Menaker M, Underwood H (1976) Extraretinal photoreception in birds. Photochem Photobiol 23: 299-306

Menaker M, Takahashi JS, Eskin A (1978) The physiology of circadian pacemakers. Annu Rev Physiol 40: 501-526

Menaker M, Hudson DJ, Takahashi JS (1981) Neural and endocrine components of circadian clocks in birds. In: Follet BK, Follett DE (eds) Biological clocks in seasonal reproductive cycles. Wright, Bristol, pp 171-183

Moore RY (1978) Central neural control of circadian rhythms. In: Ganong WF, Martini L (eds) Frontiers in neuroendocrinology, vol V. Raven Press, New York, pp 185-206

Ralph CL (1959) Some effects of hypothalamic lesions on gonadotrophin release in the hen. Anat Rec 134: 411-431

Rusak B, Zucker I (1979) Neural regulation of circadian rhythms. Physiol Rev 59: 449-526

Schwartz WJ, Davidsen LC, Smith CB (1980) In vivo metabolic activity of a putative circadian oscillator, the rat suprachiasmatic nucleus. J Comp Neurol 189: 157-167

Simpson SM, Follett BK (1980) Investigations on the possible roles of the pineal and the anterior hypothalamus in regulating circadian activity rhythms in Japanese quail. In: Nöhring R (ed) Acta XVII Congr Int Ornithol, Vol I. Dtsch Ornithol Ges, Berlin, pp 435-438

Takahashi JS (1981) Neural and endocrine regulation of avian circadian systems. Ph D dissertation, Univ Oregon, Eugene

Takahashi JS, Menaker M (1979a) Physiology of avian circadian pacemakers. Fed Proc 38: 2583-2588

Takahashi JS, Menaker M (1979b) Brain mechanisms in avian circadian systems. In: Suda M, Hayaishi O, Nakagawa H (eds) Biological rhythms and their central mechanism. Elsevier/North Holland, Amsterdam, pp 95-109

Takahashi JS, Hamm H, Menaker M (1980) Circadian rhythms of melatonin release from individual superfused chicken pineal glands in vitro. Proc Natl Acad Sci USA 77: 2319-2322

Zimmerman NH, Menaker M (1975) Neural connections of sparrow pineal: Role in circadian control of activity. Science 190: 477-479

3.7 Limits of Entrainment to Periodic Feeding in Rats with Suprachiasmatic Lesions

F.K. Stephan[1]

1 Introduction

The functional properties of entrainment of mammalian circadian rhythms by LD cycles are becoming increasingly well understood and have recently been eloquently summarized by Pittendrigh and Daan (1976). In contrast, despite occasional reports of entrainment of circadian rhythms by periodic temperature, noise or by social interaction, the importance of these stimuli as zeitgeber appears minor by comparison with LD cycles, particularly for mammals (Aschoff 1958, Bruce 1960, Regal and Connolly 1980).

One notable exception is the observation that restricted daily feeding leads to the synchronization of a large number of processes, including many enzymes, hormones, and behaviors (for review, see Boulos and Terman 1980). Of course, synchronization by itself does not necessarily imply the involvement of a circadian rhythm, and many earlier studies of the effects of restricted feeding failed to rule out passive driving, feedback mediation, temporal conditioning, or other plausible alternatives.

However, a number of studies indicate that the synchronization of rat locomotor activity by periodic feeding has some circadian properties. Rats maintained on restricted food access at 24 h intervals increase their activity several hours prior to food availability (Richter 1927). This "anticipatory" activity fails to occur on 19- or 29-h feeding schedules (Bolles and Stokes 1965). On the other hand, in otherwise constant conditions, restricted feeding generally does not appear to entrain the freerunning activity rhythm of rats (Gibbs 1979, Moore 1980). Furthermore, "anticipatory" activity (Stephan et al. 1979a, b, Phillips and Mikala 1979, Boulos et al. 1980), as well as the synchronization of corticosterone and core temperature (Krieger et al. 1977), in response to restricted feeding is not abolished by lesions of the suprachiasmatic nuclei (SCN), despite many demonstrations that such lesions severely disrupt or eliminate circadian rhythms (for review, see Rusak and Zucker 1979). Thus, two plausible alternatives are consistent with these results: (a) synchronization by feeding is mediated by a circadian pacemaker other than the SCN, or (b) it is based on some unknown alternative mechanisms which do not require circadian time cues.

Perhaps the strongest argument against the circadian hypothesis is the failure of feeding-entrained activity to freerun in ad lib feeding conditions in rats with SCN lesions (Stephan et al. 1979a, b, Boulos et al. 1980).

1 Department of Psychology, Florida State University, Tallahassee, Florida 32306, USA

Conversely, the following observations are more consistent with a circadian pacemaker hypothesis: neither intact rats, nor rats with SCN lesions entrain their activity to feedings at 18-h intervals (Stephan et al. 1979a, Boulos et al. 1980), but both are able to anticipate 23- and 24-h (Stephan et al. 1979b) as well as 30-h schedules (Boulos et al. 1980). Intact rats also anticipate 25-h schedules (Edmonds and Adler 1977). This limited range of entrainment corresponds at least roughly to the range of entrainment by LD cycles or light pulses (Pittendrigh and Daan 1976). Furthermore, these results seem to rule out mediation by feedback with an arbitrary time constant, or by responding to "subtle" 24-h cues which cannot be controlled in the laboratory.

The purpose of the present study was to explore more systematically the range of periodicities to which activity can be entrained by restricted food access. In particular, it was hoped that near the limits of entrainment, phenomena analogous to those seen with entrainment by single light pulses, such as relative coordination, changes in alpha and in the phase angle, or freerunning rhythms might be observed which could provide convincing evidence for the circadian pacemaker hypothesis. Only rats with SCN lesions were used, so that the functional properties of putative pacemakers outside the SCN could be studied in isolation.

2 Methods

Following bilateral electrolytic lesions aimed at the SCN, rats were housed in isolation boxes which enclosed an activity wheel and a small adjacent cage, equipped with a drinkometer and a food hopper. Access to food was controlled by means of a motorized gate. Wheel revolutions and licks were monitored continuously by computer and cumulated counts recorded every 10 min.

The animals were maintained in DD and water was ad lib throughout the study. Fresh food and water were provided approximately one per week using a dim red light.

Following a 13-day ad lib period, all rats were placed on a 24-h schedule (2 h access, 22 h deprivation) for 15 days. At this point, eight rats (Group A) were placed on a sequence of schedules with $T < 24$ h, followed by schedules with $T > 24$ h (see Fig. 1) and eight rats (Group B) were placed on schedules with $T > 24$ h, followed by schedules with $T < 24$ h (see Fig. 2). The duration of food access was 2 h on all schedules where $T \leqslant 24$ h, 2.5 h where $T = 27$, 3 h where $T = 29$ and 31, and 4 h where $T = 33$ h.

3 Results and Discussion

Of the 16 rats, 9 had histologically complete SCN lesions. Among these, one rat (Fig. 1, SCN 9) had a significant circadian rhythm during the first ad lib period (Chisquare periodogram) while three of seven rats with subtotal lesions had low amplitude rhythms. Since partial lesions did not systematically affect feeding-entrained activity, the data for all rats are discussed.

The results clearly demonstrate that anticipatory wheel-running develops to restricted feeding schedules only within a limited range of periods. Although the limits of entrainment cannot be defined precisely, under the conditions of the present study, anticipatory wheel-running could not be reliably obtained at schedules with $T \leqslant 22$ h, and $T \geqslant 31$ h (Figs. 1 and 2). In general, these results agree with those obtained previously in studies using non-24-h feeding schedules (Bolles and Stokes 1965, Stephan et al. 1979a, b, Edmonds and Adler 1977, Boulos et al. 1980).

The large upper boundary of entrainment (>30 h) and the substantial asymmetry of the entrainment limits with regard to 24 h was unexpected. By comparison, in a recent study with intact rats, using 2-h light "pulses", the entrainment limits were between 23 h and 26 h (Fig. 3).

The limits of entrainment to LD cycles clearly depend on both the period τ and the phase-response curve (PRC) of the underlying pacemaker (Pittendrigh and Daan 1967). However, the PRC of the putative pacemaker(s) entrainable by periodic feeding is obviously unknown. Furthermore, even estimates of τ are difficult because the activity pattern rarely free runs long enough to measure τ. The observed freeruns in ad lib following prolonged exposure to restricted feeding (e.g., Fig. 1, SCN 1, second ad lib period), or near the limits of entrainment (e.g., Fig. 2, SCN 19, 22-h schedule), are most likely confounded by after-effects, i.e., τ is influenced by the period of the previous entraining schedule. Such after-effects may also account for the observation that the limits of entrainment could be extended by a gradual change in the period of the feeding schedule. Following sequential exposure to a 24-, 27-, and 29-h schedule, most rats still entrained to the 31-h schedule (Fig. 2), although with decreasing reliability. In contrast, a direct change from 24 to 31 h did not consistently produce entrainment (Fig. 1).

After-effects of entrainment to LD cycles are well known. For example, τ in constant conditions reflects the influence of the period of the previous entraining LD cycle for several weeks or more (Fig. 3, DD condition). Furthermore, gradual changes in the period of the LD cycle extend the limits of entrainment, presumably because entrainment affects both τ and the PRC of the underlying circadian pacemakers (Pittendrigh and Daan 1976). Based on this analogy with entrainment by light, it appears that the putative pacemakers which mediate entrainment to feeding schedules have a labile PRC with a large phase delay and a small phase advance section. Of course, this hypothesis needs to be verified experimentally.

One interesting difference between light- and feeding-entrained circadian rhythms is that for the latter activity always precedes the zeitgeber (food access) throughout the range of entrainment. The phase angle of entrainment to light "pulses" clearly depends on the difference between the period of the entraining cycle (T) and τ; when $T < \tau$, activity offset entrains and when $T > \tau$ activity onset entrains (Fig. 3) (Pittendrigh and Daan 1976). This difference between light- and food-entrained rhythms possibly reflects the adaptive significance of the ability to anticipate periodic food sources, but the nature of the mechanism which prevents entrainment of activity onset is unclear. It should be noted that on certain schedules some activity followed feeding (e.g., Fig. 1, SCN 1, 24- and 21-h schedules). However, this is most probably a "passive" consequence of restricted feeding, rather than entrainment of a circadian rhythm.

Fig. 1. Double-plotted event records of daily activity for two rats with SCN lesions in Group A. *Black lines* indicate onset of access to food on restricted feeding schedules and the period of the schedule is shown at the left margin. No data were collected for 6 days during the second ad lib period. Rat SCN 9 was food-deprived for 3 days following the 22-h schedule. Rats were in DD and water was ad lib throughout.

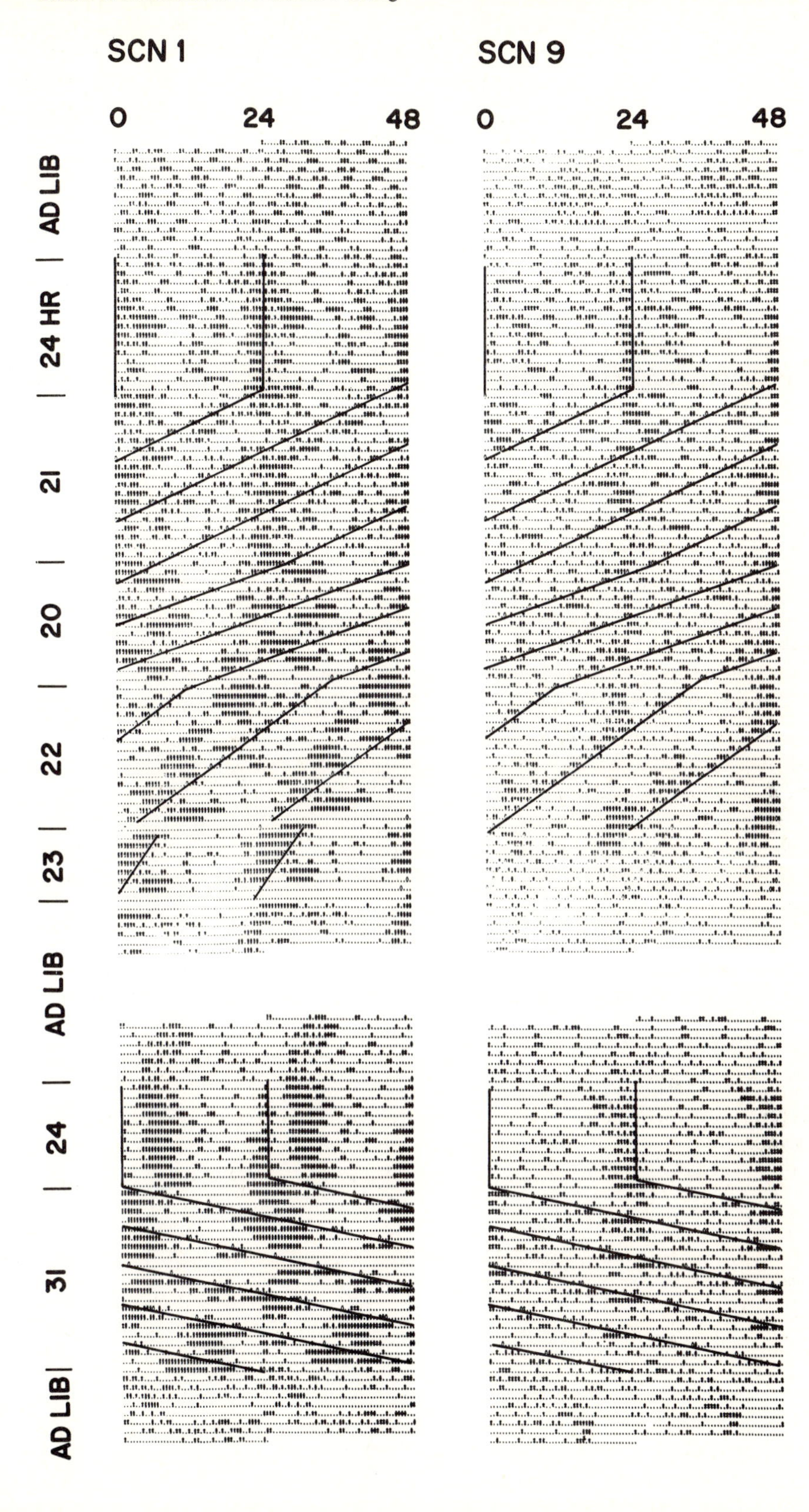

SCN 1
SCN 9
0
24
48
AD LIB
24 HR
21
20
22
23
AD LIB
24
31
AD LIB

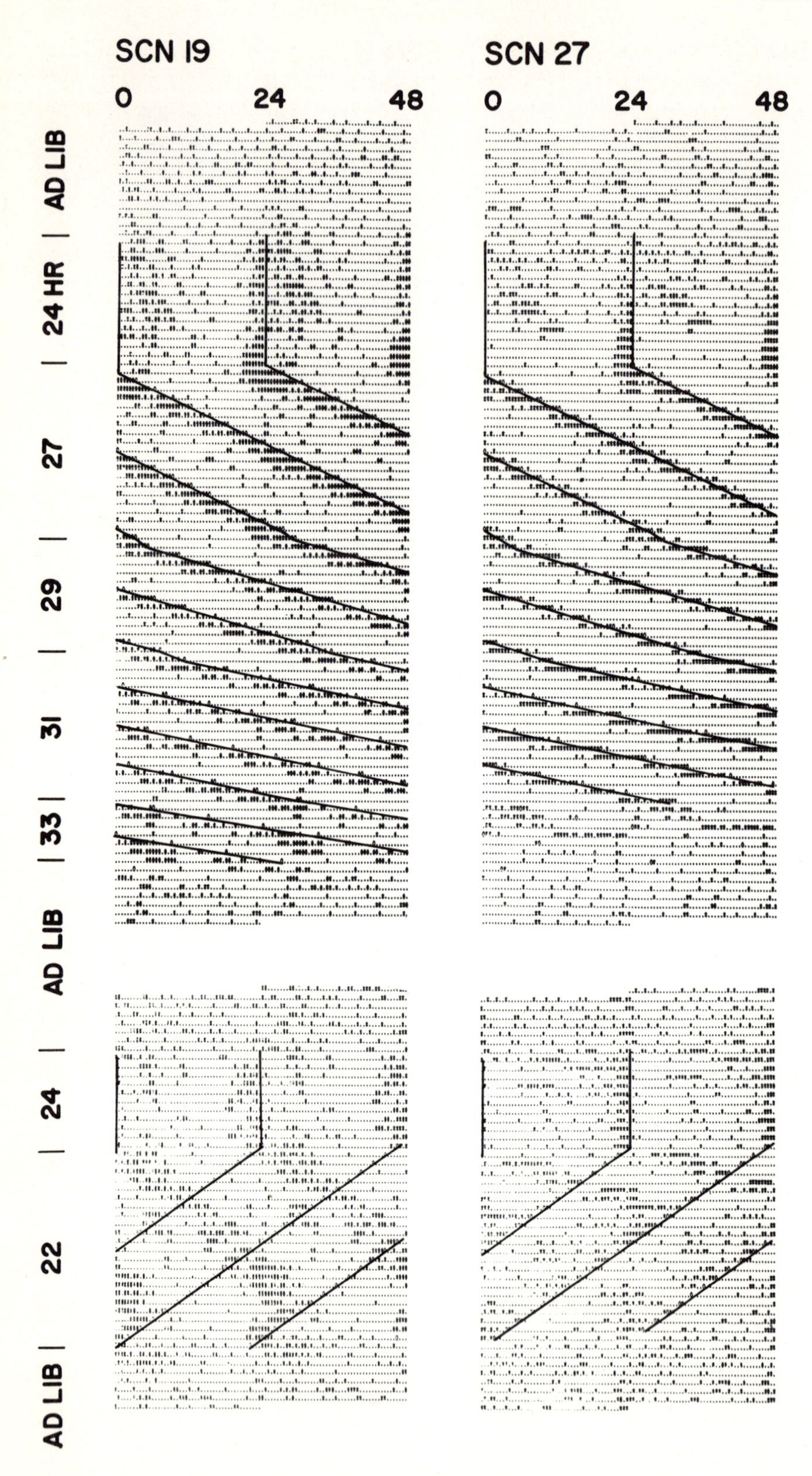
SCN 19
SCN 27
0
24
48
0
24
48
AD LIB
24 HR
27
29
31
33
AD LIB
24
22
AD LIB

Compared with circadian rhythms in intact animals, the putative pacemakers entrainable by feeding appear to have only a limited capacity to freerun. Previous results indicate that periodic activity persists in total food deprivation, but rarely in ad lib conditions (Stephan et al. 1979b, Boulos et al. 1980). In the present study, four rats each from Groups A and B were not exposed to the 23-h or 33-h feeding schedule, respectively. During 3 days of food deprivation, these rats showed a continuation of the previously entrained activity pattern (Fig. 1, SCN 9; Fig. 2, SCN 27).

However, this activity rapidly desynchronized during ad lib feeding. Unfortunately, the self-limiting nature of total food deprivation precludes the observation of prolonged freeruns.

Perhaps the most interesting findings of the present study is the behavior of the activity pattern during schedules with periods near the limits of entrainment. Since the dominant period of the activity often differed from the period of the feeding schedules by several hours, the activity appeared to be a freerunning circadian rhythm. Seven of eight rats in Group A had a weak but significant periodicity in activity (τ = 22.2-24.3 h) during the 21-h and 20-h schedules (see Fig. 1). In several cases, the period of the activity rhythm was apparently not influenced by the schedule (e.g., Fig. 2, SCN 19, 22 h schedule) while in others the activity pattern was near the period of the schedule but not completely synchronized (e.g., Fig. 1, SCN 1, 22-h schedule), and finally, "beats", i.e., waxing and waning of activity depending on the phase position of the feeding were observed (Fig. 1, 31-h schedule). Very similar phenomena (i.e., freeruns, relative coordination, and beats) have been described in intact mice and hamsters exposed to light "pulses" with non-24-h periods (Pittendrigh and Daan 1976). Figure 3 illustrates relative coordination at T = 21 and T = 27 for the rat exposed to T cycles with 2-h light pulses.

The present results, in particular the observation of prolonged freerunning rhythms near the limits of entrainment, provide strong support for the involvement of circadian pacemakers in feeding-entrained activity. There can be little doubt that the circadian system of mammals consists of multiple pacemakers (for recent reviews, see Moore-Ede et al. 1976, Block and Page 1978, Menaker et al. 1978), and that some of these oscillators are outside the SCN. Following SCN lesions, temperature rhythms have been observed in the squirrel monkey (Moore-Ede et al. 1980).

In the present study, one rat with a complete SCN lesion showed a low amplitude, but significant freerunning activity rhythm in DD (Fig. 1, SCN 9). Thus, it is not unreasonable to assume that a population of pacemakers outside the SCN is involved in the entrainment of activity by restricted feeding. Their limited capacity to drive freerunning rhythms in ad lib may be the result of internal desynchronization among such oscillators. Periodic access to food would then constitute a synchronizing signal, provided that its periodicity is within their range of entrainment. In some sense, ad lib feeding may be analogous to constant bright light. Many organisms show damping of circadian rhythms in this condition, presumably due to internal desynchronization (cf. Bunning 1973). Differences in "coupling strength" among individual pacemakers and/or their phase angle differences could explain why some animals with SCN lesions show a circadian rhythm

←

Fig. 2. Double-plotted event records of daily activity for two rats with SCN lesions in Group B. Rat SCN 27 was food deprived for 3 days following the 31-h schedule. For further explanation see Fig. 1

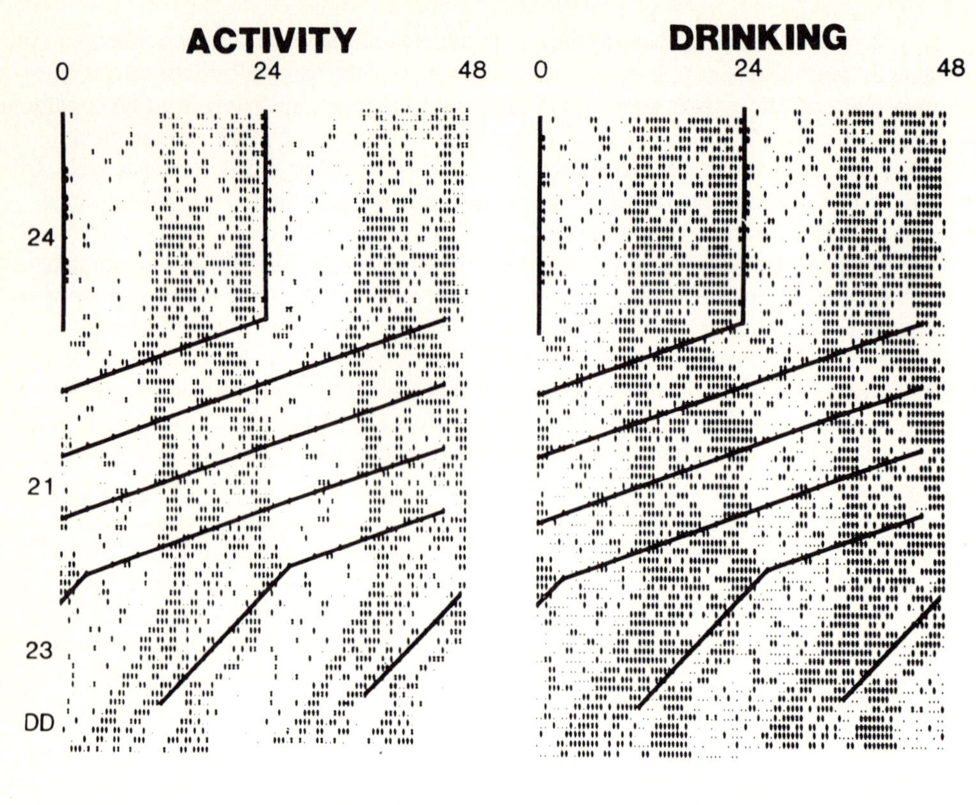
ACTIVITY
DRINKING
0
24
48
24
21
23
DD

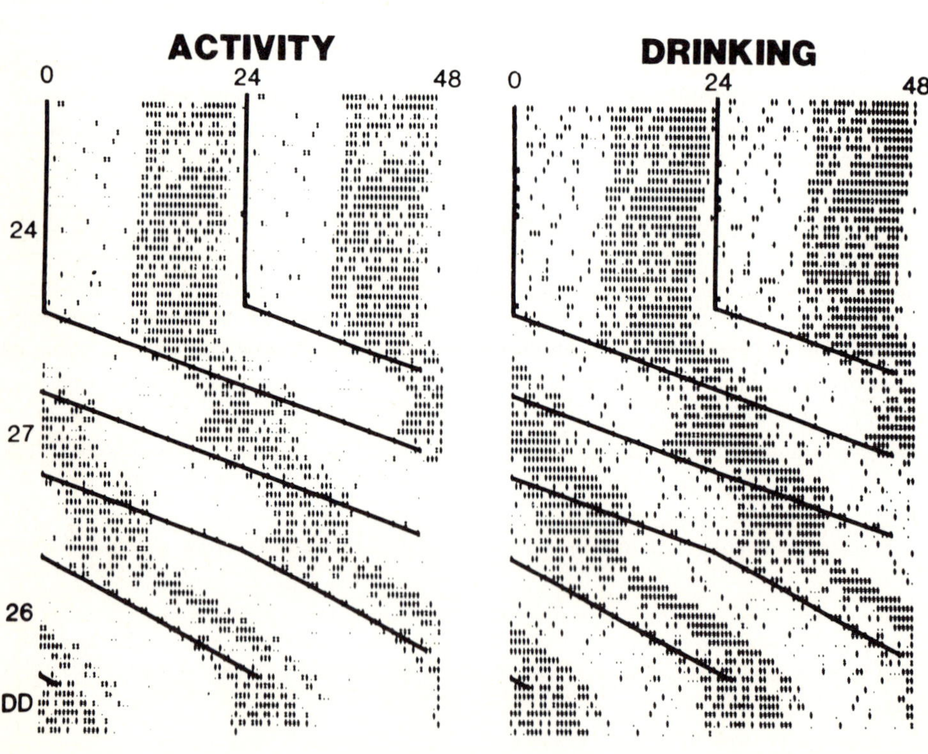
ACTIVITY
DRINKING
0
24
48
24
27
26
DD

in ad lib while others do not. Furthermore, differences in the phase position of these pacemakers, relative to introduction of the periodic feeding schedule, might explain why in some cases entrainment fails even though the schedule appears to be within the circadian range.

Finally, feeding-entrained rhythms have properties suggesting a damped oscillator (e.g., broad range of entrainment, rapid desynchronization in ad lib., rapid entrainment – often by the second cycle) as well as properties suggesting a self-sustained oscillator (e.g., freerun in food-deprived conditions and near the limits of entrainment). Current investigations in our laboratories are designed to shed more light on feeding-entrainable circadian pacemakers.

Acknowledgments. This research was supported by Grant # BNS-78-24999 from the National Science Foundation. The methods and results concerning entrainment to restricted feeding are presented more fully in paper with the same title to be published in J Comp Physiol *A*: 1981, in press.

References

Aschoff J (1958) Tierische Periodik unter dem Einfluß von Zeitgebern. Z Tierpsychol 15: 1-30

Block GD, Page TL (1978) Circadian pacemakers in the nervous system. Annu Rev Neurosci 1: 19-34

Bolles RC, Stokes LW (1965) Rat's anticipation of diurnal and adiurnal feeding. J Comp Physiol Psychol 60: 290-294

Boulos Z, Terman M (1980) Food availability and daily biological rhythms. Neuroscie Biobehav Rev 4: 119-131

Boulos Z, Rosenwasser AM, Terman M (1980) Feeding schedules and the circadian organization of behavior in the rat. Behav Brain Res 1: 39-65

Bruce VG (1960) Environmental entrainment of circadian rhythms. Cold Spring Harbor Symp Quant Biol 25: 29-48

Bunning E (1973) The physiological clock. Springer, Berlin Heidelberg New York

Edmonds SC, Adler NT (1977) The multiplicity of biological oscillators in the control of circadian running activity in the rat. Physiol Behav 18: 921-930

Gibbs FP (1979) Fixed interval feeding does not entrain the circadian pacemaker in blind rats. Am J Physiol 236: R249-253

Krieger DT, Hauser H, Krey LC (1977) Suprachiasmatic nuclear lesions do not abolish foodshifted circadian adrenal and temperature rhythmicity. Science 197: 398-399

Menaker M, Takahashi JS, Eskin A (1978) The physiology of circadian pacemakers. Annu Rev Physiol 40: 501-526

Moore RY (1980) Suprachiasmatic nucleus, secondary synchronizing stimuli and the central neural control of circadian rhythms. Brain Res 183: 13-28

Moore-Ede MC, Schmelzer WS, Kass DA, Herd JA (1976) Internal organization of the circadian timing system in multicellular animals. Fed Proc 35: 2333-2338

Moore-Ede MC, Lydic R, Czeisler CA, Tepper B, Fuller CA (1980) Characterization of separate circadian oscillators driving rest-activity and body temperature in a non-human primate. Sleep Res 9: 275

Phillips JM, Mikala PJ (1979) The effects of restricted food access upon locomotor activity in rats with suprachiasmatic nucleus lesions. Physiol Behav 23: 257-262

←

Fig. 3. Double-plotted event records of daily activity and drinking for 2 intact rats. *Black lines* indicate onset of 2-h light pulse. The period of the T cycle is shown at the *left margin*

Pittendrigh CS, Daan S (1976) A functional analysis of circadian pacemakers in nocturnal rodents. J Comp Physiol 106: 223-335
Regal JP, Connolly MS (1980) Social influences on biological rhythms. Behavior 72: 171-199
Richter CP (1927) Animal behavior and internal drives. Rev. Biol 2: 307-343
Rusak B, Zucker I (1979) Neural regulation of circadian rhythms. Physiol Rev 59: 449-526
Stephan FK, Swann JM, Sisk CL (1979a) Anticipation of 24 hr. feeding schedules in rats with lesions of the suprachiasmatic nucleus. Behav Neurol Biol 25: 346-363
Stephan FK, Swann JM, Sisk CL (1979b) Entrainment of circadian rhythms by feeding schedules in rats with suprachiasmatic lesions. Behav Neural Biol 25: 545-554

4.1 Phase Responses and Characteristics of Free-Running Activity Rhythms in the Golden Hamster: Independence of the Pineal Gland

J. Aschoff[1], U. Gerecke[1], Chr. von Goetz[1], G.A. Groos[2], and F.W. Turek[3]

1 Introduction

The pineal organ is of major importance for the circadian organization of fishes, amphibians, reptiles (Underwood 1982), and birds (Takahashi and Menaker 1979), yet there is little evidence for a similar role in mammals. As was first shown by Richter (1967), pinealectomy has no effect on the free-running activity rhythm of blinded rats. Other studies have also failed to find any difference between pinealectomized (sighted) rats and intact controls with regard to the circadian period in constant conditions and its dependence on intensity of illumination (Quay 1968), the phase-angle difference to an entraining light-dark cycle (LD) (Quay 1970a), or the pattern of activity in either natural or artificial LD (Karppanen et al. 1973). Furthermore, the periods of free-running activity rhythms were similar in intact and pinealectomized golden hamsters after 9 weeks of exposure to continuous darkness (DD) as well as immediately after the transfer from LD to DD (Morin and Cummings 1981). Circadian rhythms in several functions other than locomotor activity have also been shown to be not affected, or only marginally altered, by pinealectomy: e.g., the rhythms of food intake (Baum 1970, Morimoto and Yamamura 1979, Lynch and Wurtman 1979) and of water intake (Stephan and Zucker 1972), of plasma levels in corticosterone (Takahashi et al. 1976, Morimoto and Yamamura 1979), testosterone (Kinson and Liu 1973), and prolactin (Niles et al. 1977a), of the melanocyte stimulating hormone content of the pituitary (Tilders and Smelk 1975), of visual evoked potentials in the rabbit (cf. Chap. 2.4), and of agonistic behavior in mice (Cavalieri et al. 1980) as well as body temperature in the rat (Spencer et al. 1976).

In other reports, pinealectomy has been claimed to result in an attenuation or a phase shift of rhythms, e.g., in plasma levels of some hormones (prolactin: Rønnekleiv et al. 1973, Kizer et al. 1975, Willoughby 1980; corticosterone: Niles et al. 1977b; testosterone: Niles et al. 1979), in the activity of enzymes in various tissues (Banerji and Quay 1978; Banerji et al. 1979; Scalabrino et al. 1979), and in the rhythm of sleep stages (Kawakami et al. 1972; cf. also Chap. 6.2). It should be noted, however, that a change in amplitude, or an apparent phase shift of a rhythm following pinealectomy does not

1 Max-Planck-Institut für Verhaltensphysiologie, D-8138 Andechs, FRG
2 National Institute of Mental Health, Bethesda, Maryland 20205, USA
3 Northwestern University, Department of Neurobiology and Physiology, Evanston, I11. 60201, USA

Vertebrate Circadian Systems
Ed. by J. Aschoff, S. Daan, and G. Groos

necessarily indicate that there has been a change in the circadian organization, since removal of the pineal gland may alter the variable being monitored through a mechanism independent of the circadian system. Furthermore, the reported effects of pinealectomy on various circadian rhythms are often small and/or inconsistent; in many cases their reproducibility still has to be demonstrated.

Perhaps the strongest evidence for a role of the pineal gland in the circadian organization of mammals are the observations from three different laboratories that, after a shift of the LD-cycle, pinealectomized animals re-entrain their activity rhythm faster than intact controls (rats: Quay 1970a, 1970b, 1972; Kincl et al. 1970; hamsters: Finkelstein et al. 1978). In some of these studies significantly different rates of re-entrainment have only been seen during the initial transients after the zeitgeber shift, in others the total time needed for complete re-entrainment was different between the two groups of animals. These observations indicate that the entrainment mechanism may be affected by pinealectomy (Quay 1970a), and that pinealectomized and intact animals may differ in the shape or amplitude of their phase-response curves. To test this hypothesis, two experiments were performed with the golden hamster *(Mesocricetus auratus)*. In the first study, we determined the complete phase-response curve (PRC) in intact and pinealectomized animals by studying the effects of 1-h light pulses on the free-running rhythm of locomotor activity in DD. In an additional study we examined over a longer time span the effects of 15-min light pulses applied only at either 3 h or 9 h after the onset of wheel-running activity. The following report gives a brief summary of the results.

2 Experimental Protocols and Analytical Procedures

Study I. Nine pinealectomized and nine sham-operated male hamsters which previously had been group-housed in LD 14 : 10, were kept in DD for 240 days (January to September) in individual cages, each equipped with a running wheel. Groups of six such activity cages were enclosed in light-tight wooded boxes equipped with continuously operating ventilating fans. At intervals of about 20 days, the animals were exposed to a 1-h light pulse (intensity at cage floor: 10 to 50 lx) at either 2, 4, 7, 11, 14, 19 or 23 h after the onset of wheel-running activity. The revolutions of each animal's wheel were recorded on an Esterline-Angus event recorder, and the daily records mounted beneath each other on a chart. Phase-shifts due to the light-pulses were assessed by visual inspections of the daily onset of activity for 5 consecutive days preceding a pulse and for 5 to 10 days following a pulse once a steady state was achieved. The retrojected time of onsets after the pulse was subtracted from the projected time of onsets before the pulse, resulting in negative values for delay shifts and positive values for advance shifts (Daan and Pittendrigh 1976).

Study II. Four male hamsters, kept singly in cages equipped with running wheels, were transferred from natural LD to DD (0.02 lx) in June. Each cage was enclosed in a light-tight wooden box. At intervals of 14 days or longer, 15-min light pulses (200 lux on top of the cage) were given to each animal individually at either 3 h or 9 h after the last onset of activity. These two phases were chosen because they could be expected to correspond to the phases with maximal delay and advance shifts in the strain of animals used for this experiment (Pohl, pers commun). After each animal had been exposed to at least three pulses at each of the two phases, pinealectomy was performed in March of the following year and the presentations of light pulses at either 3 or 9 h after the onset of activity continued until October. The revolutions of each animals wheel were fed into a computer, stored on tape, and displayed in standard actogram format. In addition to assessing phase shifts as in study I, the following parameters of the activity rhythms were computed from the stored data for 10 days before and 10 days during the steady state freerun after a pulse: (a) the mean circadian period τ, based on the regression lines through onsets as well as ends of daily activity; (b) the standard deviation of consecutive onsets of activity around the regression line; (c) the mean duration of activity time α, measured between onset and end of activity; (d) the total number of wheel revolutions per 24 h.

3 Results

Typical records from two of the animals of study II, which can also be considered representative for study I, are reproduced in Fig. 1. Both records cover 45 days before pinealectomy (Px) and 40 days thereafter; they illustrate the immediate delay shifts which were induced by light pulses applied 3 h after activity onset, and the transient characterizing the advance shifts due to pulses applied 9 h after activity onset.

Study I. The effects of a total of 126 1-h light pulses, administered to 18 animals, are summarized in Fig. 2. Pulses given 2 h after activity onset induced phase delays, and pulses given 7 or 11 h after activity onset phase advances. No shifts were observed after pulses given 14 to 22 h after activity onset, while light pulses presented 4 h after the onset activity yielded either advances or delays (cf. the large standard errors). The responses of the nine pinealectomized hamsters to the light pulses were indistinguishable from those of the nine sham-operated controls. The mean phase shifts did not differ significantly (t-test for paired comparisons) between the two groups at any circadian time.

Study II. Light pulses of 15-min duration consistently resulted in phase delays when applied 3 h after activity onset, and in phase advances when applied 9 h after onset. On the average, the advance shifts were more than twice as large as the delay shifts. However, the absolute amount of the shifts changed during the course of the experiment. The magnitude of both phase delays and advances increased during the first few months of the experiment and steadily decreased thereafter. Since all four animals showed similar trends, their data were pooled, and all shifts induced within an interval of two months were averaged. The results summarized in Fig. 3 D show that from the first to the second interval, the delay shifts increased from -1.5 to -2.5 h, and the advance shifts from 4.0 to 5.1 h. Thereafter, the shifts decreased in magnitude, a trend which remained constant after pinealectomy was performed in March (In Fig. 3D, the magnitude of shift is given in real time since the difference to shifts calculated in circadian time, i.e. in reference to the period τ, is negligible). Concurrent with the change in the magnitude of the shifts there was a slight change in the period τ which reached a maximal value of 24.28 h in September/October of the first year (Fig. 3C). Pinealectomy did not alter the period of the activity rhythm.

If mean values are computed, for (1) phase shifts induced by the light pulses and (2) the periods of the free-running rhythm, from all data obtained before pinealectomy, and

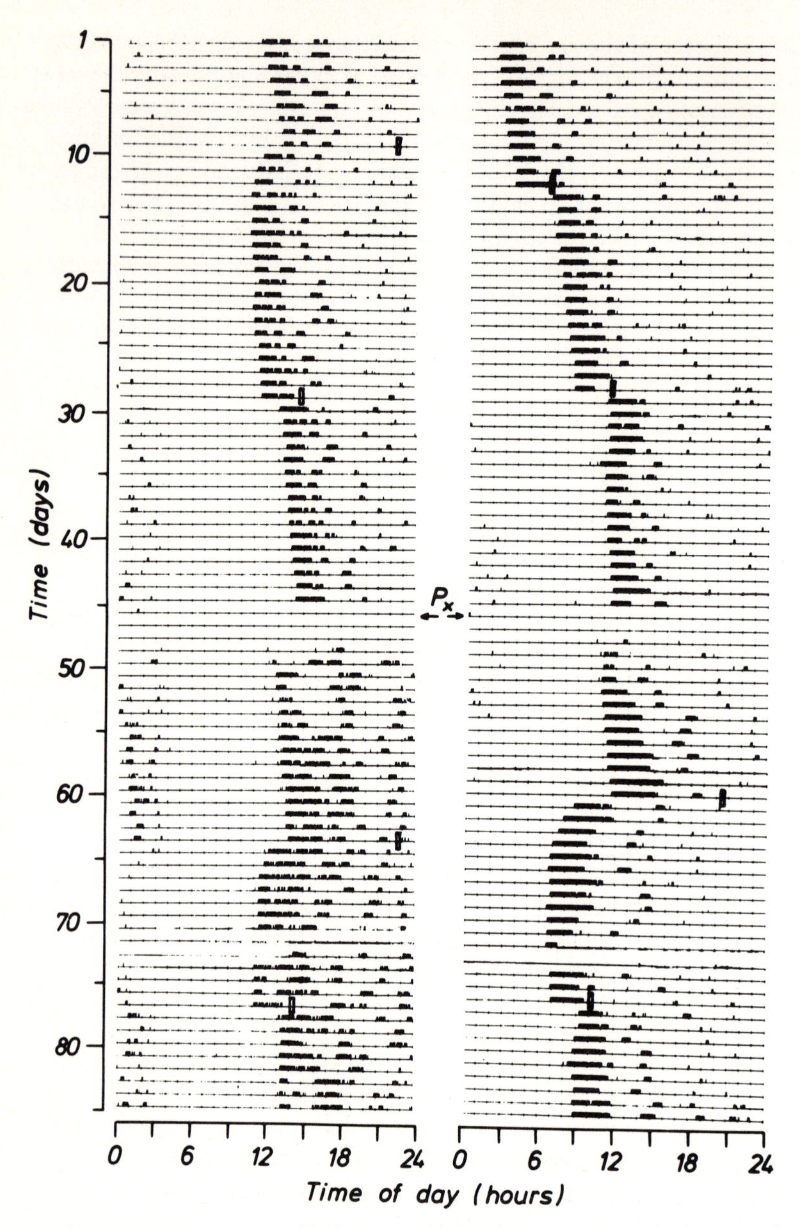

Fig. 1. Activity records of two male golden hamsters kept in constant dim illumination (0.02 lx), showing phase response to 15-min light pulses (200 lx) given at either 3 h or 9 h after onset of activity. Light pulses indicated by *vertical bars. Px* pinealectomy

from those obtained thereafter, the differences pre- and post-pinealectomy are statistically significant. These differences, however, cannot be attributed to the surgical intervention, but instead are due to the overall temporal trend that was clearly visible prior to pinealectomy (Fig. 3C and D). Thus, these data indicate that pinealectomy affected nei-

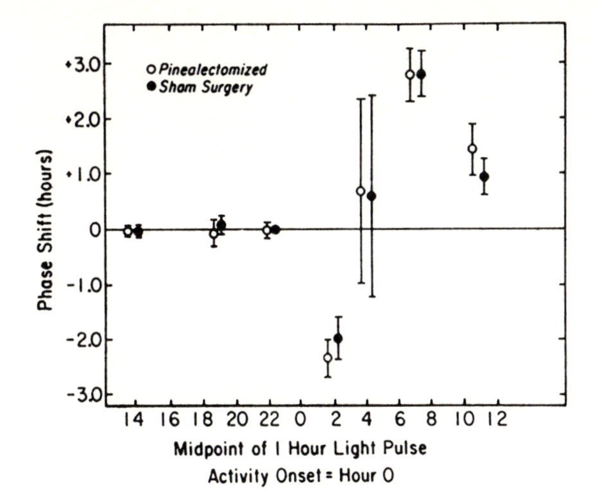

Fig. 2. Phase responses to 1-h light pulses of the circadian activity rhythm of golden hamsters kept in constant darkness. Midpoint of the light pulse drawn with reference to onset of activity (hour 0 on the abscissa). *Symbols* ± SE of nine pinealectomized and nine sham-operated animals. (Ellis et al. 1982)

ther the phase shifts induced by a 15-min pulse of light nor the circadian period of the locomotor activity rhythm.

As a further test of the possible role of the pineal gland in the circadian organization, we computed the standard deviation (SD) of τ-values from their regression line, a measure which represents the reciprocal of precision of a free-running rhythm. The mean SD of all four animals was 0.124 ± 0.017 h before pinealectomy and 0.094 ± 0.015 h thereafter. However, a closer inspection of the data reveals that this increase in precision again does not result from the pinealectomy but instead reflects the well known dependence of precision on τ (Aschoff et al. 1971, Pittendrigh and Daan 1976a). In other words, the standard deviation of τ in pinealectomized animals was smaller because their τ-values were closer to 24 h than those measured before the operation. (Mean τ of all four animals before P_X: 24.24 ± 0.11 h, after P_X: 24.02 ± 0.19 h). Importantly, the differences in τ (which are probably the causes for the differences in precision) cannot be attributed to pinealectomy, but are due to the gradual change in τ over time (Fig. 3C).

We furthermore computed the duration of the activity time α. There was an obvious trend over time similar to that seen in τ: α increased up to October of the first year and decreased thereafter (Fig. 3B). The mean values of α taken from all four animals just before and after pinealectomy were identical. Finally, we analyzed how α and τ were affected by the light pulses or the shifts, respectively. In accordance with observations made by Elliott (1981), phase shifts induced by 1-h light pulses resulted in aftereffects on both parameters: α as well as τ were lengthened after delay shifts and shortened after advance shifts (Fig. 4). After pinealectomy, these correlations were not altered, although the magnitude of aftereffects was decreased because of the smaller shifts in consequence of the temporal trend (cf. Fig. 3D).

4 Discussion

The results of study I and II indicate that there are no differences between the phase-response curves of pinealectomized and intact hamsters, at least for 1-h light pulses with

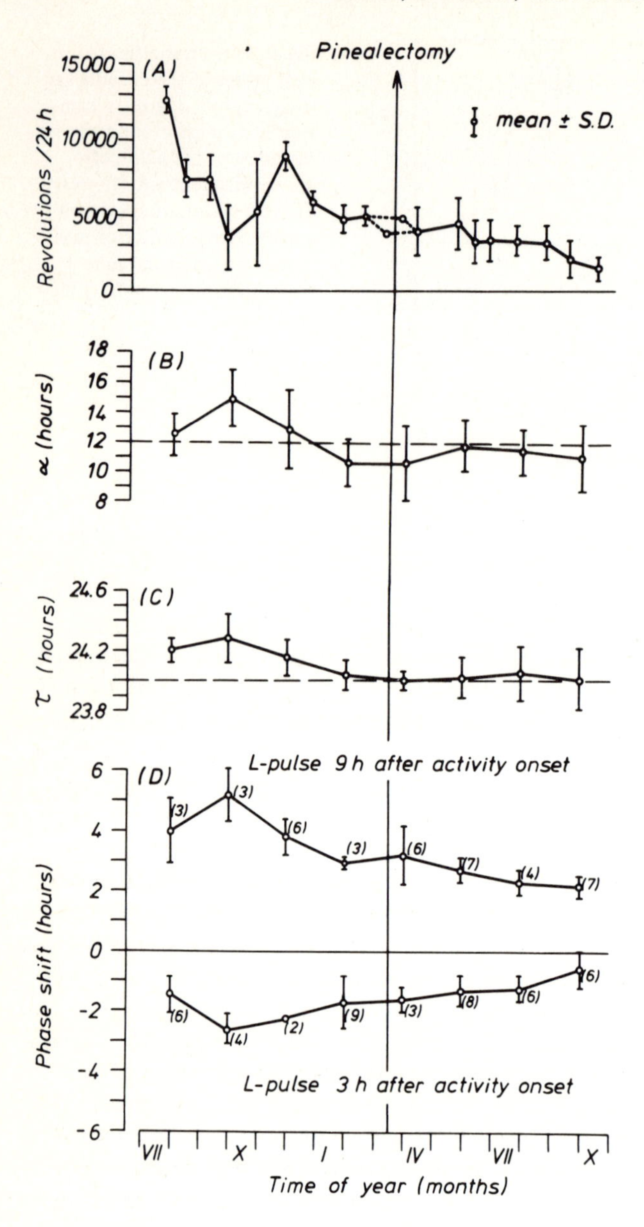

Fig. 3. Total wheel revolutions **(A)**, activity time α **(B)** and circadian period τ **(C)** throughout the course of the study, plotted together with phase shifts **(D)** induced in the activity rhythm by 15-min light pulses (200 lx) at either 3 h or 9 h after the onset of activity in four golden hamsters kept in dim constant illumination (0.02 lx). In **D** the number of shifts averaged is given in *brackets*

an intensity of 10 to 50 lx, and 15-min pulses of about 200 lx. The question arises why pinealectomized hamsters show a faster rate of re-entrainment after a shift of the LD-cycle than intact hamsters (Finkelstein et al. 1978). As outlined elsewhere in more detail (cf. Chap. 1.2), zeitgeber signals affect a circadian rhythm in at least two ways: they entrain the rhythm via phase control of the circadian pacemaker, and they directly influence the overt rhythm through the mechanism of masking. Masking is especially well documented for the effects of light on locomotor activity. In a nocturnal species, activity is often suppressed when light is turned on (negative masking), and it is released (or enhanced) when light is turned off (positive masking). Some of the data used to demonstrate faster re-entrainment of pinealectomized animals after a shift of the LD-cycle strongly suggest that the "real" shift was blurred by masking, more precisely by positive masking of the onset of activity. The justification of such an interpretation is well illustrated by one of the records published by Quay (1970b, cf. his Fig. 1) which demonstrates that, in a pinealectomized rat, onset of activity was immediately reset after the shift of the L : D transition, while the end of activity slowly drifted in D and regained its normal phase-angle difference to light-on at a rate similar to that of a sham operated and an unoperated sibling (cf. Fig. 2 and 3 in Quay 1970b). It should be noted that other activity records from pinealectomized animals show less masking (cf. Fig. 1 in Quay 1970a). However, in these experiments, the difference between the L- and D-intensity of illumination was somewhat smaller than in the above-mentioned experiments (resulting in less intense masking stimuli), and it was in these experiments that a difference in shift rate was only seen during the initial transient but not in the time needed for complete re-entrainment.

Considering, (1) the important effect masking has on the expression of the activity rhythm, and (2) our observation that pinealectomy does not alter the PRC of hamsters, we are inclined to conclude that in previous studies demonstrating an effect of pinealectomy on the rate of re-entrainment, the circadian oscillators underlying the activity rhythm were shifted at equal rates in pinealectomized and intact animals, but that pinealectomy resulted in stronger masking effects. Rusak (1982) has arrived at a similar conclusion in a recent review article which includes the following statement: "Masking and entraining effects of light are both mediated by retinal photoreceptors.... If pinealectomy affects both mechanisms it may be acting by modifying photoreceptor responses or an early stage of visual processing. If only masking is modified by pinealectomy, the locus of the effect should be sought at central targets of visual projection." Our data support Rusak's second train of thought because the entrainment mechanism, as determined by monitoring the PRC, was not altered by pinealectomy. Rusak (1982) has suggested that melatonin, a major pineal product which has a rhythm of large amplitude in the gland itself (Deguchi 1975; Tamarkin et al. 1979; Wilkinson et al. 1977), might be involved in the masking effects of light, and that reducing melatonin levels by pinealectomy could produce a central state functionally equivalent to that produced by increasing light intensity.

Rusak's hypothesis that pinealectomy alters the light input which bypasses the circadian pacemaker (cf. Chap. 1.2) can account for the apparent faster rate of re-entrainment in pinealectomized animals following a shift of the LD-cycle, and is in agreement with the observation that the PRC (Figs. 1-3), and the dependence of τ on light intensity (Quay 1968) are not altered by the removal of the pineal gland. Other properties of the activity rhythm which depend on the circadian system also do not appear to be influenced by the pineal gland. The duration of α, which is possibly controlled by the coupling forces between postulated evening and morning circadian oscillators (Pittendrigh and Daan 1976b), is not affected by pinealectomy (Fig. 3B). Furthermore, the dissociation of these oscillators (i.e. splitting), is also not influenced by pinealectomy (Ellis and Turek, unpub results), and the same applies to the precision of the free-running rhythm.

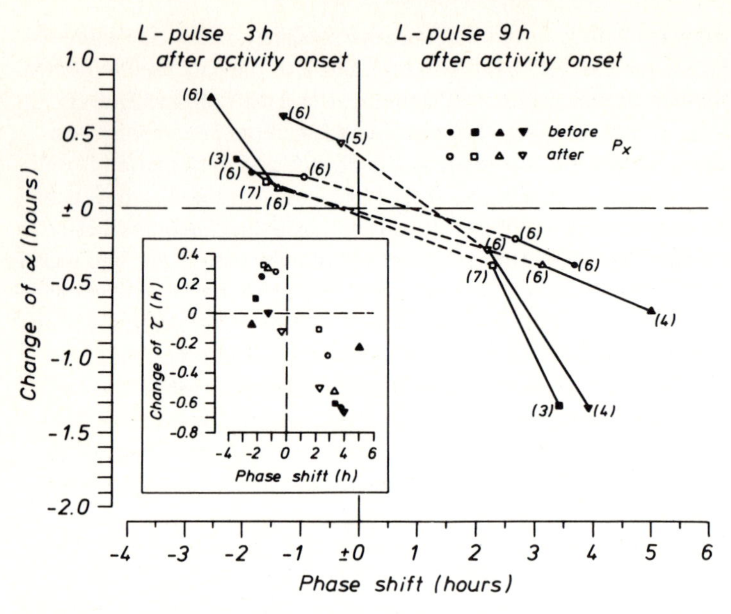

Fig. 4. Changes in activity time α, measured between onset and end of activity in free-running rhythms, and changes in period τ *(insert)*, depending on direction and magnitude of phase shifts induced by 15-min light pulses (200 lx) at 3 h or 9 h after onset of activity in four golden hamsters (4 symbols) kept in constant dim illumination (0.02). At each symbol, the number of shifts averaged is given in *brackets*. P_X pinealectomy. N.B.: the differences pre- and post-P_X do not result from the removal of the gland but are due to a temporal trend (cf. Fig. 3)

The lengthening of α after delay shifts, and its shortening after advance shifts (Fig. 4), can be interpreted as changes in the phase-angle difference between the evening and the morning oscillator. On the basis of this model, the data shown in Fig. 4 suggest that the morning oscillator is more strongly affected than the evening oscillator by delaying as well as by advancing light pulses. The concurrent changes in τ (Fig. 4, insert) agree with the observation, that, in the hamster, α and τ are positively correlated with each other (Fig. 3B and C), but the directions of the changes in τ are opposite to those seen after light pulses in diurnal rodents (cf. Fig. 2C and D in Chap. 8.3.).

It is not known why there was a change over time in the magnitude of the delay and advance shifts induced by the light pulses in study II (Fig. 3D). It should be pointed out that the initial increase in the magnitude of advances and delays, and the later decrease do not necessarily reflect changes in the "amplitude" of the PRC. Similar deviations could have been observed if there had been changes in the relationship between the PRC and the reference phase of the activity rhythm (i.e., onset of activity). However, changes in amplitude seem to be more likely, and could be related to the concurrent (although small) changes in τ (Fig. 3C). In search of causes for these phenomena, aging cannot be excluded a priori, but is less plausible than other time-related changes in the physiological state of the animals. We can safely assume that the hamsters had large testes in June, and that regression took place when they were transferred to DD in July. It is well known that castration results in a drastic reduction of wheel-running in rats (Wang et al.

1925, Slonaker 1930, Richter and Uhlenhut 1954). The decrease in activity during the first 3 months of study II and the subsequent increase in activity over the next 2 months (Fig. 3A) corresponds well so that observed by Ellis and Turek (1979) in hamsters transferred from LD 14 : 10 to LD 6 : 18. As Ellis and Turek (1979) have shown, these changes in activity are closely related to changes in testis size and plasma levels of testosterone.

If the activity of our hamsters during the first 5 months of Study II reflect changes in testosterone level due to regression and spontaneous recrudescence of testes, it still has to be explained why thereafter activity decreased steadily (aging? ; cf. Richter 1922/23, Jones et al. 1953, and Figs. 73 and 74 in Aschoff 1962). In view of the fact that the testes had probably already spontaneously recrudesced at the time of pinealectomy, removal of the gland would not be expected to cause any further change in the reproductive hormonal state of these animals since effects of pinealectomy on neuroendocrine-gonadal activity are not observed in golden hamsters with a functionally mature reproductive system (Turek and Ellis 1980). At the termination of study II, large testes were found in all hamsters. The lack of any effect of pinealectomy on the hormonal state of the animals probably accounts for why there was no change in the amount of activity due to pinealectomy per se as has been reported by others (Reiss et al. 1963, Krapp 1977).

5 Concluding Remarks

Much of the search for the oscillator(s) which drive(s) circadian rhythmicity in lower vertebrates has centered around the pineal gland (Takahashi and Menaker 1979; Underwood 1982). In contrast, attempts to identify the driving circadian oscillator(s) in mammals has centered around the suprachiasmatic nucleus (SCN) of the hypothalamus (cf. Chap. 3.1), and data presented in this paper, as well as elsewhere, indicate that the mammalian pineal gland does not function as a master oscillator. That the SCN may also be involved in the circadian organization of birds is suggested by the observation that lesions of the suprachiasmatic region abolish the rhythm of locomotor activity in three avian species (Takahashi and Menaker 1979, Ebihara and Kawamura 1980, Simpson and Follett 1981). Thus in birds, the pineal and SCN may act together to regulate circadian rhythmicity. Although there is no evidence in mammals to support a direct role for the pineal gland in the generation of circadian rhythms, it should be noted that the pineal gland does mediate the effects of light on seasonal reproductive cycles (Turek and Campbell 1979, Hoffmann 1981); an event which itself depends on the circadian system for the measurement of day length (Elliott 1981). The precise role of the pineal gland in the photoperiodic response is not known. Interestingly, the control of biochemical circadian rhythms in the mammalian pineal gland does depend upon neural input from the SCN (Moore and Klein 1974). Taken together, data from both birds and mammals suggest that there is an important functional relationship between the pineal gland and the SCN, although the nature of that relationship has clearly been altered through the course of evolution. In attempting to elucidate the physiological mechanisms which underlie the way in which the pineal gland and the SCN are involved in the overall circadian organization of higher vertebrates, it may prove extremely useful to investigate lower vertebrates where at the present time little is known about the role of the pineal gland, and nothing is known about the importance of the suprachiasmatic region in the generation of circadian rhythms.

References

Aschoff J (1962) Spontane lokomotorische Aktivität. In: Helmcke JG, Lengerken HV, Stark D (eds) Handbuch der Zoologie. Band 8 (II,4) Walter de Gruyter, Berlin.

Aschoff J, Gerecke U, Kureck A, Pohl H, Rieger P, Saint Paul U von, Wever R (1971) Interdependent parameters of circadian activity rhythms in birds and man. In: Menaker M (ed) Biochronometry. Natl Acad Sci, Washington, pp 3-27

Banerji TK, Kachi T, Quay WB (1979) Circadian changes in adrenal dopamine-β-hydroxylase activity: dependence of change at darkness onset, and the effect of pinealectomy, on animal strain and age. Chronobiologia 6: 1-7

Banerji TK, Quay WB (1978) Modification of plasma dopamine-β-hydroxylase activity by adrenal and pineal extirpations, and time of day dependency of changes. Chronobiologia 5: 379-395

Baum MJ (1970) Light-synchronization of rat feeding rhythms following sympathectomy or pinealectomy. Physiol Behav 5: 325-329

Cavalieri VM, Sollberger A, Bliss DK (1980) Rhythmicity of intermale aggression in mice: potential pineal influences. J Interdiscipl Cycle Res 11: 299-324

Daan S, Pittendrigh CS (1976) A functional analysis of circadian pacemakers in nocturnal rodents. II. The variability of phase response curves. J Comp Physiol 106: 253-266

Deguchi T (1975) Ontogenesis of a biological clock for serotonin: acetyl coenzyme A N-acetyltransferase in pineal gland of rat. Proc Natl Acad Sci USA 72: 2814-2818

Ebihara S, Kawamura H (1980) Central mechanism of circadian rhythms in birds. In: Tanabe Y, Tanaka K, Ookawa T (eds) Biological rhythms in birds. Neural and endocrine aspects. Springer, Berlin Heidelberg New York, pp 71-78

Elliott JA (1981) Circadian rhythms, entrainment and photoperiodism in the Syrian hamster. In: Follett BK and Follett DE (eds) Biological clocks in seasonal reproductive cycles. Wright, Bristol, pp 203-217

Ellis GB, Turek FW (1979) Changes in locomotor activity associated with the photoperiodic response of the testes in male golden hamsters. J Comp Physiol 132: 277-284

Ellis GB, McKleveen RE, Turek FW (1982) Dark pulses affect the circadian rhythm of activity in hamsters kept in constant light. Am J Physiol 242: R44-R50

Finkelstein JS, Baum FR, Campbell CS (1978) Entrainment of the female hamster to reversed photoperiod: Role of the pineal. Physiol Beh 21: 105-111

Jones DC, Kimeldorf DJ, Rubadeau DO, Castenera TJ (1953) Relationship between volitional activity and age in the male rat. Amer J Physiol 172: 109-114

Hoffmann K (1981) The role of the pineal gland in the photoperiodic control of seasonal cycles in hamsters. In: Follett BK, Follett DE (eds) Biological clocks in seasonal reproductive cycles. Wright, Bristol, pp 237-250

Karppanen H, Airaksinen MM, Särkimäki J (1973) Effects in rats of pinealectomy and oxypertine on spontaneous locomotor activity and blood pressure during various light schedules. Ann Med Exp Biol Fenn 51: 93-103

Kawakami M, Yamaoka S, Yamaguchi T (1972) Influence of light and hormones upon circadian rhythm of EEG slow wave and paradoxial sleep. In: Itoh S, Ogata K, Yoshimura H (eds) Advances in climatic physiology. Igaku Shoin Ltd, Tokio, pp 349-366

Kincl FA, Chang CC, Buzkova V (1970) Observation on the influence of changing photoperiod on spontaneous wheel-running activity of neonatally pinealectomized rats. Endocrinology 87: 38-42

Kinson GA, Liu CC (1973) Effects of blinding and pinealectomy on diurnal variations in plasma testosterone. Experientia 29: 1415-1416

Kizer HS, Zivin JA, Jacobowitz DM, Kopin IJ (1975) The nyctohemeral rhythm of plasma prolactin, effects of ganglionectomy, pinealectomy, constant light, constant darkness or 6-OH-dopamine administration. Endocrinology 96: 1230-1240

Krapp Ch (1977) Der Einfluß der Epiphyse auf die Lokomotionsaktivität bei Ratten. Experientia 33: 731-732

Lynch HJ, Wurtman RJ (1979) Control of rhythms in the secretion of pineal hormones in humans and experimental animals. In: Suda M, Hayashi O, Nakagawa H (eds) Biological rhythms and their central mechanism (The Naito Foundation Symp 1979) Elsevier, North-Holland Biomedical Press, Amsterdam New York Oxford, pp 117-131

Moore RY, Klein DC (1974) Visual pathways and the central neural control of a circadian rhythm in pineal serotonin N-acetyltransferase activity. Brain Research 71: 17-33

Morimoto Y, Yamamura Y (1979) Regulation of circadian adrenocortical periodicities and of eating – fasting cycles in rats under various lighting conditions. In: Suda M, Hayaishi O, Nakagawa H (eds) Biological rhythms and their central mechanism (The Naito Foundation Symp 1979). Elsevier, North-Holland Biomedial Press, Amsterdam New York Oxford, pp 176-188

Morin LP, Cummings LA (1981) Effect of surgical or photoperiodic castration, testosterone replacement or pinealectomy on male hamster running rhythmicity. Physiol Beh 26: 825-838

Niles LP, Brown GM, Grota LJ (1977a) Effects of neutralization of circulating melatonin and N-acetylserotonin on plasma prolactin levels. Neuroendocrinology 23: 14-22

Niles LP, Brown GM, Grota LJ (1977b) Endocrine effects of the pineal gland and neutralization of circulating melatonin and N-acetyl-serotonin. Can J Physiol Pharmacol 55: 537-544

Niles LP, Brown GM, Grota LJ (1979) Role of the pineal gland in diurnal endocrine secretion and rhythm regulation. Neuroendocrinology 29: 14-21

Pittendrigh CS, Daan S (1976a) A functional analysis of circadian pacemakers in nocturnal rodents. I. The stability and lability of spontaneous frequency. J Comp Physiol 106: 223-252

Pittendrigh CS, Daan S (1976b) A functional analysis of circadian pacemakers in nocturnal rodents. V. Pacemaker structure: A clock for all seasons. J Comp Physiol 106: 333-355

Quay WB (1968) Individuation and lack of pineal effect in the rat's circadian locomotor rhythm. Physiol Behav 3: 109-118

Quay WB (1970a) Physiological significance of the pineal during adaptation to shifts in photoperiod. Physiol Beh 5: 353-360

Quay WB (1970b) Precocious entrainment and associated characteristics of activity patterns following pinealectomy and reversal of photoperiod. Physiol Beh 5: 1281-1290

Quay WB (1972) Pineal homeostatic regulation of shifts in the circadian activity rhythm during maturation and aging. Trans Acad Sci, Ser. 11, 34: 239-254

Reiss,M, Davis RH, Sideman MB, Lichta ES (1963) Pineal gland and spontaneous activity of rats. J Endocrin 28: 127-128

Richter CP (1922/23) A behavioristic study of the activity of the rat. Comp Psychol Monogr 1, No. 2, pp 1-55

Richter CP (1967) Sleep and activity: their relation to the 24-hour clock. In: Kety SS, Evarts EV, Williams HC (eds) Sleep and altered states of consciousness. Williams & Williams C, Baltimore, pp 8-29

Richter CP, Uhlenhuth EH (1954) Comparison of the effects of gonadectomy on spontaneous activity of wild and domesticated Norway rats. Endocrinology 54: 311-3222

Rønnekleiv OK, Krulich L, McCann SM (1973) An early morning surge of prolactin in the male rat and its abolition by pinealectomy. Endocrinology 92: 1339-1342

Rusak B (1982) Circadian organization in mammals and birds: role of the pineal gland. In: Reiter RJ (ed) The pineal gland, Vol III Extra-reproductive effects. CRM Press, Boca Raton, Florida (in press)

Scalabrino G, Ferioli ME, Nebuloni R, Fraschini F (1979) Effects of pinealectomy on the circadian rhythms of the activities of polyamine biosynthetic decarboxylases and tyrosine aminotransferase in different organs of the rat. Endocrinology 104: 377-384

Simpson SM, Follett BK (1981) Pineal and hypothalamic pacemakers: Their role in regulating circadian rhythmicity in Japanese Quail. J Comp Physiol 144: 381-389

Slonaker JR (1930) The effect of the excision of different sexual organs on the development, growth and longevity of the albino rat. Am J Physiol 93: 307-317

Spencer F, Shirer HW, Yochim JM (1976) Core temperature in the female rat: effect of pinealectomy or altered lighting. Am J Physiol 231: 355-360

Stephan FK, Zucker I (1972) Rat drinking rhythms: Central visual pathways and endocrine factors mediating responsiveness to environmental illumination. Physiol Beh 8: 315-326

Takahashi JS, Menaker M (1979) Brain mechanisms in avian circadian systems. In: Suda M, Hayaishi O, Nakagawa H (eds) Biological rhythms and their central mechanism (The Naito Foundation Symp 1979). Elsevier, North-Holland Biomedical Press, Amsterdam New York Oxford, pp 95-109

Takahashi K, Inoue K, Takahashi Y (1976) No effect of pinealectomy on the parallel shift in circadian rhythms of adrenocortical activity and food intake in blinded rats. Endocrinol Japan 23: 417-421

Tamarkin L, Reppert SM, Klein DC (1979) Regulation of pineal melatonin in the Syrian hamster. Endrocrinology 104: 385-389

Tilders FJH, Smelk PG (1975) A diurnal rhythm in melanocytestimulating hormone content of the rat pituitary gland and its independence from the pineal gland. Neuroendocrinol 17: 296-308

Turek FW, Campbell CS (1979) Photoperiodic regulation of neuroendocrine-gonadal activity. Biol Reprod 20: 32-50

Turek FW, Ellis GB (1980) Role of the pineal gland in seasonal changes in neuroendocrine testicular function. In: Steinberger A, Steinberger E (eds) Testicular development, structure and function. Raven Press, New York, pp 389-393

Underwood H (1982) Circadian organization in fish, reptiles and amphibians: role of the pineal gland. In: Reiter RJ (ed) The pineal gland. Vol III Extra-reproductive effects. CRM Press, Boca Raton, Florida (in press)

Wang GH, Richter CP, Guttmacher AF (1925) Activity studies on male castrated rats with ovarian transplants and correlation of the activity with the histology of the grafts. Amer J Physiol 73: 581-599

Wilkinson M, Arendt J, Bradtke J, Ziegler Dde (1977) Determination of a dark-induced increase of pineal N-acetyltransferase activity and simultaneous radio-immunoassay of melatonin in pineal, serum and pituitary tissue of the male rat. J Endocr 72: 243-244

Willoughby JO (1980) Pinealectomy mildly disturbs the secretory patterns of prolactin and growth hormone in the unstressed rat. J Endocr 86: 101-107

4.2 Circadian and Infradian Activity Rhythms in the Mammalian Pineal Body

L. Vollrath[1]

1 Introduction

The neuroendocrine nature of the pineal organ is well established. Linked to the optic system via postganglionic sympathetic fibres coming from the superior cervical ganglia, the parenchymal cells of the pineal synthesize serotonin from tryptophan and convert it to melatonin, the most widely studied pineal substance, and other indoleamines. Melatonin synthesis is stimulated by noradrenaline released from intrapineal sympathetic nerve fibres, involving β-adrenergic mechanisms and the adenyl cyclase-cAMP-system. The rate-limiting enzyme of melatonin formation is serotonin-N-acetyltransferase (NAT) that converts serotonin to N-acetylserotonin. Hydroxyindole-O-methyltransferase (HIOMT) catalyzes the formation of 5-methoxy-N-acetylserotonin (melatonin) from N-acetylserotonin (cf. Vollrath 1981). A typical feature of the mammalian pineal gland is a pronounced circadian rhythm of melatonin formation, that is entrained by environmental lighting conditions. The day-night ratios of melatonin formation are 1 : 60 in rat, 1 : 3 in hamster and Mongolian gerbil and 1 : 1.5 in guinea-pig (Rudeen et al. 1975).

Circadian rhythmicity at the cellular level has recently been established by electrophysiological studies (Semm and Vollrath 1980). In these experiments the unexpected observation was made that pinealocytes differ in their electrical behaviour, when studied over periods of up to 24 h. Constantly firing cells were characterized by a lack of a clear-cut circadian rhythm. Darkness-activated cells exhibited low oscillatory activity levels during the day and high levels at night. These cells conform to the well-established pattern of melatonin formation.

A third type of cell was classified as light-activated cell, showing high activity levels at daytime and low activity levels at night. Although nothing is known about the incidence of the three cell types, it is clear that the secretory activity as well as quantitative morphological data of the gland as a whole, at a given time point, depend on the proportions of the three cell types present and on the degree of synchronization within each cell group. In this context it is relevant to note that, although there are as yet no morphological indications for the presence of functionally distinct types of parenchymal cell in the mammalian pineal gland, only some of the pinealocytes contain immunoreactive melatonin (Bubenik et al. 1974, Freund et al. 1977, Vivien-Roels et al. 1981).

1 Department of Anatomy, Johannes Gutenberg-University, Mainz, F.R.G.

2 Circadian Changes in Morphological Parameters

In contrast to a wealth of biochemical data on pineal circadian rhythmicity, morphological data are sparse (cf. Vollrath 1981). For a proper evaluation of the morphological changes observed, data obtained at all levels – gross-anatomical, histological and ultrastructural – should be considered. A complicating feature may be a perhaps mosaic-like architecture of the pineal parenchyma (cf. Vollrath 1979).

At the gross anatomical level there is insufficient evidence for circadian rhythmicity. While Axelrod et al. (1965) described rhythmic changes in pineal weight with lowest values just before lights off (LD 12 : 12; 0700-1900 h) and a clear rise after the onset of darkness, Merritt and Sulkowski (1969) were unable to observe clear-cut differences. In recent planimetric studies of serially sectioned rat pineal glands, statistically significant circadian differences in pineal volume could not be detected (Diehl 1981).

The intrinsic cells of the pineal body, the pinealocytes, show changes which suggest that they are largest during photophase. As the pinealocytes have long, slender and tortuous cytoplasmic processes, the overall size of pinealocytes is difficult to estimate, especially in light microscopical specimens. Kachi et al. (1971, 1973) assessing pinealocyte size by counting the number of pinealocyte nuclei per unit area, found that both in male and female mice nuclear density is lowest during the light phase, i.e., pinealocytes appear to be larger during the day than at night. This conclusion is in agreement with findings in rats, in which Quay and Renzoni (1966) showed that pinealocyte nuclei and nucleoli are largest at noon and reach lowest values late in the afternoon. However, in a thorough electron microscopic study, Welsh et al. (1979) could not detect clear-cut day-night differences in pinealocyte nuclear and nucleolar sizes of the Mongolian gerbil. Instead they found that the total volume of individual pinealocytes showed a clear depression during daytime.

Morphometric analyses of individual pinealocytes at the light microscopic level are hampered by the fact that the pineal body, at least in some mammals, is traversed by a prominent system of intercellular canaliculi. The canaliculi are widened intercellular spaces measuring 0.2-0.4 μm in width (Krstić 1975), which, according to injection experiments by Quay (1974), show dramatic changes in width over a period of 24 h, the greatest perfusability being observed at the end of the light phase, followed by a rapid decrease. As this parameter shows a parallel behaviour to the serotonin content of the pineal, it is interesting to note that administration of serotonin during the dark phase increased the width of the spaces. A functional significance of these canaliculi is suggested by the observation that they extend from perivascular spaces, where most of the functionally important sympathetic nerve fibres terminate, into the parenchyma where they may enmesh individual pinealocytes. Hence, it is conceivable that they may play a role in transporting substances from nerve fibres and blood vessels to pinealocytes and vice versa.

In view of the prominent day/night differences of melatonin formation the cell organelles of pinealocytes are of particular interest. However, apart from the enigmatic synaptic ribbons to be described below, very little of a definitive nature is known. The only such study has been carried out in the Mongolian gerbil (Welsh et al. 1979). Here it was found that the volume of organelle-free cytoplasm was smallest between 0700 and 1300 h, thereafter reaching a prominent peak at 1900 h and apparently a second peak at 1300 h. The mitochondrial volume is smallest at 0700 h, showing a more or less steady increase until 1300 h. A roughly similar behaviour is shown by the smooth ER and rough ER (endoplasmic reticulum) plus ribosomes. These findings taken together suggest that in the Mongolian gerbil the cell organelles of the pinealocytes are activated in the afternoon, reaching a peak early at night. In the same species, pineal serotonin-N-acetyltransferase activity has been shown to increase between 2000 and 2400 h, reaching a peak at 0400 h (Rudeen et al. 1975).

In view of the ribosomal changes mentioned above, it is relevant to enquire to what extent they are paralleled by alterations of RNA. In a histochemical, cytophotometric study Nováková et al. (1971) noted that in rats kept under normal lighting conditions the pineal RNA content was 106.0 ± 4.5 (arbitrary working units) at 1100 h and 96.7 ± 4.2 at 2300 h. These data are basically in agreement with those of one detailed biochemical study. Nir et al. (1971) noted in mature male rats that the pineal RNA content was highest at noon, decreased gradually to reach a nadir at midnight, that was followed by a renewed increase. In mature and immature female rats the rhythm was less accentuated than in males. The enhancement of the RNA content between midnight and the early morning has been observed also by Merritt and Sulkowski (1969); however, in their studies the RNA content was lowest (6.3 μg/mg pineal tissue) at noon, followed by a clear increase between noon and 1800 h. Clearly, further studies have to be carried out to solve this problem.

In view of the fact that the pineal is a neuroendocrine organ, it is of interest to study the circadian behaviour of the morphological correlate of its secretory product(s). The mammalian pinealocyte differs from most other endocrine cells in that typical secretory granules are extremely rare (Pévet 1979). This together with a non-random distribution of the granules makes quantification difficult. Besides granular vesicles, ependymal-like secretory processes may play a role, characterized by the occurrence of flocculent proteinaceous material in cisterns of the granular endoplasmic reticulum or in vacuoles (Pévet 1979). The functional significance of the two types of secretion is obscure. It is therefore pertinent to enquire as to whether a circadian study may provide clues regarding the chemical nature of the substances contained in the granular vesicles. Pévet (1979) has suggested that the ependymal-like secretory process is involved in the production of an antigonadotropic compound, whereas the granular vesicles may be related to a progonadotropic stimulatory substance. There is clear evidence that the granular vesicles originate in the Golgi apparatus. In the rabbit, the vesicles lying in the vicinity of the Golgi apparatus show a distinct increase between 0700 h and noon, followed by a prominent decline, reaching minimum numbers at midnight; in terminals of pinealocyte processes, where the vesicles accumulate, they peak 7 h later than near the Golgi apparatus (Romijn et al. 1976), suggesting a transport of vesicles from the pinealocyte perikaryon into the processes. In both mouse (Benson and Krasovich 1977) and Mongolian gerbil (Welsh et al. 1979), the granular vesicles begin to increase in number at the onset of the photoperiod, reaching a maximum at its end, followed by a decrease during scotophase. A lack of correlation of these findings with the known biochemical events in the pineal gland makes it unlikely that the vesicles contain melatonin. Their increase during the photoperiod rather suggests a relationship to serotonin, that is stored in increasing amounts during the day and decreases during scotophase as a result of an enhanced activity of serotonin-N-acetyltransferase. Whether the vesicles contain non-aminergic substances of perhaps hormonal nature cannot be deduced, as nothing is known about circadian rhythmicity of peptidergic anti- or progonadotropic pineal substances.

Clear evidence for circadian rhythmicity has been obtained with respect to the functionally enigmatic synaptic ribbons of mammalian pinealocytes. Most of these structures consist of a rod measuring 30-40 nm in width surrounded by electron-lucent vesicles of 40-60 nm in diameter. These structures fulfil synaptic functions in the retina, the pineal complex of lower vertebrates and other sense organ. In the mammalian pineal body they may link adjacent pinealocytes. In guinea-pig (Vollrath 1973), rat (Kurumado and Mori 1977, King and Dougherty 1980), baboon (Theron et al. 1979) and goldfish (McNulty 1981), synaptic ribbons are small in number during the day and large in number during the night, showing an increase already prior to the onset of darkness. Blinding is followed by a significant decrease in the amount of synaptic ribbons; 6 months after blinding normal values and clear-cut circadian rhythmicity were observed (Kurumado and Mori 1980).

Circadian rhythmicity has been established also for the dense-core vesicles of the intrapineal sympathetic nerve fibres (Mukai and Matsushima 1980), which by way of release of noradrenaline regulate pineal melatonin formation.

3 Infradian Rhythms?

In contrast to circadian and circannual rhythms, nothing precise is known about the presence of infradian rhythms in the pineal body. A 7-day cycle of pineal HIOMT activity, exhibiting a peak on Saturdays and a trough on Thursday, has been described in male and female rats (Vollrath et al. 1975). As both Diehl (1981) and U. Becker (unpubl.) from this laboratory found highly variable patterns of pinealocyte nuclear size changes over periods of 24 h as well as clear differences in the absolute sizes of pinealocyte nuclei on alternate days, this latter aspect was systematically studied in male Sprague-Dawley rats. In four 7- to 8-day-experiments carried out at 0900, 1130, 1600 and 2400 h respectively, it was invariably found that pinealocyte nuclear volume showed statistically significant differences between experimental days (ANOVA $p < 0.05$) and that the curves exhibited distinct troughs in the Friday/Saturday region, independently of whether the experiments started on a Monday or a Wednesday. Recent biochemical studies on the pineal melatonin-forming activity likewise point in the direction of a 7-day-cycle (H. Welker, unpubl.).

4 Conclusions

The present survey has shown that there is clear biochemical, electrophysiological and morphological evidence for circadian rhythmicity in the mammalian pineal gland. When the results obtained with the different techniques are compared with one another, it cannot necessarily be expected that they are in unison. Biochemical studies are dealing with the function of the pineal gland as a whole, disregarding different functional states of individual pinealocytes. The latter can be monitored by electrophysiological, extracellular single unit recordings.

With morphometric techniques it is possible to study the pineal gland at the "tissue" level and to detect a perhaps mosaic-like structure of the parenchyma.

The morphometric results mentioned in the third section of this review indicate that a search for infradian rhythms, especially in the 7-day range, is worthwhile. The 7-day rhythmicity could be due to the working routine in the animal unit, in that case the pineal would be highly susceptible to certain exogenous factors, to be clarified subsequently. Circadian rhythmicity may differ in various regions and smaller functional subunits of the pineal gland. Due to functional interactions between these, beat frequencies may occur resulting in infradian rhythms.

References

Axelrod J Wurtman RJ, Snyder SH (1965) Control of hydroxyindole O-methyltransferase activity in the rat pineal gland by environmental lighting. J Biol Chem 240: 949-954

Benson B, Krasovich M (1977) Circadian rhythm in the number of granulated vesicles in the pinealocytes of mice. Effects of sympathectomy and melatonin treatment. Cell Tissue Res 184: 499-506

Bubenik GA, Brown GM, Uhlir I, Grota LJ (1974) Immunohistological localization of N-acetylindolealkylamines in pineal gland, retina and cerebellum. Brain Res 81: 233-242

Diehl BJM (1981) Time related changes in size of nuclei of pinealocytes in rats. Cell Tissue Res 218: 427-438

Freund D, Arendt J, Vollrath L (1977) Tentative immunohistochemical demonstration of melatonin in the rat pineal gland. Cell Tissue Res 181: 239-244

Kachi T, Matsushima S, Ito T (1971) Diurnal changes in glycogen content in the pineal cells of the male mouse. A quantitative histochemical study. Z Zellforsch Mikrosk Anat 118: 310-314

Kachi T, Matsushima S, Ito T(1973) Diurnal variations in pineal glycogen content during the estrous cycle in female mice. Arch Histol Jpn 35: 153-159

King TS, Dougherty WJ (1980) Neonatal development of circadian rhythm in "synaptic" ribbon numbers in the rat pinealocyte. Am J Anat 157: 335-343

Krstić R (1975) Scanning electron microscope observations in the canaluculi in the rat pineal gland. Experientia 31: 1072-1073

Kurumado K, Mori W (1977) A morphological study of the circadian cycle of the pineal gland of the rat. Cell Tissue Res 182: 565-568

Kurumado K, Mori W (1980) Pineal synaptic ribbons in blinded rats. Cell Tissue Res 208: 229-235

McNulty JA (1981) Synaptic ribbons in the pineal organ of the goldfish: circadian rhythmicity and the effects of constant light and constant darkness. Cell Tissue Res 215: 491-497

Merritt JH, Sulkowski TS (1969) Alterations of pineal gland biorhythms by N-methyl-3-piperidyl benzilate. J Pharmacol Exp Ther 166: 119-124

Mukai S, Matsushima S (1980) Effect of continuous darkness on diurnal rhythms in small vesicles in sympathetic nerve endings of the mouse pineal – Quantitative electron microscopic observations. J Neural Transm 47: 131-143

Nir I, Hirschmann N, Sulman FG (1971) Diurnal rhythms of pineal nucleic acids and protein. Neuroendocrinology 7: 271-277

Nováková V, Sterc J, Sandritter W, Křečel J (1971) The day-night differences in total RNA content of parenchymal cells of rat pineal gland. Beitr Pathol 144: 211-215

Pévet P (1979) Secretory processes in the mammalian pinealocyte under natural and experimental conditions. Progr Brain Res 52: 149-194

Quay WB (1974) Pineal canaliculi: demonstration, twenty-four-hour rhythmicity and experimental modification. Am J Anat 139: 81-94

Quay WB, Renzoni A (1966) Twenty-four-hour rhythms in the pineal mitotic activity and nuclear dimensions. Growth 30: 315-324 (1966)

Romijn HJ, Mud MT, Wolters PS (1976) Diurnal variations in number of Golgi-dense core vesicles in light pinealocytes of the rabbit. J Neural Transm 38: 231-237

Rudeen PK, Reiter RJ, Vaughan MK (1975) Pineal serotonin-N-acetyltransferase activity in four mammalian species. Neurosci Lett 1: 225-229

Semm P, Vollrath L (1980) Electrophysiological evidence for circadian rhythmicity in a mammalian pineal organ. J Neural Transm 47: 181-190

Theron JJ, Biagia R, Meyer AC, Boekkoot S (1979) Microfilaments, the smooth endoplasmic reticulum and synaptic ribbon fields in the pinealocytes of the baboon *(Papio ursinus)*. Am J Anat 154: 151-162

Vivien-Roels B, Pévet P, Dubois MP, Arendt J, Brown GM (1981) Immunohistochemical evidence for the presence of melatonin in the pineal gland, the retina and the Harderian gland. Cell Tissue Res 217: 105-115

Vollrath L (1973) Synaptic ribbons of a mammalian pineal gland. Circadian changes. Z Zellforsch Mikrosk Anat 145: 171-183

Vollrath L (1979) Comparative morphology of the vertebrate pineal complex. Progr Brain Res 52: 25-38

Vollrath L (1981) The pineal organ. In: Oksche A, Vollrath L (eds) Hdb mikr Anat Mensch VI/7. Springer, Berlin Heidelberg New York

Vollrath L, Kantarjian A, Howe C (1975) Mammalian pineal gland: 7-day rhythmic activity? Experientia 31: 458-460

Welsh MG, Cameron IL, Reiter RJ (1979) The pineal gland of the gerbil, *Meriones unguiculatus*. II. Morphometric analysis over a 24-hour period. Cell Tissue Res 204: 95-109

4.3 Electrophysiology of the Mammalian Pineal Gland: Evidence for Rhythmical and Non-Rhythmical Elements and for Magnetic Influence on Electrical Activity

P. Semm[1,2]

1 Introduction

Although the mammalian pineal gland has been extensively studied by biochemical, pharmacological and morphological techniques, the precise nature of the intrinsic cells, the pinealocytes, was unknown for a long time. Biochemically the pinealocyte has been shown to be an endocrine cell synthesizing melatonin and other hormones. In view of the importance of the mammalian pineal as a neurotransducer, and of the limited usefulness of the methods mentioned, an attempt was made to characterize the cells electrophysiologically.

In addition to the electrophysiological properties of pinealocytes, their neuronal input is of interest. According to a current concept, which is based mainly on morphological findings in rats, peripheral sympathetic fibres from the superior cervical ganglia provide the functionally most important input to pinealocytes while an influence of central fibres reaching the pineal body by means of the habenular nuclei and the posterior commisure has also been established. Finally, as an endocrine organ, the pineal is capable of synthesizing materials which can affect the brain and particularly the activity of the neuroendocrine system. Biochemical studies have shown that transmitters from the nerve endings of the sympathetic nervous system are related to or promote the synthesis of pineal enzymes and hormones. Moreover, experimental evidence indicates that the pineal is under feedback control by its target tissues.

2 Aim of the Present Study

Insight into the influence of transmitters and hormones on electrically excitable cells has recently been gained by using microiontophoresis. This approach permits the introduction of minute quantities of charged molecules to the immediate vicinity of the cell while simultaneously monitoring electrical activity. We have carried out experiments to investigate the effects of locally applied transmitters and hormones on the electrical unit activity of pineal cells in guinea pigs. Since the synthesis of melatonin and other substances in pineal cells follows a circadian pattern, we also studied if the responses of these substances varied with time of day.

1 Dept. of Anatomy, Neurophysiol. Lab., University of Mainz, Saarstr. 19/21, D - 6500 Mainz -FRG-
2 Present address: University of Frankfurt, Dept. of Zoology, Siesmayerstr. 70, D-6000 Frankfurt -FRG-

Vertebrate Circadian Systems
Ed. by J. Aschoff, S. Daan, and G. Groos

Lastly we wanted to know whether in addition to light, an artificial magnetic field with a strength comparable to that of the earth could influence the electrical activity of pineal cells. This choice was made because it is known that artificial magnetic fields can influence circadian rhythms (Bliss and Heppner 1976, Brown and Scow 1978).

When we started our investigations, only little information was available about the electrophysiology of the mammalian pineal. Some studies had been done in rats, indicating that pineal elements exhibit a spontaneous activity which can be influenced by light given to the lateral eyes, and by electrical stimulation of the sympathetic ganglia and several parts of the primary optic and the limbic system. An important hint was the fact that light could be a stimulus even after sympathectomy indicating a double route of innervation, sympathetic and central (for review see Vollrath 1981, Semm 1981).

Guinea pigs were used, in contrast to rats, as in this species

1. the morphology of the synaptic ribbons which were considered important for intercellular communication was well documented (Vollrath 1981)
2. the pineal complex belongs to the ABC-type (Vollrath 1979) and is therefore more suitable for electrophysiological experiments and
3. a major part of the organ comes into close contact with the commissural region.

3 Results and Discussion

3.1 Electrophysiology of Rhythmical and Non-Rhythmical Cells in the Pineal

Figure 1 shows a schematic diagram of the experimental situation containing a four-barrelled micropipette in the posterior part of the organ, and a stimulation electrode in

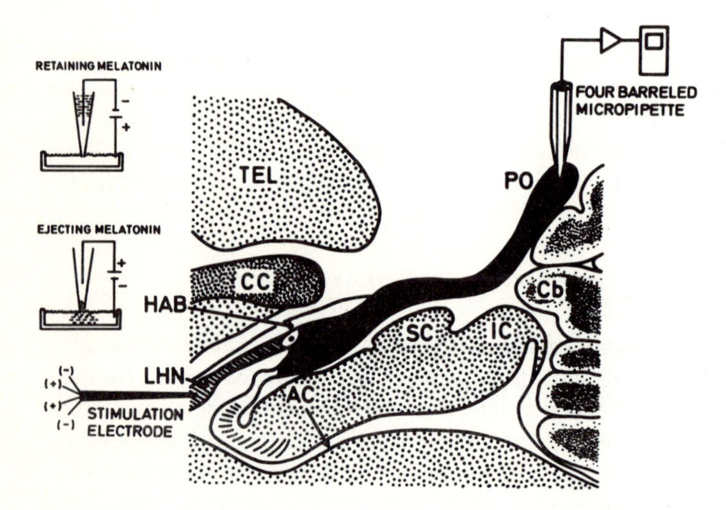

Fig. 1. Schematic diagram of the experiment showing a four-barrel glass micropipette and a stimulation electrode in a sagittal section of the brain as well as the signal to an oscilloscope. The other barrels were connected to iontophoresis programmers that can retain or eject a compound *(inserts)*. A stimulating electrode was placed in the lateral habenular nucleus. *AC* Aqueduct; *Cb* Cerebellum; *CC* Corpus callosum; *HAB* Habenula; *IC* Inferior Colliculus; *LHN* Lateral Habenular Nucleus; *PO* Pineal Organ; *SC* Superior Colliculus; *TEL* Telecenphalon

the lateral habenular nucleus, while the inserts show the mechanism of iontophoresis. The recording barrel of the electrode was connected to a preamplifier, the other barrels were connected to the iontophoresis programmers.

In extracellular recordings from the anterior, middle and posterior part of the guinea pig organ pineal spontaneously active cells and sometimes fibres have been found. Two main populations can be distinguished: (a) cells which are influenced mainly by sympathetic fibres and (b) cells which receive a central input from brain structures reaching the organ via the habenular nuclei. Sympathetically influenced cells show a spontaneous discharge which is characterized by alternating phases of high (burst) and low activity. They can be activated electrically from the optic chiasma with a relatively constant latency of about 16 ms, and they lose their spontaneous activity following sympathectomy; they differ in their reactions to light and darkness (Semm and Vollrath 1979a, b).

During exposure of the animal to darkness for 10 min in the morning light-activated cells show a low oscillatory activity. After exposure to natural or artificial light the firing rate of the cells increases. When exposed to the natural light-dark cycle, these light-activated cells show a higher activity during daytime and a depressed activity at night (Fig. 2). Dark pulses presented during daytime depress the activity gradually (Fig. 3B, C) but it seems that they are an effective stimulus only until about 1600 h. Furthermore, artificial illumination during the night had no measurable effect on these cells (Fig. 3B, C). A second category comprises cells which show the opposite behaviour. Long-term recordings reveal that the spontaneous activity of these "darkness"-activated cells is depressed during daytime and activated during the night (Fig. 2) (for further detail see Semm and Vollrath 1980).

In darkness-activated cells artificial light given for 1 min during nighttime leads to an abrupt decrease of electrical activity followed by a gradual increase in the firing rate. Artificial darkness during daytime had no measurable effect on these cells which fit best into the current concept of pineal function. It has been formerly established that melatonin formation increases during the night and, as shown by Rudeen et al. (1975), the guinea pig is no exception in this respect. The concept that these cells may be directly or indirectly involved in melatonin secretion is strengthened by three observations:

1. Darkness-activated cells can be strongly inhibited by 1 min of light given during the night to the lateral eyes, a procedure which also depresses pineal melatonin formation as assessed by N-acetyl-transferase activity.
2. The darkness-activated cells do not respond to exposure to artificial darkness during the day.
3. The different durations of cellular activation in different seasons may implicate different amounts of melatonin formation as suggested for the rat (Rudeen and Reiter 1977). Furthermore, in our microiontophoretic studies (Semm et al. 1981 a, b, c) when the responses of pineal cells to the application of melatonin during the day are compared with those occurring at night, the proportions of excitatory and inhibitory effects are markedly different (Fig. 4). While in the daytime most cells responded with an increase in impulse frequency to the application of the indole, inhibition prevailed after 1900 h (April and May experiments). Thus, the differences in the ratio of excited/inhibited cells in daytime as opposed to at night may reflect both the circadian changes in melatonin synthesis and the daily rhythms in the firing rate of cells

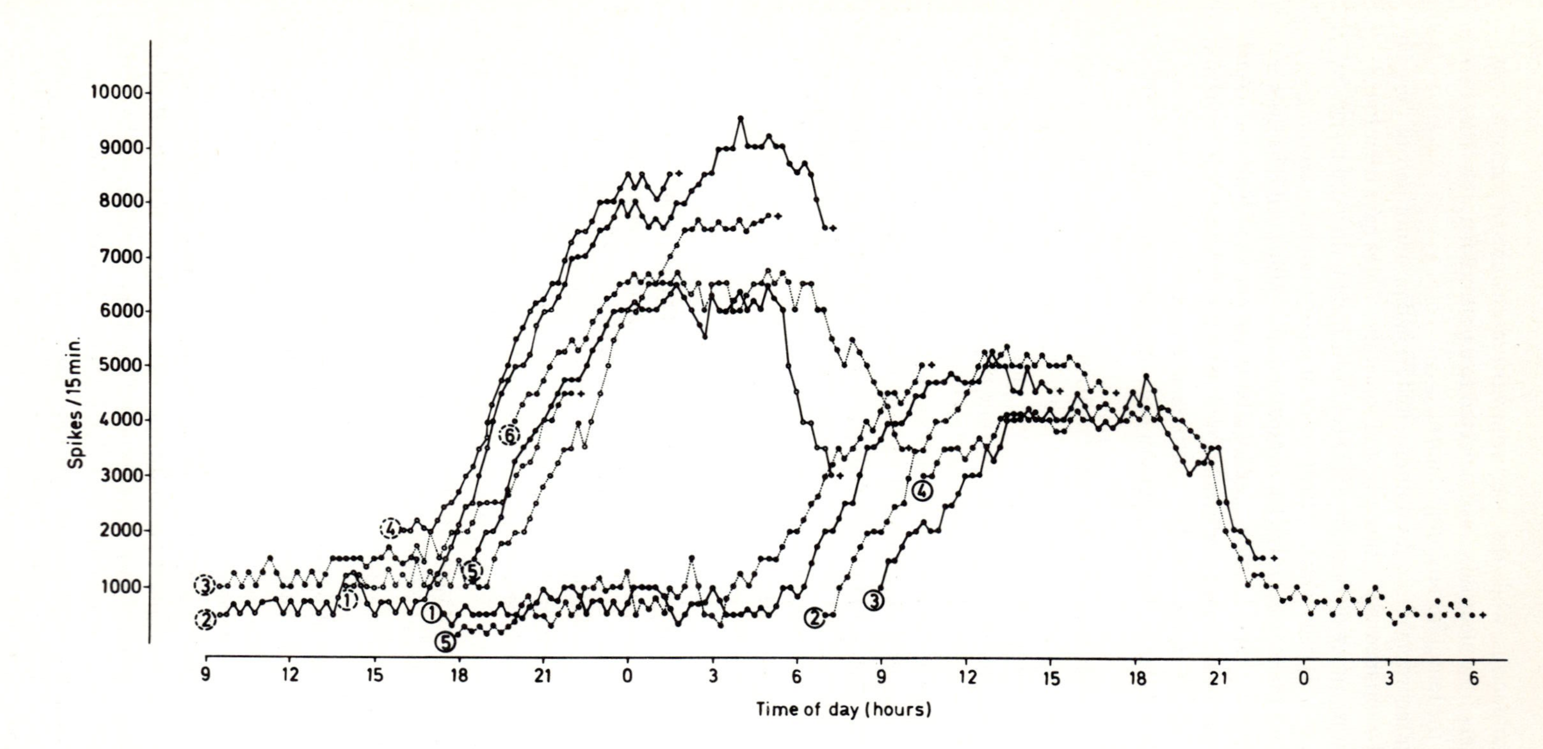

Fig. 2. Long-term recordings from light-activated (*1-5, filled circles*) and darkness-activated cells (*1-6, open circles*) in the guinea pig. The electrical activity of light-activated cells is distinctly higher during the day than during the night. Artificial illumination during night-time had no measurable effect on these cells. Darkness-activated cells show low oscillatory activity during the day, which cannot be elevated by artificial darkness, and exhibit high levels of activity during the night

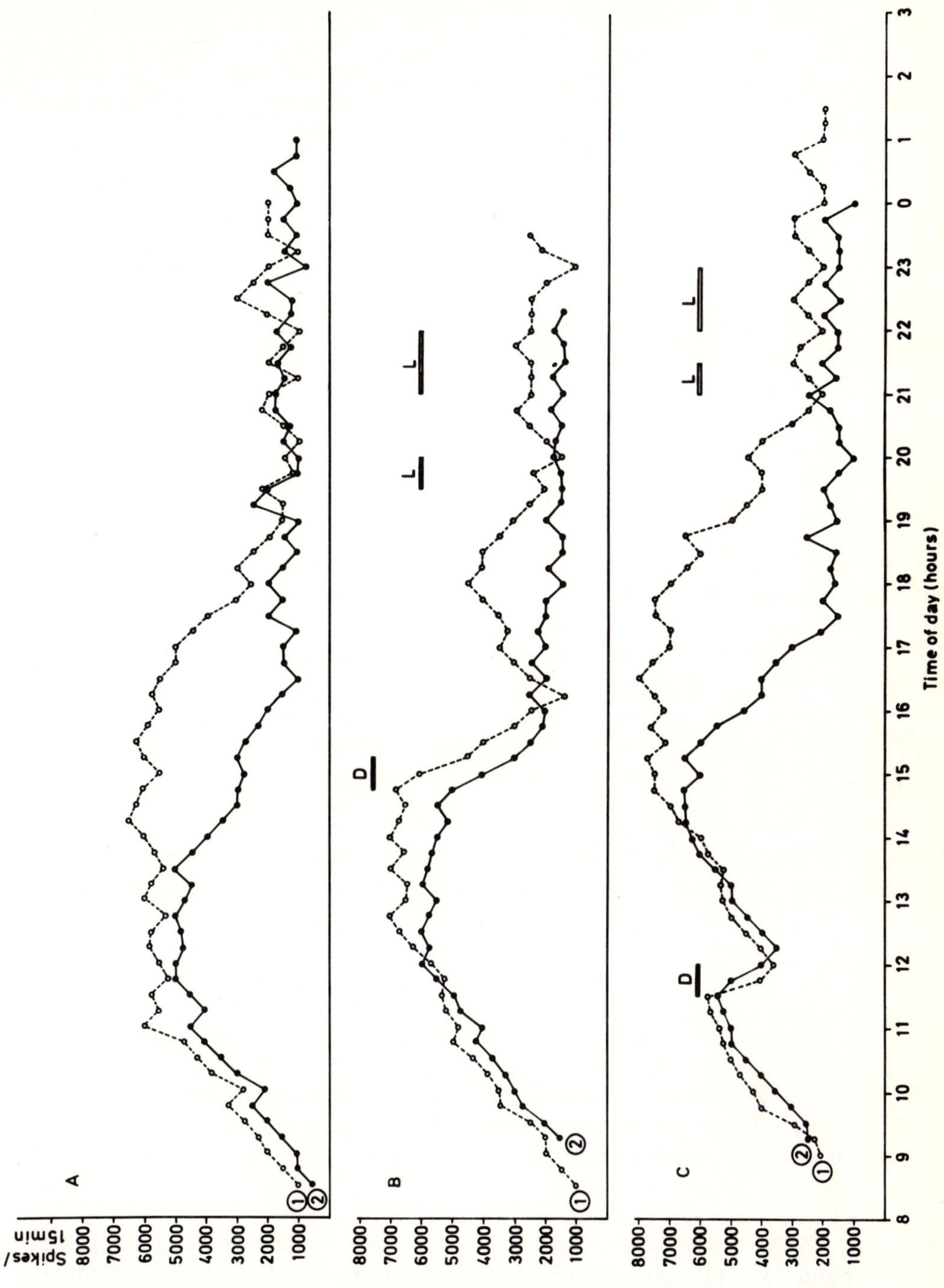

Fig. 3. A Long-term recordings from two light-activated cells in the guinea pig. Cell *1* was recorded in summertime, cell *2* was recorded in wintertime. As the experiments started at 0900 h, a difference can be seen in the evening only. Activity of the wintercell *2* is diminished at 1600 h, activity of the summercell *1* is diminished later at about 1900 h. **B** Artificial darkness at 1500 h depresses the activity of two other light-activated cells within a range of 1 h, whereas artificial light during the evening had no measurable effect. These cells were recorded during wintertime. **C** Long-term recordings of two summercells the activity of which was influenced slightly by artificial darkness given at about 1100 h. Artificial light during the evening had no measurable effect

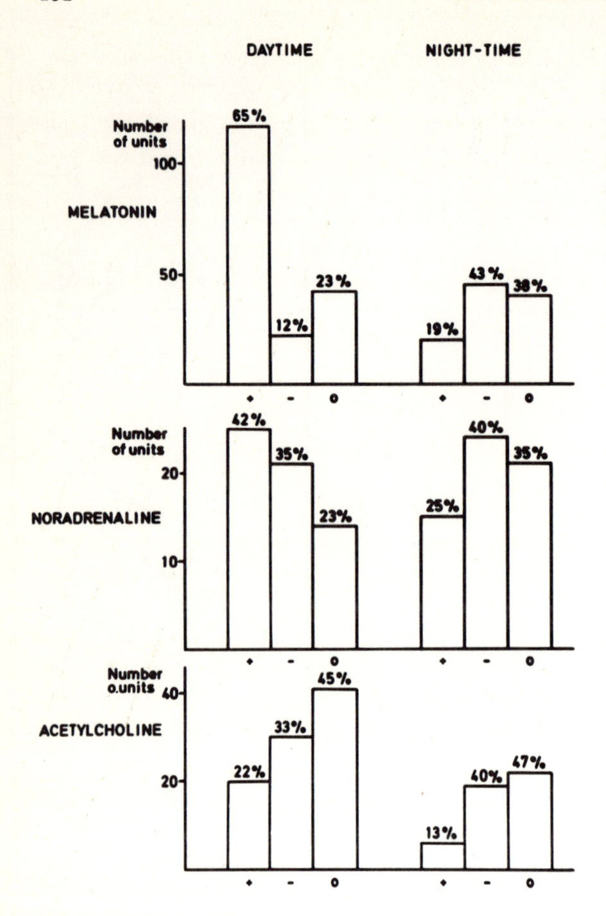

Fig. 4. A-C Comparison of the responses of pineal cells to the application of melatonin, noradrenaline and acetylcholine made during daytime (0600-1900 h) and at night (1900-0600 h) in the guinea pig. + Excitation; – Inhibition; 0 No response

in the pineal. We also observed a rhythmicity in the responses of cells to the application of noradrenaline (Fig. 4), thyroxine (in which recordings 79% of the cells responded in the same way as to melatonin) and of testosterone in male animals. A large majority of cells responded to testosterone with excitation between 1400 and 2200 h, but in the morning the predominant effect was inhibitory.

No rhythmicity could be observed in the response to the application of acetylcholine. Perhaps these acetylcholine-sensitive cells belong to a special class of pinealocytes which show a relatively constant activity pattern both at day and night and do not exhibit visual responsiveness (Semm and Vollrath 1980). After sympathectomy, the spontaneous activity of these cells is not affected. In some cases it could be demonstrated that these constantly firing cells are innervated by central fibres from or passing through the lateral habenular nucleus (Fig. 5).

Morphological and electrophysiological investigations (Semm et al. 1981d) have shown that (1) fibres from the habenular nuclei enter the anterior part of the pineal organ via the habenulae and the pineal stalk and terminate in the intercellular space where they form a dense network intermingling with the long processes of the pinealocytes, and (2) 44% of the pineal cells studied were driven orthodromically by electrical stimulation of the lateral hebanular nucleus; 80% of these were excited and 20% were inhibited. Long-term recordings from constantly firing cells show that some of these cells can be influenced by stimulation of the habenular nuclei. They respond with an increase of their mean discharge rate which outlasted the 10-min stimulation period by half an hour or more (Fig. 5).

Our microiontophoretic results show that pineal cells which are innervated by central fibres may respond in a similar manner to the application of acetylcholine (Semm et al. 1981a). In these experiments, of the nine cells activated by habenular stimulation, eight units also showed a positive response to this drug. No such relationship between the responses to central stimulation and the effects of melatonin or noradrenaline application were found. Only one cell which was inhibited by central stimulation could be tested with acetylcholine and this unit was inhibited by the neurotransmitter.

In summary, we can distinguish three different categories of cells in the guinea pig pineal organ:

1. Darkness-activated cells, innervated by the sympathetic nervous system
2. Light-activated cells, also innervated by the sympathetic nerves and
3. Constantly firing cells, probably receiving a central innervation from the limbic system

Light-activated cells and constantly firing cells cannot, as yet, be correlated with any known pineal function. As the cells of category 1 and 2 show seasonal differences, they may serve to measure daylength. Both cells together may represent a complex timekeeping device in which continuously oscillating mechanisms play an important role. The constantly firing cells could serve as a reference principle which is related to the amount of activation of the two cell types.

The insensitivity of light-activated cells to light during the night and the insensitivity of darkness-activated cells to darkness during the day poses the question whether these properties are due to intrinsic mechanisms in the pineal (e.g. rhythms in photosensitivity) or are dependent on related brain structures, in particular the suprachiasmatic nucleus.

3.2 The Effects of an Earth-Strength Magnetic Field on the Electrical Activity of Single Pineal Cells

It is now well established that magnetic fields can influence biological systems. From a theoretical point of view the vertebrate pineal organ might be a part of a magnetic sensitive system, on the following grounds:

1. The pineal organ is involved in the regulation or transmission of circadian rhythms and thereby essential for the expression of migratory restlessness ("Zugunruhe") in birds (McMillan 1972). Orientation at that time can be altered by an artificial magnetic field with a direction differing by 90° from that of the earth. Circadian rhythms can be inhibited from phase shifting by compensation of the earth's magnetic field (Bliss and Heppner 1976) and can be influenced by an artificial magnetic field (Brown and Scow 1978).
2. In birds, the pineal is a light-sensitive timekeeping organ (Binkley et al. 1978) and could form part of a combined compass-solar-clock system which has been postulated for maintaining orientation in birds (Walcott 1977).

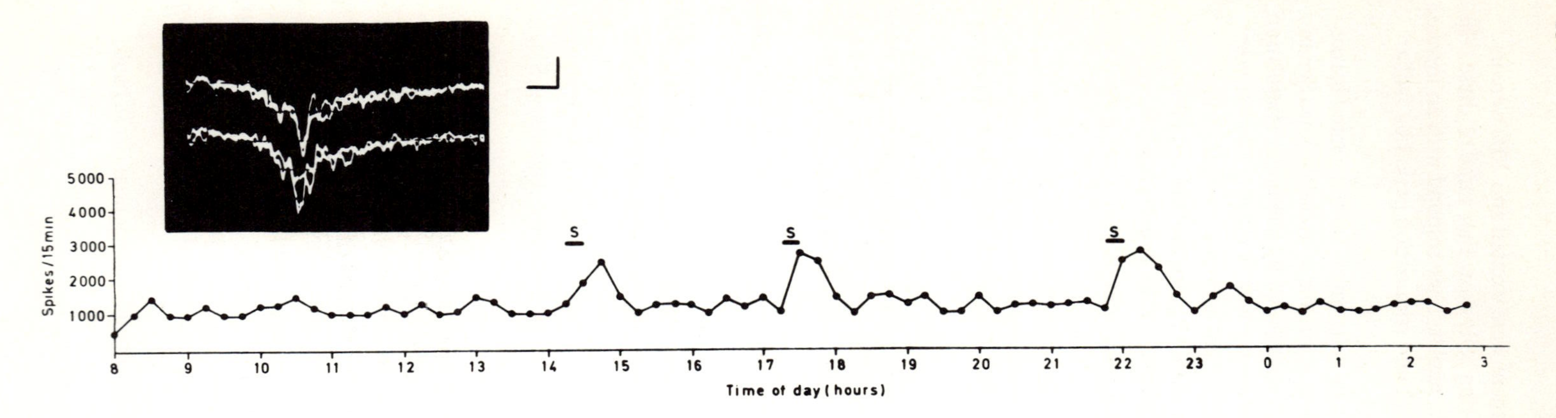

Fig. 5. Long-term recordings from a cell without circadian rhythmicity, which could be excited by electrical stimulation of the lateral habenular nuclei. The cell was recorded in the posterior part of the guinea pig pineal organ and exhibited a latency of 4.0 ms (see the oscilloscope-photograph, calibration 1 ms 2 mV). Following stimulation (S = 60 Hz for about 10 min) during the day and at night the cell showed an increase of electrical activity for ca. 30 min

3. The pineal organ is strongly dependent on its sympathetic innervation and the sympatho-adrenergic system as a whole is sensitive to magnetic stimuli (Sakharova 1977).

To test the above hypothesis, recordings were made from the pineal glands of guinea pigs and homing pigeons. For the generation of the magnetic fields we used Helmholtz coils control to alter the vertical (in guinea pigs) or the vertical and the horizontal component (in pigeons) of the local magnetic field.

3.2.1 Guinea Pigs

From a total of 71 cells 15 cells showed a marked effect to the magnetic stimulus. Activity could be diminished by more than 50% by an induced field and restored when the magnetic field was inverted (Fig. 6). Both the latency of depression and renewed activation was in the range of about 2 min. After turning off the inversed magnetic field, the cells showed their normal spontaneous activity which could be depressed again with a low magnetic stimulus (Fig. 6).

In other brain structures under identical experimental conditions no reactions to these stimuli could be measured (Semm et al. 1980).

3.2.2 Homing Pigeons

As, to our knowledge, nothing is known about effects of magnetic fields on the behaviour of guinea pigs, we turned to homing pigeons which are known to use the earth's magnetic field for directional information especially under total overcast (Walcott 1977). The electrical activity of nearly 30% of pigeon pineal cells could be altered by changing the strength of the local magnetic field and the direction of its vertical and horizontal components. Such cells responded to a rapid and a gradual change in the magnetic field with either excitation or inhibition with a latency in the range of ms. Figure 7 shows an example of a pineal cell, which was influenced by an inversion of the horizontal component of the magnetic field (Semm et al. 1982).

In conclusion, we can say that some pineal cells are sensitive to changes in the local magnetic field; the structural and functional specialities of the phenomenon remain to be established. Some of the magnetic sensitive cells in the guinea pig showed a spontaneous oscillatory activity with a pattern characteristic for that of rhythmical cells. Therefore it might be possible that these cells combine the abilities of time measurement, sensing variations in the natural magnetic field and in lighting conditions, and the synthesis of hormones.

Acknowledgments. Financial support of the Volkswagenwerk-Stiftung is gratefully acknowledged.

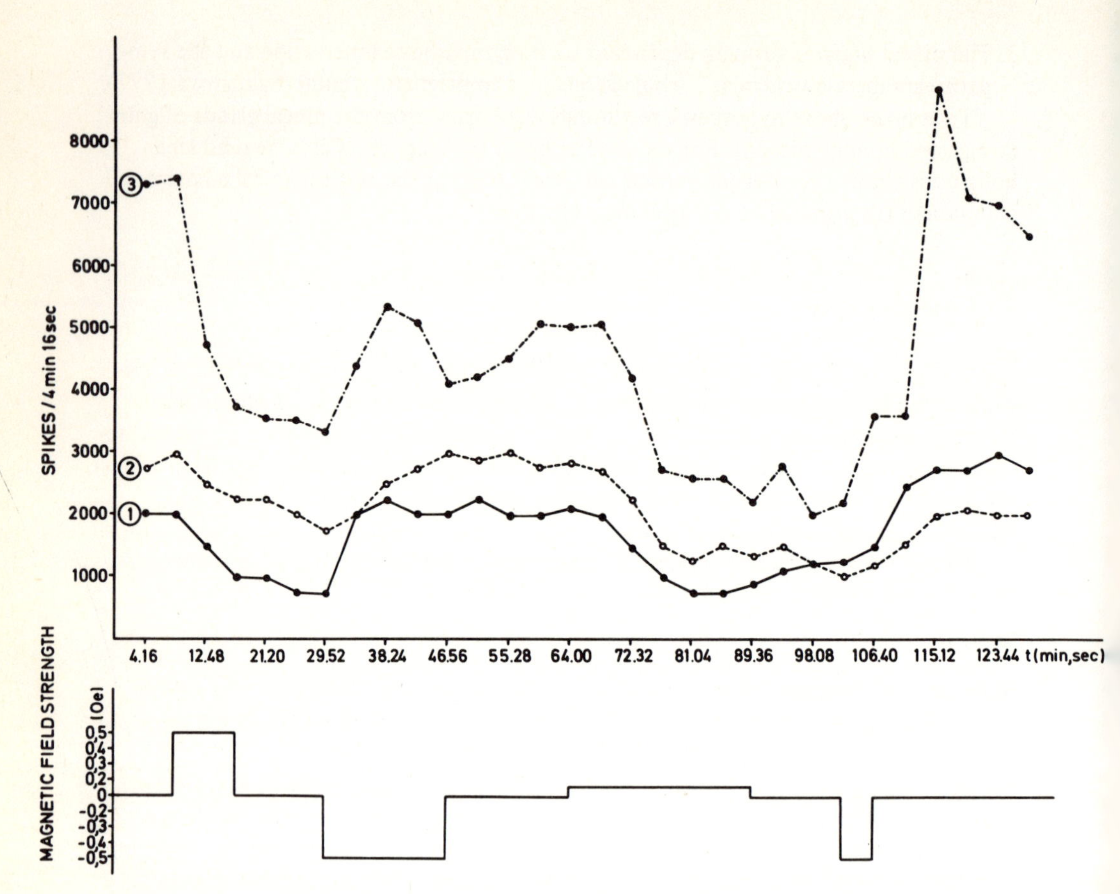

Fig. 6. The spike number of three single pineal cells recorded in the guinea pig. Cell *1* shows a clear depression of activity after a magnetic stimulus of 0.5 Oerstedt (*Oe*) and remains depressed after cessation of the stimulus. After an inverted stimulus of 0.5 Oe (the vertical component of the ambient magnetic field was inverted), cell activity returns to initial level and is depressed again by a stimulus of 0.10 Oe, south pole up. After a period of depression, activity increases again after a short stimulus of 0.5 Oe with inverted polarity of the magnetic field. Cell *2* shows a similar reaction, whereas cell *3* does not reach its initial level after the first application of the inverted magnetic field. After a second inverted magnetic stimulus, cell activity reached output level again

References

Binkley SA, Riebman JB, Reilly KB (1978) The pineal gland: a biological clock in vitro. Science 202: 1198-1201

Bliss VL, Heppner FH (1976) Circadian activity rhythm influenced by near zero magnetic field. Nature (London) 261: 411-412

Brown FA Jr, Scow KM (1978) Magnetic induction of a circadian cycle in hamsters. J Interdiscip Cycle Res 9: 137-145

McMillan JP (1972) Pinealectomy abolishes the circadian rhythm of migratory restlessness. J Comp Physiol 79: 105-112

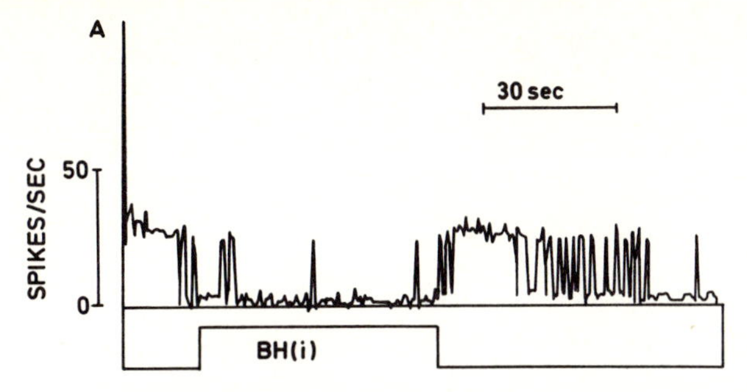

Fig. 7. Post-stimulus time histogram of a pineal cell in the homing pigeon. The spontaneous activity was clearly depressed following a rapid onset of a magnetic stimulus [BH(i) = inverted horizontal component.] Immediately after the stimulus is switched off, activity raised again reaching the original level of spontaneous activity.

Rudeen PK, Reiter RJ (1977) Effect of shortened photoperiods on pineal serotonin N-acetyltransferase activity and rhythmicity. J Interdiscip Cycle Res 8: 47-54

Rudeen PK, Reiter RJ, Vaughan MK (1975) Pineal serotonin N-acetyl-transferase activity in four mammalian species. Neurosci Lett 1: 225-229

Sakharova SA (1977) Reactions of the central and peripheral mediator links of the sympatho-adrenal system to a single exposure to an alternating magnetic field. Biol Nauki 9: 35-39

Semm P (1981) Electrophysiological aspects of the mammalian pineal gland. In: Oksche A, Pévet P (eds) The Pineal Organ. Photobiology – Biochronometry – Endocrinology. Developments in Endocrinology Vol. 14, Amsterdam, Elsevier/North-Holland Biomedical Press, pp 81-96

Semm P, Vollrath L (1979a) Electrophysiology of the guinea-pig pineal organ: sympathetically influenced cells responding differently to light and darkness. Neurosci Lett 12: 93-96

Semm P, Vollrath L (1979b) Electrophysiology of the guinea-pig pineal organ: sympathetic influence and different reactions to light and darkness. In: Kappers JA, Pévet P (eds) The pineal gland of vertebrates including man (Program in brain research, vol 52). Elsevier, Amsterdam, pp 107-111

Semm P, Vollrath L (1980) Electrophysiological evidence for circadian rhythmicity in a mammalian pineal organ. J Neural Transm 57: 181-190

Semm P, Schneider T, Vollrath L (1980) The effects of an earth-strength magnetic field on the electrical activity of pineal cells. Nature (London) 288: 607-608

Semm P, Demaine C, Vollrath L (1981a) Electrical responses of pineal cells to melatonin and putative transmitters: evidence for circadian changes in sensitivity. Exp Brain Res 43: 361-370

Semm P, Demaine C, Vollrath L (1981b) The effects of sex hormones, prolactin and chorionic gonadotropin on pineal cell electrical activity in guinea-pigs. Cell Mol Neurobiol 1 (3): 259-269

Semm P, Demaine C, Vollrath L (1981c) Electrical responses of pineal cells to thyroid hormones and parathormones: a microelectrophoretic study. Neuroendocrinology 33: 212-217

Semm P, Schneider T, Vollrath L (1981d) Morphological and electrophysiological evidence for habenular influence on the guinea-pig pineal organ. J Neural Transm 50: 247-266

Semm P, Schneider T, Vollrath L, Wiltschko W (1982) Magnetic sensitive pineal cells in pigeons. In: Papi F, Wallraff HG (eds) Avian Navigation. Proc. Life Sci., Springer-Verlag Berlin-Heidelberg-New York, pp 329-337

Vollrath L (1979) Comparative morphology of the vertebrate pineal complex. Prog Brain Res 52: 25-38

Vollrath L (1981) The pineal organ. In: Oksche A, Vollrath L (eds) Handbuch der mikroskopischen Anatomie des Menschen, vol. VI. Springer, Berlin Heidelberg New York

Walcott C (1977) Magnetic fields and the orientation of homing pigeons under sun. J Exp Biol 70: 105-123

4.4 Circadian Rhythms of the Isolated Chicken Pineal in Vitro

J.S. Takahashi[1]

1 Introduction

Behavioral experiments with passerine birds and physiological experiments with the pineal of chickens demonstrate that the avian pineal contains one or more circadian oscillators (for review see, Takahashi and Menaker 1979). In house sparrows, the pineal is necessary for the persistence of circadian rhythmicity (Gaston and Menaker 1968); pineal transplants restore circadian rhythmicity in arrhythmic pinealectomized hosts (Zimmerman and Menaker 1975, Menaker and Zimmerman 1976); and, more importantly, pineal transplants determine the phase of the restored rhythm (Zimmerman and Menaker 1979). These results argue that the pineal functions as a pacemaker within the circadian system of the sparrow.

A good deal is known about the physiology and biochemistry of the avian pineal gland. The synthesis of melatonin in the chicken pineal appears to be regulated by the activity of the enzyme, serotonin N-acetyltransferase (Binkley et al. 1973). A dramatic circadian rhythm in the activity of this enzyme persists in vivo for at least two cycles in constant darkness (Binkley and Geller 1975). This rhythm of enzyme activity is in many ways similar to the rhythm of locomotor activity of sparrows. It entrains to light cycles, it free runs in constant darkness, and it damps out in bright constant light. Pinealectomy, which in chickens reduces plasma melatonin to below detectable levels (Pelham 1975, Ralph et al. 1974) abolishes free-running locomotor rhythmicity in sparrows. Both circadian entrainment in sparrows and the inhibition of pineal N-acetyltransferase activity by light in chickens are mediated, in part, by extraretinal photoreceptors (Binkley et al. 1975). The regulation of the rhythm of pineal N-acetyltransferase – which in rats is under the control of norepinephrine from the sympathetic fibers that innervate the pineal (Axelrod and Zatz 1977, Klein 1978) – appears to be different in birds (Binkley 1976). Although it is not clear how the oscillation in N-acetyltransferase activity in the avian pineal is generated, it is consistent with the data to conclude that the controlling mechanism is within rather than external to the pineal.

Recent experiments have established several important properties that are expressed by the isolated pineal gland in organ culture. Four laboratories have documented a rise and fall in N-acetyltransferase activity in isolated pineals for at least 24 h in organ culture (Binkley et al. 1978, Deguchi 1979, Kasal et al. 1979, Wainwright and Wainwright

1 Institute of Neuroscience, Department of Biology, University of Oregon, Eugene, Oregon 97403

Vertebrate Circadian Systems
Ed. by J. Aschoff, S. Daan, and G. Groos

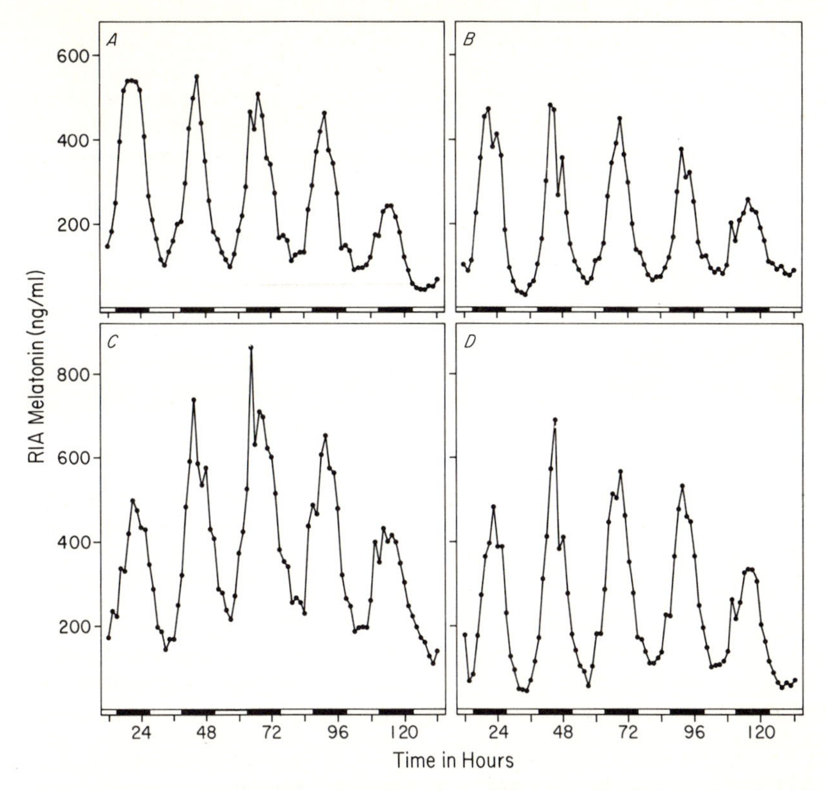

Fig. 1. Rhythms of melatonin release from four different individually isolated chicken pineals cultured in a flow-through superfusion system. The in vitro LD 12 : 12 (L = 350-500 lx cool white fluorescent: D = 0 lx) light cycle is indicated at the bottom of each panel. Samples of perfusate were collected for 90 min. Concentration of melatonin in the culture medium was determined by radioimmunoassay *(RIA)*. Each point represents a single RIA determination and is plotted at the onset of the collection interval. The flow rate in this experiment was 0.25 ml/h. (Takahashi et al. 1980)

1979). Kasal et al. (1979) have shown that the rhythm persists for two cycles in constant darkness with a circadian period in vitro. However, the difficulty in measuring rhythms in populations of isolated glands (unavoidable when N-acetyltransferase activity is used as an endpoint) and the fact that the rhythm appears damped, have precluded a definitive conclusion that the avian pineal is a self-sustained oscillator.

2 Circadian Rhythms of Melatonin Release in Cultured Pineals

We have developed a new flow-through pineal organ culture method which allows us to measure the release of melatonin continuously from individually isolated pineal glands (Takahashi et al. 1980). Chicken pineal glands maintained in the flow-through culture system release melatonin into the perfusate rhythmically. Figure 1 shows the concentration of melatonin in pineal perfusates collected continuously over a period of 5 days

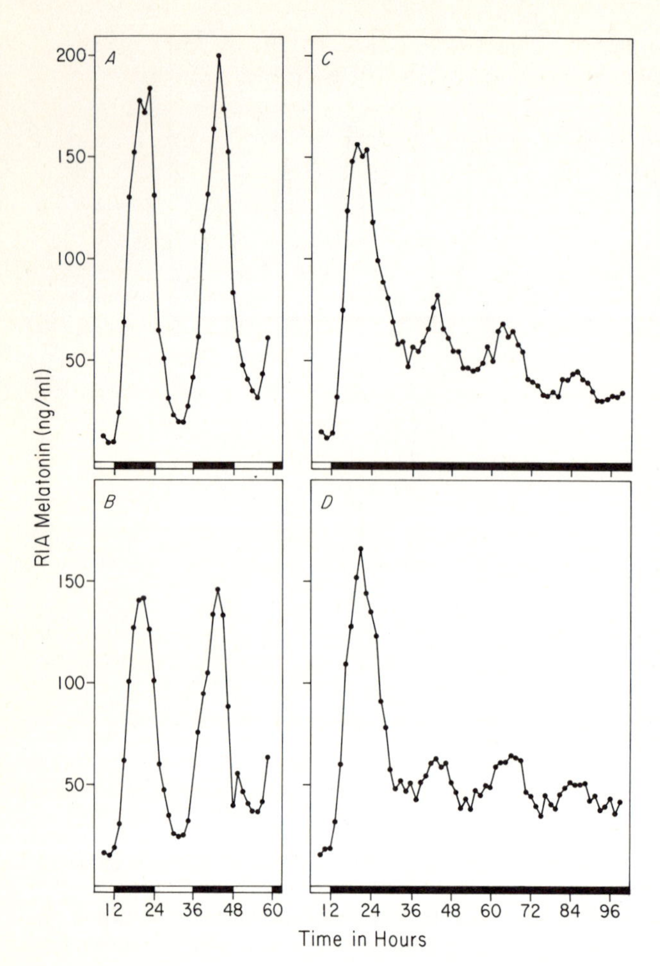

Fig. 2. Comparison of the rhythms of melatonin release in a light cycle and in constant conditions. The light treatments are indicated at the bottom of each panel. **A** and **B** are two individual pineals from a 60-h LD 12 : 12 (L = 350-500 lx cool white fluorescent: D = 1-5 lx red light) experiment. **C** and **D** are two individual pineals from a 96-h experiment in constant conditions (dim red illumination, 1-5 lx). Pineals in both experiments came from the same group of chickens. The flow rate was 0.5 ml/h. The flow rate in these experiments was twice as high as that in the experiments shown in Fig. 1, which accounts for most of the difference in the melatonin concentration in the perfusate. Collection interval and points plotted are as in Fig. 1. (Takahashi et al. 1980)

from four individually isolated pineal glands. In this experiment, an LD 12 : 12 light cycle was continued in vitro with the same phase as the light cycle the chickens had seen previously. Melatonin concentration in the superfusate reaches peak levels during the dark and minimal levels during the light (3- to 15-fold increase at night). While there are quantitative differences in the amplitude of the rhythms among glands, the general pattern of the rhythm is remarkably stable and uniform. In almost every case the fall in melatonin concentration precedes the lights-on transition and the rise in melatonin levels begins before lights off. Thus the rhythm does not appear to be entirely driven by the light cycle but rather shows anticipatory behavior which may reflect the entrainment of an oscillator. There is very little damping of the rhythm until the fourth or fifth peak. The phase of the rhythm is similar to that found in vivo in the pineal (Binkley et al. 1973).

To investigate the endogenous nature of the pineal rhythm, melatonin production was examined under constant dim red light (1-5 lx). Figure 2 shows the results of two different experiments, both performed under identical culture conditions using pineals from the same group of chickens. In a 60-h experiment in LD 12 : 12, there is a high amplitude melatonin rhythm on both days (Fig. 2A, B). In contrast, in the absence of a light-dark cycle, there is one high amplitude peak in melatonin concentration, followed by three lower amplitude peaks at about 24-h intervals (Fig. 2C, D). In constant conditions there is an oscillation in pineal melatonin which persists for at least four cycles; however, the amplitude of the rhythm is low compared to the rhythm present under light-dark conditions. The reduction in amplitude is probably not due to unfavorable culture conditions since the amplitude does not decrease when a light cycle is present.

We have also investigated the effects of light upon melatonin release at night. In three experiments we exposed chicken pineals to 1 or 2 h of light at the peak of the melatonin rhythm on the first night. Light rapidly inhibited the release of melatonin from organ-cultured pineals. The halving time of melatonin concentration in the pineal perfusate was 20 to 30 min (1.0 ml/h flow rate). The rapid inhibition of melatonin by light in our culture system is very similar to the time course of inhibition by light of N-acetyltransferase in static organ culture reported by Deguchi (1979).

3 Effects of Light on the Pineal Melatonin Rhythm

The pineal has been shown to be photoreceptive in culture. Exposure of the glands to bright constant light in vitro reduces the rise in N-acetyltransferase activity (Binkley et al. 1978, Deguchi 1979, Wainwright and Wainwright 1979, 1980). In addition, acute light exposure at night when enzyme activity is elevated rapidly inhibits N-acetyltransferase activity (Deguchi 1979, Wainwright and Wainwright 1980). The results reported in the previous section show that rhythmic input of light maintains the amplitude of the daily pineal melatonin rhythm and prevents the damping of the rhythm observed in constant conditions. Taken together, these data demonstrate that light inhibits the production of melatonin, and suggest that it synchronizes the pineal oscillation. Although the chicken pineal is clearly photoreceptive, there is little direct evidence that the pineal oscillation can be phase-shifted or entrained by light.

To explore entrainment of the pineal oscillation by light, we asked whether the phase of the melatonin rhythm could be influenced when we shifted the phase of the in vitro light cycle. Four pineals were exposed to an LD 12 : 12 cycle that was delayed 6 h relative to the light cycle the chickens had seen previously; and four pineals were exposed to an LD 12 : 12 cycle that was advanced 6 h. The phases of the melatonin rhythms were clearly shifted by the light cycles. By the second or third cycle, the phases of the oscillation were strongly correlated with the phases of the shifted light cycles and were not correlated with the phase of previous light cycle. The results from this experiment show that the pineal melatonin rhythm can be synchronized by light cycles. As long as the light cycle is present, the phase of the melatonin rhythm is determined by the phase of the light cycle.

4 Discussion

We have shown that robust rhythms of melatonin release can be measured continuously from individual pineal glands in a flow-through culture system. Using this method it is feasible to assay pineal rhythms of melatonin in vitro for long periods of time and with high resolution in the time domain. In constant conditions the rhythm persists in culture for at least four cycles with a circadian period; however, the rhythm appears to be heavily damped. It does not seem likely that damping is due to poor culture conditions because in light-dark cycles the rhythm is not damped and persists with high amplitude.

Whatever mechanism underlies the damping in constant conditions of the chicken pineal melatonin rhythm in vitro, it is clear that rhythmic light input maintains the amplitude and synchronizes the rhythm. Shifting the phase of the entraining cycle produces a similar shift in the phase of the melatonin rhythm. In addition to the synchronizing effects of light, acute light exposure at night, when melatonin levels are high, rapidly inhibits melatonin release. Thus, light acts on the isolated pineal in two ways: as a synchronizing agent and as an inhibitor of melatonin production.

Acknowledgments. I thank Michael Menaker and Heidi Hamm for their contributions to this research; and C. Norris and S. Wisner for technical assistance. Research was supported by NIH grants HD-03803 and HD-07727 to Michael Menaker; and a National Science Foundation graduate fellowship and National Institute of General Medical Sciences National Research Award 5T32 GM 07257 to J. Takahashi.
aical Sciences National Research Award 5T32 GM 07257 to J. Takahashi.

References

Axelrod J, Zatz M (1977) The β-adrenergic receptor and the regulation of circadian rhythms in the pineal gland. In: Litwack G (ed) Biochemical actions of hormones, vol IV. Academic Press, London New York, pp 249-268

Binkley S (1976) Comparative biochemistry of the pineal gland of birds and mammals. Am Zool 16: 57-65

Binkley S, Geller EB (1975) Pineal N-acetyltransferase in chickens: Rhythm persists in constant darkness. J Comp Physiol 99: 67-70

Binkley S, MacBride SE, Klein DC, Ralph C (1973) Pineal enzymes: Regulation of avian melatonin synthesis. Science 181: 273-275

Binkley S, MacBride SE, Klein DC, Ralph C (1975) Regulation of pineal rhythms in chicks: Refractory period and nonvisual light perception. Endocrinology 96: 848-853

Binkley SA, Reibman JB, Reilly KB (1978) The pineal gland: A biological clock in vitro. Science 202: 1198-1201

Deguchi T (1979) Circadian rhythm of serotonin N-acetyltransferase activity in organ culture of chicken pineal gland. Science 203: 1245-1247

Gaston S, Menaker M (1968) Pineal function: The biological clock in the sparrow? Science 160: 1125-1127

Kasal C, Menaker M, Perez-Polo R (1979) Circadian clock in culture: N-acetyltransferase activity of chick pineal glands oscillates in vitro. Science 203: 656-658

Klein DC (1978) The pineal gland: A model of neuroendocrine regulation. In: Reichlin S, Baldessarini R, Martin JB (eds) The hypothalamus. Raven, New York, pp 303-327

Menaker M, Zimmerman N (1976) Role of the pineal in the circadian system of birds. Am Zool 16: 45-55

Pelham RW (1975) A serum melatonin rhythm in chickens and its abolition by pinealectomy. Endocrinology 96: 543-546

Ralph CL, Pelham RW, MacBride SE, Reilly DP (1974) Persistent rhythms of pineal and serum melatonin in cockerels in continuous darkness. J Endocrinol 63: 319-324

Takahashi JS, Menaker M (1979) Physiology of avian circadian pacemakers. Fed Proc 38: 2583-2588

Takahashi JS, Hamm H, Menaker M (1980) Circadian rhythms of melatonin release from individual superfused chicken pineal glands in vitro. Proc Natl Acad Sci USA 77: 2319-2322

Wainwright SD, Wainwright LK (1979) Chick pineal serotonin acetyltransferase: A diurnal cycle maintained in vitro and its regulation by light. Can J Biochem 57: 700-709

Wainwright SD, Wainwright LK (1980) Regulation of the cycle in chick pineal serotonin N-acetyltransferase activity in vitro by light. J Neurochem 35: 451-457

Zimmerman NH, Menaker M (1975) Neural connections of sparrow pineal: Role in circadian control of activity. Science 190: 477-479

Zimmerman NH, Menaker M (1979) The pineal: A pacemaker within the circadian system of the house sparrow. Proc Natl Acad Sci USA 76: 999-1003

4.5 Endogenous Oscillator and Photoreceptor for Serotonin N-Acetyltransferase Rhythm in Chicken Pineal Gland

T. Deguchi[1]

1 Introduction

The pineal gland of various species of vertebrates synthesizes a specific indoleamine, melatonin, in a circadian fashion with a high production during nighttime and a low content during daytime. This circadian rhythm of melatonin synthesis is regulated by the change in the activity of serotonin N-acetyltransferase that converts serotonin to N-acetylserotonin (Weissbach et al. 1961). The day-night changes of N-acetyltransferase activity in the pineal gland persist in blinded rats (Deguchi 1975) and in blinded chickens (Binkley and Geller 1975), indicating that the enzyme rhythm is generated by an endogenous circadian oscillator. The neural mechanism that controls the circadian rhythm of N-acetyltransferase activity has been well established in rat pineal gland (Klein and Weller 1970, Klein et al. 1971, Deguchi and Axelrod 1972a, Moore and Klein 1974). The neural impulse originating in the suprachiasmatic nuclei in hypothalamus is transmitted through superior cervical ganglion to the pineal gland. The nerve endings in the pineal gland release more transmitter norepinephrine at night compared to daytime, which increases the intracellular cyclic AMP levels and stimulates the synthesis of N-acetyltransferase molecules in pinealocytes. A similar mechanism has been recently proposed for the pineal glands of other mammals.

In contrast to rat pineals, the N-acetyltransferase rhythm in chicken pineals was not abolished by superior cervical ganglionectomy (Ralph et al. 1975) or by treatment with adrenergic blocking agents (Deguchi 1978). Injection of catecholamines into animals or addition of catecholamines to culture medium markedly elevated N-acetyltransferase activity in rat pineals (Deguchi and Axelrod 1972b, Klein and Weller 1973), but not in chicken pineal glands (Binkley et al. 1975, Deguchi 1978). Binkley et al. (1975) reported that the environmental illumination prevented the nighttime increase of N-acetyltransferase activity in blinded chickens indicating the presence of a nonretinal photoreceptor. Recently we and other laboratories have demonstrated that chicken pineal gland contained an endogenous circadian oscillator and a photoreceptor, and that the N-acetyltransferase rhythm persisted in the isolated chicken pineals (Binkley et al. 1978, Deguchi 1979a, Kasal et al. 1979, Wainwright and Wainwright 1979).

1 Department of Medical Chemistry, Tokyo Metropolitan Institute for Neurosciences, 2-6 Musashidai, Fuchu-city, Tokyo 183 Japan

Vertebrate Circadian Systems
Ed. by J. Aschoff, S. Daan, and G. Groos

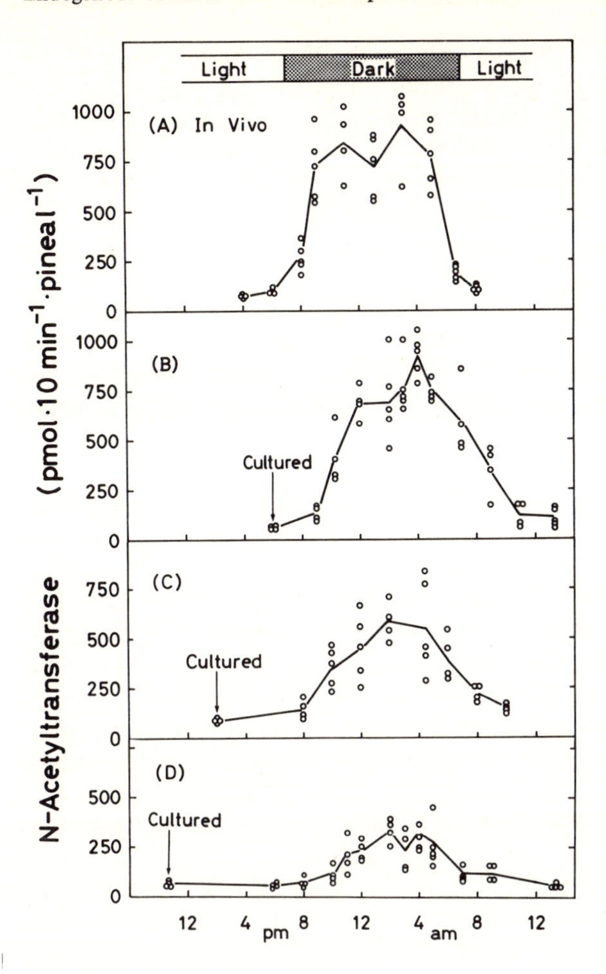

Fig. 1. Circadian change of pineal N-acetyltransferase activity in vivo **(A)** and in organ culture **(B-D)** of chicken pineal gland. Each *point* represents the enzyme activity of individual pineal. (Deguchi 1979a)

2 Rhythmic Change of N-Acetyltransferase Activity in Organ-Culture of Chicken Pineal Gland (Deguchi 1979a)

In 10- to 14-day-old chickens raised in a LD cycle (12 : 12 h), pineal N-acetyltransferase activity was low during daytime and increased tenfold at night in darkness (Fig. 1A). When chickens were killed at 1800 h and their pineals were organ-cultured in darkness, N-acetyltransferase activity increased 7- to 10-fold at midnight (Fig. 1B). The activity then declined to the daytime level on the next morning at 1000 h. The maximum level of enzyme activity was the same as that of in vivo rhythm. When chickens were killed at 1400 h or at 1100 h and their pineals were cultured in darkness, N-acetyltransferase activity remained at a low level during daytime and increased at 2100 or 2200 h, reaching a maximum level at midnight. The enzyme activity then declined to the initial level on the next morning (Fig. 1C, D). Although the amplitude of the N-acetyltransferase rhythm was somehow decreased during a long period of culture, the pattern of the increase and the decrease of the N-acetyltransferase activity was comparable to the rhythmic change of enzyme activity in vivo (Fig. 1 A). The observation indicates that the N-acetyltransferase rhythm in isolated chicken pineals is regulated by the lighting schedule under which they had been maintained before being killed, and that chicken pineal gland contains an endogenous circadian oscillator for the N-acetyltransferase rhythm. The N-acetyltransferase rhythm in organ culture was blocked by cycloheximide, a protein synthesis inhibitor, or by actinomycin D, an inhibitor of m-RNA synthesis (Deguchi 1979b).

When chicken pineals were organ-cultured in a LD cycle (12 : 12 h), N-acetyltransferase activity continued to change in oscillation with the environmental lighting schedule for at least 3 days examined. However, the daytime as well as the nighttime levels of enzyme activity gradually decreased on 2 and 3 days of culture. This decline of enzyme activity was presumably due to a necrosis of the glands.

3 Light Suppression of N-Acetyltransferase Rhythm in Isolated Chicken Pineal Gland (Deguchi 1979a)

It has previously been shown that when rats were exposed to light at midnight, N-acetyltransferase activity rapidly declined to less than 10% of the initial level within 10 min (Deguchi and Axelrod 1972a, Klein and Weller 1972). A comparable light-suppression was observed with N-acetyltransferase activity in chicken pineal gland. When chickens were exposed to continuous illumination until midnight, the nighttime increase of N-acetyltransferase activity was suppressed with a only 2-2.5-fold elevation at midnight (Fig. 2A). When chickens were illuminated at midnight after N-acetyltransferase activity had reached a maximum level, the enzyme activity decreased to

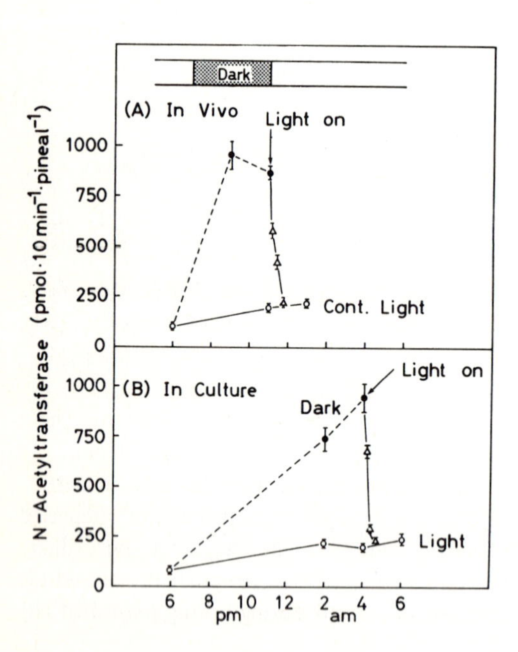

Fig. 2. Effect of illumination on the nighttime increase of N-acetyltransferase activity **(A)** and in organ culture of chicken pineal gland **(B)**. There were five samples in each group. *Vertical bars* indicate the standard errors of the means. (Deguchi 1979a)

one-quarter within 20-30 min. This decrease of enzyme activity in chicken pineals was slower than that of rat pineals. A similar light-suppression of N-acetyltransferase rhythm was observed in cultured chicken pineals (Fig. 2B). When chicken pineals were cultured in darkness, there was a 7- to 12-fold increase of enzyme activity at midnight. In the chicken pineals cultured under continuous illumination, N-acetyltransferase activity was suppressed, resulting in a 2-2.5-fold elevation at midnight. When chicken pineals were cultured in darkness until midnight and then exposed to illumination, the high level of N-acetyltransferase activity decreased to one-quarter within 20-30 min. The observation clearly indicates that chicken pineal gland contains a photoreceptor and transduces its stimulation by light to the synthesis and degradation of N-acetyltransferase molecules.

4 Rhodopsin-Like Photosensitivity of Chicken Pineal Gland (Deguchi 1981)

To study the nature of photosensitive substance, chicken pineals were cultured under various intensities of wavelengths from 400 nm to 800 nm. When pineals were cultured from 1900 to 0300 h in darkness, N-acetyltransferase activity increased from 75-125 pmol to 700-1300 pmol per 10 min per pineal. When cultured under 70 $\mu W/cm^2$ of white light, enzyme activity increased only to 150-250 pmol per 10 min per pineal. A wavelength of 500 nm was most effective in inhibiting the N-acetyltransferase rhythm. There was a significant inhibition with the intensity of 0.02 $\mu W/cm^2$ and a maximum inhibition at 0.3 $\mu W/cm^2$, which was equivalent to that caused by 70 $\mu W/cm^2$ of white light (Fig. 3A). With the wavelength of 400 nm, the maximum inhibition occurred at 1.0 $\mu W/cm^2$, while 600 nm was far less effective, showing no signif-

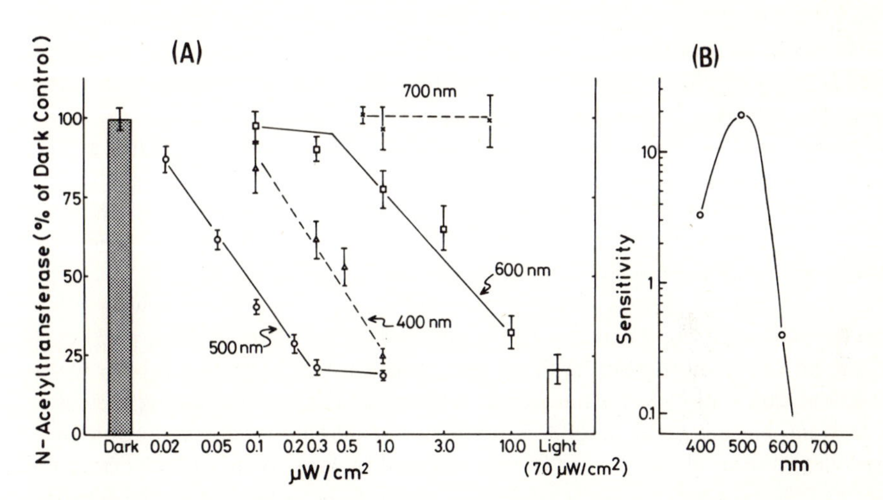

Fig. 3. Effect of various intensities of monochromatic light on the nighttime increase of N-acetyltransferase activity in organ culture of chicken pineal gland. Pineals were cultured from 1900 h to 0100 h under the intensities of monochromatic light indicated. N-Acetyltransferase activity is shown as percent of dark control. *Points* and *bars* indicate means and standard errors of the means of five samples. (Deguchi 1981)

icant inhibition with 0.3 μW/cm^2 and a submaximum inhibition with 10 μW/cm^2. The wavelength of 700 nm showed no effect on enzyme activity with 7 μW/cm^2 or less. The reciprocals of the light intensities that produce 50% inhibition of the nighttime increase of N-acetyltransferase activity were plotted as sensitivities on the ordinate in a logarithmic scale (Fig. 3B). The sensitivity curve resembled the absoprtion spectrum of rhodopsin (Wald and Brown 1956). The threshold light intensity of the chicken pineal is in the range of the threshold intensities of several photosensitive systems. It is likely that chicken pineal gland contains a rhodopsin-like photoreceptor.

5 Circadian Rhythm of N-Acetyltransferase Activity in Dispersed Cell Culture of Chicken Pineal (Deguchi 1979c)

The above observations suggest that chicken pineal gland contains a circadian oscillator, a photoreceptor and melatonin-synthesizing machinery. A central question is whether the circadian oscillation of N-acetyltransferase activity and its response to illumination are generated within the cell or are emergent properties of interaction between different types of pineal cells. To study this point, chicken pineals were dissociated by treatment with collagenase and dispersed cells were cultured as a monolayer. When dispersed cells were cultured in a LD cycle (12 : 12 h), N-acetyltransferase activity was 5- to 10-fold higher during the dark periods than in the light periods on days 2 and 3 of culture (Fig. 4A). The rhythmic change of enzyme activity continued to oscillate in phase with the LD cycle for at least 17 days of cell culture. The specific enzyme activity per mg protein in the cell culture was comparable to that of the in vivo rhythm.

To determine whether one type of pinealocytes released a substance that stimulated the synthesis of N-acetyltransferase molecules in another type of pinealocytes, conditioned medium was taken from 3-day cell culture during daytime and at night in darkness, and was added to another cell culture. The contitioned medium neither stimulated nor inhibited N-acetyltransferase activity in other cells. Taking these observations together, it is concluded that the photoreceptor, circadian oscillator and melatonin-synthesizing machinery reside in the same cell of chicken pineal gland.

The rhythmic change of N-acetyltransferase activity was not due to a synchronized cell cycle, because cytosine-1-β-D-arabinofuranoside (1 X 10^{-5}M) that inhibits cell division did not influence the N-acetyltransferase rhythm (Fig. 4B).

To examine the endogenous nature of the rhythmicity, dispersed cells were cultured under vaious lighting schedules. In Fig. 4C, cells were cultured under a LD cycle for 2 days and on the third day the lights were turned off at 1300 h, 6 h prior to the onset of usual darkness. N-Acetyltransferase activity did not increase during the subjective daytime, but increased at 2200 h, reaching a maximum level at midnight. In the experiment shown in Fig. 4D, cells were cultured in continuous darkness from the

→

Fig. 4. Circadian rhythm of N-acetyltransferase activity in the dispersed cell culture of chicken pineal glands under various lighting schedules. Each *point* represents the specific enzyme activity of an individual dish. *Ara-C* indicates the addition of cytosine-1-β-D-arabinofuranoside (1 X 10^{-5}M) to the medium. (Deguchi 1979c)

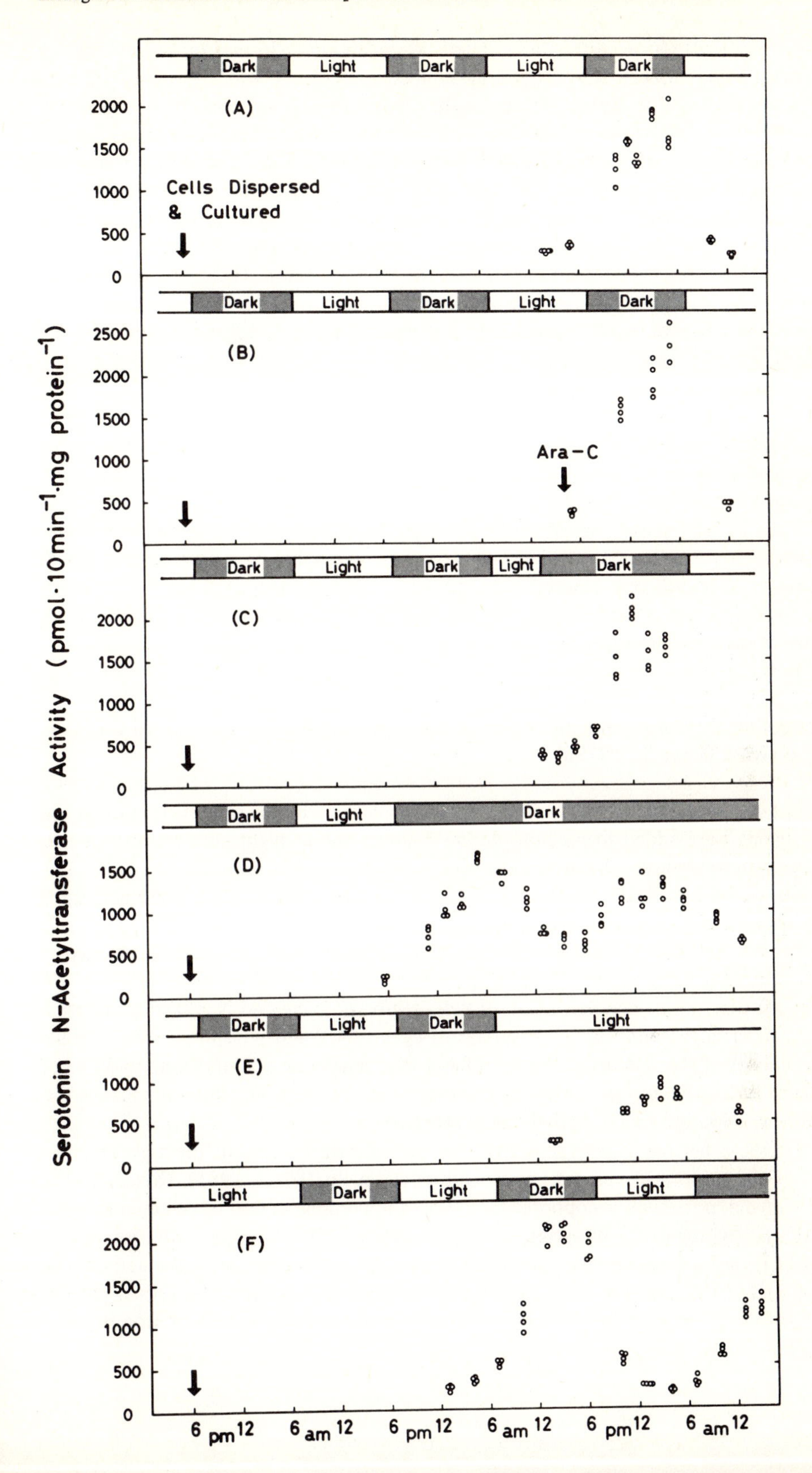
Dark
Light
Dark
Light
Dark
(A)
Cells Dispersed
& Cultured
(B)
Ara-C
(C)
(D)
(E)
(F)
Serotonin N-Acetyltransferase Activity (pmol·10min⁻¹·mg protein⁻¹)
6 pm 12
6 am 12

second day for two successive days. There were rhythmic changes of at least two cycles with a high enzyme activity during subjective nighttime and a low activity during subjective daytime. However, the day-night difference was damped out on the third day of culture: the daytime level being higher and the nighttime value being lower than those of the cells cultured in a LD cycle (Fig. 4A). This is presumably due to a desynchronization of the rhythms of individual cells. When cells were exposed to continuous illumination in the third night, the increase of N-acetyltransferase activity was suppressed (Fig. 4E). There was, however, a threefold elevation of enzyme activity during the subjective nighttime compared to the daytime value. These observations indicate that the rhythmic change of N-acetyltransferase activity in the cell culture is not induced by environmental lighting, but is generated by an endogenous oscillator in the cells.

To study whether environmental lighting can entrain N-acetyltransferase rhythm in the cells, dispersed cells were cultured in the LD cycle inverted to the lighting schedule in which the chickens had been raised. The rhythmic change of enzyme activity was inverted to be in phase with the reversed lighting schedule where the cells were cultured (Fig. 4F).

Thus environmental lighting not only suppresses the nighttime increase of N-acetyltransferase activity, but also entrains the enzyme rhythm to a new lighting schedule. Since dispersed cells responded in the same manner as organ culture or as in vivo, it is conceivable that a rhodopsin-like photoreceptor is incorporated in the cell membrane of chicken pinealocytes.

6 Effect of Various Agents on N-Acetyltransferase Rhythm in the Isolated Chicken Pineal Gland (Deguchi 1979b)

To study the membrane mechanism of the circadian oscillator of chicken pineals, various agents were added to medium during daytime and at night, and N-acetyltransferase rhythm was assayed. Cholera toxin that activated adenylate cyclase and increased intracellular cyclic AMP in a variety of tissues elevated N-acetyltransferase activity during daytime and at night under illumination. Derivatives of cyclic AMP and phosphodiesterase inhibitors, theophylline or Ro 20-1724, also caused an increase of enzyme activity during daytime and at night under illumination. A high concentration of KCl (55 mM) that causes depolarization of cell membrane increased N-acetyltransferase activity under both conditions. Derivatives of cyclic AMP and phosphodiesterase inhibitors enhanced the nocturnal rhythm of N-acetyltransferase activity in an additive or more-than-additive manner, whereas cholera toxin or a high KCl did not enhance the nighttime elevation of the activity in darkness.

On the other hand, several agents have been shown to prevent the nighttime elevation of N-acetyltransferase activity in culture (Deguchi 1979b, 1981). Nonactin or valinomycin, potassium ionophores that cause a hyperpolarization of cell membranes. prevented the nighttime elevation of enzyme activity. Cocaine, that prevents depolarization of neural cell membrane, also blocked the nighttime increase of the activity. It has been well established that light hyperpolarizes membrane potential in retinal photoreceptor cells of vertebrates (Tomita 1970) and in photoreceptor cells of frog pineal gland (Morita 1975). It is conceivable that the light-induced suppression of N-acetyltransferase rhythm in the isolated chicken pineal gland could be due to a hyperpolarization of pinealocyte membrane.

Thus depolarization of cell membrane of chicken pinealocytes increased N-acetyltransferase activity during daytime and at night under illumination, whereas hyperpolarization of membrane blocked the nighttime elevation of the activity in darkness. It is possible that depolarization and hyperpolarization of membrane are somehow involved in the generation of the circadian rhythmicity of N-acetyltransferase activity in the isolated chicken pineal gland.

7 Conclusion

There are controversial reports as to whether or not avian pineal gland is an essential part of a biological clock. Menaker and his associates (1978) have shown that the circadian rhythm of locomotor activity of house sparrows was abolished by pinealectomy and restored by transplantation of pineal gland in the anterior chamber of eyes. In other birds including chickens, however, pinealectomy has been reported to have no effect on the activity rhythm. It is also uncertain whether the avian pineal gland retains a photoreceptive capability and whether the gland is a photosensor involved in the photoperiodic control of reproductive systems and circadian rhythms in birds. The present study, however, has clearly indicated that chicken pineal gland contains an endogenous oscillator and a rhodopsin-like photoreceptor that control the circadian rhythm of the activity of serotonin N-acetyltransferase, the key enzyme of the melatonin synthesis in the pineal gland. It has also been shown that injection of melatonin modified or entrained the circadian rhythm of locomotor activity in birds (Gwinner and Benzinger 1978, Hendel and Turek 1978). Taking these observations together, it is indicated that the circadian oscillator and the photoreceptor in avian pineal gland would play an essential role in the regulation of the circadian rhythmicity of activity and other physiological conditions in birds.

Acknowledgments. This research was supported in part by Grant-in-Aid for Special Project Research on Animal Behaviors and by Grant-in-Aid for Scientific Research (No. 548121) from the Ministry of Education of Japan.

References

Binkley S, Geller EB (1975) Pineal N-acetyltransferase in chickens: Rhythm persists in constant darkness. J Comp Physiol 99: 67-70

Binkley S, MacBride SE, Klein DC, Ralph CL (1975) Regulation of pineal rhythms in chickens: Refractory period and nonvisual light perception. Endocrinology 96: 848-853

Binkley S, Riebman JB, Reilly KB (1978) The pineal gland: A biological clock in vitro. Science 202: 1198-1201

Deguchi T (1975) Shift of circadian rhythm of serotonin: acetyl coenzyme A N-acetyltransferase activity in pineal gland of rat in continuous darkness or in the blinded rat. J Neurochem 25: 91-93

Deguchi T (1978) Control of circadian rhythm of serotonin N-acetyltransferase activity in pineal gland of chicken. In: Ito M, Tsukahara N, Kubota K, Yagi K (eds) Integrative control functions of the brain, vol I. Kodansha, Tokyo/Elsevier, Amsterdam, pp 345-347

Deguchi T (1979a) Circadian rhythm of serotonin N-acetyltransferase activity in organ culture of chicken pineal gland. Science 203: 1245-1247

Deguchi T (1979b) Role of adenosine 3', 5'-monophosphate in the regulation of circadian oscillation of serotonin N-acetyltransferase activity in cultured chicken pineal gland. J Neurochem 33: 45-51

Deguchi T (1979c) A circadian oscillator in cultured cells of chicken pineal gland. Nature (London) 282: 94-96

Deguchi T (1981) Rhodopsin-like photosensitivity of isolated chicken pineal gland. Nature (London) 290: 702-704

Deguchi T, Axelrod J (1972a) Control of circadian change of serotonin N-acetyltransferase activity in the pineal organ by the β-adrenergic receptor. Proc Natl Acad Sci USA 69: 2547-2550

Deguchi T, Axelrod J (1972b) Induction and superinduction of serotonin N-acetyltransferase by adrenergic drugs and denervation in rat pineal organ. Proc Natl Acad Sci USA 69: 2208-2211

Gwinner E, Benzinger I (1978) Synchronization of a circadian rhythm in pinealectomized European Starlings by daily injection of melatonin. J Comp Physiol 127: 209-213

Hendel RC, Turek FW (1978) Suppression of locomotor activity in sparrows by treatment with melatonin. Physiol Behav 21: 275-278

Kasal CA, Menaker M, Perez-Polo JR (1979) Circadian clock in culture: N-Acetyltransferase activity of chicken pineal glands oscillates in vitro. Science 203: 656-658

Klein DC, Weller JL (1970) Indole metabolism in the pineal gland: A circadian rhythm in N-acetyltransferase. Science 169: 1093-1095

Klein DC, Weller JL (1972) Rapid light-induced decrease in pineal serotonin N-acetyltransferase activity. Science 177: 532-533

Klein DC, Weller JL (1973) Adrenergic-adenosine 3', 5'-monophosphate regulation of serotonin N-acetyltransferase activity and the temporal relationship of serotonin N-acetyltransferase activity to synthesis of ^{3}H-N-acetylserotonin and ^{3}H-melatonin in the cultured rat pineal gland. J Pharmacol Exp Ther 186: 516-527

Klein DC, Weller JL, Moore RY (1971) Melatonin metabolism: Neural regulation of pineal serotonin: acetyl coenzyme A N-acetyltransferase activity. Proc Natl Acad Sci USA 68: 3107-3110

Menaker M, Takahashi JS, Eskin A (1978) The physiology of circadian pacemakers. Annu Rev Physiol 40: 501-526

Moore RY, Klein DC,(1974) Visual pathway and the central neural control of a circadian rhythm in pineal serotonin N-acetyltransferase activity. Brain Res 71: 17-33

Morita Y (1975) Direct photosensory activity of the pineal. In: Knigge KM, Scott E, Kobayashi H, Ishii S (eds) Brain-endocrine interaction. Karger, Basel, pp 346-387

Ralph CL, Binkley S, MacBride SE, Klein DC (1975) Regulation of pineal rhythms in chickens: Effects of blinding, constant light, constant dark, and superior cervical ganglionectomy. Endocrinology 97: 1373-1378

Tomita T (1970) Electrical activity of vertebrate photoreceptor. Q Rev Biophys 3: 179-222

Wainwright SD, Wainwright LK (1979) Chicken pineal serotonin N-acetyltransferase: A diurnal cycle maintained in vitro and its regulation by light. Can J Biochem 57: 700-709

Wald G, Brown PK (1956) Synthesis and bleaching of rhodopsin. Nature (London) 177: 174-177

Weissbach H, Redfield BG, and Axelrod J (1961) The enzymic acetylation of serotonin and other naturally occurring amines. Biochim Biophys Acta 54: 190-191

5.1 Role of Hormones in the Circadian Organization of Vertebrates

F.W. Turek and E. Gwinner[1]

1 Introduction

In searching for mammalian circadian pacemakers, Richter (1965) examined the possibility that components of the rhythmic-generating processes may reside in the endocrine system. However, after observing no effect on circadian rhythmicity following the removal of the gonads, pituitary, pancreas, pineal or adrenal glands of the rat, he concluded that the driving circadian oscillator(s) must be outside the endocrine system. Indeed, most attempts to localize circadian clocks in mammals have focused on the nervous system (see Part 3, this Vol.). However, there are now a number of reports which suggest that various endocrine secretions can at least modulate circadian rhythms (for reviews see Rusak and Zucker 1979, Zucker 1979). Therefore, any attempt to establish a coherent picture of the temporal organization of vertebrates must take into account the way in which the nervous and endocrine systems interact in generating and/or modulating circadian rhythmicity.

The endocrine organs that have received most attention with respect to the role of hormones in the circadian organization of vertebrates are the pineal gland and the gonads. Since a number of chapters in this volume (see part 4) are devoted to the role of the pineal gland in the circadian system, we will not attempt to review that literature here. In this chapter we will review and discuss the evidence which suggests that gonadal hormones influence central circadian pacemakers. It should be noted that a number of other hormones, particularly those from the adrenal and thyroid glands, have been observed to influence the phase, amplitude, or sensitivity to various stimuli, of other circadian rhythms (Dunn et al. 1975, Ooka-Souda et al. 1979, Bellinger et al. 1979). However, for the most part these effects are believed to be due to the action of hormones on peripheral systems and there is little evidence that hormones from outside the pineal gland or the gonads alter the circadian pacemaker itself.

1 Department of Neurobiology and Physiology, Northwestern University, Evanston, Illinois 60201 USA and Max-Planck-Institut für Verhaltensphysiologie, Vogelwarte Radolfzell, Radolfzell and Andechs, D-8138 Andechs

Vertebrate Circadian Systems
Ed. by J. Aschoff, S. Daan, and G. Groos

2 Effect of Gonadal Hormones on Circadian Rhythmicity

2.1 Females

Evidence that ovarian hormones are involved in the regulation of circadian rhythmicity has been obtained almost exclusively from studies carried out with golden hamsters *(Mesocricetus auratus)* and rats *(Rattus norvegicus)*. In an elegant series of experiments, Morin et al. (1977a, b) demonstrated that estradiol could alter the circadian rhythm of wheel running activity in hamsters maintained under a variety of different conditions. The impetus to look for a role of estradiol came from the observation that the phase angle difference (ψ) between lights off and the beginning of activity varied systematically with the days of the estrous cycle (Fig. 1). When estrogen levels were high (on proestrus) the onset of activity occurred earlier than on those days when circulating estrogen levels were lower. This phenomenon has been observed in intact hamsters maintained on LD 12 : 12, LD 14 : 10 and in intact and blind animals maintained in constant light (Morin et al. 1977a, Finkelstein et al. 1978). Furthermore, the continuous administration of estradiol benzoate via hormone-filled Silastic capsules produces a significant advance in the onset of wheel-running in ovariectomized hamsters maintained on LD 12 : 12 (Morin et al. 1977a). One interpretation of the estrogen-induced change in ψ is that estrogen shortens the underlying period τ (Morin et al. 1977a). This hypothesis is supported by the observation that the administration of estradiol benzoate to blind ovariectomized hamsters leads to a shortening of τ (Morin et al. 1977a).

In addition to affecting the period of the rhythm and its phase relationship to the LD cycle, estrogen also alters other properties of the locomotor activity rhythm. Decreased estrogen levels are associated with a decrease in the total amount of activity, an increase in the variability of day-to-day onsets of activity, and a change in the pattern of activity expressed (Finkelstein et al. 1978, Widmaier and Campbell 1980) Takahashi and Menaker 1980).

In contrast to the pronounced effect of estrogens on the rhythm of locomotor activity in ovariectomized hamsters, progesterone treatment alone does not influence the rhythm (Morin et al. 1977a, Takahashi and Menaker 1980). However, it appears that progesterone can antagonize the effects of estrogen. Takahashi and Menaker (1980) observed that progesterone-filled Silastic capsules blocked all the estrogen-induced effects on the activity rhythm.They concluded that the interaction of estrogens and progesterone may modulate the circadian rhythm of activity in female hamsters on a day-to-day basis.

Recent studies in rats (Albers et al. 1981, Albers 1981) indicate that estrogens alter the circadian activity in this species in a manner similar to that observed in hamsters. It should be noted that although effects of estrogens on basic circadian properties, such as τ, have only been well documented in rats and hamsters, few attempts have been made to study this action of estrogens in other species. Indeed, a proestrous peak in locomotor activity has laso been documented in cats and mice (see Campbell and Turek 1981, for review), and it is likely that estrogens influence locomotor activity in a wide variety of mammalian, and perhaps nonmammalian species. The adaptive func-

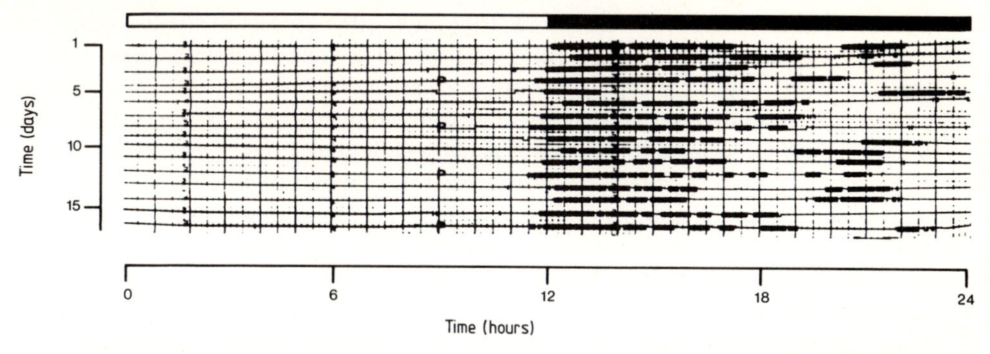

Fig. 1. Activity record of a female hamster housed in LD 12 : 12 *(black bar at the top* indicates the hours of darkness) showing the estrous rhythm of wheel-running. The time scale runs from left to right with successive days represented from top to bottom. The day of proestrus, when estrogen levels reach their peak in the blood stream, is indicated by the symbol *P.* (Adapted from Campbell and Turek 1981)

tion of these estrogen effects on circadian activity parameters is not yet known. However, it seems reasonable to assume that the earlier onset of activity of female hamsters and rats on the evening of proestrous, as well as the general increase of the level of activity during that night, increases the chances of the female coming in contact with a potential mate (Morin et al. 1977b).

2.2 Males

Experiments on both birds and rodents indicate that gonadal secretions, principially testosterone, can modulate the circadian system of male animals. As in the female, the evidence for gonadal secretions influencing rhythmicity comes almost exclusively from the analysis of locomotor activity rhythms.

Daan et al. (1975) observed that castration in mice *(Mus musculus)* resulted in a lengthening of the circadian rhythm in wheel-running behavior and that continuous testosterone treatment reversed the effects of castration. A golden-mantled Ground Squirrel *(Citellus lateralis)* maintained in constant light for over a year showed a circannual cycle in the period of the activity rhythm, with the period being shortest when the animals were in an active reproductive condition and circulating testosterone levels were presumably high (Mrosovsky et al. 1976).

In contrast to the above studies, testosterone did not appear to alter the period of the circadian pacemaker in male hamsters (Ellis and Turek 1979, Morin and Cummings 1981). Transfer of male hamsters to short days (e.g., LD 6 : 18) initially induced gonadal atrophy and a decrease in circulating testosterone levels; but after a prolonged exposure to short days (i.e., 20-25 weeks) there was a spontaneous increase in the size of the testes and serum testosterone levels (Ellis and Turek 1979). However, neither the duration of the daily active phase nor the phase relationship between activity onset and lights off was correlated with changing testosterone levels. A change in the phase relationship was expected if the period of the endogenous oscillator was affected by testosterone (Aschoff 1979). Interestingly, two other aspects of the locomotor activity rhythm, the lability of the daily onset of activity and the total number of wheel revolutions per day, were correlated with circulating testosterone levels (Ellis and Turek 1979). A later study revealed that testosterone treatment of castrated hamsters could indeed induce an increase in locomotor activity (Ellis and Turek 1981). It is important to note that the effect of testosterone on the total amount of wheel-running activity was influenced by the photoperiod, since exposure to short days rendered castrated hamsters less sensitive to the effects of testosterone on activity (Ellis and Turek 1981).

Experiments with male birds also indicate that testosterone has only little or no effect on the period of circadian activity rhythms. Still, there is some limited evidence suggesting that in the common redpoll *(Carduelis flammea)* and in the European starling *(Sturnus vulgaris)*, τ measured in LL is shorter during the breeding season than at other times of the year (Pohl 1974, Gwinner 1975). However, other studies have failed to find such a relationship (Gwinner and Turek 1971, Pohl 1972). Moreover, in starlings maintained in LL, neither castration nor testosterone treatment of castrated males changes the period of the circadian activity rhythm (Gwinner 1975).

While testosterone has little, if any, effect on τ, it does have major effects on the duration of activity (α) in birds. Birds transferred to LL or DD during the reproductive season showed a longer α than birds moved to constant conditions during the non-breeding season (Gwinner and Turek 1971, Pohl 1972, Gwinner 1975). Furthermore, while α increased steadily in intact birds whose testes grew in LL, α remained the same in castrated birds (Gwinner 1975). Finally, injections of testosterone in castrated birds induced clear increases in α. These effects of testosterone on α can fully explain the well-documented fact that the daily activity time of free-living male birds of several species is considerably longer and often starts considerably earlier during the spring breeding season than on days with similar photoperiods in autumn (e.g., Daan and Aschoff 1975). An early onset of activity in spring may be of adaptive significance in these birds when they establish or defend a territory, activities known to be concentrated to the early morning hours (cf. Chap. 8.1, this Vol.).

An important role for testosterone in the circadian organization of birds is suggested by the observation that under the influence of testosterone the activity rhythm of starlings tended to "split" into two components which temporarily free-ran with different circadian frequencies (Gwinner 1974). Splitting was associated with testicular growth in intact birds (Fig. 2A) and in castrated birds injected with testosterone during exposure to dim LL (Fig. 2B). Since splitting is presumed to occur when two circadian oscillators uncouple from each other (see Turek et al. Chap. 5.1, this Vol.), these results suggest that testosterone may be involved in the coupling of circadian oscillators which control the rhythm of locomotor activity. A recent report on female hamsters also indicated that gonadal steroids may play a role in synchronizing a multi-oscillatory circadian system although in this study splitting was associated with low estrogen levels (Morin 1980).

In European starlings a phenomenon related to splitting can be observed even in birds synchronized with a light-dark cycle. Male starlings transferred to a long 12-h photoperiod in early spring not only began to show gonadal development, but simultaneously develop nocturnal restlessness in addition to their normal daytime activity. This nocturnal activity most probably corresponds to the well-known migratory rest-

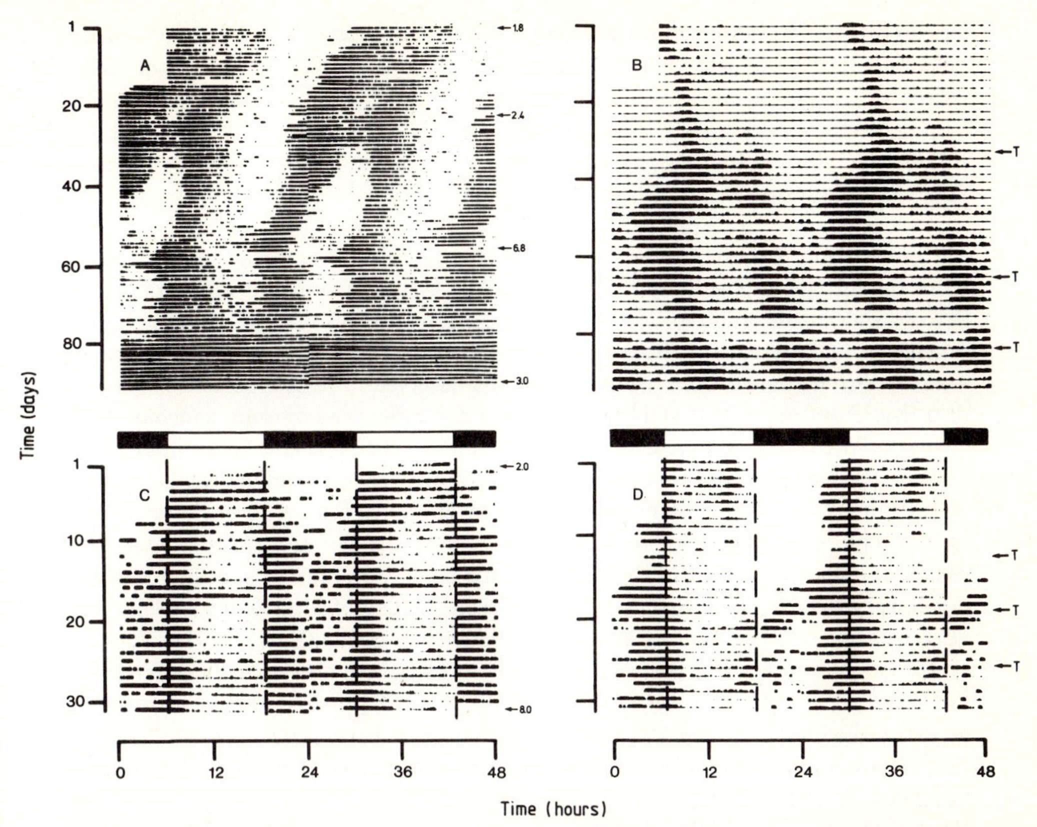

Fig. 2. Activity recordings of four male European starlings kept either in continuous dim light of 0.7 lx **(A, B)** or in LD 12 : 12, 200 : 0.05 lux **(C, D)**; the *black bars* at the top of **C** and **D** indicate the hours of darkness. To fascilitate interpretation of the data, the records have been double plotted on a 48-h time scale. Birds **A** and **C** were normal photosensitive birds whose testes began to develop after transfer to experimental conditions. *Arrows* in the right-hand margin of **A** and **C** indicate days on which laparotomies were performed; the numbers give the testicular width (in millimeters) measured that day. Birds **B** and **D** were castrates. They were injected with 2.5 mg testosterone *(T)* dissolved in 0.1 ml of sesame oil on the days indicated by the *arrows*. **(A** and **B** after Gwinner 1974)

lessness (Zugunruhe) shown by many nocturnally migrating birds during the migratory seasons. Nocturnal activity in captive starlings often appears to result from a shift of the morning and/or evening components of locomotor activity into the night; i.e., from a change in the phase relationship between the morning and evening components of activity (Fig. 2C). A similar "splitting" of activity components under synchronized conditions could be induced in castrated birds by injection of testosterone (Fig. 2D). These preliminary data (Gwinner, unpub.) suggest that the development of nocturnal activity in the starling and possibly other migratory birds may be the result, at least in part, of steroid effects on the coupling of circadian activity oscillators, leading to a shifting of one or more of these components into the night (Gwinner 1975). Although gonadal steroids have long been associated with the development of the migratory condition (Rowan 1925, Stetson and McMillan 1974), no mechanism has been suggested as to how hormonal effects on the circadian system may bring about nocturnal activity.

2.3 Sex Differences

The observation that gonadal hormones may play a role in the circadian organization raises the intriguing possibility that the circadian system may be organized differently in male and female animals. Few studies have addressed this question, although a recent paper (Zucker et al. 1980a) indicates that the circadian system in the hamster is sexually differentiated. Contrary to the results observed in female hamsters, the circadian rhythm of locomotor activity in males is not influenced by subcutaneously placed estrogen-filled capsules. Furthermore, as with many other sexually differentiated behaviors, the hormonal environment during a critical stage of development influences how the adult circadian system will respond to hormonal signals. A single injection of testosterone propionate given to female hamsters on the day of birth prevented estrogen from inducing changes in the period of the activity rhythm in adulthood, while preliminary data in males that were castrated on the day of birth indicated that they were sensitive to the effects of estrogen on activity as adults (Zucker 1979, Zucker et al. 1980a). It appears that perinatal exposure to androgens also sexually differentiates the circadian system of the rat (Albers 1981), although perinatal androgen treatment does not eliminate estrogen sensitivity of the rat circadian system.

There are preliminary reports indicating that the range and pattern of entrainment to light-dark cycles differs between male and female hamsters (Alvis et al. 1978, Davis et al. 1979). These differences did not appear to be dependent on the adult gonadal steroid environment, but were dependent on the perinatal hormonal environment. Sex differences in the circadian pattern of locomotor activity have also been reported from several species of birds; under natural environmental conditions the males tend to wake up earlier and roost later than the females, particularly during the spring reproductive season (Aschoff 1979). Further studies are needed to determine the extent to which the circadian system of male and female animals may be organized and/or activated by sex steroid hormones.

3 Possible Mechanism by Which Hormones Influence the Circadian Pacemaker

Although it has been conclusively established that sex steroid hormones can modulate circadian rhythmicity, the mechanism by which these effects take place is not known. Various authors have pointed out that there are a variety of direct and indirect ways steroid hormones could influence the circadian system (Gwinner 1974, Daan et al. 1975, Morin et al. 1977a, Zucker 1979). A direct action of estradiol on the clock system itself has been questioned, but not ruled out, since the suprachiasmatic nuclei (SCN) (an important component of the circadian system in mammals, see Rusak, Chap. 3.1, this Vol.) is not a major tissue for the uptake or binding of estrogens (Zucker 1979). A variety of indirect influences have been proposed, including the possibility that steroid hormones act on steroid sensitive neurons which have input to the circadian pacemaker. In addition, due to the many hormonal changes which occur in response to alteractions in circulating steroid levels (e.g., changes in gonadotropin and gonadotropin-releasing hormone activity), it may be that other hormones mediate the effects of steroids in circadian rhythmicity.

Implicit in almost all the published literature on the effects of steroid hormones on circadian rhythmicity is the assumption that gonadal secretions excert specific effects on the circadian organization of animals by directly (or indirectly by altering the activity of other hormones or neural tissue which have a direct input to circadian oscillators) affecting circadian oscillators or the coupling among them. Only little attention has so far been given to an alternative hypothesis: that steroid hormones, at least in some cases, may not have a specific effect on the circadian system, but rather alter certain circadian properties by changing the general physiological state of the organism.

Aschoff (1960) suggested a number of years ago that the "level of excitement" of an organism is somehow related to the frequency of the circadian pacemaker. Indeed it is well documented that changes in τ of the activity rhythm which are induced by variations in temperature or light intensity are strongly correlated with changes in the level of activity (Aschoff 1960, 1979). Very little is known about the cause and effect relationships between the "arousal" state (Aschoff 1979) of the organism and the frequency of the circadian oscillator. However, there is some indication that access to a running wheel can influence the period of the activity rhythm possibly because the level of activity has some feedback effect on the circadian system (Aschoff et al. 1973). Some findings suggesting that animals kept in groups have shorter circadian periods than conspecifics kept singly might also be interpreted as resulting from an increase in the level of arousal (Aschoff 1979). In general, however, literally no investigations have been carried out as yet to analyse the possible effects of the "arousal" state of the organism on its own circadian system.

One of the most consistent effects of steroid hormones on the rhythm of locomotor activity is a dramatic increase in activity (Gwinner 1975, Ellis and Turek 1979). Could such changes in the level of activity be responsible for some of the small changes in τ and other rhythm parameters which are observed following castration or treatment with steroids? In addition to the activity level, steroid hormones also influence food intake, body weight and a variety of autonomic functions in rodents (Morin and Fleming 1978, McCabe et al. 1981). For finches, Pohl (1977) has demonstrated that apart from the circadian period, the daily duration of high metabolic activity and the resting metabolic rate during the rest time varied seasonally along with changes in the reproductive state. In addition, castration has been shown to have conspicuous effects on several parameters of body temperature and energy metabolism of male chickens (*Gallus domesticus,* Mitchell et al. 1927), Greenfinches *(Chloris chloris),* Bramblings (*Fringilla montifringilla,* Rautenberg 1952) and quail (Feuerbacher und Prinzinger 1981). Some of the effects of castration could be abolished by testosterone injections (Feuerbacher and Prinzinger 1981). Thus, changes in the steroid hormone environment are associated with a vast array of physiological variations and at the present time it is not clear if these physiological changes may in fact be mediating the effects of steroids on the circadian system. That physiological changes can influence the period of a circadian rhythm has been conclusively demonstrated by Menaker (1961) who found that the body temperature rhythm of bats *(Myotis lucifugus)* was much shorter in summer ($\simeq$22.5 h) than winter ($\simeq$25 h). It is important to note that this 2 1/2 h alteration in period is much longer than has ever been affected by changing steroid levels, and is presumably due to other physiological variations that are associated with hibernation in this species.

If it is true that steroid hormones in at least some cases affect circadian rhythms indirectly by affecting the physiological state of an organism, then one should expect that other hormones which induce major physiological changes (e.g., hormones of adrenal or thyroidal origin) should influence circadian rhythms as well. To our knowledge detailed studies comparable to those performed with steroids have not been carried out. However, recently Zucker et al. (1980b) demonstrated that removal of the pituitary gland lengthened the free-running period of the activity rhythm in hamsters of both sexes. Since there is no evidence that gonadal steroids alter τ of male hamsters (Ellis and Turek 1979, Morin and Cummings 1981), these results indicate that the effect of hypophysectomy was not mediated, at least in males, by an alteration in gonadal function.

A fundamental property of circadian systems is their general homeostasis (Pittendrigh and Caldarola 1973). However, it is also clear that within rather narrow limits, the circadian system does show a great deal of variability. Indeed, many of the variations in circadian properties that have been observed (e.g., after effects, spontaneous changes in τ, changes in τ with age, splitting) may be due to changes in the general physiological state of the organism and are not due to intrinsic changes in the circadian system itself. In order to define the actual role of hormones in the circadian organization it is necessary that experiments be carried out to determine if hormones have access to the circadian oscillator. Do hormones have specific effects on the circadian system, or do they simply alter the general metabolic and physiological state of the organism which in turn induces changes in the fundamental properties of the circadian pacemaker? Answers to these questions are needed before the role of hormones in the overall circadian organization can be defined.

References

Albers HE (1981) Gonadal hormones organize and modulate the circadian system of the rat. Am J Physiol 241: R62-R66

Albers HE, Gerall AA, Axelson JF (1981) Effect of reproductive state on circadian periodicity in the rat. Physiol Behav 26: 21-25

Alvis JD, Davis FC, Menaker M (1978) Sexual dimorphism of the biological clock in golden hamsters. Physiologist 21: 2

Aschoff J (1960) Exogenous and endogenous components in circadian rhythms. Cold Spring Harbor Symp Quant Biol 25: 11-28

Aschoff J (1979) Circadian rhythms: influences of internal and external factors on the period measured in constant conditions. Z Tierpsychol 49: 225-249

Aschoff J, Figala J, Pöppel E (1973) Circadian rhythms of locomotor activity in the golden hamster *(Mesocricetus auratus)* measured with two different techniques. J Comp Physiol Psychol 85: 20-28

Bellinger LL, Williams FE, Bernadis LL (1979) Effect of hypophysectomy, thyroidectomy, castration, and adrenalectomy on diurnal food and water intake in rats. Proc Soc Exp Biol Med 161: 162-166

Campbell CS, Turek FW (1981) Cyclic functions of the mammalian ovary. In: Aschoff J (ed) Handbook of behavioral neurobiology, vol IV. Biological rhythms, Plenum Press, New York, pp 523-545

Daan S, Aschoff J (1975) Circadian rhythms of locomotor activity in captive birds and mammals: their variations with season and latitude. Oecologie 18: 269-316

Daan S, Damassa D, Pittendrigh CS, Smith ER (1975) An effect of castration and testosterone replacement on a circadian pacemaker in mice *(Mus musculus)*. Proc Natl Acad Sci USA 72: 3744-3747

Davis FC, Alvis JD, Menaker M (1979) Effects of gonadal steroids on the development of the hamster circadian pacemaker. Soc Neurosci Abstr 5: 441

Dunn JD, Hess M, Johnson DC (1975) Effect of thyroidectomy on rhythmic gonadotropin release. Proc Soc Exp Biol Med 151: 22-27

Ellis GB, Turek FW (1979) Changes in locomotor activity associated with the photoperiodic response of the testes in male golden hamsters. J Comp Physiol 132: 277-284

Ellis GB, Turek FW (1981) Testosterone and the photoperiod interact to regulate daily locomotor activity in male golden hamsters. Fed Proc 40: 307

Feuerbacher I, Prinzinger R (1981) The effects of the male sex-hormone testosterone on body temperature and energy metabolism in male Japanese quail *(Coturnix coturnix Japonica)*. Comp Biochem Physiol 70A: 247-250

Finkelstein JS, Baum FR, Campbell CS (1978) Entrainment of the female hamster to reversed photoperiod: role of the pineal. Physiol Behav 21: 105-111

Gwinner E (1974) Testosterone induces "splitting" of circadian locomotor activity rhythms in birds. Science 185: 72-74

Gwinner E (1975) Effects of season and external testosterone on the freerunning circadian activity rhythm of European starlings *(Sturnus vulgaris)*. J Comp Physiol 103: 315-328

Gwinner E, Turek F (1971) Effects of season on circadian activity rhythms of the starling. Naturwissenschaften 12: 627-628

McCabe PM, Porges SW, Carter CS (1981) Heart period variability during estrogen exposure and withdrawal in female rats. Physiol Behav 26: 535-538

Menaker M (1961) The freerunning period of the bat clock: seasonal variations at low body temperature. J Cell Comp Physiol 57: 81-86

Mitchell HH, Card LE, Haines WT (1927) The effect of age, sex and castration on the basal heat production of chickens. J Agric Res 34: 945-961

Morin LP (1980) Effect of ovarian hormones on synchrony of hamster circadian rhythms. Physiol Behav 24: 741-749

Morin LP, Cummings LA (1981) Effect of surgical or photoperiodic castration, testosterone replacement or pinealectomy on male hamsters running rhythmicity. Physiol Behav 26: 825-838

Morin LP, Fleming AS (1978) Variation of food intake and body weight with estrous cycle, ovariectomy, and estradiol benzoate treatment in hamsters *(Mesocricetus auratus)*. J Comp Physiol Psychol 92: 1-6

Morin LP, Fitzgerald KM, Zucker I (1977a) Estradiol shortens the period of hamster circadian rhythms. Science 196: 305-307

Morin LP, Fitzgerald KM, Rusak B, Zucker I (1977b) Circadian organization and neural mediation of hamster reproductive rhythms. Psychoneuroendocrinology 2: 73-98

Mrosovsky N, Boches M, Hallonquist JD, Lang K (1976) Circannual cycle of circadian cycles in a Golden-mantled ground squirrel. Naturwissenschaften 63: 298-299

Ooka-Souda A, Draves DJ, Timiras PS (1979) Diurnal rhythm of pituitary-thyroid axis in male rats and the effect of adrenalectomy. Endocr Res Commun 6: 43-56

Pittendrigh CS, Caldarola PC (1973) General homeostasis of the frequency of circadian oscillations. Proc Natl Acad Sci USA 70: 2697-2701

Pohl H (1972) Seasonal change in light sensitivity in *Carduelis flammea.* Naturwissenschaften 11: 518-519

Pohl H (1974) Interaction of effects of light, temperature, and season on the circadian period of *Carduelis flammea.* Naturwissenschaften 9: 406

Pohl H (1977) Circadian rhythms of metabolism in cardueline finches as function of light intensity and season. Comp Biochem Physiol 56A: 145-153

Rautenberg W (1952/53) Körpergewicht und Grundumsatz beim kastrierten männlichen Vogel. Wiss Z Univ Greifswald II: 230-236

Richter CP (1965) Biological clocks in medicine and psychiatry. CC Thomas, Springfield, IL

Rowan W (1925) Relation of light to bird migration and developmental changes. Nature (London) 115: 494-495

Rusak B, Zucker I (1979) Neural regulation of circadian rhythms. Physiol Rev 59: 449-526

Stetson MH, McMillan JP (1974) Some neuroendocrine and endocrine correlates in the timing of bird migration. In: Scheving LE, Halberg F, Pauly JE (eds) Chronobiology. Igaku Shoin, Tokyo, pp 630-635

Takahashi JS, Menaker M (1980) Interaction of estradiol and progesterone: effects on circadian locomotor rhythm of female golden hamsters. Am J Physiol 239: R497-R504

Widmaier EP, Campbell CS (1980) Interaction of estradiol and photoperiod on activity patterns in the female hamster. Physiol Behav 24: 923-930

Zucker I (1979) Hormones and hamster circadian organization. In: Suda M, Hayaishi O, Nakagawa H (eds) Biological rhythms and their central mechanism. Elsevier/North-Holland, Biomedical Press, pp 369-381

Zucker I, Fitzgerald KM, Morin LP (1980a) Sex differentiation of the circadian system in the golden hamster. Am J Physiol 238: R97-R101

Zucker I, Cramer CP, Bittman EL (1980b) Regulation by the pituitary gland of circadian rhythms in the hamster. J Endocrinol 85: 17-25

5.2 The Neuropharmacology of Circadian Timekeeping in Mammals

A. Wirz-Justice[1], G.A. Groos[2], and T.A. Wehr[3]

"How can this clock be speeded up; slowed down; reset; or stopped? Blinded rats were subjected to almost every conceivable kind of metabolic, endocrinologic and neurologic interference . . . to no effect." C.P. Richter (1965)

1 Introduction

Richter's observation, that the homeostasis of circadian oscillations is reflected in their resistance to most chemical manipulations of the internal milieu, may be less generally valid today. For many years the only substance known to have a reproducible effect on circadian rhythms in mammals was deuterium oxide. Heavy water has been found to lengthen the freerunning period as an increasing function of its concentration in the body (Daan and Pittendrigh 1976). Under light-dark (LD) cycles of different periods, heavy water has also been observed to alter the range of entrainment (Hayes and Palmer 1976). Unfortunately, the mechanism by which deuterium oxide alters rhythm characteristics in mammals is completely unknown.

More recently, a growing body of evidence indicates that changes in the endocrine milieu can modulate the frequency of circadian rhythm parameters (for a review, see Chap. 5.1, this Vol.). Several hormones – oestradiol, testosterone, melatonin, thyroid hormone – have been studied. For example, administration of oestradiol shortens the freerunning period (τ) of the hamster rest-activity cycle. Moreover, both under constant conditions and under LD cycles, there is a consistent change of phase with respect to day of the oestrus cycle in female rodents (Morin et al. 1977a). These observations led to the conclusion that the circadian pacemaker is not independent of the concentration of circulating steroid hormones. It had also been shown that the suprachiasmatic nuclei (SCN), which have been identified as a putative pacemaker of the mammalian circadian system, are essential components of the mechanism that drives the oestrus cycle (see Rusak and Zucker 1979, for review). The modulation from the periphery of the central

1 Fellow of the Swiss Foundation for Biomedical Research. Present address: Psychiatrische Universitätsklinik, Wilhelm Klein Straße 27, 4025 Basel, Switzerland
2 National Institute of Mental Health, Bethesda, Maryland 20205, USA , USA
3 Department of Physiology and Physiological Physics, University of Leiden, The Netherlands

Vertebrate Circadian Systems
Ed: by J. Aschoff, S. Daan, and G. Groos

pacemaker controlling oestrus cyclicity probably occurs via brain regions intimately related to the SCN – the anterior and preoptic hypothalamus – that show high affinity for oestradiol binding (Krieger et al. 1976). Thus in contrast to heavy water, the effects of sex hormones on circadian timekeeping can be readily related to specific brain structure, neuroendocrine mechanisms, and even possibly function, in terms of optimal timing of oestrus behaviour (Morin et al. 1977b).

In addition to hormones, certain neurophamacologically active substances have recently been shown to affect the frequency of circadian rhythms. The three drugs that have been investigated, the antidepressants lithium, imipramine, and clorgyline, will be discussed in the present chapter. As in the case of hormones, their action may be expected to be related to specific brain structures and neurochemical function. We shall discuss their effects on neurotransmitter systems in the CNS and in particular those known to be present in the SCN, the major neural centre regulating circadian rhythms in mammals.

2 The Effects of Antidepressant Drugs on Circadian Rhythms

Clinicians have long observed that many of the striking symptoms of affective illness can be considered in terms of altered circadian timekeeping. Depression may involve an abnormal phase-advance of the circadian rhythms of body temperature, rapid-eye-movement sleep, and cortisol with respect to the sleep-wake or day-night cycle (for a review of these hypotheses, see Wehr and Goodwin 1981).

One experimental approach to testing theories of disordered circadian timekeeping in manic-depressive illness, is to examine the effect of chronic administration of those drugs used to treat the illness on the freerunning and entrained circadian rhythms in animals. Three such drugs have now been investigated: lithium, which is used clinically as an antimanic and prophylactic antidepressant; the tricyclic antidepressant imipramine; and the monoamine-oxidase inhibitor antidepressant clorgyline.

Of these substances, lithium has been most extensively studied. It has been found to slow free-running rhythms in two invertebrate species (*Aplysia* – Strumwasser and Viele 1980, cockroach: Hofmann et al. 1978). Lengthening of the freerunning locomotor activity period has been documented in various rodent species: the desert rat (Engelmann 1973), the albino rat (Kripke and Wyborney 1980), and the golden hamster (Zucker, Engelmann, unpublished data). Figure 1 illustrates this effect on τ for the circadian rhythm of food intake in the blinded rat (Groos, unpublished data).

Lithium also appears to alter the range of entrainment: rats on a lithium diet have been found to entrain to 28-h days, whereas control animals did not (McEachron et al. 1981).

In freerunning healthy human subjects, lithium can also slow the body temperature rhythm and sleep-wake cycle (Johnsson et al. 1979). Furthermore, lithium delays the phase-position of the sleep-wake cycle in normal subjects under entrained conditions (Kripke et al. 1979, Pflug, Sitaram and Gillin, pers. commun.). The only studies of the effects of lithium on phase-position in rodents are biochemical: lithium delays many neuroendocrine, plasma electrolyte, and CNS neurotransmitter receptor rhythms (Mc Eachron et al. 1982, Kafka et al. 1982).

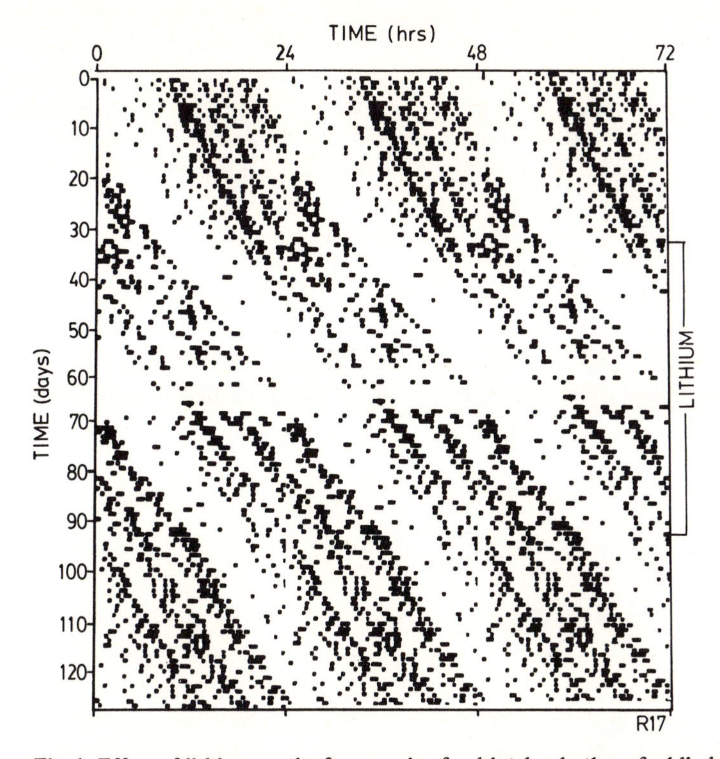

Fig. 1. Effect of lithium on the free-running food intake rhythm of a blinded rat. During days 1-31 and 96-130 this animal was fed on food pellets to which sodium carbonate was added (2.5 mg/g food) while during days 32-89 was substituted by one containing lithium carbonate (3 mg/g food). The free-running periods before, during and after lithium administration were 24.29, 24.53 and 24.37 h respectively. In eight animals subjected to a similar protocol the free-running period increased from a baseline mean of 24.26 h to a mean of 24.62 h during lithium administration. In none of these cases was splitting observed when the animals were fed on a lithium diet

It is a matter of controversy whether lithium facilitates splitting, as has been reported by Kripke and Wyborney (1980). Other workers have not been able to substantiate this claim (Engelmann, Groos, unpubl.). However, the evidence for the lengthening of τ by lithium is substantial and convincing.

Clorgyline and possibly also imipramine lengthen the free-running circadian rest-activity cycle in hamsters (Wirz-Justice et al. 1980a, 1982b, Wirz-Justice and Wehr 1981). Furthermore, both drugs appear to promote dissociation of different components of the circadian activity rhythm. Examples of the patterns of drug-induced lengthening of τ and dissociation of activity components are shown in Fig. 2.

The effects of these two antidepressant drugs are complex, varying in latency and persistence, often showing long-lasting after-effects, as has been observed with other agents affecting the circadian system, such as oestradiol.

A further study followed up these findings by measuring the phase-position of the rest-activity rhythm under entrained conditions. Clorgyline, administered chronically to hamsters, clearly delayed activity onset in a dose-dependent manner under an LD cycle (Craig et al. 1981).

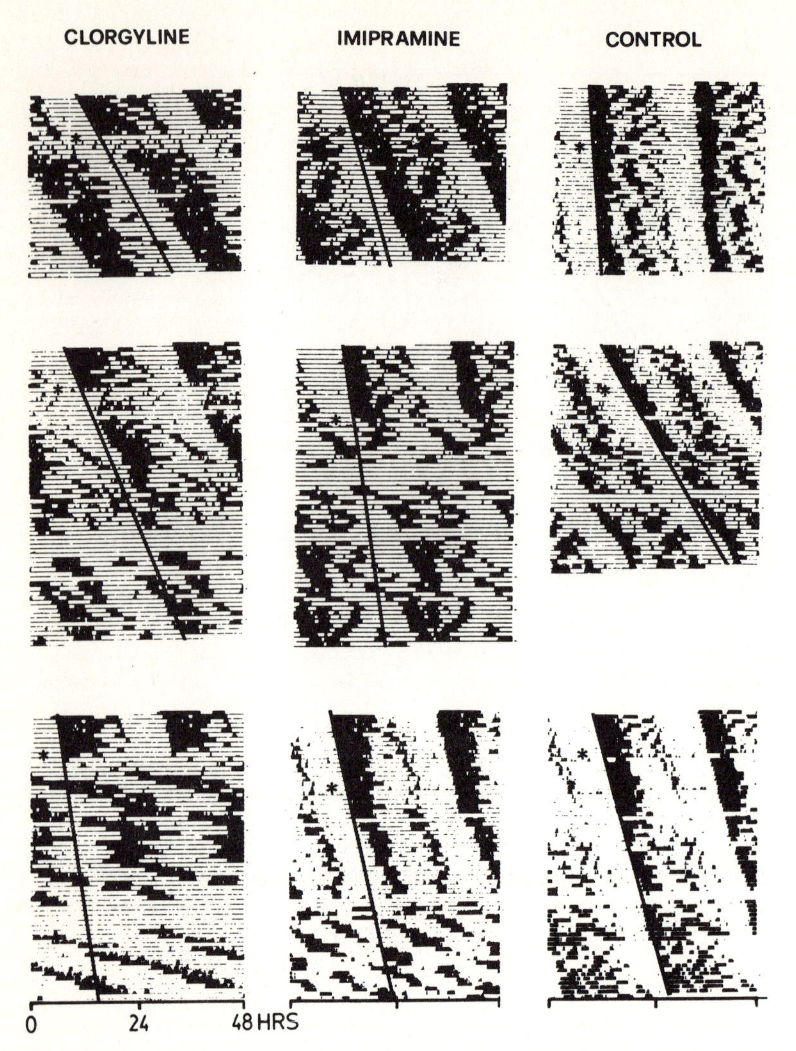

Fig. 2. Representative running-wheel activity records of female hamsters free-running in constant dim red light. On the days indicated by an *asterisk* the animals were implanted subcutaneously at CT 7-9 with osmotic minpumps. The minipumps contained solutions of clorgyline (adminstered at 2 mg/kg/day, for 2 weeks), imipramine (20 mg/kg/day, for 2 weeks) or were empty (control). The free-running period of the baseline activity rhythm is indicated by the *oblique line* in each actogram

Thus for antidepressant drugs from two different classes, there is evidence that they modify circadian rhythm characteristics in a manner possibly implicating a direct effect on the frequency of the circadian pacemaker. Biochemical studies also show that both clorgyline and imipramine can delay the phase-position of many CNS neurotransmitter receptor rhythms (Wirz-Justice et al. 1980a, b; Naber et al. 1980, Kafka et al. 1981b, Wirz-Justice et al. 1982a).

To establish whether clorgyline and imipramine exert their effects on circadian rhythms directly through their action on the neurones of the SCN, both drugs were

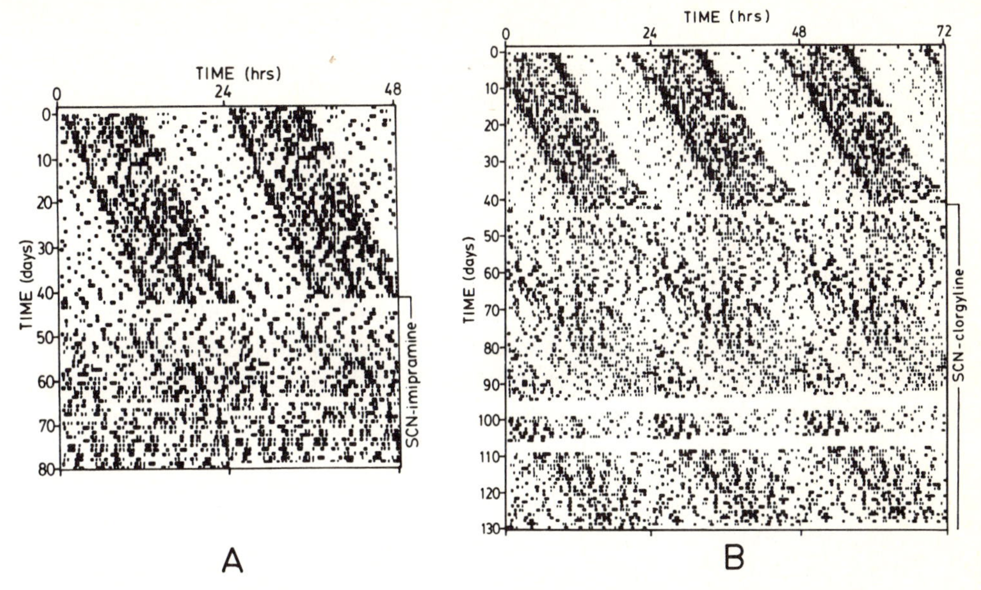

Fig. 3. Effects of local administration of imipramine **(A)** or clorgyline **(B)** in the SCN on the freerunning food intake rhythm of the blinded rat. After imipramine implantation in the SCN the rhythm is abolished **(A)**. Local administration of clorgyline in the SCN of the animal in **B** resulted in an amplitude decrease of the food intake rhythm while the freerunning period increased from 24.39 to 24.58 h. In neither animal did the SCN sustain significant damage due to implantation of the cannulas

chronically applied to the suprachiasmatic region. For this purpose, each compound was administered from a cannula implanted in the brain of blinded rats (Groos, unpublished data). The freerunning rhythms of food intake were consistently abolished after implantation of imipramine cannulas in or very close to the SCN (Fig. 3A). Implantation of empty cannulas in the SCN or imipramine cannulas 1 mm lateral or dorsal to the SCN did not modify rhythmicity in any respect. Clorgyline implantation near the SCN was also found to lead to a similar arrhythmicity. In some animals, however, the rhythm amplitude was slightly reduced with a significant lengthening of τ (Fig. 3B). These findings support the hypothesis that imipramine and clorgyline modify circadian rhythm parameters by pharmacological modulation of the pacemaker in the SCN rather than by diffuse action on a number of brain structures. Dose-response curves of the concentrations of antidepressants required to lengthen τ or induce arrhythmicity when applied to the SCN itself, will be important in relating these direct effects to the concentrations of antidepressants required to induce these effects when applied peripherally.

3 The Neuropharmacology of Antidepressants in the SCN

It would be of great interest to relate the effects of antidepressant drugs on circadian rhythms to their neuropharmacological effects in the SCN. In attempting to classify the known neurochemical changes induced by these drugs with respect to their common effect of slowing circadian rhythms, one faces the problem that lithium, clorgyline and imipramine modify the synthesis, release, re-uptake, or metabolism of many different neurotransmitters.

The first questions are, specifically, which neurotransmitters/neuromodulators are present in the SCN, what are the effects on circadian timekeeping of these neurotransmitters directly administered in the SCN, and of drugs that selectively modify neurotransmitter metabolism? This topic will be briefly reviewed before discussing the neuropharmacology of the antidepressants themselves.

The strongest evidence for any neurotransmitter in the SCN exists for serotonin (5-HT). The SCN are rich in 5HT-terminals, the cell bodies of which are located in the midbrain raphé complex (Azmitia and Segal 1978). The 5HT-synthesising enzyme tryptophan hydroxylase shows high concentrations in the SCN (Brownstein et al. 1975), as does the metabolising enzyme monoamine oxidase (Saavedra et al. 1976, Kamase 1980), and the amine itself (Saavedra et al. 1974). The 5HT-terminals of the SCN are capable of 5HT uptake in vitro, a process which shows a circadian rhythm (Meyer and Quay 1976), and SCN cells respond to microiontophoretic application of 5HT (Nishino and Kolzumi 1977). In contrast to these clear demonstrations of the presence of high concentrations of 5HT in the SCN, there is a generally observed lack of effect of 5HT depletion on circadian timekeeping. Raphe lesions, administration of PCPA, 5,6- or 5,7-dihydroxytryptophan have no effect on circadian rhythm parameters other than mean and amplitude (Rusak and Zucker 1979, Honma et al. 1979, Szafarczyk et al. 1979, 1980a).

In addition to 5HT, acetylcholine may also be a transmitter in the SCN. The synthesising enzyme choline acetyltransferase is present (Palkovits et al. 1974), as is the inactivating enzyme acetylcholine-esterase (Parent and Butcher 1976). Postsynaptic nicotinic acetylcholine receptors have been localized on SCN cells (Segal et al. 1978), and shown to be highly concentrated in these nuclei (Block and Billiar 1981). In contrast, muscarinic acetylcholine receptors show no specific pattern of enrichment in any hypothalamic nuclei (Block and Billiar 1981). Carbachol, a cholinergic agonist, mimics the effect of light on the circadian rhythm of rat pineal indoleamine metabolism (Zatz and Brownstein 1979) and on the circadian rhythm of wheel-running activity in the mouse (Zatz and Herkenham 1981). Alpha-bungarotoxin, a nicotinic receptor blocker, prevents the effect of light on the circadian rhythm of rat pineal indoleamine metabolism (Zatz and Brownstein 1981). These results suggest that acetylcholine is involved in the effects of light on circadian rhythms, via a nicotinic rather than a muscarinic receptor.

Catecholamine-synthesising enzymes are found in low concentrations in the SCN as compared to other hypothalamic nuclei (Saavedra 1975), and recently a sensitive histofluorescence technique has been used to demonstrate the absence of catecholaminergic terminals (Szafarczyk et al. 1980a). Nevertheless, a high proportion of SCN neurons respond to microiontophoretic application of the catecholamines dopamine and noradrenaline (Nishino and Koizumi 1977). Catecholamine depletion with chemical lesions (6-hydroxy-DOPA) does not affect circadian rhythmicity (Honma and Hiroshige 1979). Thus on the basis of the available evidence it is not yet possible to implicate the catecholamines unequivocally in circadian timekeeping in the SCN.

High concentrations of vasopressin are also present in cell bodies in the SCN (Sofroniev and Weindl, Chap. 3.2, this Vol.). Lack of this peptide in the genetically deficient Brattleboro is associated with longer τ than in heterozygons controls; however replacement vasopressin has no effect (Groblewski et al. 1981). Moderate concentrations of histamine, substance P, TRH, CRF, LHRH, somatostatin, vasoactive intestinal peptide, prolactin, beta-endorphin and enkephalins have been identified in the SCN (reviewed in Palkovits 1980). None of these substances has been studied with respect to a possible role in the circadian function of the SCN.

The next question is, how may the wide neuropharmacological profile of antidepressants be related to these transmitters in the SCN? All three drugs have their most potent, although not exclusive, effects on the serotonergic system.

Lithium poses a problem in its ubiquitous effects on many aspects of neurotransmitter metabolism and membrane function. It increases the intracellular concentration of calcium ions (Aldenhoff and Lux 1982), and can therefore be expected to facilitate the release of many transmitters from the pre-synaptic terminal as well as alter post-synaptic cyclic-AMP concentrations. Many studies document its consistent effects on serotonin metabolism: increasing uptake of the precursor amino acid, 5HT concentration and turnover, decreasing serotonin receptor binding (Knapp and Mandell 1973, 1975, Sangdee and Franz 1980, Maggi and Enna 1980). Lithium also enhances noradrenaline receptor sensitivity, whether measured electrophysiologically or biochemically (Schultz et al. 1981, Kafka et al. 1982). Even though there are possibly no noradrenergic terminals in the SCN, there is a known interaction between noradrenergic and serotonergic neurotransmitter systems, which may modulate SCN serotonergic function indirectly.

The tricyclic imipramine primarily inhibits re-uptake of 5HT into the presynaptic terminal (for review of imipramine pharmacology, see Langer and Briley 1981), which, after chronic treatment, leads to a decrease in binding to both serotonin$_2$ and beta-adrenergic receptors (Peroutka and Snyder 1980). It appears that the one consistent result after chronic antidepressant drug treatment is a parallel decrease in binding to both these receptors, supporting the idea of a functional interaction between serotonergic and adrenergic systems (Peroutka and Snyder 1980). Imipramine itself also binds with high affinity to sites which have a very high density in the SCN and show a circadian rhythm (Wirz-Justice et al., unpublished data). The regional distribution and characteristics of this binding site indicate that it may be identical to the 5HT pre-synaptic re-uptake site. Raphé lesions reduce both 5HT concentrations in the hypothalamus and imipramine binding (Langer and Briley 1981). Therefore it seems likely that a high local accumulation of imipramine in the SCN may interfere with serotonergic transmission. Since 5HT is mainly an inhibitory transmitter in the SCN (Nishino and Koisumi 1977), this increase in 5HT after acute imipramine administration will result in an increased inhibition of those SCN cells innervated by 5HT fibres from the raphé complex. That such inhibition can indeed occur is demonstrated by microiontophoretic application of imipramine to SCN cells in the rat (Fig. 5). This mechanism may underlie the arrhythmicity in the circadian rhythm of food intake after imipramine cannulas in the SCN of the rat (Fig. 3A). The resulting high local concentration of 5HT will induce massive inhibition of SCN cells receiving a serotonergic input which could be related to the disruption of the neural timekeeping process in this nucleus.

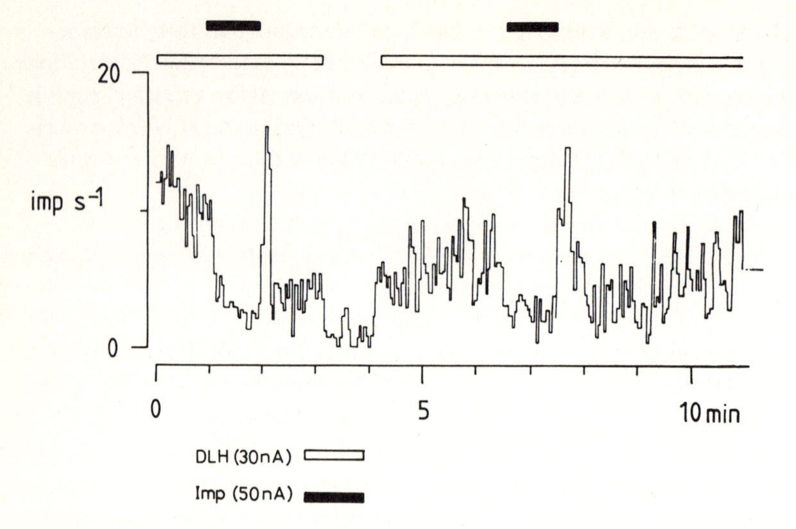

Fig. 4. The effect of microiontophoretic administration of imipramine on the discharge rate of a single cell in the rat SCN. Since the firing rate of SCN cells is generally low the discharge of this neurone was increased by microiontophoresis of DLH. The firing rate of this cell was consistently decreased by application of imipramine *(imp)*. This suppressive effect was followed by an excitatory rebound when imipramine administration was stopped. (Mason and Groos, unpubl. observ.)

Clorgyline is a selective inhibitor of monoamine-oxidase A (MAO-A) activity in the brain; chronic treatment rapidly inhibits the enzyme, increases both serotonin and noradrenaline concentrations, and subsequently decreases serotonergic and beta-adrenergic receptor binding (Campbell et al. 1979, Cohen et al. 1982, Savage et al. 1979).

Thus for the three antidepressants, there are certain neurochemical similarities in that serotonergic transmission is enhanced. Adrenergic transmission is reduced after imipramine and clorgyline, but increased after lithium. To explain our observations of arrhythmicity after application of imipramine and clorgyline in the SCN, we propose that it is the elevated levels of 5HT that can interfere with the function of this nucleus in circadian timekeeping. It may be that only increases, not decreases, of 5HT above a given threshold are functionally important. This idea is not inconsistent with the lack of effect of 5HT depletion on circadian rhythm parameters. In this regard, Szafarczyk et al. (1980b) found that the circadian rhythm of ACTH, which could be abolished with PCPA, was re-established with the 5HT precursor, 5-hydroxytryptophan, only when given at a specific circadian phase.

The antidepressant drugs can therefore be considered as new tools to investigate questions of circadian physiology: what common neurochemical mode of action underlies those drugs that lengthen τ, and what is the chemical substrate of the clock? In contrast to the once generally accepted view that the circadian system of mammals is sensitive only to very few substances, it has now become apparent that circadian timekeeping can be altered by pharmacological manipulation. Three clinically effective antidepressants lengthen the freerunning period in mammals, and two have been shown to delay the phase-position under entrainment to LD cycles as well. For these drugs there exists a great deal of evidence implicating their effects on 5HT metabolism in the brain, and hence also in the SCN. However, it is difficult to relate their pharmacology to any single neurotransmitter; moreover, many putative neurotransmitter substances have been identified in the SCN and the drugs may affect any or all of these. Nevertheless, in the present state of ignorance of their role in circadian timekeeping, it seems parsimonious as well as realistic to relate their action to the serotonergic terminals of raphé fibres in the SCN, and design specific experiments to test these assumptions.

References

Aldenhoff JB, Lux HD (1982) Effects of lithium on calcium-dependent membrane properties and on intracellular calcium-concentration in Helix neurons. In: Emrich HM, Aldenhoff JB, Lux HD (eds) Basic mechanisms in the action of lithium. Excerpta Medica, Amsterdam, pp. 50-63

Azmitia EC, Segal M (1978) An autoradiographic analysis of the differential projections of the dorsal and medial raphé nuclei in the rat. J Comp Neurol 179: 641-668

Block GA, Billiar RB (1981) Properties and regional distribution of nicotinic cholinergic receptors in the rat hypothalamus. Brain Res 212: 152-158

Brownstein MJ, Palkovits M, Saavedra JM, Kizer KS (1975) Tryptophan hydroxylase in the rat brain. Brain Res 97: 163-166

Campbell IC, Robinson DS, Lovenberg W, Murphy DL (1979) The effects of chronic regimens of clorgyline and pargyline on monoamine metabolism in the rat brain. J Neurochem 32: 49-55

Cohen RM, Campbell IC, Dauphin M, Tallman JF, Murphy DL (1982) Changes in alpha and beta-receptor densities in rat brain as a result of treatment with monoamine oxidase inhibiting antidepressants. Neuropharmacology 21: 293-298

Craig C, Tamarkin L, Garrick N, Wehr TA (1981) Long-term and short-term effects of clorgyline (a monoamine oxidase type A inhibitor) on locomotor activity and on pineal melatonin in the hamster. Abstract # 229.14 Soc Neurosci 11th Ann Meeting.

Daan S, Pittendrigh CS (1976) A functional analysis of circadian pacemakers in nocturnal rodents. III. Heavy water and constant light: homeostasis of frequency? J Comp Physiol 106: 267-290

Engelmann W (1973) A slowing down of circadian rhythms by lithium ions. Z Naturforsch 28c: 733-736

Groblewski TA, Nunez AA, Gold RM (1981) Circadian rhythms in vasopressin deficient rats. Brain Res Bull 6: 125-130

Hayes CJ, Palmer JD (1976) The chronomutagenic effect of deuterium oxide on the period and entrainment of a biological rhythm. II. The reestablishment of lost entrainment by artificial LD cycles. Int J Chronobiol 4: 63-69

Hofmann K, Günderoth-Palmowski M, Wiedemann G, Engelmann W (1978) Further evidence for period lengthening effect of Li^+ on circadian rhythms. Z Naturforsch 33c: 231-233

Honma KI, Hiroshige T (1979) Participation of brain catecholaminergic neurons in a self-sustained circadian oscillation of plasma corticosterone in the rat. Brain Res 169: 519-529

Honma KI, Watanabe K, Hiroshige T (1979) Effects of parachlorophenylalanine and 5,6-dihydroxytryptamine on the freerunning rhythms of locomotor activity and plasma corticosterone in the rat exposed to continuous light. Brain Res 169: 531-455

Johnsson A, Pflug B, Engelmann W, Klemke W (1979) Effect of lithium carbonate on circadian periodicity in humans. Pharmacopsychiatria 12: 423-425

Kafka MS, Wirz-Justice A, Naber D, Wehr TA (1981) Circadian acetylcholine receptor rhythm in rat brain and its modification by imipramine. Neuropharmacology 20: 421-425

Kafka MS, Wirz-Justice A, Naber D, Marangos PJ, O'Donahue T, Wehr TA (1982) Effect of lithium on circadian neurotransmitter receptor rhythms. Neuropsychobiology 8: 41-50

Kamase H (1980) The diurnal variations of monoamine oxidase activity in discrete nuclei of rat brain. Folia Psychiat et Neurolog Jap 34: 481-492

Knapp S, Mandell AJ (1973) Short and long-term lithium administration: effects on the brain's serotonergic biosynthetic systems. Science 180: 645-647

Knapp S, Mandell AJ (1975) Effects of lithium chloride on parameters of biosynthetic capacity for 5-HT in rat brain. J Pharmac Exp Ther 198: 123-132

Krieger MS, Morrell JI, Pfaff DW (1976) Autoradiographic localization of estradiol concentrating cells in the female hamster brain. Neuroendocrinology 22: 193-205

Kripke DF, Wyborney VG (1980) Lithium slows rat circadian activity rhythms. Life Sci 26: 1319-1321

Kripke DF, Judd LL, Hubbard B, Janowsky DS, Huey LY (1979) The effect of lithium carbonate on the circadian rhythm of sleep in normal human subjects. Biol Psychiat 14: 545-548

Langer SZ, Briley M (1981) High-affinity ^{3}H-imipramine binding: a new biological tool for studies in depression. TINS 4: 28-31

Maggi A, Enna SJ (1980) Regional alterations in rat brain neurotransmitter systems following chronic lithium treatment. J Neurochem 34: 888-892

McEachron DL, Kripke DF, Wyborney VG (1981a) Lithium promotes entrainment of rats to long circadian light-dark cycles. Psychiatry Res 2: 511-9

McEachron DL, Kripke DF, Hawkins R, Haus E, Pavlinac D, Deftos L (1982) Lithium delays biochemical circadian rhythms in rats. Neuropharmacology 8: 12-29

Meyer DC, Quay WB (1976) Hypothalamic and suprachiasmatic uptake of serotonin in vitro: twenty-four hour changes in male and proestrous female rats. Endocrinology 98: 1160-1165

Morin LP, Fitzgerald KM, Zucker I (1977a) Estradiol shortens the period of hamster circadian rhythms. Science 196: 305-307

Morin LP, Fitzgerald KM, Rusak B, Zucker I (1977b) Circadian organization and neural mediation of hamster reproductive rhythms. Psychoneuroendocrinology 2: 73-98

Naber D, Wirz-Justice A, Kafka MS, Wehr TA (1980) Dopamine receptor binding in rat striatum: ultradian rhythm and its modification by chronic imipramine. Psychopharmacology 68: 1-5

Nishino H, Koizumi K (1977) Responses of neurons in the suprachiasmatic nuclei of the hypothalamus to putative transmitters. Brain Res 120: 167-172

Palkovitz M (1980) Topography of chemically identified neurons in the central nervous system: progress in 1977-1979. Med Biol 58: 188-227

Palkovits M, Saavedra JM, Kobayashi RM, Brownstein M (1974) Choline acetyltransferase content of limbic nuclei of the rat. Brain Res 79: 443-450

Parent A, Butcher LL (1976) Organization and morphologies of acetylcholinesterase-containing neurons in the thalamus and hypothalamus of the rat. J Comp Neurol 170: 205-226

Peroutka SJ, Snyder SH (1980) Long-term antidepressant treatment decreases spiroperidol-labelled receptor binding. Science 210: 88-90

Richter CP (1965) Biological clocks in medicine and psychiatry. Thomas, Springfield, p 21

Rusak B, Zucker I (1979) Neural regulation of circadian rhythms. Physiol Rev 59: 449-526

Saavedra JM (1975) Localization of biogenic amine-synthesizing enzymes in discrete hypothalmic nuclei. In: Anatomical neuroendocrinology. Karger, Basel, p 397

Saavedra JM, Brownstein MT, Palkovits M (1976) Distribution of catechol-O-methyltransferase, histamine N-methyltransferase and monoamine oxidase in specific areas of the rat brain. Brain Res 118: 152-156

Saavedra JM, Palkovits M, Brownstein MJ, Axelrod J (1974) Serotonin distribution in the nuclei of the rat hypothalamus and preoptic region. Brain Res 77: 157-165

Sangdee C, Franz DN (1980) Lithium enhancement of central 5-HT transmission induced by 5-HT precursors. Biol Psychiatry 15: 59-75

Savage DD, Frazer A, Mendels J (1979) Differential effects of monoamine oxidase inhibitors and serotonin reuptake inhibitors on ^{3}H-serotonin receptor binding in rat brain. Eur J Pharmacol 58: 87-88

Schultz JE, Siggins GR, Schocker FW, Türck M, Bloom FE (1981) Effects of prolonged treatment with lithium and tricyclic antidepressants on discharge frequency, norepinephrine responses and beta receptor binding in rat cerebellum: electrophysiological and biochemical comparison. J Pharmacol Exp Ther 216: 28-38

Segal M, Dudai Y, Amsterdam A (1978) Distribution of an α-bungarotoxin-binding cholinergic nicotinic receptor in rat brain. Brain Res 148: 105-119

Strumwasser F, Viele DP (1980) Lithium increases the period of a neuronal circadian oscillator. Soc Neurosci Abstr 6: 707

Szafarczyk A Ixart G. Malaval F, Nouguier-Soulé J, Assenmacher I (1979) Effects of lesions of the suprachiasmatic nuclei and of p-chlorophenylalanine on the circadian rhythms of adrenocorticotrophic hormone and corticosterone in the plasma, and on locomotor activity of rats. J Endocrinol 83: 1-16

Szafarczyk A, Alonso G, Ixart G, Malaval F, Nouguier-Soulé J, Assenmacher I (1980a) Serotoninergic system and circadian rhythms of ACTH and corticosterone in rats. Am J Physiol 239: E482-E489

Szafarczyk A, Ixart G. Malaval F, Nouguier-Soulé J, Assenmacher I (1980b) The influence of the time of administration of 5-hydroxytryptophan on its restoring effect on the circadian rhythm of plasma ACTH in rats under pcpa treatment. C R Soc Biol 174: 170-175

Wehr TA, Goodwin FK (1981) Biological rhythms and psychiatry. In: Arieti S, Brodie HKH (eds) American handbook of psychiatry, vol VII. Basic Books, N.Y. pp 46-74

Wirz-Justice A, Wehr TA (1981) Uncoupling of circadian rhythms in hamsters and man. In: Koella WP (ed) Sleep 1980, Karger, Basel, pp. 69-72

Wirz-Justice A, Wehr TA, Goodwin FK, Kafka MS, Naber D, Marangos PJ, Campbell IC (1980a) Antidepressant drugs slow circadian rhythms in behaviour and brain neurotransmitter receptors. Psychopharmacol Bull 16: 45-47

Wirz-Justice A, Kafka MS, Naber D, Wehr TA (1980b) Circadian rhythms in rat brain alpha and beta-adrenergic receptors are modified by chronic imipramine. Life Sci 27: 341-347

Wirz-Justice A, Kafka MS, Naber D, Campbell IC, Marangos PJ, Tamarkin L, Wehr TA (1982a) Clorgyline delays the phase-position of circadian neurotransmitter receptor rhythms. Brain Res 241: 115-122

Wirz-Justice A, Wehr TA, Goodwin FK, Campbell IC (1982b) Antidepressant drugs can slow or dissociate circadian rhythms. Experientia (in press)

Zatz M, Brownstein MJ (1979) Intraventricular carbachol mimics the effects of light on the circadian rhythm in the rat pineal gland. Science 203: 358-360

Zatz M, Brownstein MJ (1981) Injection of alpha-bungarotoxin near the suprachiasmatic nucleus blocks the effects of light on nocturnal pineal enzyme activity. Brain Res 213: 438-442

Zatz M, Herkenham MA (1981) Intraventricular carbochol mimics the phase-shifting effect of light on the circadian rhythm of wheel-running activity. Brain Res 212: 234-238

5.3 Entraining Agents for the Circadian Adrenocortical Rhythm in the Rat

K. Takahashi and N. Murakami[1]

1 Introduction

Circadian variation of adrenocortical function has been well documented, both in man and in many animal species. The factors entraining the adrenocortical rhythm in the rat, both light and food, have been extensively studied. However, only a limited number or studies are available on the maternal influence in the development of this rhythm (Levin et al. 1976, Takahashi et al. 1979, Hiroshige and Honma 1979, Miyabo et al. 1980). Moreover, the influence of social cues has not been sufficiently studied in this animal, while its importance has been well established for man (Aschoff et al. 1975). The present study is mainly concerned with these two subjects; i.e. the maternal influence and the social influence on the corticosterone rhythm.

We have developed a new method for the determination of corticosterone rhythm in individual pups as young as 2 weeks of age as reported previously (Takahashi et al. 1979). Using a microsampling technique we measured blood corticosterone levels at 4-h intervals in individual animals for 2 days. Because of the insufficient number of a data points for statistical analysis of the individual rhythm, the animals were considered to have a distinct rhythm only when the highest peak elevation of blood corticosterone levels on both days occurred at the same time of day, plus or minus 4 h.

2 The Maternal Influence on the Development of the Circadian Adrenocortical Rhythm in Blinded Pups

2.1 Effect of LD Reversal in Blinded Mothers and Pups

In previous papers pups, optically enucleated on the day of birth and raised by an intact mother under LD, have been reported to manifest an overt circadian corticosterone rhythm at 4 weeks of age. The acrophase of this rhythm was near the light-dark transition (Levin et al. 1976, Takahashi et al. 1979). In the present study, similar findings were obtained in both blinded pups groups born and raised under LD and under an inverted lighting regimen (DL) (Fig. 1). These facts suggested that either the light or the mother entrained the blinded pups' rhythm.

1 Department of Medical Chemistry, Tokyo Metropolitan Institute for Neuroscience, 2-6 Musashi-dai, Fuchu-city, Tokyo, Japan (183)

Vertebrate Circadian Systems
Ed. by J. Aschoff, S. Daan, and G. Groos

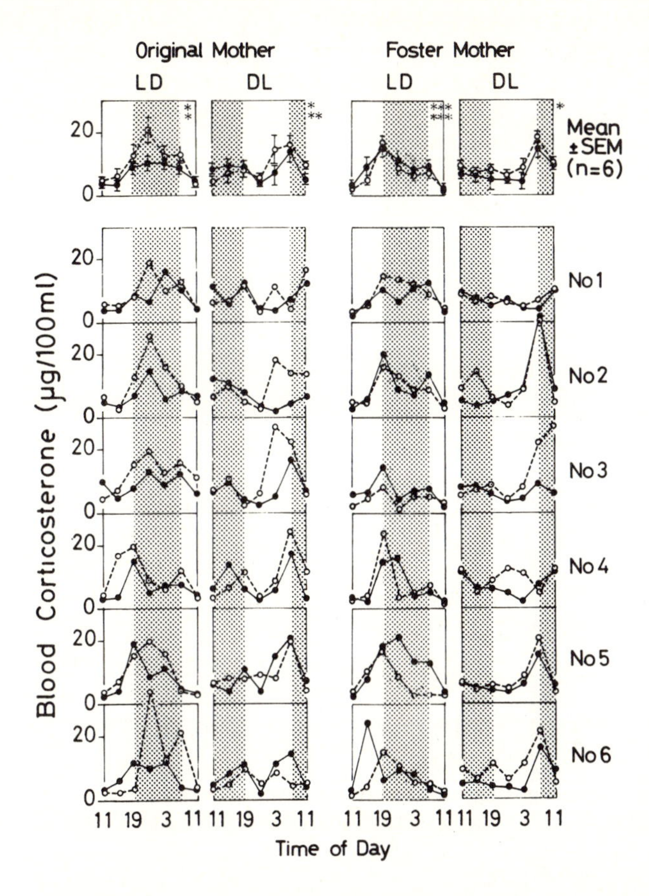

Fig. 1. Two-day patterns of blood corticosterone levels in blinded pups 4-5 weeks of age. *Left two columns* show the blinded pups born and raised by an original mother under LD *(left)* or DL *(right)*. *Right two columns* show the blinded pups subjected to the mother exchange experiment. Blinded pups were raised by a foster mother with a rhythm inverted to that of the original mother under LD *(left)* or DL *(right)*. Individual 2-day patterns of six pups are shown in the *lower part* and the *panel in the top* shows the mean of them in each column. Each *circle* and *bar* in the top panel represents the mean ± SEM of six pups. For this and the subsequent figures except Fig. 7, the *solid lines* connect values on the 1st day and the *dotted lines* connect values on the 2nd day. *Shaded areas* represent the dark periods. *Upper* and *lower asterisks* indicate a significant difference between peak and trough values on the 1st and 2nd day, respectively. ***$p < 0.01$, **$p < 0.02$, *$p < 0.05$. (P-value on the 2nd day in the pups raised by a foster mother under DL is less than 0.1 but larger than 0.05).

To determine the influential factor in entraining the pup's rhythm, both the mother and the pups were simultaneously blinded on the day pups were born and transferred to a room with an DL cycle. Blinded mothers were expected to maintain their previous rhythm after transfer, although a small phase shift would occur during the nursing period. Accordingly, the blinded pups were raised under conditions in which the phase angle difference between mother and light was approximately 180°.

Figure 2 illustrates the representative mothers and litters. At 4 weeks a distinct rhythm appeared in four blinded pups (PIV-1, PVI-3, PVI-1, PVI-3), whose acrophase were located between 2300 h and 0300 h. As peak elevation of blood corticosterone levels took place around 0300 h in both mothers, pups' rhythms appeared to be in phase with those of their mothers. In both pups and mothers a similar pattern of free-run of the corticosterone rhythm was observed throughout the observation period of 8-10 weeks.

Figure 3 shows all acrophases observed at each examination week in mothers and pups who manifested a distinct rhythm. At 4 weeks the majority of mothers and pups

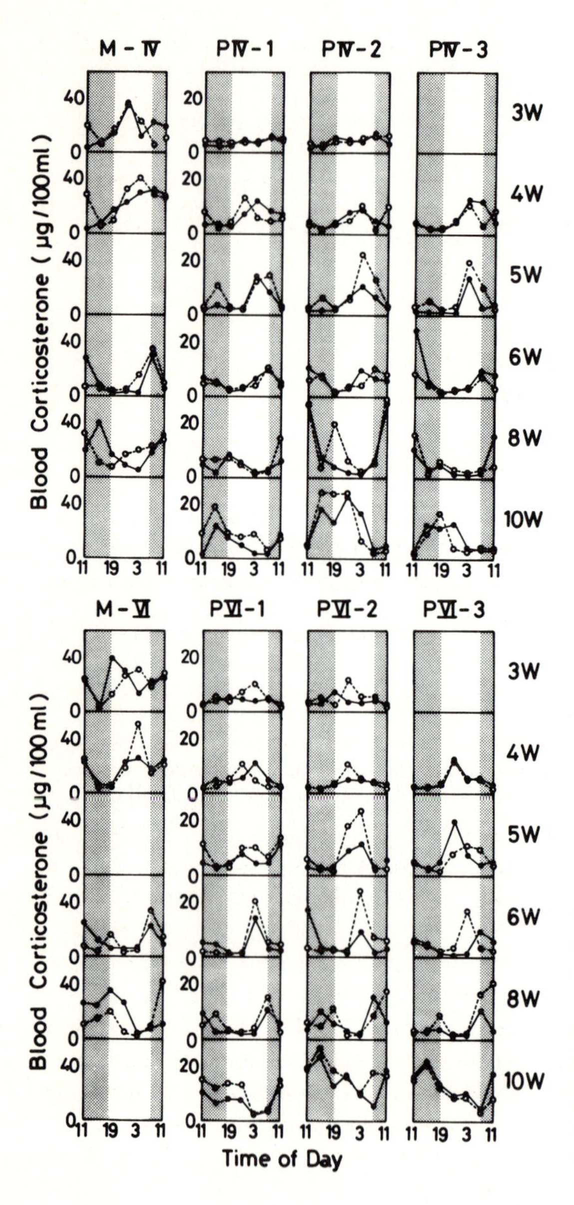

Fig. 2. Two-day patterns of blood corticosterone levels in blinded pups born under LD and raised under DL. Two representative litters and their mothers are illustrated. *M* and *P* identify the mothers and pups, respectively. For this figure, Fig. 3 and Fig. 4, the *numbers along the right edge* of the illustration identify the number of weeks after birth

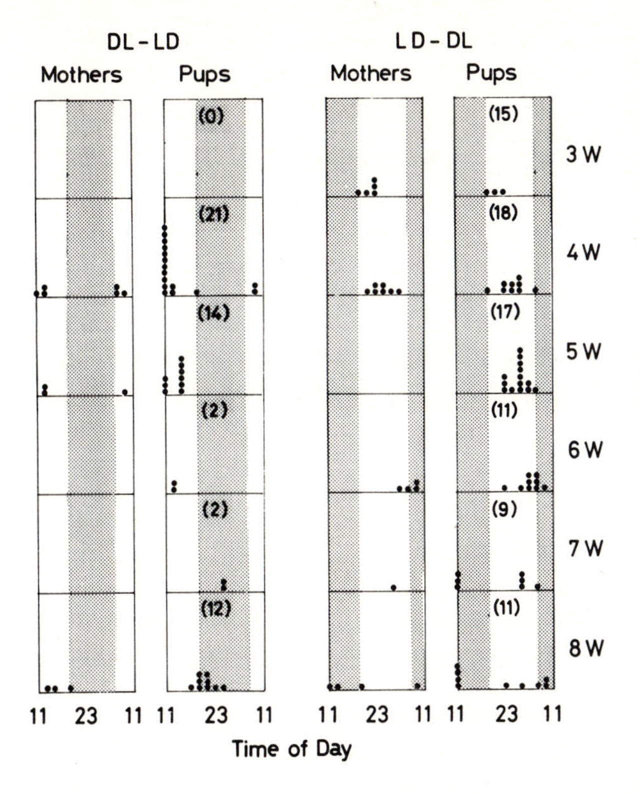

Fig. 3. Acrophase of circadian corticosterone rhythm in blinded mothers and pups. Mothers and pups under DL or LD just prior to reversal of the lighting condition are illustrated in the *left* and *right columns,* respectively. Each *circle* represents the acrophase of the corticosterone rhythm in animals with a distinct rhythm, as defined in the text. When the temporal difference in the highest peak elevation of blood corticosterone levels between two days was 4 h, acrophase was determined at the midpoint between the peaks of each day. *Numbers in parentheses* show those of blinded pups examined at each week

showed their acrophase of the rhythm in the first half or in the middle of the light phase, not near the light-dark transition. The fact that the corticosterone rhythm of blinded pups was in phase with that of the mother would seem to indicate that mother rats, rather than the light cycle, entrained the rhythm of blinded pups.

2.2 Effect of Maternal Exchange and Light Reversal on Blinded Pups

To determine whether the phase was established prior to birth or during the nursing period, we conducted a mother-exchange experiment. Pups born under a LD or a DL schedule, blinded on the day of birth, were raised by intact foster mother under a DL or a LD schedule. Two-day patterns of blood corticosterone levels of a representative litter at 4-5 weeks are illustrated in the right half of Fig. 1. The majority of blinded pups raised by a foster mother manifested a distinct rhythm, acrophase being observed near the light-dark transition in both groups born under LD and under DL. In our laboratory intact mother rats always showed peak elevation of corticosterone levels near the light-dark transition. Accordingly, the rhythm in blinded pups, who were reared by an intact foster mother with a rhythm inverted to that of the original mother, was presumed to be in phase with that of the foster mother before they were weaned at 21 days of age.

Figure 4 shows individual free-running patterns of corticosterone rhythm obtained from another litter born under LD and raised by a foster mother under DL. All blinded

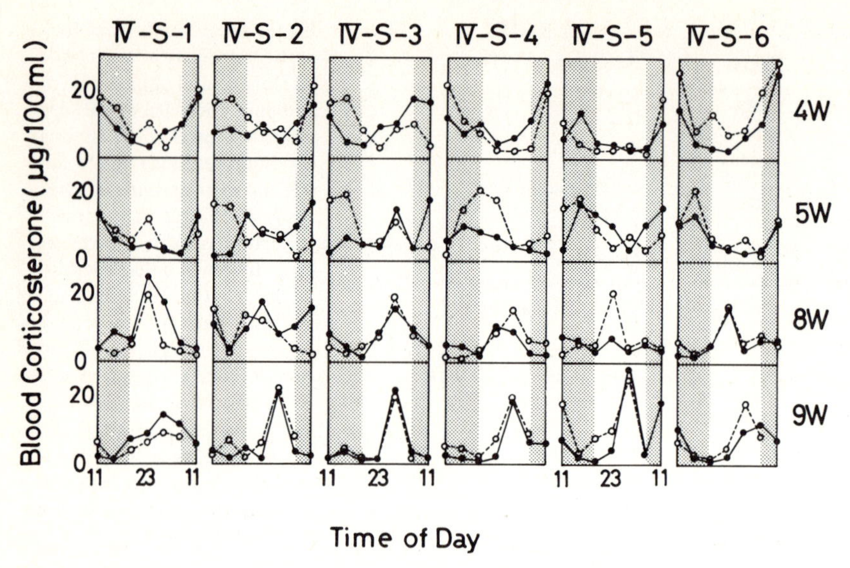

Fig. 4. Two-day patterns of blood corticosterone levels in individual six blinded pups born under LD and raised by a foster mother with a rhythm inverted to that of the original mother under DL. Each *number on the top* identifies the rat number

pups except IV-S-3 manifested a distinct rhythm at 4 weeks and the acrophase of the rhythm was in the first half of the dark period at this week. At 8 and 9 weeks, the highest peak elevation of corticosterone levels occurred in the middle of the light phase. This phase shift indicates that the corticosterone rhythm observed in the blinded pups was an endogenous one.

These results suggested that the pups' rhythm was transiently entrained by the mother during the nursing period and subsequently free-ran. Aschoff and Meyer-Lohmann (1954) showed that chicks, which are not affected by their mother after birth, manifested a circadian locomotor activity rhythm on the 1st day of birth under LL. In mammals, also, the circadian locomotor rhythm was demonstrated to persist for several generations when exposed to LL (Aschoff 1955). Thus, it is evident that the endogenous clock mechanism is of inborn nature. However, little is known about the factor setting the phase of the endogenous rhythm. The results of the present study showed the predominant role of the mother as the entrainer of the rhythm.

Our findings confirmed the previous report by Deguchi, who demonstrated the importance of the mother's role in entraining the pups' rhythm (Deguchi 1977). He clearly showed that serotonin N-acetyltransferase activity in the pineal gland of the intact pups was in phase with that of the foster mother, when the pups were raised under an ultradian lighting condition (L:D = 6:6). Moreover, he found that for entrainment of the rhythm by a foster mother the pups must be raised by the foster mother for more than 2 days, but less than 2 weeks. Presumably, some circadian factors in the mother's behavior may entrain the pups' rhythm. Miyabo et al. (1980) reported that neonatally blinded pups raised by periodic 12-h suckling and feeding, beginning on day 5, established circadian corticosterone rhythm fully synchronized to their feeding patterns by day 25. Taking this into consideration, it seems that maternal behavior related to nurture is a most likely factor in entraining the corticosterone rhythm in blinded pups.

3 Social Influence on the Circadian Corticosterone Rhythm

In 1976 Wilson et al. reported that blinded rats kept with intact ones in the same cage under LD were not entrained by the intact rats 10 weeks after blinding. We also demonstrated that intact and blinded rats housed in the same cage did not entrain each other under both LD and LL. Under LL both intact and blinded rats showed free-running rhythms of their own period throughout the observation period of 8 weeks (Takahashi et al. 1978).

These facts suggest that social influences seem not to be potent for entraining the corticosterone rhythm in the rat. However, to investigate the possible effects of social influence on the adrenocortical rhythm, the experiment should have the group size as the only variable. In the present study, differences in the free-running period of blinded rats housed individually and housed in groups was studied.

Two-day patterns of corticosterone levels were determined every 2 to 3 weeks, for 13 weeks, in blinded male adult rats who were housed either singly or grouped with three rats per cage. The animals had been kept under LD (light: 0700 h – 1900 h) and were exposed to LL after blinding. On both days the peak elevation in each rat, housed either singly or in a group, occurred at the same time of day plus or minus 4 h at each of the study week. As shown in Fig. 5, rats kept in groups exhibited a rhythm synchronized at both the 7th and 13th week. Acrophase was observed between 0700 h and 1500 h in the 7th week and between 0700 h and 1100 h in the 13th week. Rats kept singly did not exhibit a synchronized pattern at the 13th week, although the rhythm was better synchronized in the 7th week, as shown in Fig. 6.

Figure 7 illustrates the free-running pattern of individual rats, housed either singly or in groups. All blinded rats manifested a rhythm with a free-running period longer than 24 h. Until the 13th week, the rats in groups exhibited a synchronized pattern, while singly housed rats lacked a synchronized pattern after the 10th week. Essentially similar findings were obtained with sighted rats under LL. Intact rats housed in groups exhibited a better synchronization of free-running rhythm than the rats housed singly (data not shown).

These results reveal that when the only variable is group size, rats housed in groups entrain each other better than singly housed rats. This suggests some social factors may act as a synchronizer for the circadian free-running rhythm as reported in the human. There is ample evidence showing the importance of the social environmeter the entraining the circadian rhythm in man (Aschoff et al. 1975). Vernikos-Danellis and Winget (1979) demonstrated that social cues have a profound effect on the synchronization of the plasma cortisol rhythm in human subjects exposed to LL for 21 days.

However, when blinded and sighted rats were housed together in the same cage, social factors did not act as a synchronizer as reported previously (Takahashi et al. 1978). Although the reason for this difference is unknown, there may be two possible explanations. One is that there may be differences in some of the many physiological

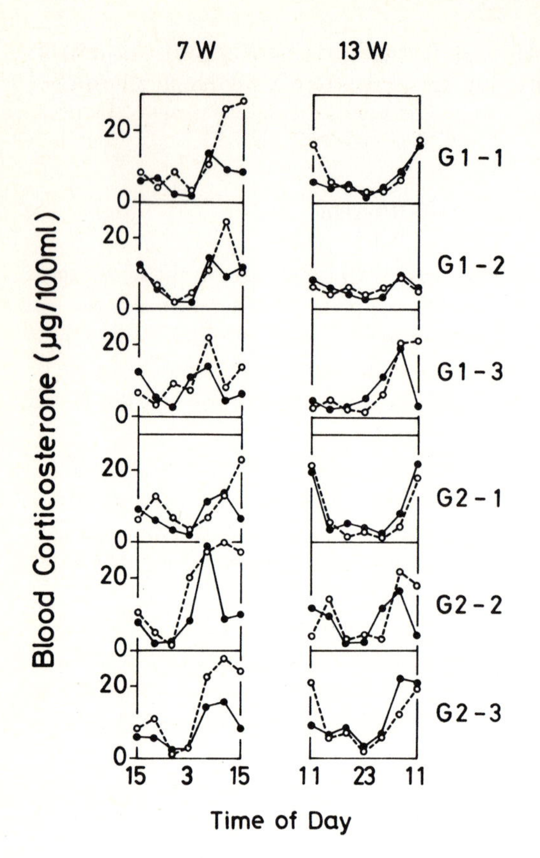

Fig. 5. Two-day patterns of blood corticosterone levels in blinded rats housed in groups. Each group *(G1* and *G2)* contained three rats in the same cage (18 cm x 40 cm x 30 cm). The rhythm was determined at 7 *(left column)* and 13 *(right column)* weeks after blinding

conditions which couple with the differences in the free-running period in such a way that entrainment is prevented. Another possibility is that the difference in the free-running period between the two types, intact and blinded, of rats is too great for one type to entrain the other type. Under constant light with an intensity of 500-700 lx, the mean free-running period of sighted rats was about 25.0 h, while the mean period of blinded rats was about 24.3 h. Social factors might be insufficient to entrain those rats having large differences in the free-running period. Conversely, when the difference in free-running period is not large, social influence can act as a synchronizer. Of these two possibilities the latter one seems more probable, since a potent synchronizer, such as restricted feeding, entrains both blinded and sighted rats under constant conditions (Takahashi et al. 1977, Morimoto et al. 1979).

Acknowledgements. The authors are very grateful to Dr. R. Rowland for preparation of the manuscript. The present study was supported in part by a Ministry of Education Grant.

References

Aschoff J (1955) Tagesperiodik bei Mäusestämmen unter konstanten Umgebungsbedingungen. Pflügers Arch 262: 51-59

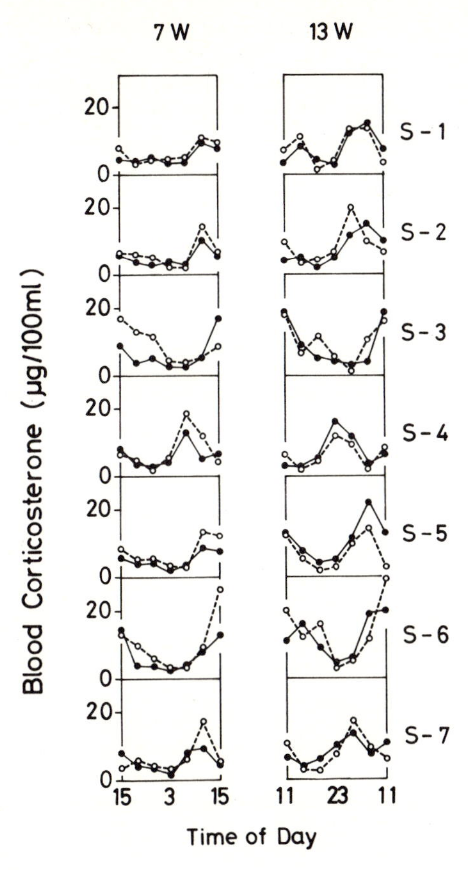

Fig. 6. Two-day patterns of blood corticosterone levels in blinded rats housed singly. The rhythm was determined at 7 *(left column)* and 13 *(right column)* weeks after blinding

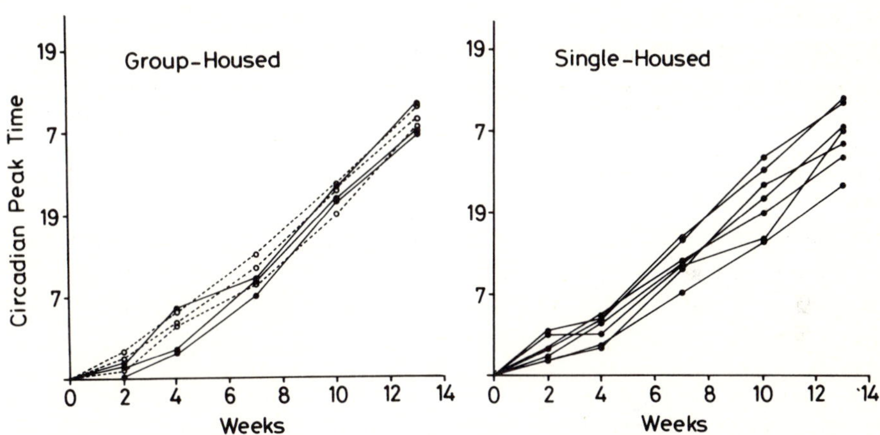

Fig. 7. Free-running patterns of the circadian adrenocortical rhythm in blinded rats housed either singly *(right)* or in groups *(left)* under LL. Circadian peak time is shown on the ordinate and the number of weeks after blinding is shown on the abscissa. Each *circle* represents the acrophase of the rhythm in each animal at the examination week. In the group-housed illustration, the *solid lines* represents the pattern for each of the rats in one group. The *three dotted lines* represent the pattern for each of the rats in another group

Aschoff J, Meyer-Lohmann J (1954) Angeborene 24-Stunden-Periodik beim Kücken. Pfluegers Arch 260: 170-176

Aschoff J, Hoffmann K, Pohl H, Wever R (1975) Re-entrainment of circadian rhythms after phase-shifts of the Zeitgeber. Chronobiologia 2: 23-78

Deguchi T (1977) Circadian rhythms of enzyme and running activity under ultradian lighting schedule. Am J Physiol 232: E375-E381

Hiroshige T, Honma K (1979) Internal and external synchronization of endogenous rhythms in the rat and involvement of brain biogenic amines. In: Suda M, Hayaishi O, Nakagawa H (eds) Biological rhythms and their central mechanism. Elsevier, Amsterdam, pp 233-245

Levin R, Fitzpatrick KM, Levine S (1976) Maternal influence on the ontogeny of basal levels of plasma corticosterone in the rat. Horm Behav 7: 41-48

Miyabo S, Yanagisawa K, Ooya E, Hisada T, Kishida S (1980) Ontogeny of circadian corticosterone rhythm in female rats: Effects of periodic maternal deprivation and food restriction. Endocrinology 106: 636-673

Morimoto Y, Oishi T, Arisue K, Yamamura Y (1979) Effect of food restriction and its withdrawal on the corcadian adrenocortical rhythm in rats under constant dark or constant lighting condition. Neuroendocrinology 29: 77-83

Takahashi K, Inoue K, Kobayshi K, Hayfuji C, Nakamura Y, Takahshi Y (1977) Effects of food restriction on circadian adrenocortical rhythm in rats under constant lighting conditions. Neuroendocrinology 23: 193-199

Takahashi K, Inoue K, Kobayashi K, Hayafuji C, Nakamura Y, Takahashi Y (1978) Mutual influence of rats having different circadian rhythm of adrenocortical activity. Am J Physiol 234: E515-E520

Takahashi K, Hanada K, Kobayashi K, Hayafuji C, Otani S, Takahashi Y (1979) Development of the circadian adrenocortical rhythm in rats: Studied by determination of 24- or 48-hour patterns of blood corticosterone levels in individual pups. Endocrinology 104: 954-961

Vernikos-Danellis J, Winget CM (1979). The importance of light, postural and social cues in the regulation of the plasma cortisol rhythm in man. In: Reinberg A, Halberg F (eds) Chronopharmacology. Pergamon Press, Oxford, pp 101-106

Wilson MM, Rice RW, Critchlow V (1976) Evidence for a free-running circadian rhythm in pituitary-adrenal function in blinded adult female rats. Neuroendocrinology 20: 289-295

5.4 Splitting of the Circadian Rhythm of Activity in Hamsters

F.W. Turek, D.J. Earnest, and J. Swann[1]

1 Introduction

There is now unequivocal evidence, obtained from a variety of different experimental approaches, that the circadian system of multicellular organisms is composed of a population of circadian oscillators (Pittendrigh 1974, Aschoff and Wever 1976, Moore-Ede et al. 1976, Menaker et al. 1978, Block and Page 1978, Jacklet 1981). Indeed, it appears that even a single measurable circadian rhythm may in fact be regulated by more than one circadian oscillator. Strong evidence for this proposition is the observation that the circadian rhythm of locomotor activity can dissociate into two distinct components. An important feature of this dissociation is that for at least a period of time the two components, or bouts of activity, can free-run with clearly distinct periods which result in a series of changing phase relationships between the two components. Usually these components become recoupled some 12-h out of phase with each other and thereafter assume an identical free-running period. This "splitting" phenomenon is difficult to explain in terms of a single oscillator-regulating activity, but instead indicates that at least two mutually coupled circadian pacemakers underlie the circadian rhythm of activity (Pittendrigh 1974, Pittendrigh and Daan 1976, Daan and Berde 1978).

Aschoff (1954) noted a number of years ago that the activity rhythm of many vertebrate species appeared to be made up of two components; a peak of activity often occurs during the first few hours of the active phase and a second peak at the end of the active phase. It is important to note that a two-oscillator model is not necessary to explain such a bimodal distribution of activity, since each component could be coupled to different phase points of the same underlying circadian pacemaker. Therefore, we reserve the use of the term "splitting" for only those instances in which there is evidence that two parts of the same rhythm have become dissociated from each other such that the phase relationship between the two components is altered, and where there is evidence to suggest that the two components are capable of free-running with different circadian periods.

Splitting of circadian rhythms into two components has now been observed to occur in a variety of different vertebrate species, including hamsters, rats, monkeys, tree shrews, mice, squirrels, starlings and lizards (Pittendrigh 1960, Hoffmann 1971, Gwin-

1 Department of Neurobiology and Physiology, Northwestern University, Evanston, IL, 60201

Vertebrate Circadian Systems
Ed. by J. Aschoff, S. Daan, and G. Groos

ner 1974, Pittendrigh and Daan 1976, Underwood 1977, Boulos and Terman 1979, Fuller et al. 1979, Ellis et al. 1982). Therefore, it appears that the circadian system in vertebrates is organized such that a single circadian rhythm may be driven by at least two endogenous circadian pacemakers. The purpose of this brief review is to report on some of our recent findings on splitting in golden hamsters *(Mesocricetus auratus)* and to relate this work to previously published studies on the splitting phenomenon.

2 Formal Properties of the Split Circadian System

We have recorded the rhythm of wheel-running activity from over 200 individually housed hamsters maintained in constant darkness (DD) and over 150 animals maintained in constant light (LL) for long periods of time (i.e., between 4 and 15 months). While we have never observed splitting to occur in any animal housed in DD, we have found that in about 70% of the animals maintained in LL (light intensity = 50-300 lx), the activity rhythm will split into two components after about 40-100 days in LL. Although there is some suggestion that surgical intervention (see Part 3) or brief periods of darkness can induce splitting (Ellis et al. 1982), splitting in hamsters maintained in LL can occur spontaneously with no external intervention.

The basis of a theoretical explanation for why splitting occurs regularly in LL but not in DD has been provided by Pittendrigh and Daan (1976). They suggested that the period (τ) of one of the circadian pacemakers may be a positive while τ of the other pacemaker is a negative function of light intensity. Thus, in DD the intrinsic periods of the two pacemakers may be very similar and the normal coupling mode between the two oscillators is easily maintained. However, during exposure to LL the difference in the intrinsic periods of the two oscillators may become so great that the normal coupling mode can no longer be maintained and splitting occurs.

In hamsters, one of two patterns is usually observed when the activity rhythm spontaneously changes from the nonsplit to the split condition. In about 75% of our animals in which the activity rhythm dissociates into two components, splitting occurs gradually; the two activity bouts slowly drift apart from each other (Fig. 1). Pittendrigh (1974) has labeled these two bouts of activity in the nocturnal hamster the evening (E) and morning (M) components, based on whether they appear to be derived from the early (evening) or late (morning) portion of the active phase prior to the occurrence of splitting. When splitting occurs gradually, one component free-runs with a period less than the period of the previous nonsplit rhythm and activity which normally occurred late in the active phase of the nonsplit rhythm often disappears. In the nocturnal hamster, this bout of activity is labeled the M component. The second bout of activity con-

→

Fig. 1. Daily wheel-running activity record of a male hamster maintained in constant light *(LL)* and constant darkness *(DD)*. Successive days are plotted from top to bottom and the activity record has been double plotted over a 48-h time interval to facilitate visualization of the rhythm. On about day 92 the morning *(M)* component begins to drift away from the evening *(E)* component until the two components recouple approximately 12 h out of phase with each other. On day 217 of LL the animal was transferred into DD at the onset of the M component *(arrow)*. After 26 days in DD the animal was transferred back into LL *(black circle)* approximately 4-5 h after the onset of activity. (Earnest and Turek 1982)

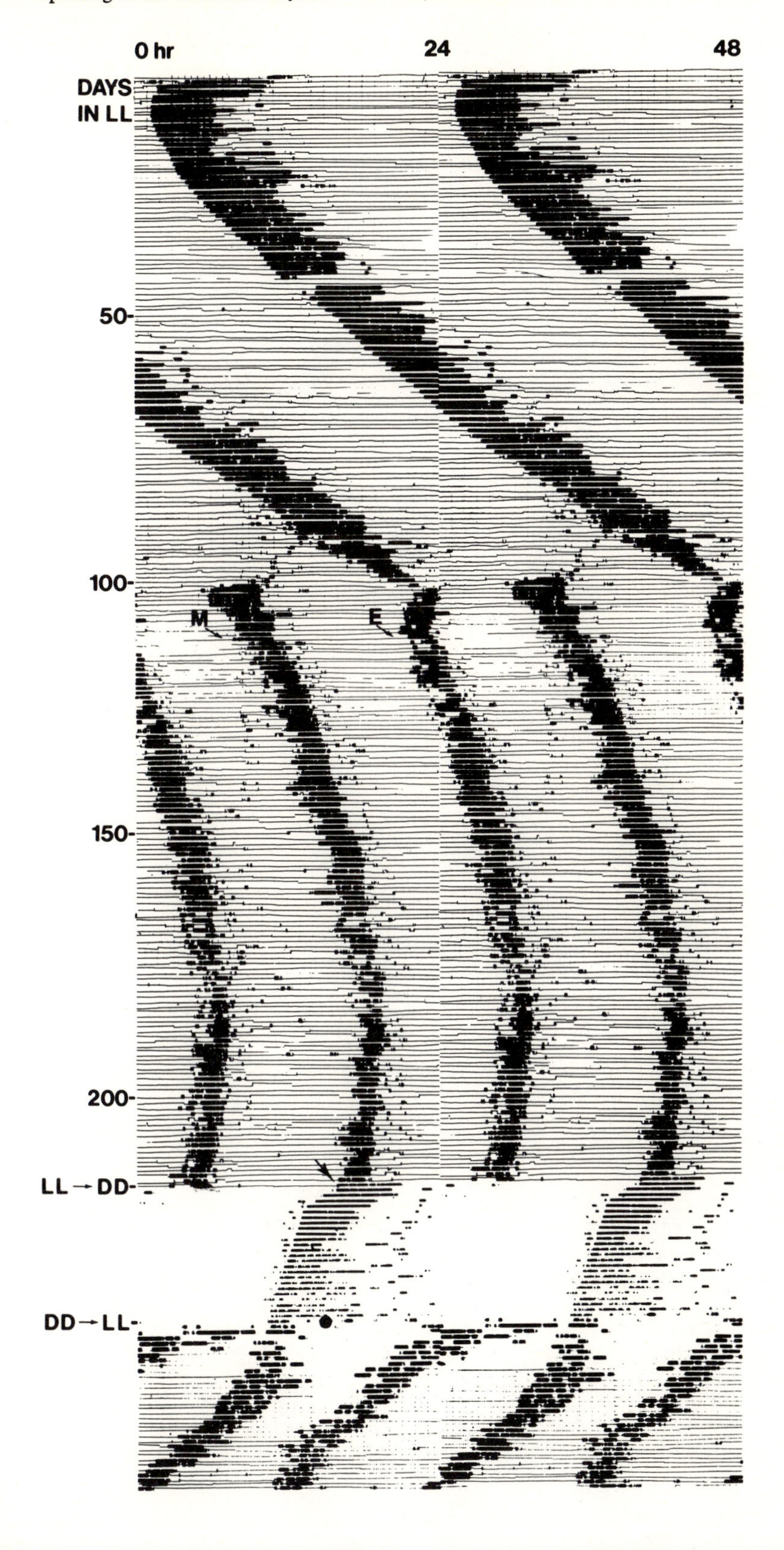
0 hr
24
48
DAYS IN LL
50
100
M
E
150
200
LL → DD
DD → LL

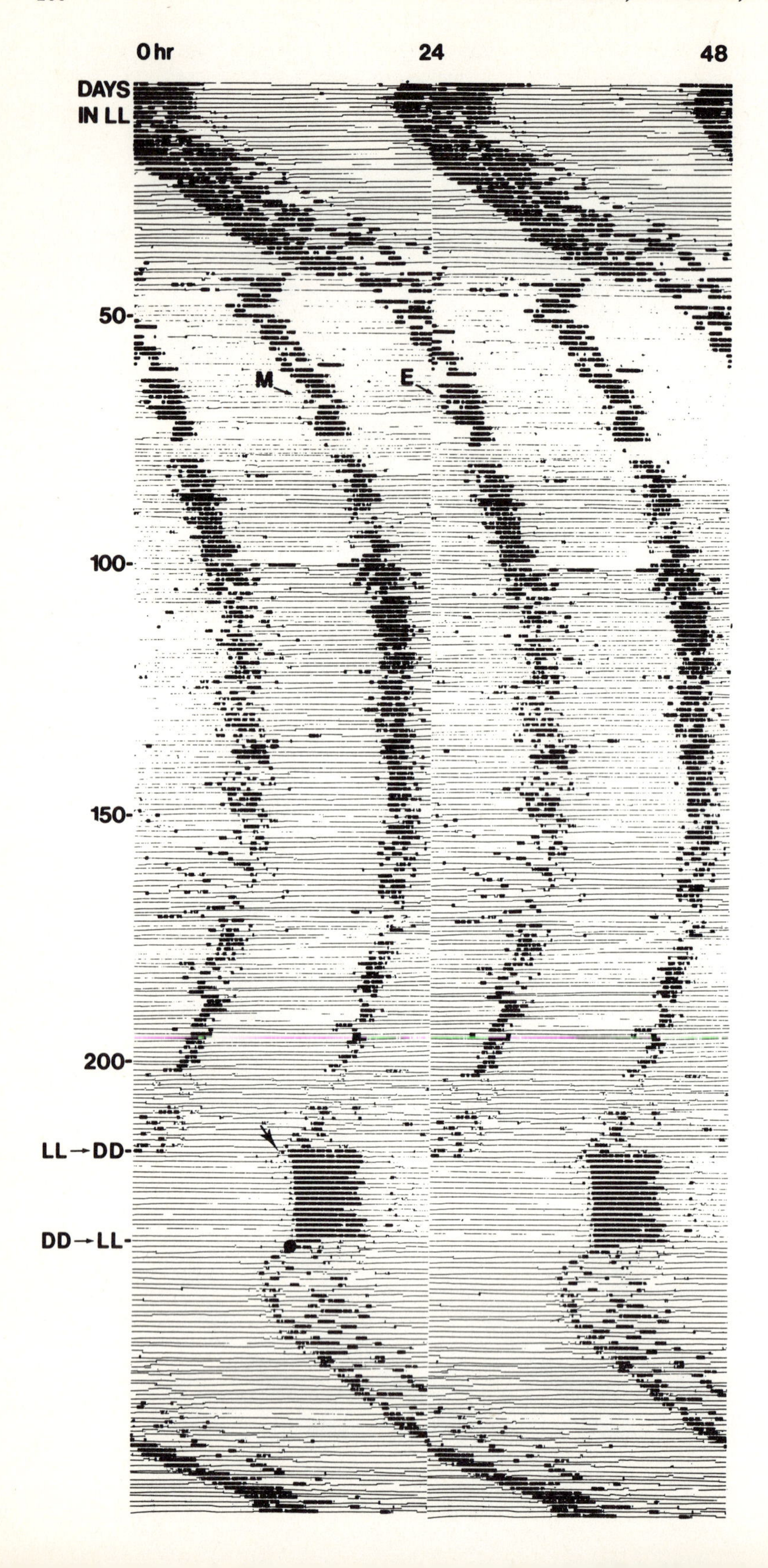
0 hr
24
48
DAYS
IN LL
50
M
E
100
150
200
LL→DD
DD→LL

tinues to free-run in phase with the onset of activity of the previous nonsplit rhythm, and is labeled the E component. Thus, when splitting occurs gradually, the M component is observed to slowly phase advance relative to the E component until they are about 12 h apart, at which time the two components recouple and free-run with the same period.

In about 25% of our animals splitting occurs rapidly with no apparent transients. The rapid occurrence of splitting is not characterized by the gradual drifting apart of the two activity bouts, instead, one day the activity rhythm is intact and on the following day two activity components appear about 12 h out of phase with each other. These two components then free-run with the same period. When splitting does occur rapidly, one activity component remains in phase with the onset of activity of the previous nonsplit rhythm and therefore we have also designated this the E component. The second activity component which occurs 12 h out of phase with the E component is designated the M component since this is the same phase relationship that is eventually assumed by the M component during the gradual development of splitting. Whether splitting occurs gradually or rapidly, τ of the split activity rhythm is always less than τ of the previous nonsplit rhythm (Pittendrigh and Daan 1976, Ellis et al. 1982, Earnest and Turek 1982).

The split activity pattern appears to be a very stable one, since less than 5% of our splitters ever show a spontaneous refusion of the two activity bouts while maintained in LL. Pittendrigh and Daan (1976) reported a few instances of refusion of the split components and suggested that this may have occurred in response to a reduction in light intensity due to aging of the lights. Boulos and Terman (1979) have reported that the transfer of a single rat with a split drinking rhythm from LL to DD abolished splitting. In view of the fact that (1) splitting is never observed to develop in hamsters maintained in DD, and (2) there is some suggestion that a reduction in light intensity may abolish the split condition, we have recently completed a detailed analysis of the effect of transferring split hamsters from LL to DD.

Transfer of split hamsters from LL to DD (N = 38) at a variety of different circadian times (e.g., at the onset of either E or M or between the two components) always results in the rapid (1-4 days) abolishment of splitting; the activity components refuse and assume the phase relationship that is characteristic of hamsters maintained in DD (Earnest and Turek 1982). A similar rapid loss of splitting also follows bilateral enucleation (N = 4), indicating that the eyes are necessary for splitting to be maintained. In order to analyze the formal properties of refusion we transferred 16 hamsters with a split-activity rhythm from LL to DD at either the onset of E (N = 9) or the onset of M (N = 7). In 14 of 16 animals transferred from LL to DD, the onset of activity in DD appeared to be derived from that component of the split activity pattern that coincided with the LL to DD transition (Figs. 1 and 2; Earnest and Turek 1982). In the other two animals the onset of activity in DD occurred in phase with that component of the split activity rhythm which was not coincident with the LL to DD transition. This find-

Fig. 2. Daily wheel-running activity record of a male hamster maintained in LL and DD in which splitting occurred spontaneously after about 40 days in LL. On day 218 of LL the animal was transferred into DD at the onset of the M component *(arrow)*. After 18 days in DD the animal was transferred back into LL near the onset of activity *(black circle)*. See Fig. 1 for further details. (Earnest and Turek 1982)

ing suggests that either the E or the M component can be associated with either the beginning or the end of the nonsplit activity phase, and that either of the two oscillators which underlie the intact activity rhythm can be responsible for the early or late part of the active phase. Alternatively, each oscillator may play a role in regulating all of the activity which occurs throughout the active phase.

Transfer of 16 hamsters with a split activity rhythm from LL to DD resulted in an increase in τ in 14 and a decrease in τ in 2 animals (see Fig. 1 for a decrease and Fig. 2 for an increase in τ). When splitting occurs spontaneously in LL, τ is always shortened (Pittendrigh and Daan 1976), thus it might be expected that a return to the nonsplit state would lengthen τ. On the other hand, in hamsters, as well as in other nocturnal species, a transfer from LL to DD usually shortens the period of the activity rhythm if splitting has not occurred in LL (Aschoff 1979). The change in period which occurs when split hamsters are transferred from LL to DD is presumably the sum of the two effects of DD. Apparently, the change in period (lengthening) which is due to a shift from the split to the nonsplit coupling mode is usually greater than the change in τ (shortening) that occurs following a reduction in light intensity.

The pattern of recoupling of the two activity bouts which was induced by the transition from LL to DD was not predictable. Sometimes the onset of the refused activity rhythm was phase-advanced (10/16 animals; Fig. 1), and at other times it was phase-delayed (6/16 animals; Fig. 2) relative to the activity component that was coincident with the LL to DD transition. Although a transfer from LL into DD usually resulted in an immediate return to the normal coupling mode, there were often 3-5 days of transients before a new steady state τ for the refused rhythm was achieved (Fig. 1). These transients suggest that it takes the two oscillators at least a few cycles to go from the split to the nonsplit coupling mode, just as it usually takes a number of cycles to complete the transition from the nonsplit to the split coupling mode (Fig. 1).

When previously split hamsters were maintained in DD for 15-30 days, transfer back into LL resulted in the reinitiation of splitting (7 of 9 animals) within 1-4 days if the transfer occurred 4-5 hours after the onset of activity (Fig. 1; Earnest and Turek 1982). This was unexpected since splitting in hamsters normally takes 30-60 days to develop in LL (Pittendrigh and Daan 1976). Thirteen previously split animals that were transferred from DD back into LL at other circadian times (i.e., 0, 1, 2, 3, 6, 7, 17, 18, or 20 h after activity onset) did not show a rapid reinitiation of splitting (Fig. 2). The fact that the rapid reinitiation of splitting could only be induced if the transfer to LL occurred near the middle of the subjective night suggests that this phenomenon may be related to the phase response curve (PRC) of the hamster. Light pulses presented about 4-5 h after the onset of the activity fall within the transition region or the hamster's PRC; exposure to light during this period may induce either phase delays of phase advances (Daan and Pittendrigh 1976, Ellis et al. 1982). It is not known if each of the two oscillators which underlie the circadian activity rhythm have a complete PRC with both phase advances and phase delays. Pittendrigh and Daan (1976) have raised the possibility that the PRC of the E oscillator has only phase delays while the PRC of the M oscillator has only phase advances. Perhaps the rapid onset of splitting observed in hamsters transferred from DD into LL near the middle of the subjective night is due to the abrupt onset of light delaying one oscillator while the other is phase-advanced, with rapidly leads to the oscillators being 180° out of phase with each other.

The rapid induction of splitting that is induced by the transfer from DD to LL near the middle of the subjective night appears to be a history-dependent effect of the previous split condition. The same DD to LL transition does not induce splitting (0/10 trials) if the animals had not previously shown the split conditions during a prior lengthy exposure to LL (Earnest and Turek 1982). Importantly, during the second prolonged exposure to LL, the activity rhythm of half of these animals spontaneously split into two components, indicating that splitting could occur in these animals. This history-dependent effect of splitting suggests that there are some "after-effects" of splitting at a physiological level. Presumably, some change in the physiology of the circadian system which accompanies splitting continues even after the system has refused in constant darkness.

3 Physiological Basis of Splitting

Over the last 10 years a number of studies have demonstrated that both the nervous and endocrine systems play a major role in the physiology of the circadian oscillators of vertebrates (Menaker et al. 1978, Rusak and Zucker 1979, Chap. 5.1, this Vol.). However, at the present time only a few studies have been directed at determining the physiological basis for splitting or what alterations in the neural and/or endocrine environment could lead to a dissociation of circadian oscillators which regulate a single rhythm.

The observations that the removal of the pineal gland or the gonads, as well as the administration of various hormones produced by these organs (e.g., melatonin, testosterone, estrogens), can alter the circadian rhythm of activity raises the possibility that these organs may play a functional role in splitting (Morin et al. 1977, Menaker et al. 1978, Chap. 5.1, this Vol.). Indeed, both pineal and gonadal hormones have been hypothesized to be involved in the coupling of circadian oscillators to each other as well as to the rhythms they drive (Gwinner 1974, Fitzgerald and Zucker 1976, Morin 1980). Support for a role of gonadal hormones in splitting was obtained in starlings where splitting of the locomotor rhythm was associated with testicular growth, and injections of testosterone induced splitting in castrated birds (Gwinner 1974). An opposite result was observed in female hamsters where the administration of estradiol benzoate appeared to reduce the incidence of splitting in ovariectomized animals, although treatment with estradiol benzoate did not restore synchrony in already split ovariectomized animals (Morin 1980). Evidence for a role of the pineal gland in splitting has been obtained in lizards, where it was found that pinealectomy could induce splitting of the circadian activity rhythm during exposure to LL (Underwood 1977). Although pinealectomy only induced splitting in 1 of 12 animals (the other 11 animals showed either a marked change in period or arrhythmicity), splitting has never been observed in intact lizards under any lighting conditions.

Despite the fact that hormones from the pineal gland and the gonads may be able to facilitate or inhibit the incidence of splitting, data from hamsters indicate that neither the presence (or absence) of these organs is necessary for splitting. Splitting has been observed to develop in both castrated and pinealectomized hamsters exposed to LL and once splitting has occurred, neither surgical treatment has a noticeable effect on the activity rhythm (Turek and Ellis 1979, Ellis and Turek, unpubl. results). Furthermore, the administration of melatonin or testosterone, either continuously via hormone-filled Silastic capsules or via daily injections, fails to alter the split condition (Fig. 3; Ellis and Turek, unpubl. results).

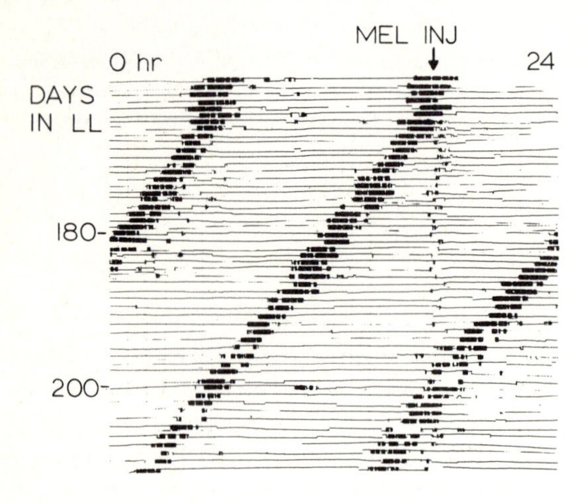

Fig. 3. Daily wheel-running activity record of a male hamster that was maintained in LL for 210 days. The activity rhythm had split into two components during the first 100 days of LL (data not shown). From days 158 to 210 of LL the animal received an injection of melatonin *(arrow)* at the same real clock time each day. No noticeable disturbance of the split activity pattern by either the daily injection or the handling of the animal was discernible. (Ellis and Turek, unpubl. results)

In searching for a neural basis for the phenomenon of splitting, we have focused our attention on the suprachiasmatic nuclei (SCN) of the hypothalamus because of the major role these nuclei play in the generation of circadian rhythms in vertebrates (see Part 3, this Vol.). We have recently observed that following the ablation of a single SCN in hamsters, the split activity pattern disappears and a single bout of activity is established (Pickard and Turek 1982). In view of the fact that the SCN are reciprocally innervated (Silverman and Pickard 1980) these results suggest that an interaction between the two SCN may play a role in the splitting phenomenon. One interpretation of these results is that the bilaterally paired SCN might function as separate circadian oscillators in a manner similar to mutually coupled bilaterally paired oscillators in invertebrates (Block and Page 1978, Hudson and Lickey 1980, Jacklet 1981). Alternatively, the split activity pattern may be abolished following the destruction of a critical amount of SCN tissue; whether this destruction is bilateral or unilateral may not be important.

4 Effect of Splitting on the Cycle of Lordosis Behavior

Until recently, all the reported cases of splitting in vertebrates involved only the circadian rhythm of locomotor activity. Although the data are not extensive, it appears that many other rhythms may also be capable of splitting into two components. Shibuya et al. (1980) found that the circadian rhythm of drinking in a single hamster dissociated into two components at about the same time as the rhythm of wheel-running activity. Furthermore, when the running wheel was removed from the cage, the split drinking rhythm persisted, indicating that splitting of this rhythm was not dependent on the expression of the rhythm of wheel-running. In rats there is evidence that splitting can occur in the circadian rhythms of feeding, drinking, and electrical brain self-stimulation (Boulos and Terman 1979). Fuller et al. (1979) have demonstrated that splitting of the colonic temperature rhythm in squirrel monkeys can occur even though the skin temperature rhythm continued to show a unimodal circadian rhythm. Taken together, these data indicate that a variety of circadian functions are under the control of at least two circadian pacemakers which can be uncoupled from each other.

In an attempt to develop a better understanding of how other circadian rhythms may be affected when the rhythm of locomotor activity has dissociated into two components, we have recently analyzed the rhythm of lordosis behavior in hamsters after splitting has occurred (Swann and Turek 1982). While its 4-day period length prevents the lordosis rhythm from being considered a true circadian rhythm, it appears that the onset of lordosis does have a circadian component, since it occurs at or just prior to the onset of activity in hamsters entrained to a light-dark cycle as well as in animals maintained in DD or LL (Alleva et al. 1971, Stetson and Anderson 1980, Swann and Turek, unpubl. results). Furthermore, the signal which controls the precise timing of lordosis behavior, the large amount of luteinizing hormone (LH) released by the pituitary gland on proestrous, has a circadian basis. Although the LH surge occurs only at a precise time once every 4 days in the intact animal (Stetson and Gibson 1977), a similar release of LH occurs daily in ovariectomized animals at the same time each day (Shander and Goldman 1978, Stetson et al. 1978).

We have found that the rhythm of lordosis behavior (as determined by exposing the female to a male hamster at 2-h intervals) is maintained in hamsters showing a split activity rhythm. Among 21 female hamsters with a split rhythm of activity 18 animals showed regular (i.e., 4-day) or irregular (i.e., non-4 day) cycles in lordosis behavior (Swann and Turek 1982). Only two animals were observed to be in constant behavioral estrus, while lordosis was never observed in one animal. Among the 18 animals showing regular or irregular cycles in lordosis behavior, seven showed lordosis onsets only near the onset of the E component of the split activity rhythm, five began lordosis behavior only near the onset of the M component, and one animal initiated lordosis between the two components. Importantly, in five animals the onset of lordosis was associated on different days with either the E or M component (Fig. 4).

These data indicate that cycles in lordosis behavior persist even after the circadian rhythm of activity has dissociated into two components and that the onset of lordosis still occurs near the onset of one of the activity bouts. The observation that either the E or M activity component could be associated with the onset of lordosis behavior was unexpected since the onset of lordosis always occurs near the onset of locomotor activity when splitting has not occurred. Thus when the activity rhythm is intact, the onset of lordosis is always associated with what is presumably the E oscillator. An examination of the temporal pattern of the hormones which underlie lordosis behavior in females with a split activity pattern is needed to elucidate the relationship between the estrous cycle and the split circadian rhythm of activity.

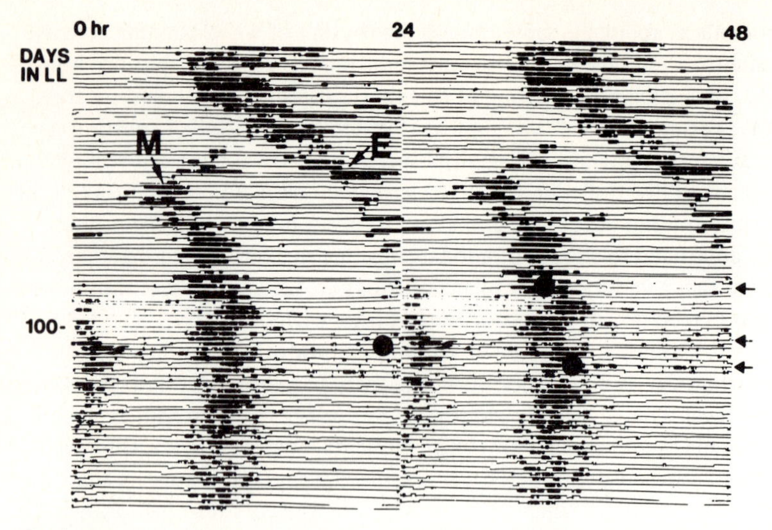

Fig. 4. Daily wheel-running activity record of a female hamster that was maintained in LL for 134 days (first 35 days have been omitted from record). Periodically the female was exposed to a male hamster at 2-h intervals to determine the time of the onset of lordosis behavior. *Black circles,* indicated by *arrows in the right hand border,* represent the time on that day when lordosis behavior was first observed. Note that on two days the onset of lordosis was associated with the M component of the split activity rhythm, while on another day it was associated with the E component. (Swann and Turek 1982)

5 Conclusions

The observation that a single circadian rhythm can dissociate into two distinct components, and that each of these two components can free run for a period of time with clearly different frequencies, indicates that circadian rhythms in vertebrates are generated by the interaction of two coupled circadian pacemakers. While recent studies have elucidated some of the formal properties of the split system, and have offered some insight into the physiological basis for the splitting phenomenon, a number of fundamental questions remain unanswered. For example, it is not known how the two split activity components relate to the normal nonsplit activity rhythm. The observations that (1) following dark-induced refusion of the split system either the E or the M component can be associated with either the beginning or the end of the refused activity rhythm, and (2) that the onset of lordosis behavior can be associated with either the E or M activity component, indicate that the two split activity bouts may not simply represent the evening and morning portions of the nonsplit activity rhythm. It may be that either of the two oscillators underlying the nonsplit activity rhythm can be responsible for the early or late part of the active phase, or that each oscillator plays a role in regulating all the activity which occurs through the active phase of the intact rhythm. At the physiological level, questions about how neural tissue is organized so that oscillators can be coupled in two different modes, or how oscillators interact to drive circadian rhythms remain to be answered. It is anticipated that the "split circadian system" will prove to be an extremely useful model for developing a more complete understanding of the fundamental properties and the physiology of the circadian system in vertebrates.

Acknowledgments. This work was supported by a grant from the Whitehall Foundation, NIH Grants HD-09885, HD-12622 and Research Career Development Award HD-00249 to F.W.T.

References

Alleva JJ, Waleski MV, Alleva FR (1971) A biological clock controlling the estrus cycle of the hamster. Endocrinology 88: 1368-1379

Aschoff J (1954) Zeitgeber der tierischen Jahresperiodik. Naturwissenschaften 41: 49-56

Aschoff J (1979) Circadian rhythms: Influences of internal and external factors on the period measured in constant conditions. Z Tierpsychol 49: 225-249

Aschoff J, Wever R (1976) Human circadian rhythms: a multioscillatory system. Fed Proc 35:2326-2332

Block GD, Page TL (1978) Circadian pacemakers in the nervous system. Annu Rev Neurosci 1: 19-34

Boulos Z, Terman M (1979) Splitting of circadian rhythms in the rat. J Comp Physiol 134: 75-83

Daan S, Berde C (1978) Two coupled oscillators: simulations of the circadian pacemaker in mammalian activity rhythms. J Theor Biol 70: 297-313

Daan S, Pittendrigh CS (1976) A functional analysis of circadian pacemakers in nocturnal rodents. II. The variability of phase response curves. J Comp Physiol 106: 253-266

Earnest DJ, Turek FW (1982) Splitting of the circadian rhythm of activity in hamsters: effects of exposure to constant darkness and subsequent re-exposure to constant light. J Comp Physiol 145: 405-411

Ellis GB, McKlveen RE, Turek FW (1982) Dark pulses affect the circadian rhythm of activity in hamsters kept in constant light. Am J Physiol 242: R44-R50

Fitzgerald KM, Zucker I (1976) Circadian organization of the estrous cycle of the golden hamster. Proc Natl Acad Sci USA 73: 2923-2927

Fuller CA, Sulzman FM, Moore-Ede MC (1979) Circadian control of thermoregulation in the squirrel monkey, *Saimiri sciureus.* Am J Physiol 236: R153-R161

Gwinner E (1974) Testosterone induces "splitting" of circadian locomotor activity rhythms in birds. Science 185: 72-74

Hoffmann K (1971) Splitting of the circadian rhythm as a function of light intensity. In: Menaker M (ed) Biochronometry. Washington DC, Natl Acad Sci, pp 134-150

Hudson DJ, Lickey ME (1980) Internal desynchronization between two identified circadian oscillators in Aplysia. Brain Res 183: 481-485

Jacklet JW (1981) Circadian timing by endogenous oscillators in the nervous system: toward cellular mechanisms. Biol Bull 160: 199-227

Menaker M, Takahashi JS, Eskin A (1978) The physiology of circadian pacemakers. Annu Rev Physiol 40: 501-526

Moore-Ede MC, Schmelzer WS, Kass DA, Herd JA (1976) Internal organization of the circadian timing system in multicellular animals. Fed Proc 35: 2333-2338

Morin LP (1980) Effect of ovarian hormones on synchrony of hamster circadian rhythms. Physiol Behav 24: 741-749

Morin LP, Fitzgerald KM, Rusak B, Zucker I (1977) Circadian organization and neural mediation of hamster reproductive rhythms. Psychoneuroendocrinology 2: 73-98

Pickard GE, Turek FW (1982) Splitting of the circadian rhythm of activity is abolished by unilateral lesions of the suprachiasmatic nuclei. Science 215: 1119-1121

Pittendrigh CS (1960) Circadian rhythms and the circadian organization of living systems. Cold Spring Harbor Symp Quant Biol 25: 159-184

Pittendrigh CS (1974) Circadian oscillations in cells and the circadian organization of multicellular systems. In: Schmitt FO, Worden FG (eds) The neuroscience: Third study program. MIT Press, Cambridge, pp 437-458

Pittendrigh CS, Daan S (1976) A functional analysis of circadian pacemakers in nocturnal rodents. V. Pacemaker structure: a clock for all seasons. J Comp Physiol 106: 333-355

Rusak B, Zucker I (1979) Neural regulation of circadian rhythms. Physiol Rev 59: 449-526

Shander D, Goldman B (1978) Ovarian steroid modulation of gonadotropin secretion and pituitary responsiveness to luteinizing hormone-releasing hormone in the female hamster. Endocrinology 103: 1383-1393

Shibyua CA, Melnyk RB, Mrosovsky N (1980) Simultaneous splitting of drinking and locomotor activity rhythms in a golden hamster. Naturwissenschaften 67: 45-46

Silverman AJ, Pickard GE (1980) Retinal and CNS input to the suprachiasmatic nucleus of the golden hamster. Soc Neurosci Abstr 6: 266

Stetson MH, Anderson PJ (1980) Circadian pacemaker times gonadotropin release in free-running female hamsters. Am J Physiol 238: R23-R27

Stetson MH, Gibson JT (1977) The estrous cycle in golden hamsters: a circadian pacemaker times preovulatory gonadotropin release. J Exp Zool 201: 289-294

Stetson MH, Watson-Whitmyre M, Matt KS (1978) Cyclic gonadotropin release in the presence and absence of estrogenic feedback in ovariectomized golden hamsters. Biol Reprod 19: 40-50

Swann J, Turek FW (1982) The cycle of lordosis behavior in female hamsters whose circadian activity rhythm has split into two components. Am J Physiol (in press)

Turek FW, Ellis GB (1979) The effect of dark pulses on the circadian rhythm of locomotor activity in hamsters maintained in constant light. Soc Neurosci Abstr 5: 462

Underwood H (1977) Circadian organization in lizards: the role of the pineal organ. Science 195: 587-589

5.5 Phase-Response Curves and the Dual-Oscillator Model of Circadian Pacemakers

Z. Boulos and B. Rusak[1]

1 Introduction

Phase-response curves (PRC's) and the splitting phenomenon have given rise to two convincing but thus far parallel interpretations of entrainment and other photic effects on circadian pacemakers.

On the one hand, the PRC shows that the responsiveness of circadian pacemakers to light is phase-dependent. In the hamster, as in other nocturnal animals, a single light pulse presented in otherwise constant darkness (DD) around the time of daily onset of locomotor activity causes a delaying phase shift of the activity rhythm, while a pulse presented toward the end of daily activity causes a phase advance. Pulses presented during much of the inactive phase of the rhythm have little or no effect (Burchard 1958, DeCoursey 1964, Elliott 1974, Daan and Pittendrigh 1976). Entrainment of a circadian rhythm by daily light-dark (LD) cycles can thus be accounted for by a daily phase shift, caused by periodic light exposure, which corrects for the difference between the free-running period (τ) measured in DD and the 24-h period of the LD cycle (Elliott 1974, Enright 1965, Eskin 1971, Pittendrigh 1965, Pittendrigh and Daan 1976b).

On the other hand, the occurrence of splitting – the dissociation of a circadian rhythm into two separate components that can temporarily free-run at different frequencies – is evidence that the pacemaker generating such rhythms consists of two mutually coupled oscillators, as described in a model by Pittendrigh and Daan (1976c). The model assumes that light exerts a differential effect on the periods of the two oscillators, and accounts for entrainment by their separate coupling to dusk (or lights-off) and dawn (or lights-on).

But the relationship between the PRC and the proposed dual-oscillator structure of the pacemaker remains to be determined. Pittendrigh and Daan (1976c) have suggested two possibilities. One is that light can only advance one oscillator and only delay the other; the former would therefore give rise to the advance portion of the PRC, the latter to the delay. Alternatively, each oscillator may be capable of both advances and delays and thus have a complete bidirectional PRC. We have attempted to distinguish between these two alternatives in the hamster by comparing the responses of split activity components to dark pulses presented on a background of constant light (LL).

1 Department of Psychology, Dalhousie University, Halifax, Nova Scotia, Canada B3H4J1

Vertebrate Circadian Systems
Ed. by J. Aschoff, S. Daan, and G. Groos

2 Dark Pulse PRC's for Normal Rhythms

Our use of dark rather than light pulses was based on the fact that almost all reported cases of splitting were obtained in animals maintained in LL (Gwinner 1974, Hoffmann 1971, Morin 1980, Pittendrigh 1960, 1974, Pittendrigh and Daan 1976c, Pohl 1972, Shibuya et al. 1980, Swade and Pittendrigh 1967), and that split rhythms were shown to re-fuse upon transfer to DD (Boulos and Terman 1979). At the time this study was initiated, however, the only published PRC's for dark pulses in vertebrates had been obtained in *Taphozus melanopogon*, a nocturnal tropical bat (Subbaraj and Chandrashekaran 1978). We therefore began by examining the effects of 2- and 6-h dark pulses presented in LL (80 lx), in hamsters showing normal free-running rhythms (Boulos and Rusak 1982). Two PRC's were obtained with each pulse duration. One represented the immediate phase shifts caused by the pulses, measured on the first day post-pulse, the other represented the steady-state phase shifts, measured after all transient cycles had ended. We obtained a measure of this transient activity by subtracting the immediate from the steady-state phase shifts, and calculated changes in steady-state τ.

The effects of 6-h dark pulses are summarized in Fig. 1. The pulses caused phase-dependent phase shifts of the activity rhythms, in a direction opposite to those caused

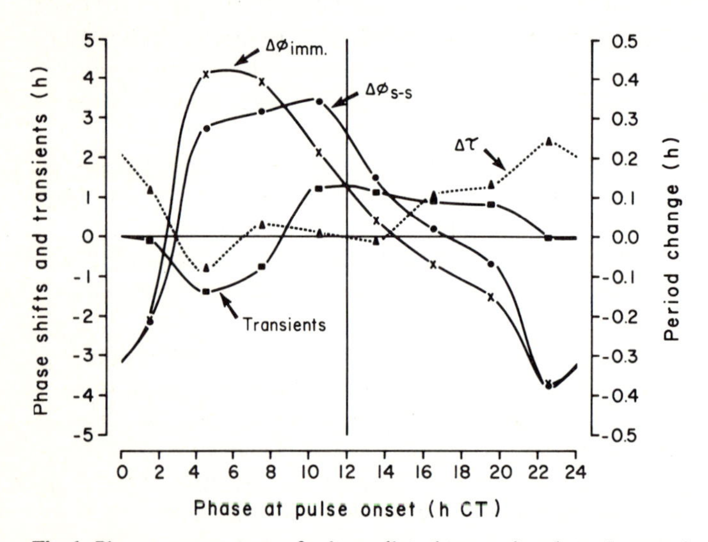

Fig. 1. Phase-response curves for immediate ($\Delta\phi_{imm.}$) and steady-state ($\Delta\phi_{s\text{-}s}$) phase shifts, transients, and changes in steady-state period ($\Delta\tau$; note expanded scale) caused by 6-h dark pulses. Phase advances are shown as positive and phase delays as negative values. Each *point* represents the average effect of several pulses (38 in all) grouped in 3-h blocks according to the phase of the rhythm at dark pulse onset. Circadian time (CT) is calculated by multiplying real time values by $\tau/24$ h. The *vertical line* at CT 12 h represents activity onset. (Redrawn from Boulos and Rusak 1982)

by light pulses of comparable duration in DD at equivalent circadian phases (cf. Burchard 1958). The PRC's for 2-h pulses were similar, but showed smaller phase shifts and were displaced slightly to the right relative to those for 6-h pulses (Boulos and Rusak 1982).

Many pulses were followed by transient changes in τ that lasted up to 13 days. Both advancing and delaying transients were observed, depending on the phase of the rhythm at dark pulse onset (Fig. 1). The pulses also caused phase-dependent changes in steady-state τ, with large increases accompanying phase delays, and fewer and smaller decreases accompanying some of the phase advances (Fig. 1). Transient as well as long-lasting changes in τ are well known to accompany phase shifts induced by light pulses (Pittendrigh and Daan 1976a).

3 Effects of Dark Pulses on Split Activity Rhythms

Having established that dark pulse PRC's were mirror images of, but otherwise similar to those for light, we set out to examine the effects of dark pulses on split activity rhythms. The results reported here were obtained from six hamsters (five intact males and one ovariectomized female) which received a combined total of 15 6-h pulses while in the split condition. Splitting occurred spontaneously in three of these animals following long-term exposure to LL (80 lx). The remaining three had received one or more dark pulses before they showed splitting (see Boulos and Rusak 1982).

Portions of the activity records of four hamsters are presented in Fig. 2. They show that dark pulses can cause concurrent but unequal, even opposite, phase shifts of the two split components, and that the direction in which each is shifted depends on its phase at pulse onset. Furthermore, this phase-response relationship for individual components appears similar to that observed with normal rhythms. For example, the first pulse in Fig. 2D, presented 3 h prior to the onset of one component, caused a phase advance of that component. That same pulse preceded the second component by about 11 h, causing it to be phase-delayed. Both of these effects are consistent with the PRC for immediate phase shifts shown in Fig. 1. The results also show that both split components can be phase-advanced (Fig. 2A, B) as well as phase-delayed by dark pulses (Fig. 2B).

One complication in these data is that several pulses resulted in a merging of the split components, making it impossible to determine the extent to which each had been shifted. In some cases the components separated again several days later (Fig. 2D, pulses 2 and 3), while in others the effect was more durable (Fig. 2B and 2C).

A more serious problem, however, is the occurrence of several instances in which phase delays were expected but could not be documented in the activity records (Fig. 2A, pulses 3-5). The absence of observable phase delays in such cases may be due to a combination of two factors: first, the phase delays that were observed were always followed by rapidly advancing transients lasting 2-7 cycles. As a result, the final steady-state phase was either in line with, or showed an advance relative to the pre-pulse phase (Fig. 2B, pulses 1 and 2; Fig. 2D, pulse 1). Phase advances, on the other hand, were followed by either delaying (Fig. 2A, pulse 3; Fig. 2D, pulse 1) or no transients (Fig. 2A, pulse 5; Fig. 2B, pulse 1). Second, a component that was separated from dark pulse onset by about 8-14 h – the phase at which the PRC for normal rhythms predicts a phase delay – was often completely suppressed for the next 1-3 days (Fig. 2A, pulses 3-5; Fig. 2D, pulse 1); in contrast, the other component (the second member of the pair) often showed increased running activity on the first day post-pulse (Fig. 2A,

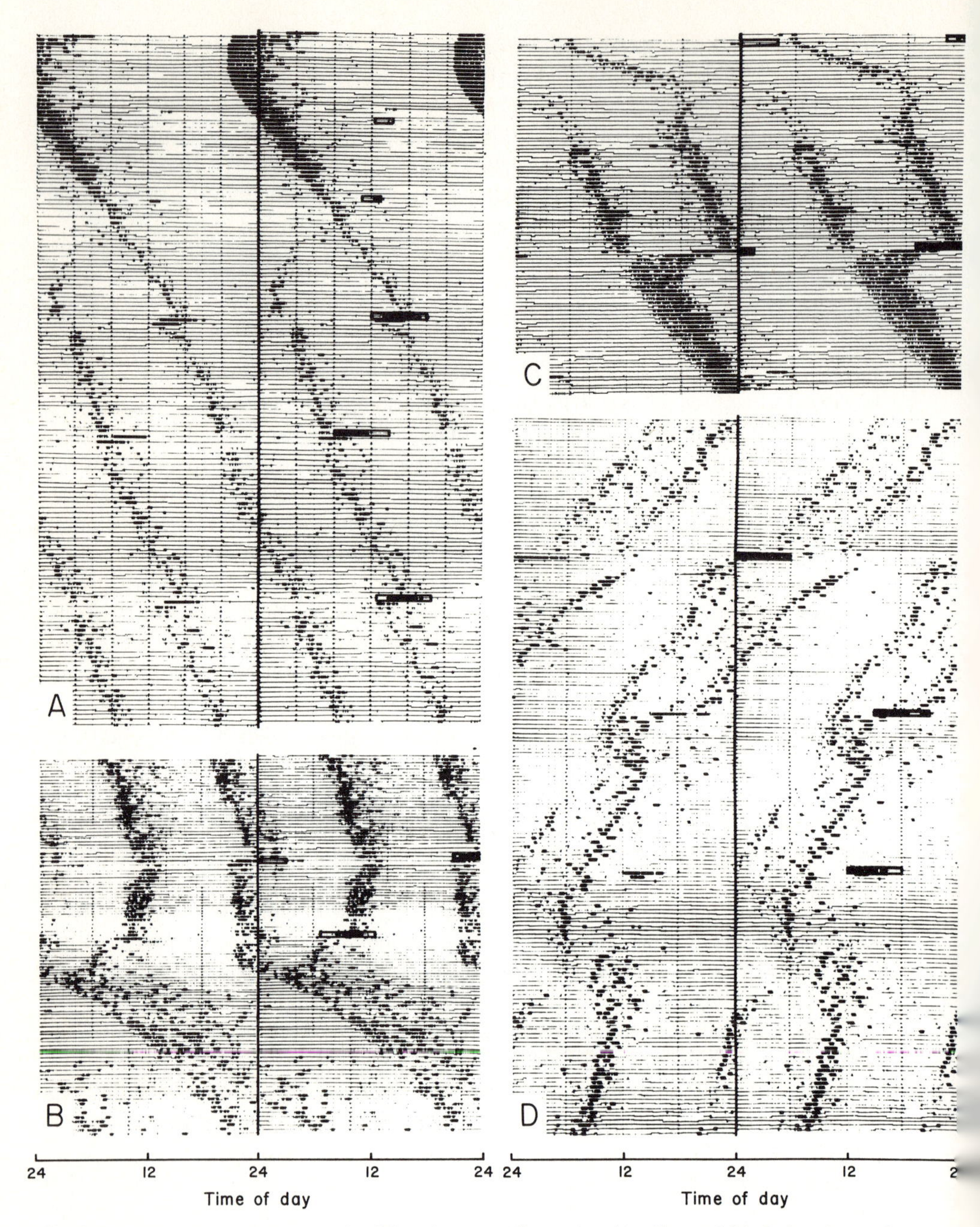

Fig. 2. Wheel-running activity records of four hamsters illustrating the effects of 6-h dark pulses on split rhythms. Three of the hamsters (**A-C**) were males, the fourth (**D**) an ovariectomized female; two (**A** and **C**) were also exposed to 2- or 6-h dark pulses before they showed splitting. In **A** the first day of the record corresponds to day 1 of LL, and in **B-D** to day 92 or more. The pulses are indicated by *rectangles* on the right half of the double-plotted records

pulses 3-5). Thus, it is possible that one of the components in Fig. 2A was in fact delayed, but that by the time it reappeared 3-4 days later, it had already advanced back to or beyond its pre-pulse phase.

Note that the size and direction of the transients that followed dark pulse presentation were such as to result either in the re-establishment of the prior phase relation between the split components (about 180° in all but the record of Fig. 2D), or in their merging to form a single daily activity band. We have also observed four cases in which dark pulses appeared to trigger splitting, and this was followed by transients in one or both components until they reached a 180° phase relation, at which point the two became synchronized and free-ran with a common period (Fig. 2A; see Boulos and Rusak 1982). These observations suggest that all transients – those shown by normal as well as split rhythms – may be attributable to the same cause: a differential effect of a light or dark pulse on two coupled oscillators temporarily altering their phase relation. The transients would then reflect the motion of one or both oscillators as they gradually regain a more stable phase relation. A similar interpretation was proposed many years ago by Pittendrigh et al. (1958) to account for the behavior of the eclosion rhythm of *Drosophila*, and more recently for that of the locomotor activity rhythm of hamsters (Pittendrigh 1981).

We have not yet examined the effects of 2-h dark pulses on split rhythms, but have obtained preliminary results in three hamsters, using a daily LD cycle with a 2-h dark segment. In two of these animals (Fig. 3A, B), both split components appeared to be synchronized by the LD cycle. Much of the running activity that occurred during the daily dark segment no doubt represents a direct or masking effect of darkness, but both animals showed at least some running activity immediately preceding dark onset, suggesting that one of the components was entrained by, and showed a small phase lead relative to the dark segment. Such a phase angle is consistent with the PRC for normal rhythms, since it is at this phase that a 2-h dark pulse causes the small phase advance necessary to correct for the difference between the period of the split component and that of the LD cycle (Boulos and Rusak 1982). The apparent entrainment of the second component at a phase angle of about 12 h relative to dark onset was more unexpected, since at this phase the PRC predicts a phase delay. It is possible that the dark segment did have a delaying effect, but that this was counteracted by the coupling force that keeps the two oscillators synchronized.

The responses of the third hamster (Fig. 3C) are more difficult to interpret. It is not clear, for example, whether running in the dark in this case was entirely a masking effect or whether it also represented entrainment of one component. The record does suggest that the second component was initially entrained by the LD cycle, but later broke loose and free-ran with a period longer than 24 h.

Many of the observations reported here clearly require confirmation before any definitive conclusions can be reached. Nevertheless, we feel that the data obtained thus far provide additional evidence that the circadian pacemaker in the hamster consists of two coupled oscillators, and suggest that these oscillators respond to dark pulses in a qualitatively similar manner, each behaving according to its own bidirectional PRC. This, of course, does not rule out the possibility that there are quantitative differences between the two PRC's. As mentioned earlier, a basic premise of Pittendrigh and Daan's model is that the two oscillators are differentially sensitive to light, and this could be

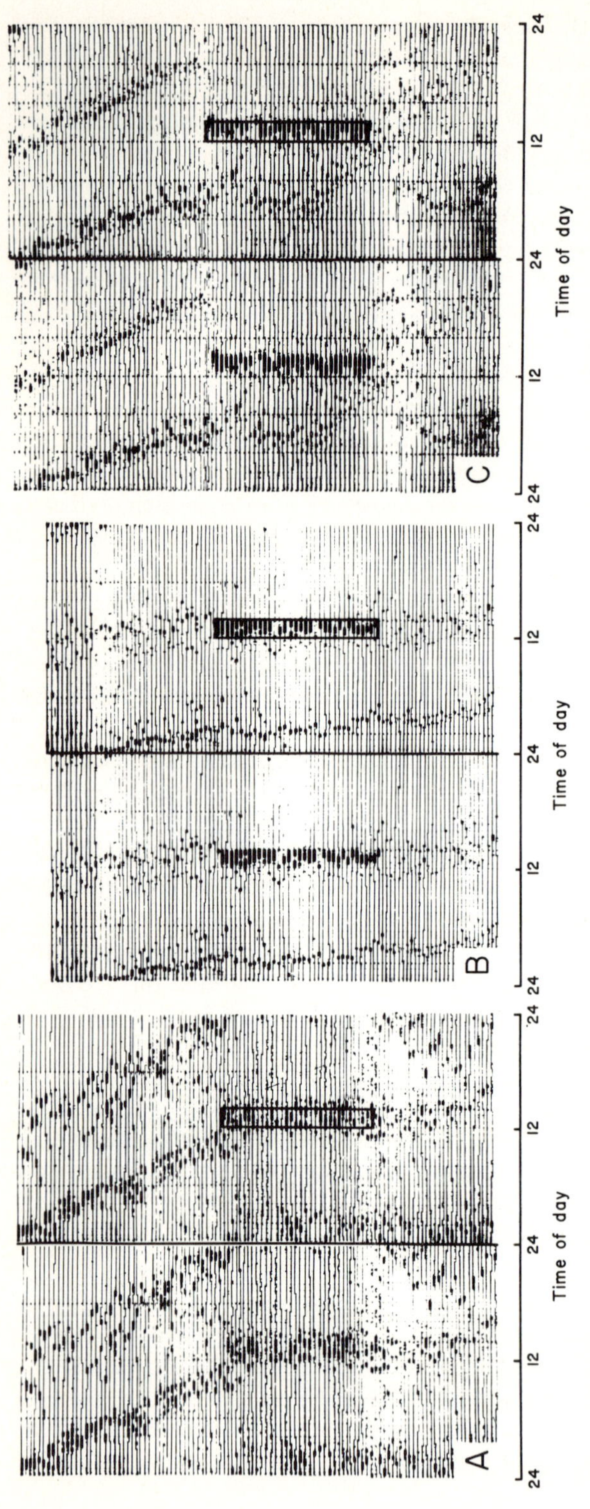

Fig. 3. Wheel-running activity records of three male hamsters that were exposed to daily LD cycles with 2-h dark segments while showing split rhythms. The hours of darkness are enclosed by *rectangles* on the right half of the records

reflected, for example, in the relative sizes of the advance and delay portions of their PRC's.

4 Concluding Remarks

The appeal of the dual-oscillator model described by Pittendrigh and Daan lies in its providing a unitary framework which accounts for many different phenomena, some of which – including splitting, transients, and τ and PRC lability – are difficult to account for by the actions of a single oscillator. The interpretation of these phenomena, however, rests to a large extent on certain assumptions about the phase relation between the two oscillators, which is only directly observable under split conditions. Pittendrigh and Daan (1976c) have suggested that for normal rhythms, this phase relation (Ψ) is reflected by the duration of daily activity, or α. But α is not always well defined in activity records, and even when it is, this does not allow the identification of equivalent phases of the two oscillators.

If each of these oscillators is in fact endowed with its own bidirectional PRC, as our data suggest, then it may be possible to obtain at least a rough estimate of their phase relation by taking into consideration the totality of the effects of light or dark pulses, i.e., immediate and steady-state phase shifts as well as τ changes. This is illustrated in the following examples, based on the data of Fig. 1. The terminology used is from Pittendrigh and Daan (1976c), where E (for evening) and M (for morning) represent the two coupled oscillators, and Ψ represents the phase relation between E and M. The E oscillator is thought to be primarily responsible for determining activity onset. We also assume, as do Pittendrigh and Daan, that changes in Ψ are reflected by changes in steady-state τ.

1. A 6-h dark pulse presented around circadian time 5 h (CT 5) causes a large immediate phase advance, followed by delaying transients. This would represent a larger phase advance of E than of M, followed by delaying transients by E (and possibly advancing transients by M) until a stable Ψ is re-established. The steady-state Ψ that is attained is somewhat greater than before, resulting in a small decrease in steady-state τ.
2. Similarly, the effects of a pulse presented around CT 12 would represent a larger phase advance of M than of E, followed by advancing transients by E until the prior Ψ is re-established.
3. Around CT 19 a pulse causes a larger phase delay of E than of M; E then advances gradually until a new steady-state Ψ is reached which, judging by the increase in steady-state τ, is smaller than before.
4. A pulse presented around CT 23 also causes a larger phase delay of E than of M, thereby decreasing Ψ, but this new Ψ is maintained and results in a large increase in steady-state τ.
5. One additional outcome – an immediate phase delay followed by delaying transients – was observed with 2-h dark pulses presented between CT 22 and CT 1, but not with 6-h pulses. This would represent a larger phase delay of M than of E, followed by delaying transients by E.

Thus, we suggest that the curve for immediate phase shifts shown in Fig. 1 is a more or less accurate representation of the PRC of the E oscillator, while the curve for steady-state phase shifts reflects the responses of both E and M. This implies that under LL, the two oscillators (and their PRC's) are phase-displaced by about 6 h, which is the phase difference between the peaks of the advance portions of the two PRC's of Fig. 1. We conclude by noting that Pittendrigh (1981, p. 120) has recently proposed a similar scheme, in which the behavior of the circadian pacemaker under LD is seen as reflecting the interaction of two oscillators – each with its own bidirectional PRC – phase-displaced by a few hours.

Acknowledgments. Supported by grants from the NSERC of Canada to B. Rusak and to G.V. Goddard. We are grateful to Dr. Goddard for his continued support and encouragement, to G. Eskes for various contributions to this research, and to M. Terman for comments on an earlier version of the manuscript.

References

Boulos Z, Rusak B (1982) Circadian phase response curves for dark pulses in the hamster. J Comp Physiol 146: 411-417

Boulos Z, Terman M (1979) Splitting of circadian rhythms in the rat. J Comp Physiol 134: 75-83

Burchard JE Jr. (1958) "Re-setting" a biological clock. Ph D thesis. Princeton Univ

Daan S, Pittendrigh CS (1976) A functional analysis of circadian pacemakers in nocturnal rodents. II. The variability of phase response curves. J Comp Physiol 106: 253-266

DeCoursey PJ (1964) Function of a light response rhythm in hamsters. J Cell Comp Physiol 63: 189-196

Elliott JA (1974) Photoperiodic regulation of testis function in the golden hamster: relation to the circadian system. Ph D thesis, Univ Texas, Austin

Enright JT (1965) Synchronization and ranges of entrainment. In: Aschoff J (ed) Circadian clocks. North-Holland, Amsterdam, pp 112-124

Eskin A (1971) Some properties of the system controlling the circadian activity rhythms of sparrows. In: Menaker M (ed) Biochronometry, Natl Acad Sci, Washington DC, pp 55-80

Gwinner E (1974) Testosterone induces "splitting" of circadian locomotor activity in birds. Science 185: 72-74

Hoffmann K (1971) Splitting of the circadian rhythm as a function of light intensity. In: Menaker M (ed) Biochronometry. Natl Acad Sci, Washington DC, pp 134-151

Morin LP (1980) Effect of ovarian hormones on synchrony of hamster circadian rhythms. Physiol Behav 24: 741-749

Pittendrigh CS (1960) Circadian rhythms and the circadian organization of living systems. Cold Spring Harbor Symp Quant Biol 25: 155-184

Pittendrigh CS (1965) On the mechanism of entrainment of a circadian rhythm by light cycles. In: Aschoff J (ed) Circadian clocks. North-Holland, Amsterdam, pp 277-297

Pittendrigh CS (1974) Circadian oscillations in cells and the circadian organization of multicellular systems. In: Schmitt FO, Worden FG (eds) The neurosciences: Third study program. MIT Press, Cambridge, Mass. pp 437-458

Pittendrigh CS (1981) Circadian systems: entrainment. In: Aschoff J (ed) Handbook of behavioral neurobiology, vol IV. Biological rhythms. Plenum Press, New York, pp 95-124

Pittendrigh CS, Daan S (1976a) A functional analysis of circadian pacemakers in nocturnal rodents. I. The stability and lability of spontaneous frequency. J Comp Physiol 106: 223-252

Pittendrigh CS, Daan S (1976b) A functional analysis of circadian pacemakers in nocturnal rodents. IV. Entrainment: pacemaker as clock. J Comp Physiol 106: 291-331

Pittendrigh CS, Daan S (1976c) A functional analysis of circadian pacemakers in nocturnal rodents. V. Pacemaker structure: a clock for all seasons. J Comp Physiol 106: 333-355

Pittendrigh CS, Bruce V, Kaus P (1958) On the significance of transients in daily rhythms. Proc Natl Acad Sci USA 44: 965-973

Pohl H (1972) Die Aktivitätsperiodik von zwei tagaktiven Nagern, *Funambulus palmarum* und *Eutamias sibiricus,* unter Dauerlichtbedingungen. J Comp Physiol 78: 60-74

Shibuya CA, Melnyk RB, Mrosovsky N (1980) Simultaneous splitting of drinking and locomotor activity rhythms in a golden hamster. Naturwissenschaften 67: 45-47

Subbaraj R, Chandrashekaran MK (1978) Pulses of darkness shift the phase of a circadian rhythm in an insectivorous bat. J Comp Physiol 127: 239-243

Swade RH, Pittendrigh CS (1967) Circadian locomotor rhythms of rodents in the arctic. Am Nat 104: 431-466

5.6 Circadian Control of Body Temperature in Primates

C.A. Fuller[1] and F.M. Sulzman[2]

1 Introduction

As homeotherms, primates are capable of maintaining a stable body temperature in a wide variety of thermal environments. In addition to showing a stable 24-h mean, primates also display a prominent diurnal variation in body temperature. The purpose of this review is to examine the physiological regulation and control of this rhythm in primates.

The temperature at any point in the body is a function of the heat content of that site. In general the thermoregulatory system operates to maintain core body temperature within relatively narrow limits. To do this mammals have evolved several effector mechanisms by which the heat content of the body can be altered (see Cabanac 1975 for review). These mechanisms include control of: metabolic heat production, evaporative heat loss, convective and radiative heat loss, internal heat conductance, and behavior. These effectors are coordinated by the thermoregulatory system so as to maintain a specific body temperature. One way this can be accomplished is by a negative feedback control system (Fig. 1). Here neural signals representing the body temperatures sensed by diverse thermoreceptors are compared with reference levels in the neural controllers, and modify neural outputs which then drive effectors to correct the sensed temperatures to appropriate levels (Satinoff 1978). The primary effectors which are utilized for thermoregulation depend on ambient temperature, since in certain environments some effectors may already be maximally utilized (Kleiber 1975).

Other physiological systems in the animal also influence body temperature regulation. For example, increases in heat production resulting from changes in activity levels may not be matched by increases in heat loss, resulting in an increase in body temperature. Influences such as age, sleep-state, eating and reproductive cycles can also influence body temperature regulation (Fig. 1).

Two examples of primate temperature rhythms are shown in Fig. 2. The top graph shows the average body temperature rhythm of the squirrel monkey *(Saimiri sciureus)* entrained to a light-dark (LD) 12:12 cycle (L:600 lx, D: < 1 lx). The 24-h mean is 37.4°C with a 1.9°C range about this mean. The body temperature was maintained above the mean when the lights were on, and fell progressively throughout the night

1 Division of Biomedical Sciences, University of California, Riverside, California 92521
2 Department of Biological Sciences, State University of New York, Binghamton, New York 13901

Vertebrate Circadian Systems
Ed. by J. Aschoff, S. Daan, and G. Groos

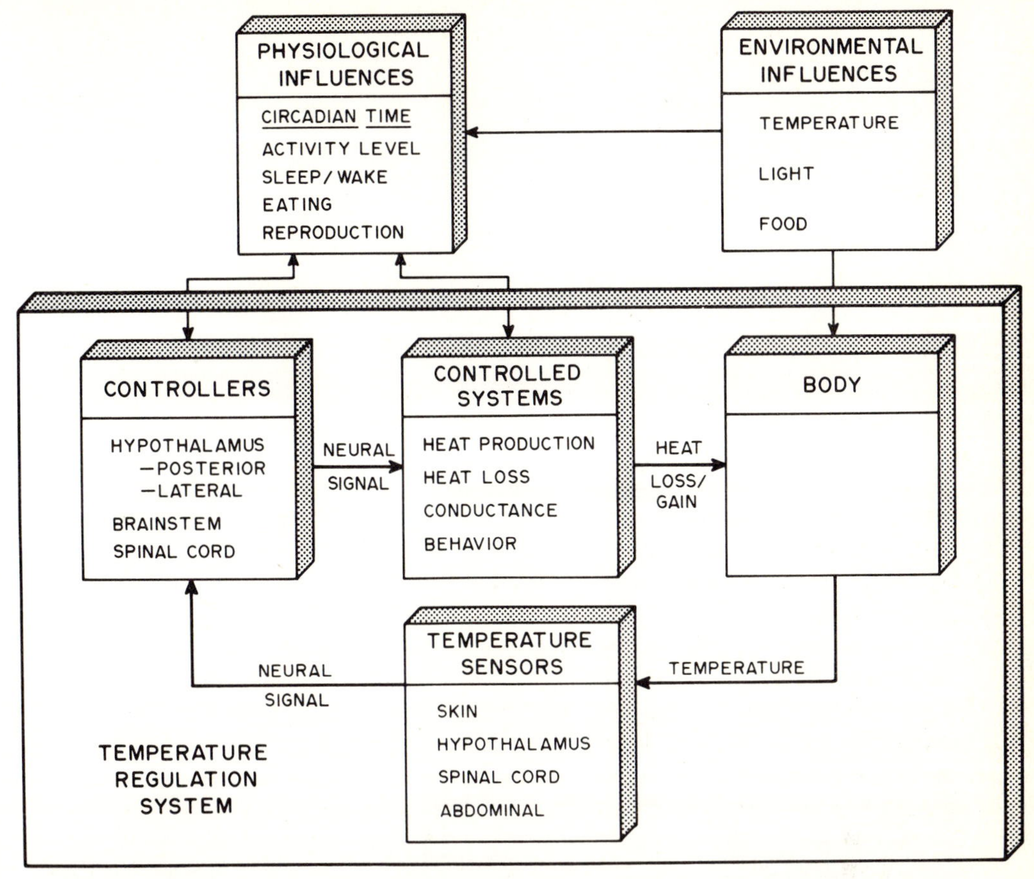

Fig. 1. Block diagram of the temperature regulation system showing the main components and the flow of information between them. Also shown are some influential factors on this control system

until it began to rise again 2 h before the lights were scheduled to come on. Conversely, the bottom graph shows the average body temperature rhythm of the nocturnal owl monkey *(Aotus trivergatus)* in an LD 12:12 cycle. In this species the 24-h average is about 37.6°C with a range of about 1.4°C. In these night-active primates, the body temperature is depressed throughout the day and begins to rise before the lights go out. During the night the temperature is elevated. The major factor contributing to the difference in the rhythm amplitude of the two species is the minimum reached during the inactive period. The temperature during the active period is similar in both species (approx. 38.3°C).

In a recent review, Aschoff (1982) noted that the range of the body temperature rhythm in primates was inversely related to body size. The daytime temperature of the diurnal primates examined in that review was very similar (38.6°C on average), while the night-time temperature increased with increasing body mass. In man the body temperature rhythm is only 1°C, compared to the 2°C range of the smaller squirrel monkey.

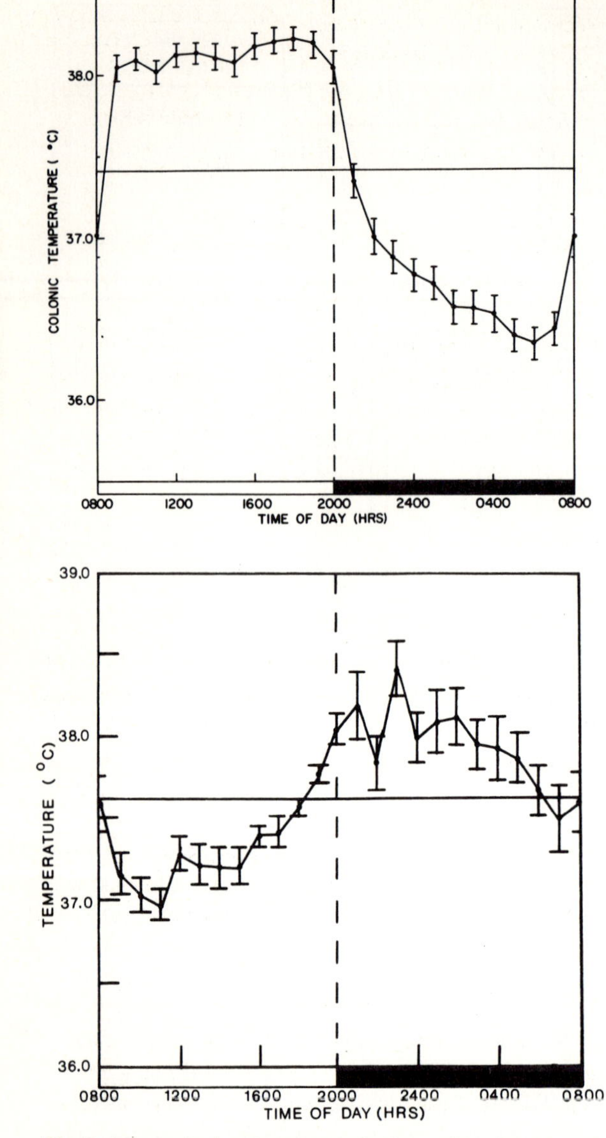

Fig. 2. Average body temperature rhythms of two primate species, squirrel monkey *(top)* and owl monkey *(bottom)*, synchronized by 24-h LD cycles. (Top from Fuller et al. 1977b, Bottom from Hoban and Sulzman 1982)

Since primates have such a pronounced rest-activity cycle, it has often been assumed that these activity changes cause the body temperature rhythm. There is no doubt that changes in the activity levels can influence body temperature. However, it should be emphasized that activity rhythms do not generate the body temperature rhythm (Aschoff 1970, Aschoff et al. 1974, Fuller et al. 1979a, 1979c). For example, the body temperature rhythm in man has been demonstrated to persist during continuous bed rest (with either normal meals or during complete fasting) and during complete sleep deprivation (Kleitman 1963, Wever 1979). Thus, although activity levels, food intake and sleep have specific effects on temperature regulation, they do not generate the body temperature rhythm but rather act to modulate the regulated temperature.

2 Environmental Influences

The endogenous nature of the body temperature rhythm is demonstrated by its persistence in constant conditions with a non-24-h free-running period and by its inability to entrain to environmental synchronizers with periods outside the circadian range (Kleitman 1963, Wever 1979). Although the rhythm is endogenous, it can be modulated by some environmental factors. In this section, we will examine the influence of light, ambient temperature and food on the timing and regulation of the primate body temperature rhythm.

2.1 Light

In nature, LD cycles provide the main form of synchronization of the circadian timing system, so that characteristic phase relationships are achieved between the temperature rhythm, the environment and other rhythms of the organism. When the LD cycle is phase-shifted, the body temperature rhythm resynchronizes along with the other rhythms. The rate of resynchronization of the temperature rhythm varies from that of other rhythms causing transient internal desynchronization (Moore-Ede et al. 1977, Fuller et al. 1981a). Several investigators have also noted a splitting or two-phase resynchronization of the body temperature rhythm itself (Winget et al. 1969, Fuller et al. 1979a). This two-phase resynchronization may be a result of multiple circadian oscillators within the thermoregulatory system resynchronizing at different rates.

Light intensity also influences body temperature levels quite separate from the normal phase control of the LD cycle (Fuller et al. 1978a, 1981a). As shown in Fig. 3, increased light intensity causes a higher body temperature. Figure 3A illustrates the average body temperature rhythm of squirrel monkeys maintained in either LD 12:12 or LD 2:2 cycles. When the lights (600 lx) are on, the body temperature is higher and when the light are off, the temperature is lower. This response is greater during the subjective night – 600 lx increased body temperature by more than 1°C during the night, while the change was as little as 0.2°C during the day.

These results help explain the reduced amplitude of the body temperature rhythm in constant light (Fuller et al. 1979a, Sulzman et al. 1979). Figure 3B shows the animals placed in LL (600 lx) had an elevated body temperature rhythm which followed the upper boundary of the ranges of body temperature seen during the LD 2:2 cycle (Fig. 3A). Thus, when light was applied during the subjective night in the LD 2:2 regimen, body temperature elevated to the same level at that phase as it did with the animals free-running in LL of the same intensity. This suggests that the range of the temperature rhythm in an LD 12:12 cycle is due to approximately equal contributions of the

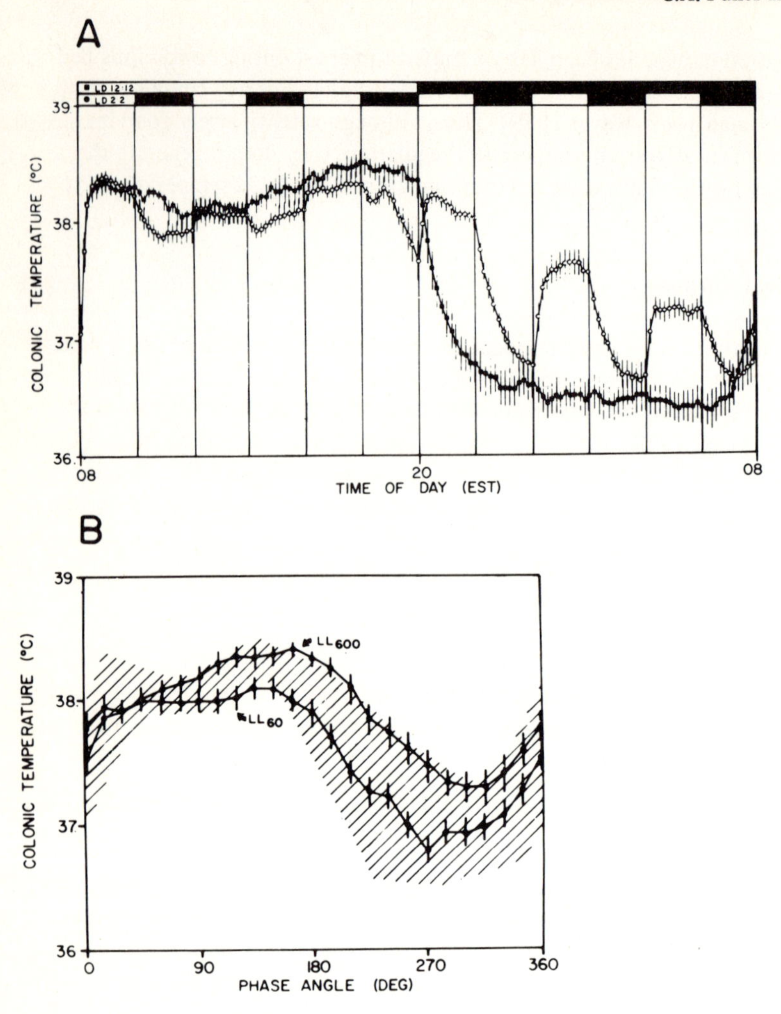

Fig. 3. A shows the average body temperature waveforms of a group of monkeys initially exposed to an LD 12:12 (•) cycle or an LD 2:2 (○) cycle (lighting conditions are shown at the top). **B** shows the average body temperature waveforms in LL of either 600 or 60 lx. The *hatched area* represents the upper and lower limits of the body temperature waveforms in A. (A from Fuller et al. 1981a)

endogenous rhythm and the passive effects of light intensity. These light intensity responses are also responsible for a lower body temperature and larger amplitude rhythm in dimmer LL (60 lx), as shown in Fig. 3B.

Other influences of constant light on the circadian timing system of primates are well known (Sulzman et al. 1979). Primates in general, do not appear to adhere to Aschoff's rule (Tokura and Aschoff 1978, Sulzman et al. 1979), and this is true also for the body temperature rhythm (unpubl. observ.). For example, in constant bright light the free-running period of the temperature rhythm is longer than the period in constant dim light.

Although the LD cycle is a strong synchronizer, it is only capable of synchronizing the body temperature rhythm within specific period ranges (Kleitman 1963). Once outside of the range of entrainment, the circadian system will free-run and the organism will then be exposed to changes in the LD cycle which are no longer in phase with the body temperature rhythm. An example of this can be seen in Fig. 4, in which a squirrel monkey was exposed to an LD 9:9 cycle – a period outside the range of entrainment (unpubl. observ.). The complex pattern of body temperature seen in Fig. 4A has two separate rhythmic components. The first is a free-running circadian temperature rhythm as shown by the phase plot in Fig. 4B. The second is the result of the light intensity influ-

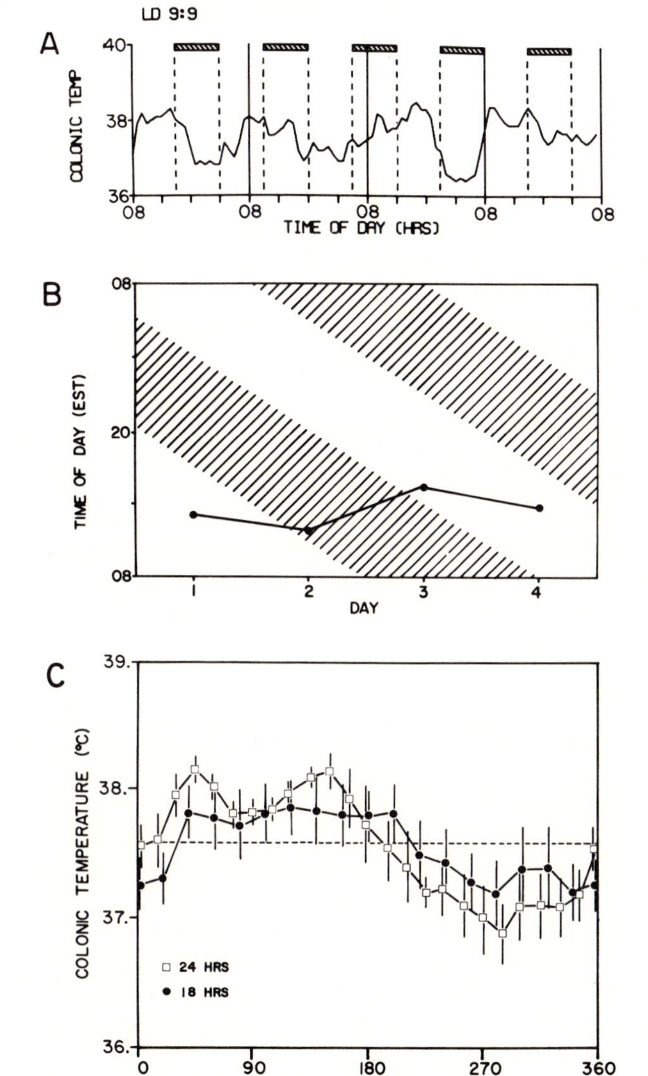

Fig. 4. A shows the colonic temperature of a monkey exposed to an 18-h LD 9:9 cycle. **B** is a phase plot of the circadian rhythm in **A**. The phase points (•) are calculated acrophases of fitted sine waves and show the rhythm persists with a period independent of the LD cycle *(clear areas* for light and *hatched areas* for dark). Waveform eduction analysis (**C**) of the data in **A** shows that the temperature data contains rhythmic components with periodicities of 18 h and 24 h. (Fuller et al. 1981a)

ences of the 18-h cycle superimposed on the circadian temperature rhythm. The rhythmic influence of each of these components on the temperature data is plotted in Fig. 4C. Both the 18-h and 24-h waveform eductions show significant oscillations in body temperature.

2.2 Ambient Temperature

Ambient temperature has been demonstrated to have some very specific effects on the regulation of the body temperature rhythm of man and primates. Human body temperature rhythms of individuals exposed to constant ambient temperatures ranging from 20°C to 32°C show a 24-h temperature mean which is positively correlated with ambient temperature with the greatest increase occurring during the subjective night (Aschoff et al. 1974). The range of the temperature rhythm is maximal in a thermoneutral environment and declines as the environment becomes both cooler and warmer. We have observed similar responses in both the body (Fuller et al. 1980) and brain (Fuller 1981) temperature rhythms in squirrel monkeys at different ambient temperatures.

The contribution of each effector in the regulation of body temperature depends on ambient temperature (Kleiber 1975). However, continuing rhythmicity can also be seen in all effectors (including heat production, conductance, and heat loss) at both warm and cool ambient temperature although the amplitude and phase-relationships between these variables shifts (Aschoff and Heise 1972, Fuller et al. 1979c, 1980).

2.3 Food

Although the ingestion of food does influence the regulation of body temperature by its specific dynamic action on heat production (Kleiber 1975), meals per se do not directly generate the body temperature rhythm. However, cycles of food availability (EF = Eat-Fast) are important synchronizers for the body temperature rhythm. We have shown that 24-h EF 3:21 cycles are capable of synchronizing the body temperature rhythm of the squirrel monkey (Sulzman et al. 1977a). Further, the rate of resynchronization of the body temperature rhythm to 8 h phase shifts of such EF cycles is slower than similar LD phase shifts, suggesting that light is the stronger synchronizer (Sulzman et al. 1978a).

3 Endogenous Regulation

We noted above that the controlled variables used to regulate body temperature always appear to be rhythmic. We also know that the circadian timekeeping system of primates appears to be composed of multiple autonomous self-sustained oscillators. The purpose of this section is to summarize the current understanding of the endogenous regulation of circadian rhythms within primates. We will first consider the circadian timing system and then the temperature regulation system.

3.1 Circadian Timing System

Several key observations have led investigators to recognize that the circadian timekeeping system is composed of multiple oscillators (Pittendrigh 1974). One piece of evidence is the transient internal desynchronization of the circadian timekeeping system which occurs during resynchronization following either LD (Moore-Ede et al. 1977, Fuller et al. 1981a) or EF (Sulzman et al. 1978a) phase shifts. Spontaneous internal desynchronization in constant conditions has been observed in both human and nonhuman primates (Aschoff et al. 1967, Sulzman et al. 1977b, 1979, Wever 1979, Czeisler et al. 1980). Further evidence comes from the observation of splitting of activity patterns (Pittendrigh 1974). The existence of two functional groups of oscillators in squirrel monkeys has also been demonstrated by contrasting LD and EF cycles (Sulzman et al. 1978b). These results show that with either conflicting phase or periods, the animals showed a synchronization of some rhythms preferentially by the EF cycle and the body temperature rhythm preferentially by the LD cycle.

Finally, lesions of the suprachiasmatic (SCN) nucleus of the hypothalamus in squirrel monkeys, one of the key pacemakers in the circadian timekeeping system of mammals, result in persisting rhythmicity in the body temperature rhythm even in the absence of a rest/activity rhythm (Fuller et al. 1981b). Figure 5 shows the temperature rhythms of two SCN-lesioned squirrel monkeys. It can be seen that body temperature rhythm (left) persists, as confirmed by spectral analysis (right). Thus, prominent circadian rhythms in the body temperature can continue in the absence of circadian rhythms in drinking.

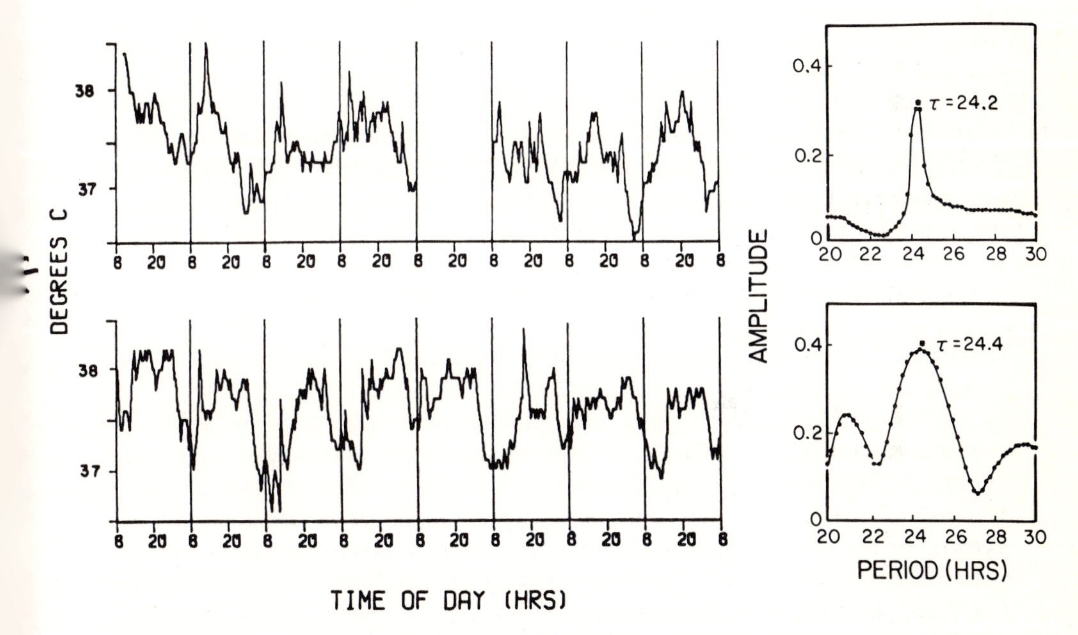

Fig. 5. Persisting body temperature rhythms *(left)* in two SCN-lesioned squirrel monkeys in LL as indicated by spectral analysis of the temperature data *(right)* which shows a significant circadian periodicity in both animals. (Fuller et al. 1981b)

3.2 Temperature Regulation System

Normally, the rhythm in body temperature occurs as a result of oscillation in temperature effectors (e.g., heat production and heat loss) which are precisely phased with each other (Aschoff 1982). The phase angle between heat production and heat loss is such that a temperature rhythm is generated by allowing heat storage to occur in the morning and heat dissipation to occur in the evening.

Several observations suggest that these effectors may be self-sustained oscillators. Squirrel monkeys in an LD cycle have diurnal rhythms in body and skin temperature (an indication of heat loss) that are precisely phased (Fuller et al. 1979a). However, when monitored in LL, both phase and period relationships between these rhythms can be markedly altered. Figure 6 (left) shows examples of the daily changes in phase between the colonic and skin temperature rhythms of four squirrel monkeys in LL, The variance of the phase angle is increased over 13-fold when the animals are in LL. and the period relationships are such that short episodes of dissociation between the two phases of the rhythms can occur.

Another alteration in the temperature rhythm can be seen in LL. The monkey in Fig. 6 (right) shows dissociation of the body temperature rhythm into higher frequency components. The skin temperature (heat loss) simultaneously exhibited a persisting circadian rhythm with none of the ultradian period components seen in the body temperature rhythm. One explanation for this is that other effectors may have had free-running circadian rhythms with different phase or period relationships, thereby causing the breakdown in the body temperature rhythm.

4 Concluding Remarks

What are the consequences of these interactions between body temperature regulation, the circadian timekeeping system, and other physiological systems? This question can be examined at three different levels: (1) influences within the thermoregulatory system, (2) influences between temperature regulation and other physiological systems, and (3) interactions between the organism and the environment. We have only begun to answer these questions, yet it appears the consequences are substantial.

We have previously reported (Fuller et al. 1978b, 1979b) that LD-entrained squirrel monkeys can maintain body temperature at any phase of the circadian cycle when exposed to mild cold. However, when these aniamls are free-running in LL they frequently cannot maintain body temperature effectively during acute cold exposure. Examples of this can be seen in Fig. 7. We have also shown that when animals were maintained in LL but synchronized by 24-h EF cycles, they could maintain body temperature during cold exposure. When internal desynchronization was forced upon these animals, the response was more severe and reproducible. This suggests that internal synchronization has a profound influence upon the ability of the temperature regulation system to maintain body temperature.

The circadian control of body temperature is a complex interaction between various physiological systems and the temperature regulation system (Fig. 1). Although body temperature has been examined by many individuals over the last century, we

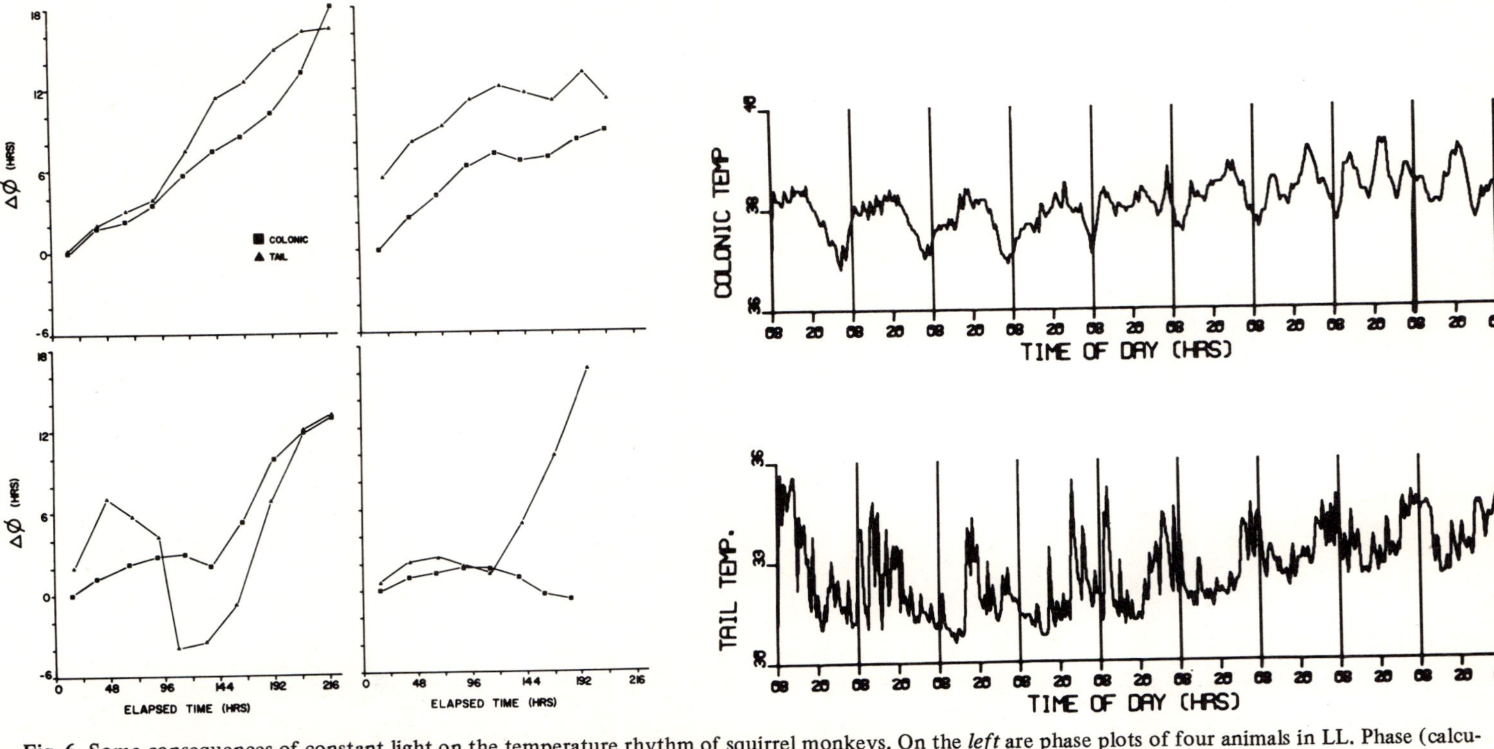

Fig. 6. Some consequences of constant light on the temperature rhythm of squirrel monkeys. On the *left* are phase plots of four animals in LL. Phase (calculated acrophase of a fitted sine wave) of the colonic temperature on day 1 was normalized to 0 h. Change in phase ($\Delta\phi$, hours) was plotted versus elapsed time for both the colonic and tail skin temperature rhythms. Plotted on the *right* are 9 days of colonic temperature *(upper)* and tail skin temperature *(lower)* of an animal in LL demonstrating dissociation of the colonic temperature rhythm. (Fuller et al. 1979a)

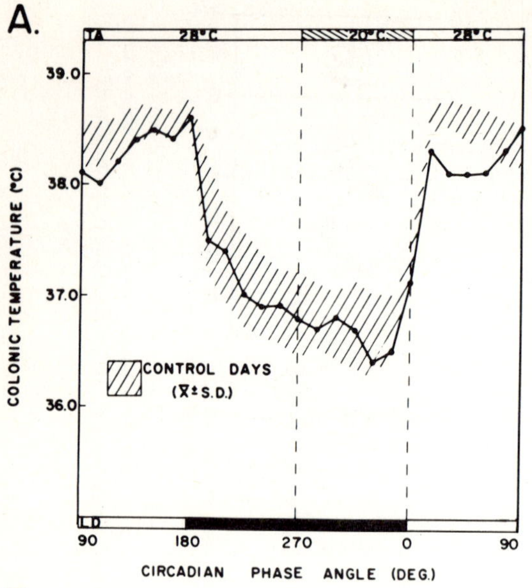

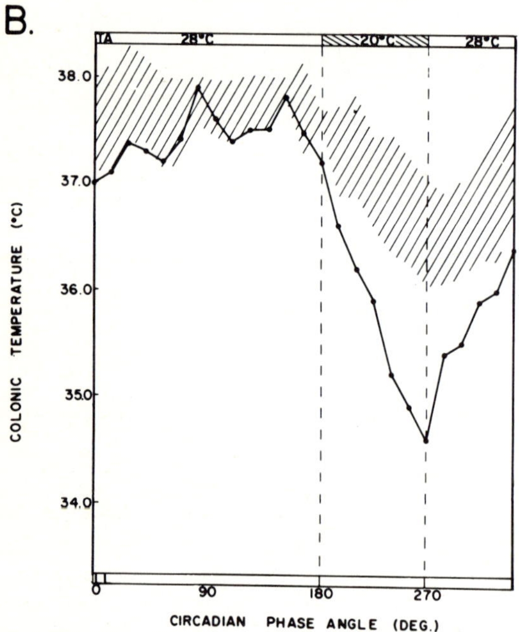

Fig. 7. Effects of 6-h cold exposures on body temperature in a squirrel monkey entrained to an LD 12:12 cycle **(A)** or a monkey free-running in LL **(B)**. At the *top* of each graph the ambient temperature (normally 28°C) is indicated on the day the cold was instituted; lighting conditions are indicated at the *bottom* of each graph. The *shading* represents the mean ± S.D. of the previour three cycles. (Fuller et al. 1978b, Copyright 1978 by the American Association for the Advancement of Science)

still do not fully understand its control and regulation. Future goals in the study of temperature regulation will hopefully not only continue to detail the formal properties of the temperature rhythm, the effectors that generate it and its neural control, but also to begin to examine where the multiple oscillators driving this rhythm are located. Are they in any or all of the temperature components described in Fig. 1, or do they reside elsewhere? What are the interactions of the temperature system with other physiological systems such as sleep? Considering the ease of 24-h temperature measurements, more effort should be devoted to examining clinical measures of this rhythm to expand its usefulness as a diagnostic tool.

Acknowledgments. We thank Ms. Gretchen Thompson and Ms. Nancy Carwile for assistance in preparing the manuscript and Ms. Tana Hoban and Ms. Sharon Sickles for technical assistance. Supported in part by NSF Grant 79-24412, PHS Grant BRD RR09070, PHS Grant BRS RR05816, PHS Grant GM28265-02 and SUNY Grant UAP/JAC 240-7447A.

References

Aschoff J (1970) Circadian rhythm of activity and body temperature. In: Hardy JD, Gagge AP, Stolwijk JAJ (eds) Physiological and behavioral temperature regulation. CC Thomas, Springfield, pp 905-919

Aschoff J (1982) The circadian rhythm of body temperature as a function of body size. In: Taylor CR, Johnsen A, Bolis L (eds) A companion to animal physiology. Cambridge Oxford (in press)

Aschoff J, Heise A (1972) Thermal conductance in man: Its dependence on time of day and on ambient temperature. In: Itoh S, Ogata K, Yoshimura H (eds) Advances in climatic physiology. Igaku Shoin Ltd, Tokyo, pp 334-348

Aschoff J, Gerecke U, Wever R (1976) Desynchronization of human circadian rhythms. Jpn J Physiol 17: 450-457

Aschoff J, Biebach H, Heise A, Schmidt T (1974) Day-night variation of heat balance. In: Monteith JL, Mount LE (eds) Heat loss from animals and man. Butterworths, London, pp 147-172

Cabanac M (1975) Temperature regulation. Annu Rev Physiol 37: 415-439

Czeisler CA, Zimmerman JC, Ronda J, Moore-Ede MC, Weitzman ED (1980) Timing of REM sleep is coupled to the circadian rhythm of body temperature in man. Sleep 2: 329-346

Fuller CA (1981) Primate circadian rhythms in hypothalamic and body temperature. Neurosci Soc Abstr 7: 719

Fuller CA, Sulzman FM, Moore-Ede MC (1978a) Active and passive responses of circadian rhythms in body temperature to light-dark cycles. Fed Proc 37: 832

Fuller CA, Sulzman FM, Moore-Ede MC (1978b) Thermoregulation is impaired in an environment without circadian time cues. Science 199: 794-796

Fuller CA, Sulzman FM, Moore-Ede MC (1979a) Circadian control of thermoregulation in the squirrel monkey *(Saimiri sciureus)*. Am J Physiol 236: R153-R161

Fuller CA, Sulzman FM, Moore-Ede MC (1979b) Effective thermoregulation in primates depends upon internal circadian synchronization. Comp Biochem Physiol 63A: 207-212

Fuller CA, Sulzman FM, Moore-Ede MC (1979c) Variations in heat production cannot account for the circadian rhythm of body temperature in the squirrel monkey. Fed Proc 38: 1053

Fuller CA, Sulzman FM, Moore-Ede MC (1980) Circadian body temperature rhythms of primates in warm and cool environments. Fed Proc 39: 989

Fuller CA, Sulzman FM, Moore-Ede MC (1981a) Shift-work and the jet-lag syndrome: Conflicts between environmental and body time. In: Johnson LC, Tepas DI, Colquhoun WP, Colligan MJ (eds) The 24-hour workday. A symposium on variations in work-sleep schedules. National institute of occupational safety and health. Washington DC 9 DHHS No 81-127

Fuller CA, Lydic R, Sulzman FM, Albers HE, Tepper B, Moore-Ede MC (1981b) Circadian rhythm of body temperature persists after suprachiasmatic lesions in the squirrel monkey. Am J Physiol 241: R385-391

Hoban T, Sulzman FM (1982) Effect of light intensity in owl monkey *(Aotus trivergatus)* circadian rhythms. (in preparation)

Kleiber M (1975) The fire of life: An introduction to animal energetics. Krieger, New York

Kleitman N (1963) Sleep and Wakefulness. Univ Chicago Press, Chicago

Moore-Ede MC, Kass DA, Herd JA (1977) Transient circadian internal desynchronization after light-dark phase shift in monkeys. Am J Physiol 232: R31-R37

Pittendrigh CS (1974) Circadian oscillation in cells and the circadian organization of multicellular systems. In: Schmitt FO, Worden FG (eds) The neurosciences: Third study program. MIT Press, Cambridge, pp 437-458

Satinoff E (1978) Neural organization and evolution of thermal regulation in mammals. Science 201: 16-22

Sulzman FM, Fuller CA, Moore-Ede MC (1977a) Feeding time synchronizes primate circadian rhythms. Physiol Behav 18: 775-779

Sulzman FM, Fuller CA, Moore-Ede MC (1977b) Spontaneous internal desynchronization of circadian rhythms in the squirrel monkey. Comp Biochem Physiol 58A: 63-67

Sulzman FM, Fuller CA, Moore-Ede MC (1978a) Comparison of synchronization of primate circadian rhythms by light and food. Am J Physiol 234: R130-R135

Sulzman FM, Fuller CA, Hiles LG, Moore-Ede MC (1978b) Circadian rhythm dissociation in an environment with conflicting temporal information. Am J Physiol 235: R175-R180

Sulzman FM, Fuller CA, Moore-Ede MC (1979) Tonic effects of light on the circadian system of the squirrel monkey. J Comp Physiol 129: 43-50

Tokura H, Aschoff J (1978) Circadian activity rhythm of the pigtailed macaque (*Macaca nemestrina*) under constant illumination. Pfluegers Arch 376:231-243

Wever RA (1979) The circadian system of man. Results of experiments under temporal isolation. Springer, Berlin Heidelberg New York

Winget CM, DeRoshia CW, Hetherington NW (1969) Response of monkey biorhythms to changing photoperiods. Comp Biochem Physiol 30: 621-630

6.1 Circadian and Sleep-Dependent Processes in Sleep Regulation

A.A. Borbély[1]

1 Sleep as a Circadian Rhythm

When man or animals live in an environment without time-cues, their sleep-wake cycle continues to exhibit a well-defined rhythm whose period (τ) is usually somewhat different from 24 h (e.g. rat: Borbély and Neuhaus 1978, man: Wever 1979). The sleep-wake cycle consitutes therefore a true circadian rhythm which is controlled by an oscillator. Research in circadian rest-activity rhythms is based heavily on motor activity recordings, whereas long-term measurements of sleep have been rarely undertaken. Waking is a prerequisite for motor activity, although motor activity is not always present during waking. Particularly when wheel-running is used as the activity measure, inactivity cannot be equalled to rest or sleep. Long-term recordings of movement activity with a transducer under the animal's cage provide a close approximation of the sleep-wake rhythm (Borbély and Neuhaus 1978).

The circadian rest-activity rhythm is insensitive to manipulations which have a profound effect on sleep. Thus a 24-h sleep-deprivation period had marked effects on sleep states in the rat (Borbély and Neuhaus 1979), whereas neither τ nor the phase-relationship of the circadian rest-activity rhythm was affected by a 24-h vigil (Borbély and Tobler, unpubl. results). Figure 1 provides an illustration for the striking insensitivity of the rest-activity rhythm to social influences. The animals were first kept in individual cages under entrained conditions (LD 12:12). After release into DD they exhibited a typical free-running circadian rhythm with individually different τ. To test for interaction between the rhythms during social contact, the two animals were placed into the same cage for a 10-day period (day 40-50). Then they were separated and recorded again in individual cages during the subsequent weeks. It is obvious from Fig. 1 that the social contact period neither affected τ nor induced a phase-shift in the circadian activity rhythm of the two animals. The experiment was performed in 14 rats and revealed no significant effect of social interaction on the rest-activity rhythm (Borbély et al., in prep.).

2 Sleep-Dependent Processes in Sleep Regulation

After a prolonged period of sleep deprivation, the duration of recovery sleep is in general longer than the usual sleep period. However, excess sleep time is modest in compar-

1 Institute of Pharmacology, University of Zürich, Zürich Switzerland

Vertebrate Circadian Systems
Ed. by J. Aschoff, S. Daan, and G. Groos

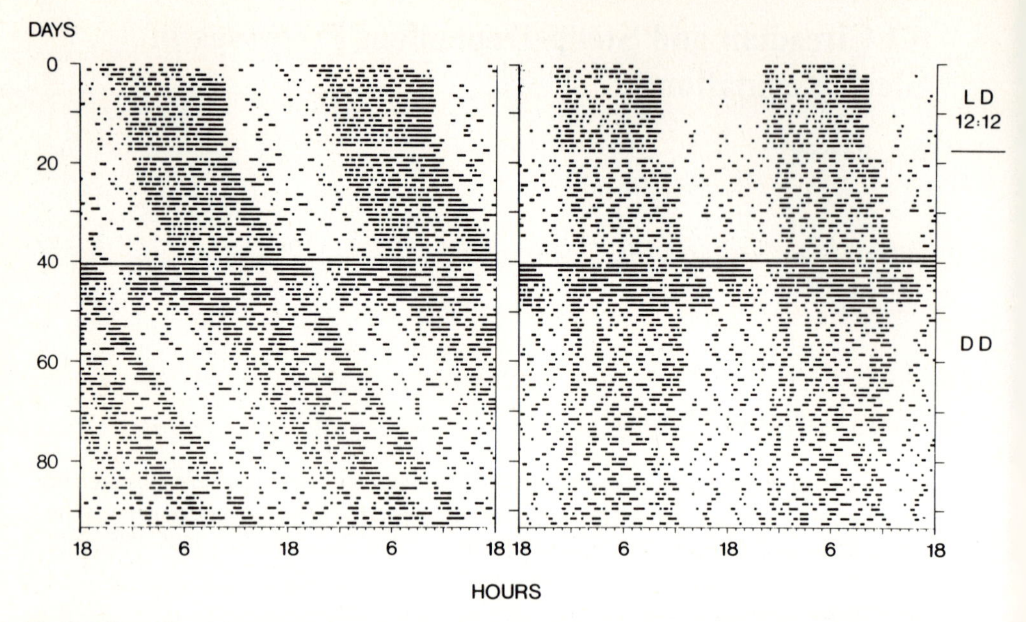

Fig. 1. Effect of social interaction on the circadian rest-activity rhythm. Activity records of two rats maintained in separate cages under entrained conditions (LD 12:12) and then in continuous darkness (DD) up to day 40. Day 40-50: Social interaction period in which both animals were maintained in the same cage. Day 51-88: Recordings in separate cages. Suprathreshold activity values recorded by a mechano-electric transducer under the cage are plotted for consecutive 15-min periods. Each day has been plotted twice to facilitate inspection of the circadian rhythm. (Borbély 1982)

ison to the sleep loss (e.g., Gulevich et al. 1966). This raises the question whether the sleep deficit can be compensated by an increase in "sleep intensity". Slow wave sleep (SWS; stages 3 and 4 of nonREM sleep) may represent a high-intensity fraction of nonREM sleep (NREMS). In accordance with this assumption, SWS predominates in the first part of the sleep period when "sleep pressure" would be expected to be highest. Figure 2 shows that stage 4 is limited to the first two NREM-REM sleep cycles, while the later cycles are dominated by stage 2 and REMS. The power density plots shown in Fig. 2 emphasize the continuous changes of the sleep process which are not evident from the sequence of discrete sleep stages (top of Fig. 2). The high peaks in the low frequency bands correspond to SWS. With the exception of the 0-1 Hz band, the peak in the first sleep cycle typically exceeds the peak in the second cycle, which in turn exceeds the power density in the subsequent cycles. The decline in EEG power density in the course of the sleep period can be closely approximated by an exponential function (Borbély et al. 1981). In the same study sleep-deprivation resulted in an increase in slow wave activity which was most prominent in the first cycle sleep. The EEG parameters seem to reflect a sleep-dependent process whose initial level at sleep onset is determined by the length of prior waking.

Studies in the rat have yielded essentially similar results (Borbély and Neuhaus 1979). Thus a SWS fraction of NREMS which was defined on the basis of EEG frequency criteria, predominated at the beginning of the daily sleep period and showed

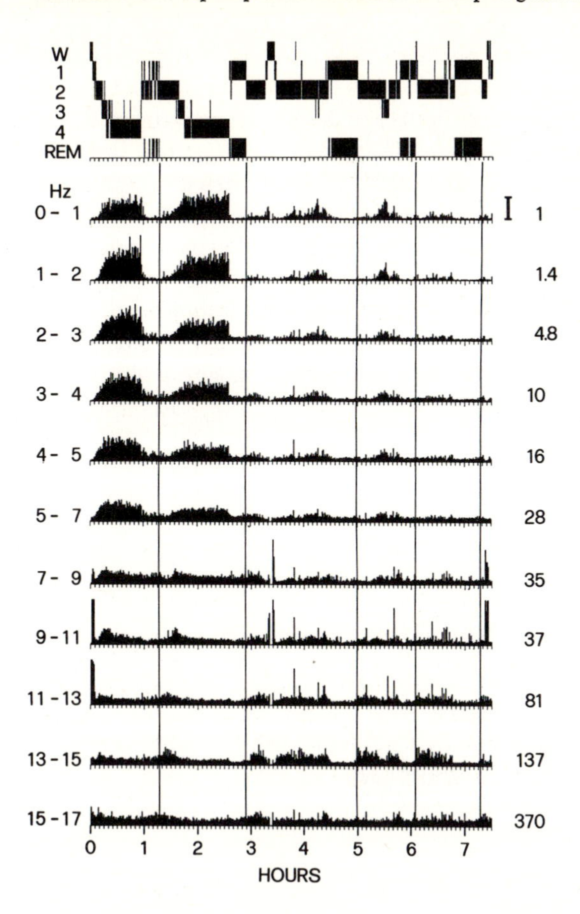

Fig. 2. Sleep states and EEG power density plots recorded from an adult young human subject. Sleep states *(top)* are plotted for 30-s epochs *(W* waking; *1-4* nonREM sleep stages; *REM* REM sleep; note that REM sleep epochs are also plotted as stage 1 sleep). Power density of the various frequency bands is plotted for 1-min segments. The lower limits of the frequency bands are 0.25 Hz higher than indicated on the left (e.g., 0-1 Hz corresponds to 0.25-1.00 Hz). The calibration mark at the *right* of the top frequency band corresponds to 200 $(\mu V)^2/0.25$ Hz for scale factor 1. The other scale factors indicated on the *right* represent relative values which were chosen to make the area of the plots equal. The nonREM-REM sleep cycles are delimited by *vertical lines.* (unpublished data from Borbély et al. 1981)

then a progressive decline. SWS but not total NREMS, was massively enhanced after a 24-h sleep-deprivation period. Since sleep-deprivation was achieved by forced locomotion, it was mandatory to investigate the effect of locomotion per se. Variations in the rotation rate of the sleep-deprivation apparatus had no effect on sleep. When rats were given access to a running wheel and ran up to 7.6 km per day, SWS was not enhanced in comparison to the no-access condition (Hanagasioglu and Borbély, submitted for publication). The results show that the enhancement of SWS is a consequence of sleep itself. Moreover, the dependence of slow wave activity on prior waking time is not restricted to human sleep, but constitutes a general feature of sleep in various mammalian species (see Borbély and Neuhaus 1979 for references).

3 The "Conflict Experiment"

When, in the rat, the 24-h sleep-deprivation period is made to terminate at dark onset, a conflict arises between the enhanced sleep propensity due to the 24-h vigil, and the tendency for waking and motor activity as a consequence of the circadian rhythm and the lighting condition. Under these conditions, a sleep rebound was observed in the first hours of the dark period, whereas the usual dark-time motor activity pattern prevailed in the subsequent hours (Borbély and Neuhaus 1979). SWS exhibited a two-stage compensatory response which consisted of an initial rise during the first dark-hours, and a delayed increase during the first light-hours. These results indicate that the compensatory response to a sleep deficit may be impaired if recovery sleep coincides with the circadian phase of waking. Although the recovery from sleep-deprivation as a function of the circadian rhythm has not yet been investigated in detail in man, fatigue (Åkerstedt and Fröberg 1977) and performance (Aschoff et al. 1972) are known to exhibit persistent rhythms during prolonged sleep deprivation. The question whether recovery from sleep loss is adversely affected by circadian parameters has considerable practical importance for occupational medicine (e.g., shift-work, time-zone shifts of air-flight personnel) and deserves further investigation.

4 Outline of a Model

In the model illustrated in Fig. 3, the sleep-wake cycle is shown as a combination of a clock-triggered circadian rhythm and a sleep-dependent process. A circadian oscillator is assumed to determine two levels of the sleep threshold: a high level in the waking

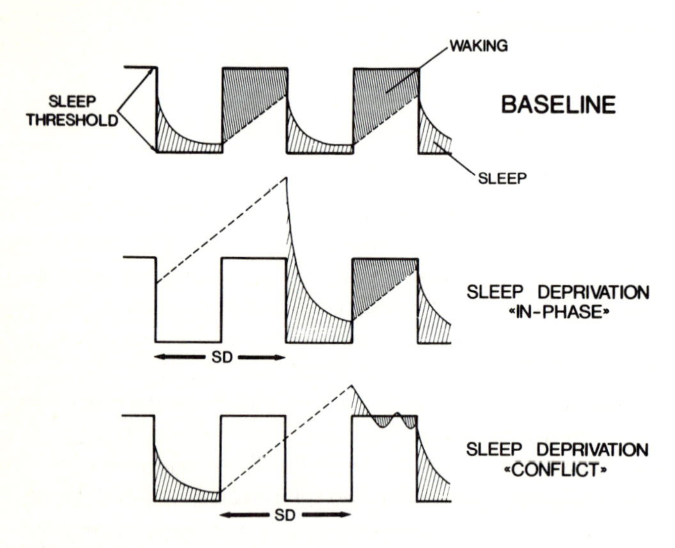

Fig. 3. Model of sleep regulation based on the combination of a circadian oscillator and a sleep-dependent process. (Modified from Borbély 1982)

phase and a low level in the sleep phase. The sleep-dependent process is represented by a variable whose level increases as a function of waking time and decreases as a function of sleep time. The propensity for sleep or waking is assumed to depend on the level of the regulatory variable relative to the sleep threshold (Fig. 3, hatched areas). During the usual sleep-wake cycle, sleep propensity is highest at the beginning of the daily sleep phase, and declines gradually with the progression of sleep. The decreasing trend of slow wave activity during sleep may reflect the progressive decline in sleep propensity. The curvilinear decrease of the sleep-dependent process in the model corresponds to the exponential decline of slow wave activity in the rat and in man (Borbély 1981, Borbély et al. 1981).

The middle and bottom diagrams in Fig. 3 illustrate the sleep-wake cycle during sleep-deprivation schedules. The period of forced waking, extending over one sleep-wake cycle, is parallel by a progressive increase in the level of the sleep-dependent process. The recovery period coinciding with the usual circadian sleep phase (Fig. 3, "in-phase") therefore shows a high initial sleep propensity which is assumed to be reflected by the high level of slow wave activity. Due to the curvilinear decline, sleep propensity at the end of the sleep phase is only slightly higher than under baseline conditions. Nevertheless, the difference may be sufficient to cause a residual effect of sleep-deprivation in the subsequent sleep phase (Borbély et al. 1981).

If the end of sleep-deprivation coincides with the beginning of the circadian waking phase (Fig. 3, "conflict"), the level of the sleep-dependent process is only slightly above the sleep threshold. Consequently, a relatively short initial sleep period is followed by a period of frequent waking episodes. However, due to the persisting high level of the sleep-dependent process, a delayed increase in sleep propensity is seen during the circadian sleep phase (Borbély and Neuhaus 1979).

5 Conclusions

The model outlined in this paper is based on the combination of two separate regulatory processes. The circadian oscillator allows sleep and waking to occur at predictable intervals in synchrony with the environmental day-night cycle. The circadian facet of sleep may be homologous with the circadian rest-phase which may be regarded as the phylogenetic percursor of sleep. However, the very stability of the circadian rhythm (Fig. 1) may impair an organism's capacity to adjust to unexpected environmental challenges. The evolution of a sleep-dependent regulatory process may have remedied this shortcoming by rendering sleep contingent upon the duration of prior waking. However, synchrony with the environmental day-night cycle is not maintained by such a process. The combination of the two mechanisms, as outlined in the model, may therefore provide an optimal solution to the requirement of stability and flexibility.

The question arises whether the circadian and the sleep-dependent facet of sleep regulation constitute independent processes. To investigate this problem, bilateral lesions of the suprachiasmatic nucleus (SCN) were carried out in enucleated rats (Borbély, Tobler and Groos, in preparation). SCN-lesioned animals are known to lose their sleep-wake rhythm (Ibuka and Kawamura 1975). First results indicate that sleep-deprivation enhances SWS and REMS in SCN-lesioned rats. Thus the intact circadian sleep-wake rhythm does not seem to be a prerequisite for the sleep-dependent processes subserving sleep regulation.

The two types of regulatory processes proposed in this article are based on a phenomenological approach which leaves the underlying neurobiological mechanisms largely unspecified. In particular, the notion of sleep-dependent processes rests heavily on EEG parameters whose generating mechanisms are still largely obscure. It would be therefore desirable to validate the concept on the basis of other parameters. Such a possibility may arise from recent developments indicating that sleep propensity in animals and man may be regulated by a humoral factor (Krueger et al. 1980, see Borbély and Tobler 1980 for a review). Particularly relevant to the present considerations are reports that the CSF concentration of the still unidentified sleep factor is a function of prior waking time (Fencl et al. 1971), and that the factor induces the same type of slow wave activity in NREMS as is observed after sleep-deprivation (Pappenheimer et al. 1975). These results suggest the intriguing possibility that the sleep-dependent process postulated in the present model may be controlled by an endogenous, sleep-promoting compound.

Acknowledgment. The experimental work reported in the paper was supported by the Swiss National Science Foundation, grants no. 3.254-0.77 and 3.561-0.79.

References

Åkerstedt T, Fröberg JE (1977) Psychophysiological circadian rhythm in women during 72 h sleep deprivation. Waking Sleep 1: 378-394

Aschoff J, Giedke H, Pöppel H, Wever R (1979) The influence of sleep-interruption and of sleep-deprivation on circadian rhythm in human performance. In: Colquhoun WP (ed) Aspects of human efficiency. English Univ Press, London pp 135-150

Borbély AA (1981) The sleep process: circadian and homeostatic aspects. In: Obál F, Benedek G (eds) Environmental physiology, Advances in physiological sciences, vol 18. Pergamon Press, Oxford, pp 85-91

Borbély AA (1982) Sleep regulation: circadian rhythm and homeostasis. In: Ganten D, Pfaff D (eds) Current topics in neuroendocrinology, vol 1. Springer, Berlin Heidelberg New York pp 83-103

Borbély AA, Neuhaus HU (1978) Circadian rhythm of sleep and motor activity in the rat during skeleton photoperiod, continuous darkness and continuous light. J Comp Physiol.128: 37-46

Borbély AA, Neuhaus HU (1979) Sleep-deprivation: effects on sleep and EEG in the rat. J Comp Physiol 133: 71-87

Borbély AA, Tobler I (1980) The search for an endogenous "sleep-substance". Trends Pharmacol Sci 1: 356-358

Borbély AA, Baumann F, Brandeis D, Strauch I, Lehmann D (1981) Sleep-deprivation: effect on sleep stages and EEG power density in man. Electroencephalogr Clin Neurophysiol 51: 483-493

Fencl V, Koski G, Pappenheimer JR (1971) Factors in cerebrospinal fluid from goats that affect sleep and activity in rats. J Physiol 216: 569-589

Gulevich G, Dement W, Johnson L (1966) Psychiatric and EEG observations on a case of prolonged (264 h) wakefulness. Arch Gen Psychiatry 15: 29-35

Ibuka N, Kawamura H (1975) Loss of circadian rhythm in sleep-wakefulness cycle in the rat by suprachiasmatic nucleus lesions. Brain Res 96: 76-81

Krueger JM, Bacsik J, Garcia-Arraras J (1980) Sleep-promoting material from human urine and its relation to factor S from brain. Am J Physiol 283: E116-E123

Pappenheimer JR, Koski G, Fencl V, Kanovsky ML, Krueger J (1975) Extraction of sleep-promoting factor S from cerobrospinal fluid and from brains of sleep-deprived animals. J Neurophysiol 38: 1299-1311

Wever RA (1979) The circadian system of man. Springer, Berlin Heidelberg New York

6.2 Sleep Circadian Rhythms in the Rat: One or Two Clocks?

J. Mouret[1]

1 Introduction

Even though it is obvious that a period of inactivity may correspond with completely different processes, e.g., quiet wakefulness, Slow Wave Sleep (SWS), Paradoxical Sleep (PS) of focused attention, most of the studies on circadian rhythms are based only on the pattern of activity of experimental animals and the epochs of inactivity are looked upon as a negative state. Moreover, in these experiments, the withdrawal and/or changing of the known zeitgebers is quite often the most commonly used procedure.

By contrast we shall report here on the PS and SWS circadian patterns of rats kept under strict light-dark cycle (LD) before and after lesions or removal of central or more peripheral structures, some of which were already known for their role in the genesis of sleep and activity rhythms (Coindet et al. 1975, Kawakami et al. 1972, Mouret et al. 1974, Ibuka and Kawamura 1975), as well as in the control of some enzymatic activities in the pineal body (Klein 1974, Moore et al. 1967, Reiter 1977, Wurtman et al. 1968).

2 Material and Methods

Fifty-two male Wistar rats, 230/250 g each at the onset of the experiments, were used in these studies. After a 15-day adaptation period to the laboratory environment (LD 12:12, light on at 0700 h temperature 21°C ± 1°C, food and water ad lib), they were chronically implanted with cortical and neck muscle electrodes (Mouret et al. 1974), and thereafter placed for 8 days under the recording conditions (single plastic jars, recording cable plugged in). At the end of this habituation period 5 days of continuous polygraphic recordings (24 h a day, paper speed 5 mm/s) were taken and scored visually to the nearest minute for the presence of Wakefulness (W), SWS or PS.

The rats were then submitted to lesions of ablations aimed at some central structures: supra chiasmatic nuclei (SCN:11 rats), raphe nuclei (RN:16 rats), or at more peripheral organs: pineal body (P:7 rats), superior cervical ganglia (SCG:4 rats), according to our already published procedures (Coindet et al. 1975, Mouret et al. 1974, Mouret and Coindet 1980), and replaced under their previous recording conditions. The

1 Neurophysiology Clinique, Hôpital Neurologique, B.P. Lyon Montchat, 69394 Lyon Cedex 3

Vertebrate Circadian Systems
Ed. by J. Aschoff, S. Daan, and G. Groos

continuous recordings were immediately resumed for 20 to 30 days, at the end of which the animals were killed for histological verification of lesion placements.

Four rats, which had not been recorded previously, were submitted to a ligature of the longitudinal venous sinus (S:4 rats) without removal of the pineal body, chronically implanted and then recorded continuously.

In this paper we shall report on the data of the last 5 days of the experiment period: the hourly amounts, expressed in minutes, of each sleep stage were submitted to a mathematical analysis, based on the least-squares method of Anderson (1971) and developed by CHOUVET (Coindet et al. 1975) in order to determine the best-fitting sine curve for a 24 h periodicity.

3 Results

As shown in Fig. 1 on days 20 to 25 after pinealectomy, the PS circadian rhythm is clearly altered, whereas that of SWS is unaffected. These alterations consist in a reduction of the amplitude of the circaidian variations together with a slight phase shift toward a later time in L. There are no quantitative modifications of either sleep stage on a 24-h time base, whereas the anticipatory increase in PS in the hour preceding the i llumination period disappears, while that of SWS persists.

Our results on sleep rhythms after SCN lesions are in keeping with those of Ibuka and Kawamura (1975): at the end of the experiment (days 22 to 27), no quantitative changes in either PS or SWS amounts per day are observed, whereas the amplitude of the 24-h variations of both sleep stages are strongly reduced (Fig. 1), the phase of that of PS being delayed and that of SWS slightly advanced. The anticipatory changes in the hour preceding the light persist for both sleep stages (Mouret et al. 1978).

Among the 25 rats with lesions aimed at the raphe nuclei, 6 had strong reduction of the amplitude of the circadian variations of both sleep stages which persisted on days 17-22 after the surgery (Fig. 1). The common lesion of these rats was found to be located at the level of the rostro-dorso-axial part of nucleus raphe dorsalis (NRD) (Fig. 2). This reduced amplitude is accompanied by a phase delay of PS, and the anticipation to the light is abolished. The small changes in PS and SWS amounts per day are apparently dependent upon extensions of the lesions in some areas which have been shown to affect the production of these sleep stages without altering their circadian rhythms (Mouret and Coindet 1980).

From the representation in Fig. 3, the sleep modifications seen on days 15 to 20 after occlusion of the longitudianl venous sinus are quite similar to those observed after pinealectomy. No quantitative alterations of either SWS or PS occur, whereas the amplitude of PS circadian rhythms is strongly decreased and its phase shifted to the beginning of the dark half of the LD cycle. The PS anticipatory increase is quite reduced if still present.

Except for a moderate increase in the amplitude of the PS circadian rhythm and a significant increase of SWS amounts per day, no obvious alterations are observed on the records taken from day 15 to 20 after superior cervical ganglioectomy. The anticipatory increase in PS is completely abolished, whereas that of SWS is still present (Fig. 3, Fig. 4).

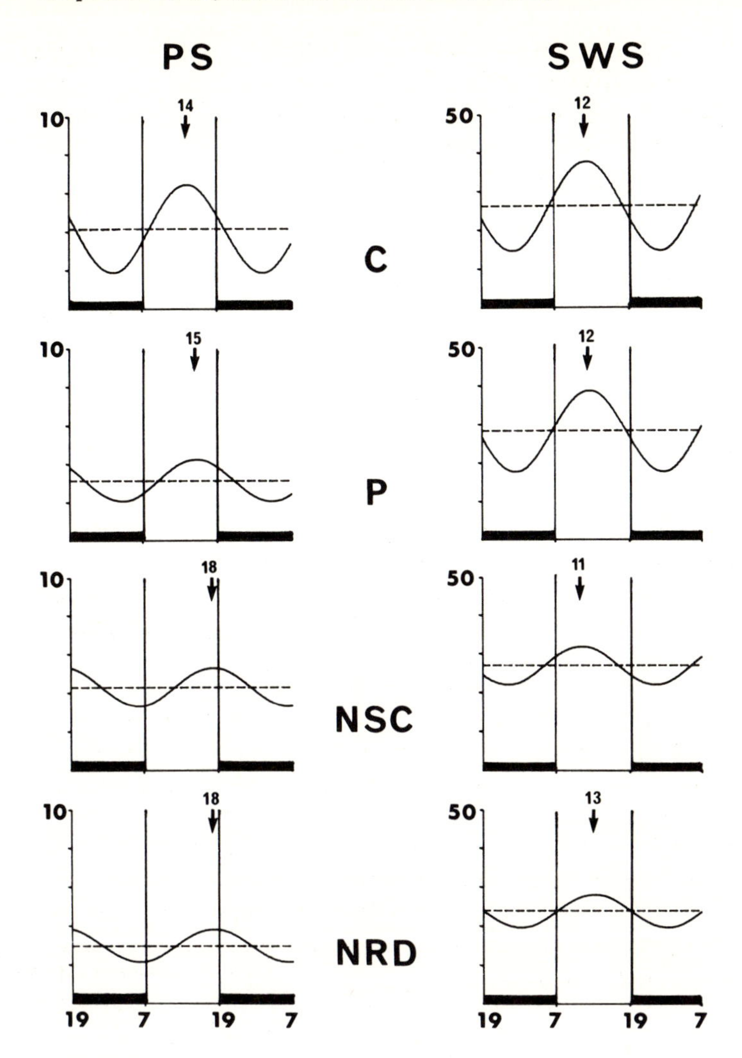

Fig. 1. Best-fitting sine curve for the 24-h periodicity of Paradoxical Sleep (PS) and Slow Wave Sleep (SWS).
C Control conditions; *P* Pinealectomy; *SCN* Supra Chiasmatic Nuclei lesions; *NRD* Nucleus Raphe Dorsalis lesions.
The *curves* were calculated from the hourly data of 5 consecutive days (see text). Ordinates expressed in minutes, the *dotted lines* indicate the median level (on a hourly base) of each individual rhythm and the *arrows* the acrophase (in local time). Light on at 0700, light off at 1900 h

4 Discussion

The similarities between the effects of SCN and NRD lesions are striking. It appears that both structures are necessary for the sleep circadian rhythms to be present. The role of some serotoninergic projections to the SCN (Aghajanian et al. 1968, Nojyo and Sano 1978) in the functional role of the latter must thus be questioned by more spe-

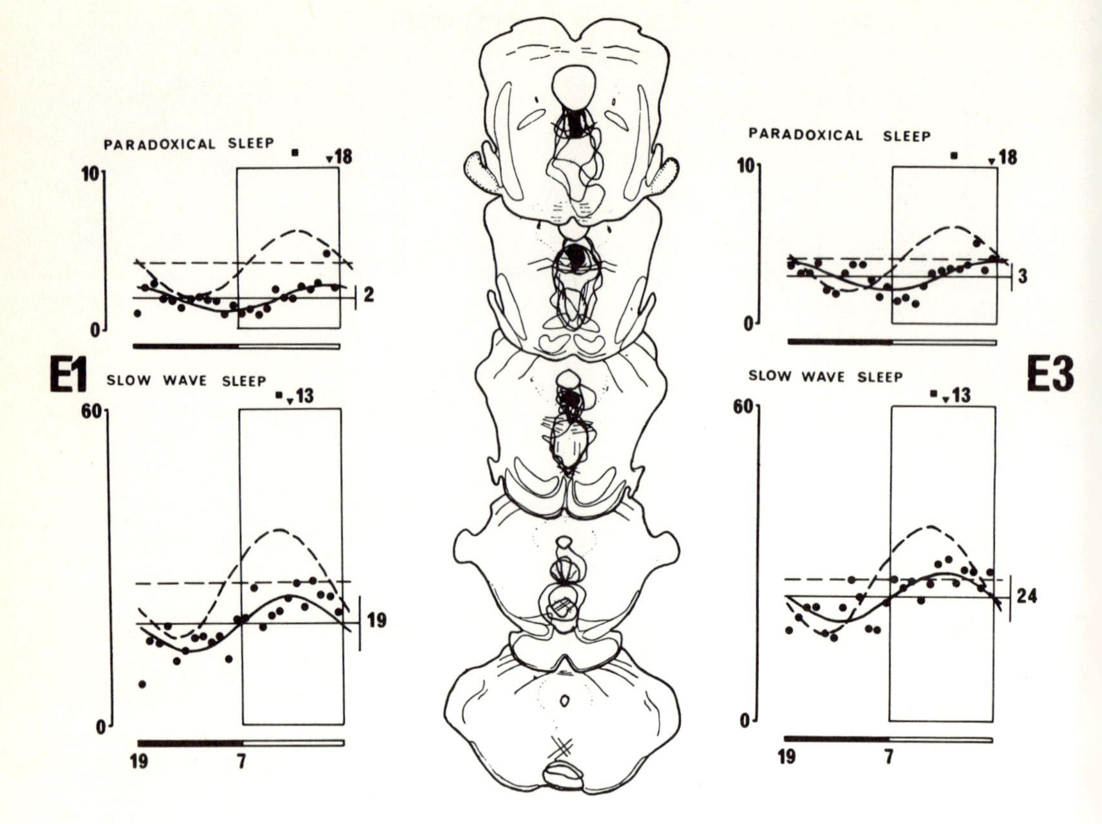

Fig. 2. Circadian rhythms of PS and SWS of six rats whose lesions are illustrated *(circled area)* on the anatomical planes. The *dark area* represents the common part of the lesions.
E1 Days 1 to 6 after the lesion; *E3* days 17-22 after the lesion. *Dashed lines* median level and circadian variations during the control period. *Solid lines* median level and circadian variations for the two experimental periods. Each *point* represents the average of 30 determinations

cific experiments. Due to the existence of these two projections and to the sensitivity of NRD to light influences (Foote et al. 1978), a gating effect of this nucleus on the reacitivity of SCN cells to direct light information should be considered.

Another heuristic point is the fact that the removal of the SCG does not reproduce the effects seen after pinealectomy or after occlusion of the venous sinus. The possibility that the catecholaminergic innervation of the pineal may come from a source other than the SCG is suggested by the fact that destroying both nuclei locus coeruleus (LC) of the superior cervical gangliectomized rats leads to sleep modifications similar to those observed after pinealectomy (Mouret, unpubl.). The dual innervation of cerebral vessels (Edvinsson et al. 1973, Itakura et al. 1977, Raichle et al. 1975) possibly suggests that this second source of pineal innervation comes from the brain vessels.

From a more general point of view, the specific effects of some lesions or ablations on PS and/or SWS circadian rhythms, without any modification of their daily quantities, favors a dissociation between the structures involved in the homeostatic control

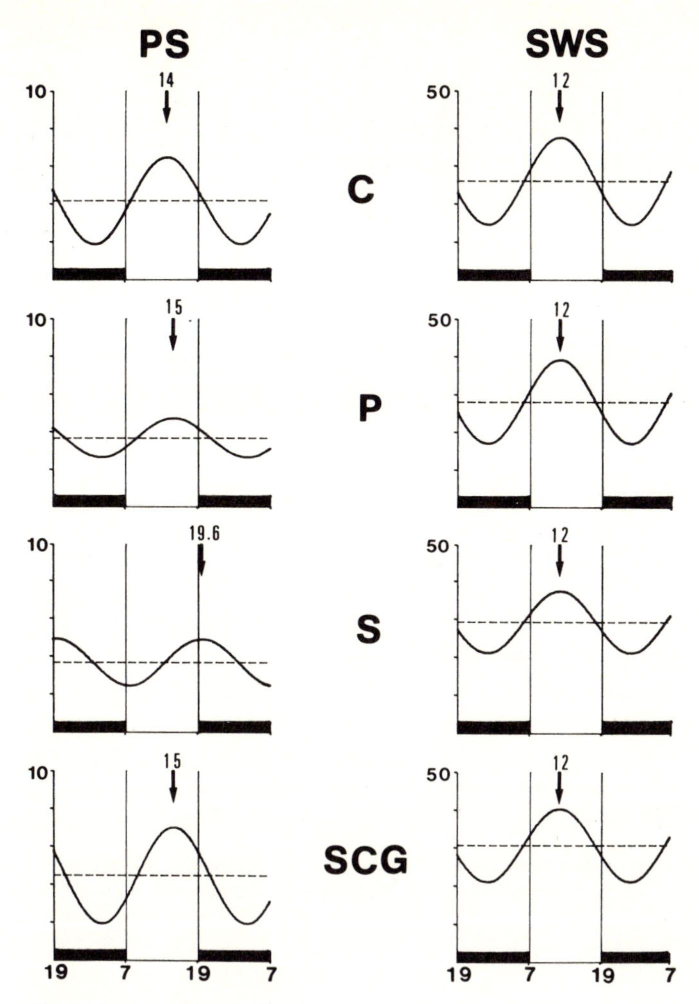

Fig. 3. Same legend as in Fig. 1.
C Control recordings; *P* pinealectomized animals; *S* ligature of the longitudinal venous sinuses; *SCG* Superior Cervical Gangliectomy

of amount of PS and SWS (cf. Borbély, Chap. 6.1, this Vol.) and those responsible for their circadian organization. The usual synchronization between the two sleep stages may also disappear, suggesting that each of them may have its own endogenous or exogenous control system. Since such a dissociation between PS and SWS rhythm is particularly prominent after pinealectomy, one should also look at this structure in some human pathological phenomena, e.g., narcolepsy, where such a dissociation is also observed.

Acknowledgments; This work has been done with the technical assistance of Mrs. J. Coindet (Department of Experimental Medicine, University Claude Bernard) and supported by grants from INSERM (U.52), CNRS (LA 56).

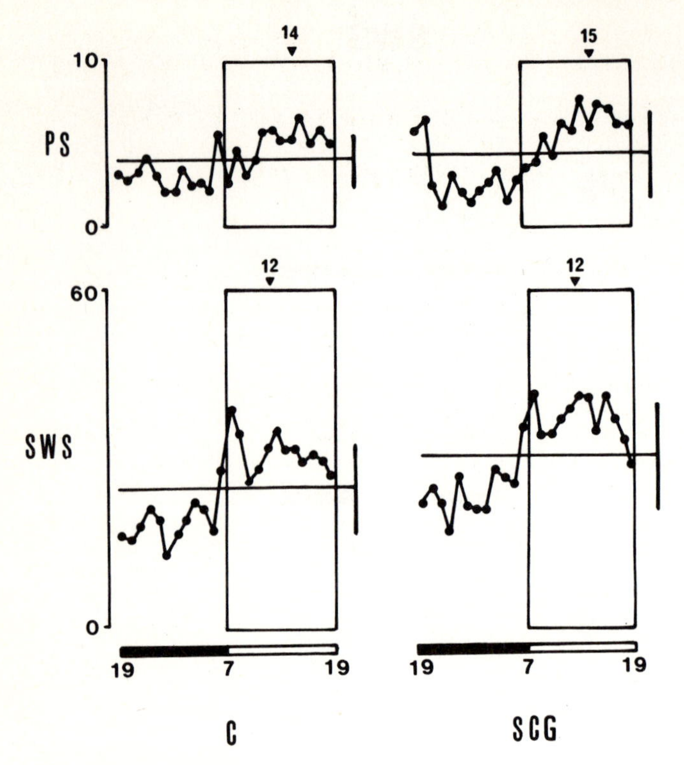

Fig. 4. Effect of Superior Cervical Gangliectomy *(SCG)* on the 24-h variations of Paradoxical Sleep *(PS)* and Slow Wave Sleep *(SWS)*.
C Control conditions.
Each *point* is the average of 20 determinations. Notice the disappearance of the anticipatory increase in PS in the hour preceding the illumination and the increase in SWS mean level *(horizontal line)*

References

Aghajanian G, Bloom FE, Sheard MH (1969) Electron microscopy of degeneration within serotonin pathway of rat brain. Brain Res 13: 266-273

Anderson TW (1971) The statistical analysis of time series. John Wiley and Sons Inc, New York

Coindet J, Chouvet G, Mouret J (1975) Effects of lesions of the suprachiasmatic nuclei on paradoxical sleep and slow wave circadian rhythms in the rat. Neurosci Lett 1: 243-247

Edvinsson L, Lindvall M, Nielsen KC, Owman CH (1973) Are brain vessels innervated also by central (non sympathetic) adrenergic neurons? Brain Res 63: 496-499

Foote WE, Taber-Pierce E, Edwards L (1978) Evidence for a retinal projection to the midbrain raphe of the cat. Brain Res 156: 135-140

Ibuka N, Kawamura H (1975) Loss of circadian rhythm in sleep-wakefulness cycle in the rat by suprachiasmatic nucleus lesions. Brain Res 96: 76-81

Itakura T, Yamamoto K, Toyama M, Shimizu N (1977) Central dual innervation of arterioles and capillaries in the brain. Stroke 8,3: 360-365

Kawakami M, Yamaoka S, Yamaguchi T (1972) Influence of light and hormones upon circadian rhythm of EEG slow wave and paradoxical sleep. In: Itoh S, Ogata K, Yoshimura H (eds) Advances in clinical physiology. Springer, Berlin Heidelberg New York, pp 349-366

Klein DC (1974) Circadian rhythm in indole metabolism in the rat pineal gland. In: Schmitt FO, Worden FG (eds) The neurosciences: Third study program. MIT Press, Cambridge, Mass, pp 509-516

Moore RY, Heller A, Wurtman RJ, Axelrod J (1967) Visual pathway mediating pineal reponse to environmental light. Science 155: 220-223

Mouret J, Coindet J (1980) Polygraphic evidence against a cirtical role of the raphe nuclei in sleep in the rat. Brain Res 186: 273-287

Mouret J, Coindet J, Chouvet G (1974) Effect de la pinealectomie sur les états de sommeli du rat mâle. Brain Res 81: 97-105

Mouret J, Coindet J, Debilly G, Chouvet G (1975) Suprachiasmatic nuclei lesions in the rat: alterations in sleep circadian rhythms. EEG Clin Neurophysiol 45: 402-408

Nijyo Y, Sano Y (1978) Ultrastructure of the serotoninergic nerve terminals in the suprachiasmatic and interpreduncular nuclei of rat brain. Brain Res 149: 482-488

Raichle M, Hartman BK, Eichling JO, Sharpe LG (1975) Central noradrenergic regulation of cerebral blood flow and vascular permeability. Proc Natl Acad Sci USA 72: 3726-3730

Reiter RT (1977) The pineal. In: Horrobin DF (ed) Annual Research Review, vol II. Eden Press

Wurtman RJ, Axelrod J, Kelly DE (1968) The pineal. Academic Press, London New York

6.3 Sleep ECoG Rhythm in the High Mesencephalic Rat

H. Kawamura and Y. Hanada[1]

The sleep-waking circadian rhythm in the rat is supposed to be influenced by the SCN. However, how the SCN influences the sleep-waking circadian rhythm is largely unknown. We performed an experiment which eliminates the serotonergic input from the raphe nuclei to the SCN. In this particular preparation, the midbrain was transected at a rostral level so that a part of the mesencephalic reticular formation was still within the forebrain, whereas the raphe and locus coeruleus were totally isolated from the forebrain (Hanada and Kawamura 1981). In this high mesencephalic preparation the appearance of a desynchronized electrocorticographic (ECoG) pattern is earlier than with the transection described by Batsel (1960) and Villablanca (1962). After such a mesencephalic transection, Bremer (1935) first described the appearance of a continuous slow sleep pattern in his acute experiment. Later, Batsel and Villablanca described the reappearance of a ECoG desynchronization pattern 2-3 weeks after transection in chronic cerveau isolé cats and dogs. Unfortunately they did not record the ECoG continuously for 24 h per day. We recorded the ECoG of cerveau isolé rats continuously for periods up to 3 weeks. Rats were fed through a stomach cannula. A desynchronized pattern appeared in the ECoG 2 to 4 days after transection (Fig. 1). Further continuous recording revealed that 4 to 7 days after surgery a circadian sleep-wakefulness rhythm reappeared (Figs. 1, 3). The circadian rhythm of sleep and wakefulness in the cerveau isolé rat can be eliminated by bilateral lesions of the SCN. After lesions, no circadian

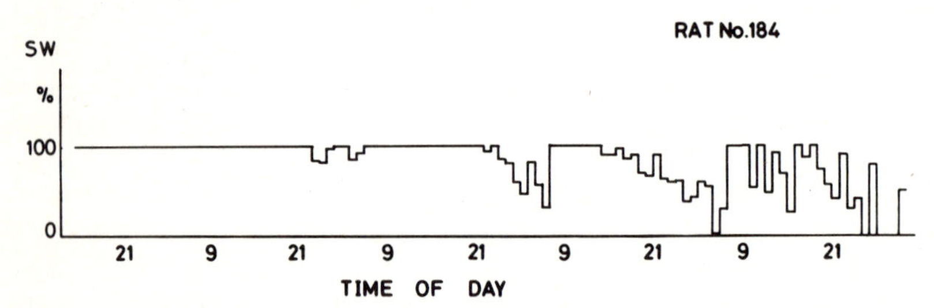

Fig. 1. Appearance of an ECoG desynchronization pattern in a blind, chronic high mesencephalic rat. *Ordinate* percentage of slow wave sleep pattern per hour. *Abscissa* time of day. A gradual increase of desynchronization is seen with a concomitant decrease of percentage of slow wave

1 Mitsubishi-Kasei Institute of Life Sciences

Vertebrate Circadian Systems
Ed. by J. Aschoff, S. Daan, and G. Groos

rhythm was observed in the forebrain ECoG. Jouvet (1972) suggested that ECoG changes in the forebrain of brain stem transected animals may be due to the humoral influence from a pacemaker in the hindbrain. However, as can be seen from the records in Fig. 2 in the brain stem transected rats, changes in ECoG and bodily movements

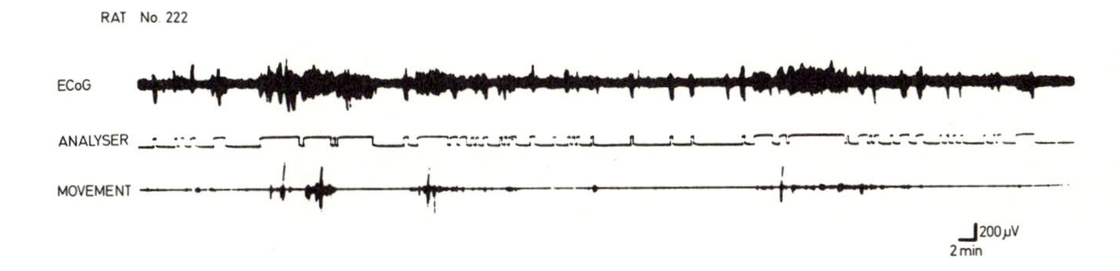

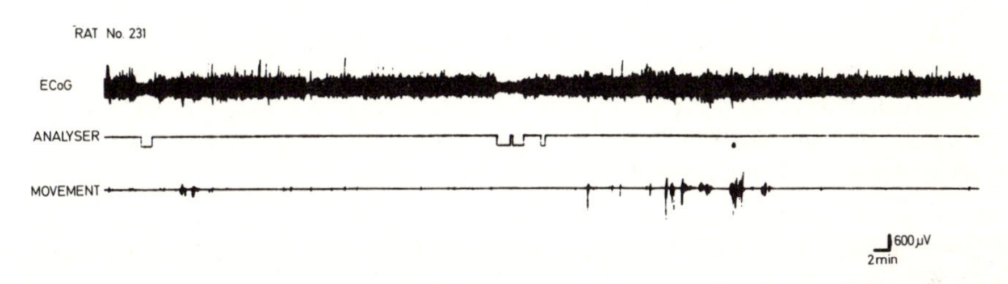

Fig. 2. Records from two rats demonstrating independent changes of ECoG and bodily movements. Vigorous movements *(bottom traces)* appear, when ECoG show a high amplitude slow sleep pattern *(top traces)*. *Middle traces* show data of automatic sleep stage analyzer of ECoG

are quite independent. There are bodily movements which do not influence the forebrain ECoG activity. Whenever there was influence of bodily movement on forebrain ECoG, the transection was incomplete and there remained some neural connection between hindbrain and forebrain. Therefore, the mechanisms inducing the forebrain sleep-waking rhythm and movement rhythm in the hindbrain are totally independent. Finally, Fig. 3 illustrates the similarity of the sleep-waking rhythms taken from 4 intact and 4 cerveau isolé rats, respectively. Data of continuous recordings for 2 days with clear circadian rhythms are shown. In these experiments rats were enucleated in both eyes.

In the cerveau isole rats, the ultradian rhythms were less marked compared to the intact rats. Records in cerveau isole rats were taken when the animal recovered sufficiently without any visible deterioration of the ECoG. These results indicate that the SCN shows an enormous effect on sleep and wakefulness via the remaining parts of the activating system in the forebrain, presumably via the hypothalamus and a small portion of the reticular formation in the mesencephalon. Apparently the raphe complex and locus coeruleus have nothing to do with the rhythm. The circadian pacemaker and

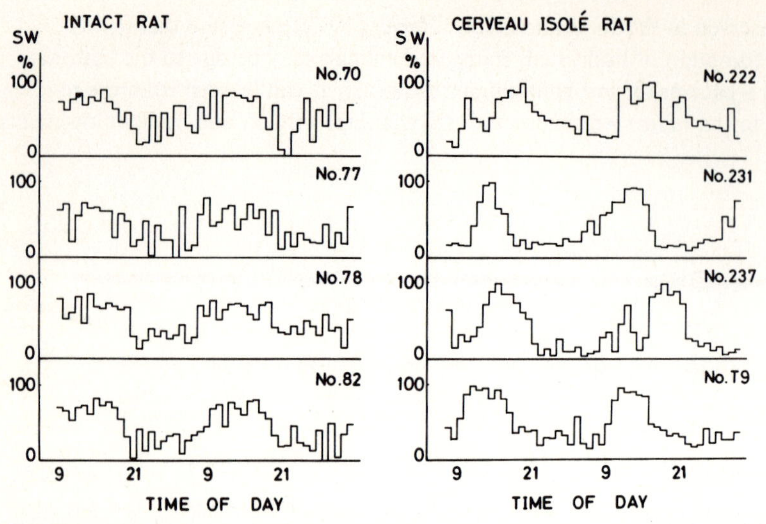

Fig. 3. Circadian rhythms of sleep-waking ECoG taken from four intact *(left)* and four cerveau isolé *(right)* rats. All rats were blinded. Intact rats show more ultradian rhythms

a pacemaker for sleep-wakefulness inducing changes in the ECoG from an awake pattern to a sleep pattern and vice versa is doubtless located in the forebrain. Of course for induction of paradoxical sleep, the ponto-mesencephalic area containing the locus coeruleus as well as pontine reticular formation may be crucial. But it is not the locus coeruleus or raphe which produces the pacemaker activity for the sleep-waking rhythm. The neurophysiological bases of the interaction between the forebrain slow wave sleep-waking mechanism and hindbrain mechanism for induction of paradoxical sleep await further study.

References

Batsel HL (1960) Electroencephalographic synchronization and desynchronization in the chronic "cerveau isolé" of the dog. Electroencephalogr Clin Neurophysiol 12: 421-430

Bremer G (1935) Cerveau isolé et physiologic du sommeil. C R Soc Biol 118: 1235-1241

Hanada Y, Kawamura H (1981) Sleep-waking electrocorticographic rhythms in chronic cerveau isolé rats. Physiol Behav 26: 725-728

Jouvet M (1972) The role of monoamines and acetylcholine-containing neurons in the regulation of the sleep-waking cycle. Ergeb Physiol 64: 166-307

Villablanca J (1962) Electroencephalogram in the permanently isolated forebrain of the cat. Science 138: 44-46

6.4 Interaction Between the Sleep-Wake Cycle and the Rhythm of Rectal Temperature

J. Zulley and R.A. Wever[1]

1 Introduction

The human circadian system has been considered as controlled by two – or even more – oscillators which are represented by the sleep-wake cycle and the rhythm of rectal temperature (Wever 1975). Both rhythms are influenced by environmental time cues but also affect each other mutually. The interaction between the two rhythms is especially obvious in the absence of external stimuli, i.e., under constant conditions, where it usually results in mutual synchronization between the two rhythms. However, in about 30% of subjects examined under constant conditions, the two rhythms show different average periods in the steady state; nevertheless, they continue to interact mutually. It is only in this state of internal desynchronization that the direction of the interaction between the two rhythms can be evaluated.

2 Method

Human subjects lived in complete isolation from environmental time cues and, hence, showed free-running circadian rhythms. The illumination in the experimental room was constant in some experiments and self-selected (i.e., light during wakefulness and dark during sleep) in others. Beside a variety of physiological and psychological variables, the state of activity and rectal temperature was recorded continuously (Wever 1979). In some of these studies, sleep was recorded polygraphically by measuring EEG, EOG, and EMG (Zulley 1979). In all experiments under consideration, the subjects lived singly isolated for, at least, 4 weeks.

Out of the larger sample of data, free-running rhythms were analyzed from ten subjects, who all showed internal desynchronization with considerably lengthened sleep-wake cycles. The temporal relationship between the rhythms of activity and rectal temperature (temporal positions of sleep episodes and of minimum values of rectal temperature), and the range of the temperature rhythm (amplitude) will be considered here. In this paper, the term "sleep" is throughout used synonymous to "bedrest" (cf. Glossary in Sleep, 2:287-288, 1980); in the two out of the ten subjects where sleep was recorded polygraphically, the difference between sleep and bedrest was less than half an hour and, hence, negligible with regard to the present investigations.

1 Max-Planck-Institut für Psychiatrie, Aussenstelle Andechs, Arbeitsgruppe Chronobiologie

Vertebrate Circadian Systems
Ed. by J. Aschoff, S. Daan, and G. Groos

3 Results

To illustrate the phenomenon of internal desynchronization, Fig. 1 presents an experiment where the rhythms of sleep-wake and rectal temperature took different periods during the whole experiment of 29 days duration. The activity rhythm had a mean period of 30.2 h and the rhythm of rectal temperature one of 25.0 h. Consequently, the phase relationship between both rhythms varied from cycle to cycle.

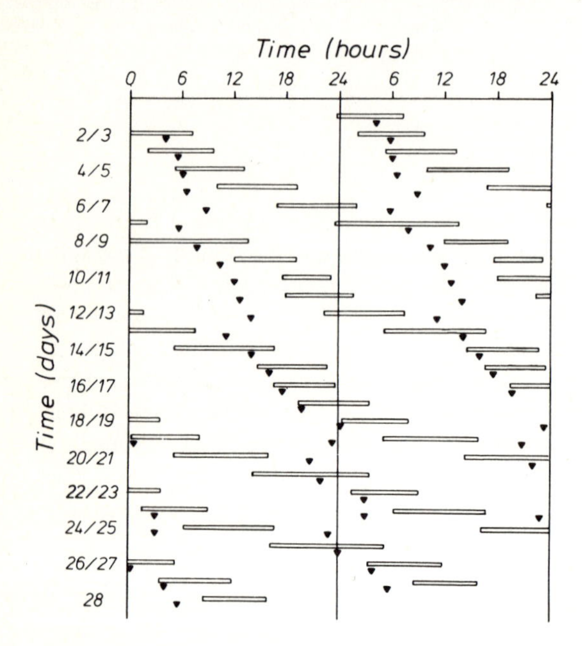

Fig. 1. Consecutive sleep episodes *(bars* bedrest) and minima of rectal temperature *(triangles)* of a subject living alone in isolation from environmental time cues. Double plot

For further analyses, the time of every single sleep episodes was recorded with reference to the actual minimum of rectal temperature. Figure 2 shows, in its upper part, the frequency distribution of sleep onsets relative to the temperature minimum. According to the state of internal desynchronization, the phase relationships cover the full cycle. They are, however, not randomly distributed, but show several clearly separated peaks: for the present analysis, the data are divided in two major groups with the centers of gravity of sleep onset at 6.70 and 0.98 h before the actual temperature minima, and minor groups with the centers of gravity at 4.8 and 11.5 h after the actual minima (cf. Table 1). The lower part of Fig. 2 shows the dependency of the duration of sleep on its position within the temperature cycle. It demonstrates that longer sleep episodes are typically placed, with their midpoints, before the minimum, and shorter sleep episodes after the minimum of rectal temperature.

A more detailed analysis of this dependency is presented in Fig. 3. It shows the correlation between the duration of the single sleep episodes and the position of sleep within the temperature cycle, separately for onset and end of every sleep episode. As Fig. 3 shows especially for sleep onset, the wave form of this rhythmic relationship is not symmetric, but its ascending slope is much steeper than its descending slope.

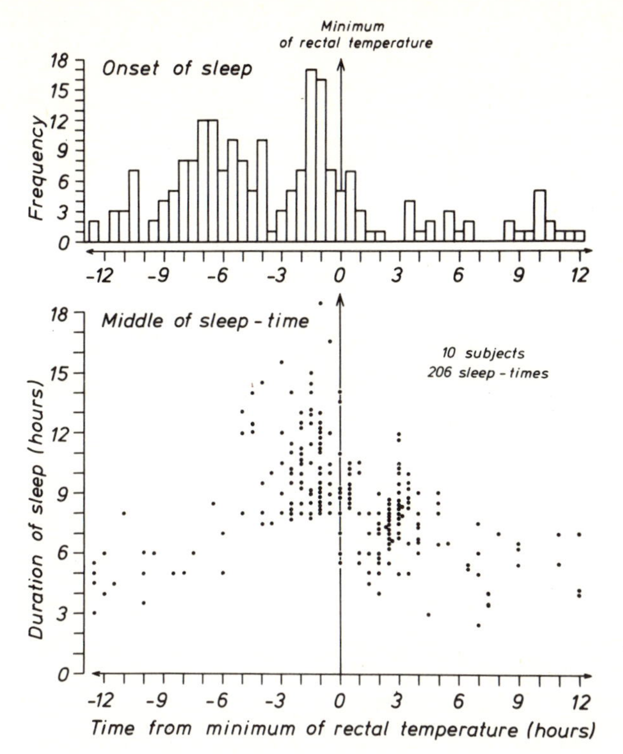

Fig. 2. *Upper part* Frequency distribution for onset of sleep referred to the time of the actual minimum of rectal temperature. *Lower part* Duration of sleep correlated to the phase position of the middle of sleep within the temperature cycle (reference: actual minimum of rectal temperature)

Around the temperature maximum, there seems even to be an ambiguity with long and short durations of sleep beginning at the same phase. This ambiguity, however, is simulated by the superposition of data from many subjects within one plot. The data from single subjects do not show such an overlap of sleep durations. The meaning is, therefore, that the steep increase in the relationship mentioned occurs in the different subjects at slightly different phases.

In Fig. 3, again, a clustering of the sleep episodes in four groups is recognizable. To elucidate this ordering, the four groups (cf. Fig. 2, upper diagram) are indicated by envelops; average data of all groups are given in Table 1. The comparison of the arrangements of the different groups in the two diagrams concerning onset and end of sleep elucidates the fact that the different groups are recognizable in the distribution of sleep onset but not in that of sleep termination (Zulley et al. 1981): with regard to the phases of sleep onset, for instance, group A is clearly separated from group B; for end of sleep, both groups overlap in their phases so that a monomodal distribution of phases results which is advanced in comparison to that of the sole consideration of group B.

In summary, Fig. 3 demonstrates the dependency of the duration of a sleep episode on its temporal position within the temperature cycle and, hence, the influence of the temperature rhythm on the activity rhythm. Moreover, Fig. 3 demonstrates again the multimodal distribution of mutual phase relationships. Whereas in the state of internal synchronization sleep onset occurs preferably at one certain position with-

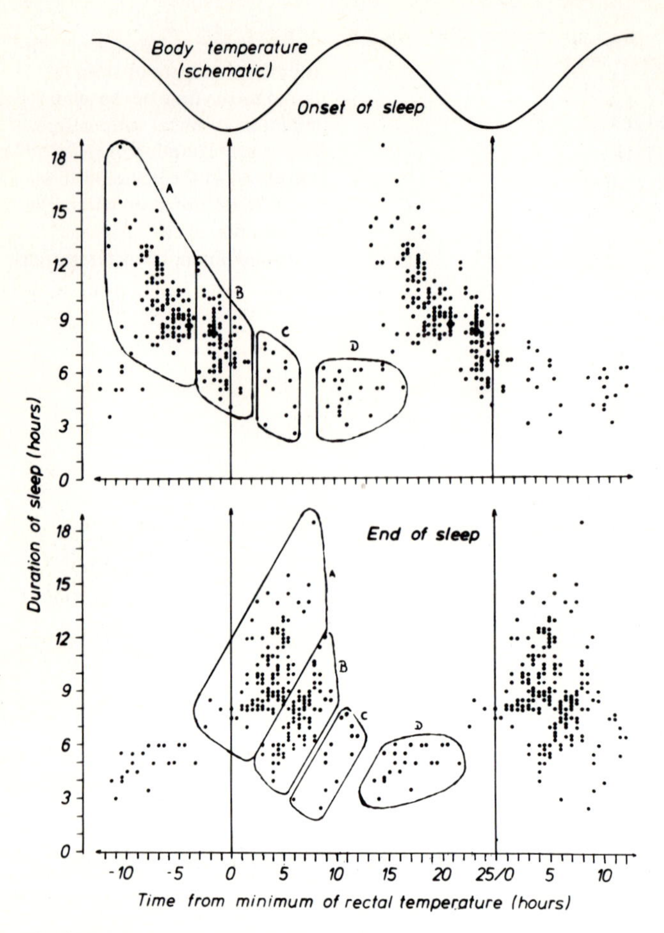

Fig. 3. Duration of sleep correlated to the phase position of sleep onset *(above)* and end of sleep *(below)* within the temperature cycle (reference: actual minimum of rectal temperature); double plot. *Upper border* schematic representation of rectal temperature, for comparison. In the two diagrams corresponding groups of data are indicated each once in the two plots; denotations of the groups corresponds to Table 1

Table 1. Relation between sleep episodes and minima of rectal temperature. (Data from 206 sleep episodes originating from ten subjects)

Group (cf. Fig. 3)	Portion out of all sleep episodes	Onset of sleep [hours before (–) or after (+) the closest temperature minimum] Range	Onset of sleep Mean ± Stand.-Dev.	End of sleep Mean ± Stand.-Dev.	Duration of sleep (h)
A	48%	–10 to – 3.5	– 6.70 ± 1.88	+ 3.68 ± 1.90	10.38 ± 2.24
B	35%	– 3 to + 2	– 0.98 ± 1.09	+ 5.98 ± 1.53	6.96 ± 1.76
C	6%	+ 3 to + 6	+ 4.8 ± 1.7	+10.3 ± 1.6	5.5 ± 1.7
D	11%	+ 8 to +16	+11.5 ± 2.3	+16.4 ± 2.3	4.9 ± 1.0

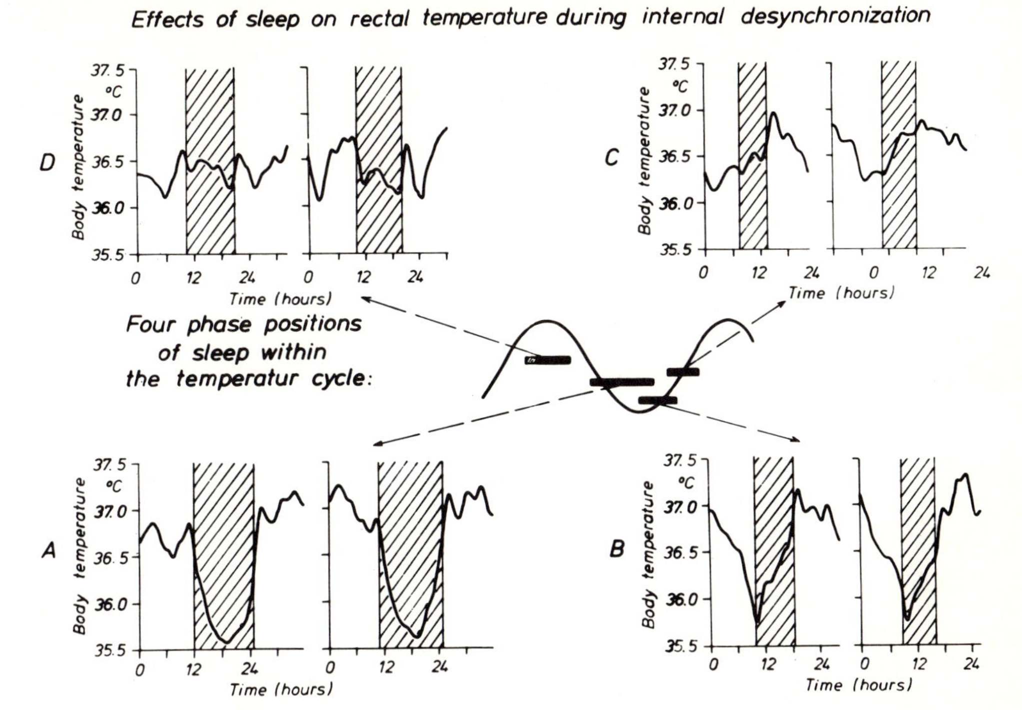

Fig. 4. Two examples from each of the four groups according to Fig. 3, presenting the course of rectal temperature around a sleep episode

in the cycle of deep body temperature, it accumulates, in the state of internal desynchronization, at several preferential positions.

Apart from phase of the rhythm of rectal temperature also the range of this rhythm is of interest. This parameter is recognizable from Fig. 4 where examples of temperature cycles in relation to the sleep-wake cycle are presented; from each of the preferred phase relationships (A to D in Fig. 3) two cycles are drawn originating from the same subject (with polygraphic sleep recordings) as shown in Fig. 1. Figure 4 demonstrates that the wave form and the range of the rectal temperature rhythm are influenced to a considerable amount by its temporal relationship to the sleep-wake cycle: during the sleep episodes, rectal temperature is always reduced, an effect which has been described as "masking effect" of activity on temperature (Aschoff 1970, Wever 1982). As a consequence, the temperature amplitude is large when sleep coincides with a temperature minimum (A), and it is small (or obscured) when sleep coincides with a temperature maximum (D).

A summarizing analysis of the relation between range of the temperature rhythm and the internal phase relationship from all ten subjects is given in Fig. 5. It shows a correlation between the two parameters; in particular, it shows that the range of the temperature rhythm is largest when sleep onset occurs shortly before the temperature minimum, and it is smallest when sleep onset occurs around the temperature maximum. Consequently, Fig. 5 demonstrates the influence of the sleep-wake rhythm on the rhythm of rectal temperature.

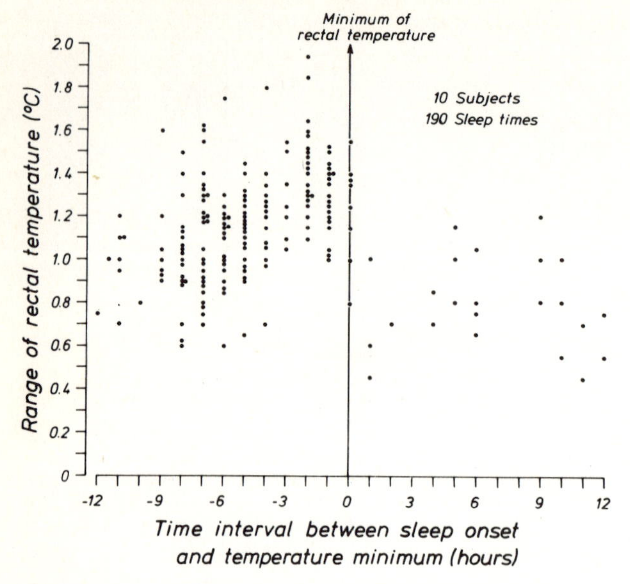

Fig. 5. Range of rectal temperature correlated to the phase position of sleep onset within the temperature cycle (reference: actual minimum of rectal temperature)

4 Discussion

The results obtained in experiments with internally desynchronized circadian rhythms show, on the one hand, a dependency of the sleep episodes on the rhythm of deep body temperature. The probability of the occurrence of a sleep episode varies with the phase of the temperature rhythm, and the duration of a sleep episode depends on its temporal position within the temperature cycle (cf. Fig. 3). On the other hand, the results also show a dependency of the rhythm of deep body temperature on the sleep-wake cycle. The range of the temperature rhythm depends on the phase relationship betwen the rhythms of temperature and sleep-wake (Fig. 5).

These significant results lead to the conclusion that there is mutual interaction between the circadian rhythms of activity and deep body temperature. This interaction is bidirectional and is present in spite of the missing mutual synchronization. As long as the rhythms run internally synchronized, only the mutual synchronization proves unambiguously mutual interaction; however, it cannot be differentiated whether this synchronization is due to an unidirectional or a bidirectional interaction. It is only from internally desynchronized rhythms that the additional conclusions to a bidirectional interaction can be drawn.

The mutual interaction between the two rhythms under consideration is based on, at least, three mechanisms which can be separated in the theoretical analysis.
1. The oscillatory interaction, which leads to internal synchronization inside and to internal relative coordination outside the mutual ranges of entrainment (Aschoff et al. 1967, Wever 1968).

2. Beat phenomena, which are based on the collective and mutually superimposing contribution of two (or more) basic oscillators to every single overt rhythm (Wever 1979).
3. The masking effect, which is due to directly evoced responses of one overt rhythm to variations in the course of the other rhythm (see above).

In addition, the results demonstrates the clustering of the internal phase relationship between the rhythms of activity and rectal temperature in separated groups, representing preferred adjustment to some few discrete values. In internally synchronized rhythms, this phase relationship covers only a small range, corresponding to group B in Fig. 3 and Table 1 (Zulley et al. 1981). In internally desynchronized rhythms, in general, all phase relationships between 0° and 360° can occur. In particular, however, there is not an equal distribution of phase relationships, and not even a monomodal distribution. Instead, there are several preferred phase relationships between the two rhythms which are represented by the groups A through D (cf. Fig. 3 and Table 1). Apart from the two secondary groups C and D, there is mainly group A with the longest sleep durations which differs in its position from that of internally synchronized rhythms (these are characterized by a position similar to that of group B; Zulley et al. 1981). Phase relationships corresponding to the two groups A and B can also be observed in rhythms which are synchronized to an external zeitgeber with varying period. Under such an influence, the temporal relationship between sleep and minimum of rectal temperature does not change steadily, but jumps unsteadily between two discrete phase relationships, with the minimum near onset and near end of sleep (Wever 1981); these preferred phase relationships correspond to those in free-running internally desynchronized rhythms which are characterized by groups A and B.

In all analyses discussed in this paper, actual values of single sleep episodes and single minima in rectal temperature are compared; this approach had been introduced by Zulley (1976, 1980) and Zulley and Schulz (1980). In addition, another type of analysis had been applied which compares single sleep episodes with the minimum of an educed shape of the temperature rhythm, i.e., with an average cycle of rectal temperature (Czeisler 1978, Czeisler et al. 1980a, 1980b). The results of this type of analysis deviate from results discussed in this paper in several details. For instance, a sleep episode begins preferably several hours before an actual minimum of rectal temperature (cf. Fig. 2, upper diagram) but several hours after the "educed" minimum (cf. Czeisler et al. 1980b). This difference is due to the systematic influence of sleep on the position of the temperature minima which results in systematic deviations between actual and "educed" minimum values. These deviations become obvious in Fig. 6 where the total course of rectal temperature (solid line) measured in an experiment with internally desynchronized rhythms (same experiment as in Fig. 1).is drawn together with a sinusoid which represents the educed rhythm of rectal temperature (constant period of 25.0 h); in addition, the sleep episodes of the subject are indicated. As can be seen in Fig. 6, the actual minimum temperature is mostly later than the educed minimum when it occurs near the onset of a sleep episode, and it is mostly earlier than the educed minimum when it occurs near the end of a sleep episode. This systematic shifting is another evidence for the influence of the sleep-wake rhythm on the temperature rhythm.

The variability of the rectal temperature rhythm in phase and amplitude could produce the impression of a random process. This impression, however, can be cor-

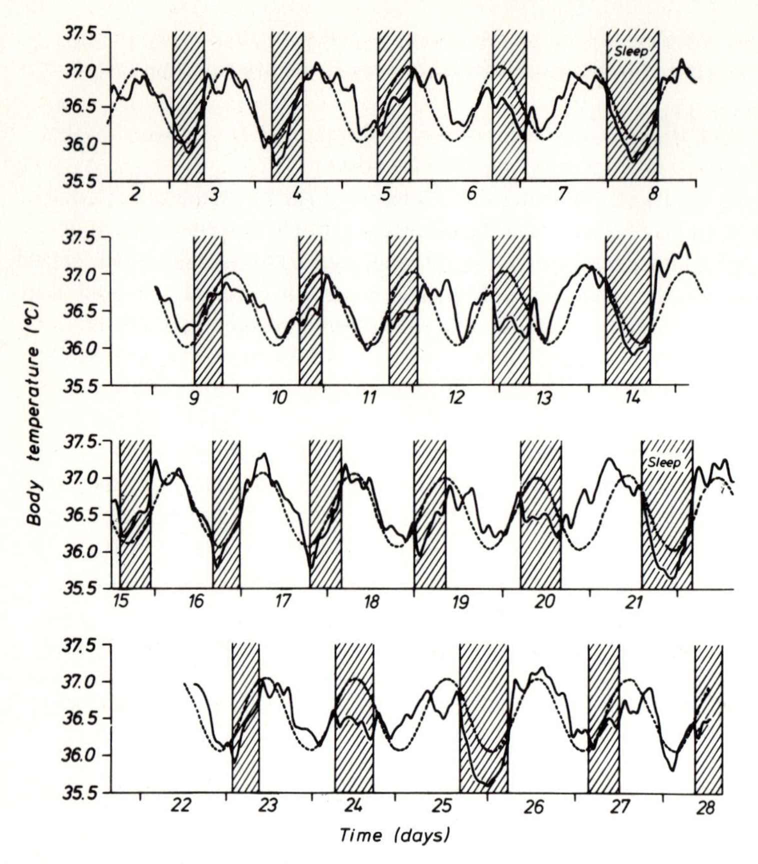

Fig. 6. Course of rectal temperature *(solid line)* measured in a 4-week experiment under constant conditions, presented in four consecutive sections (same experiment as in Fig. 1). *Dotted line* sine wave with constant parameters, best fitting the measured data. Sleep episodes are indicated by *shaded areas*

rected by the inspection of Fig. 6. In this figure, the dotted line (sinusoid) can be considered to represent the course of a hypothetical "temperature oscillator' which runs with constant period and without any random fluctuations. As can be seen, the measured temperature values (constituting the overt temperature rhythm) fluctuate around the hypothetical oscillator course; for instance, the actual temperature minima are sometimes earlier and sometimes later than the hypothetical minima but return always to the values of the hypothetical oscillator. The deviations between actual and hypothetical temperature values depend, apart from random fluctuations, regularly on the temporal relation to the sleep episodes. This picture is hardly compatible with the assumption of a random generation of the temperature rhythm; in this case, the deviations from a long-term average should increase cumulatively. The picture of Fig. 6 is compatible with the assumption that the overt temperature rhythm is, in fact, influenced by the alteration between sleep and wake, but mainly controlled by an oscillator which must not necessarily interact with the sleep-wake rhythm.

The results discussed in this paper are of theoretical interest because they give insight into the structure of the circadian multi-oscillator system. In addition, however, they are of practical interest because they enable to predict the sleeping behavior of a subject from the previously measured course of deep body temperature. In particular, they give the possibility to predict the duration of a sleep episode when the actual position of the temperature minimum is known. This holds true even during the state of internal desynchronization where the variations in sleep duration are remarkably large. On the other hand, the range of the temperature rhythm can be predicted from the phase relationship between sleep-wake and the temperature rhythm; in particular, this means the actual value of a temperature minimum can be predicted.

References

Aschoff J (1970) Circadian rhythms of activity and of body temperature. In: Hardy JD, Gagge AP, Stolwijk JA (eds) Physiological and behavioral temperature regulation. CC Thomas, Springfield, III, pp 905-919

Aschoff J, Gerecke U, Wever R (1967) Desynchronization of human circadian rhythms. Jpn J Physiol 17: 450-457

Czeisler CA (1978) Human circadian physiology: internal organization of temperature, sleep-wake, and neuroendocrine rhythms monitored in an environment free of time cues. Ph D thesis, Stanford Univ

Czeisler CA, Weitzman ED, Moore-Ede MC, Zimmerman JC, Knauer RS (1980a) Human sleep: its duration and organization depend on its circadian phase. Science 210: 1264-1267

Czeisler CA, Zimmerman JC, Ronda JM, Moore-Ede MC, Weitzman ED (1980b) Timing of REM sleep is coupled to the circadian rhythm of body temperature in man. Sleep 2: 329-346

Wever RA (1968) Einfluss schwacher elektro-magnetischer Felder auf die circadiane Periodik des Menschen. Naturwissenschaften 55: 29-32

Wever RA (1975) The circadian multi-oscillator system of man. Int J Chronobiol 3: 19-55

Wever RA (1979) The circadian system of man. Springer, Berlin Heidelberg New York

Wever RA (1981) On varying work-sleep schedules: the biological rhythm perspective. In: Johnson LC, Tepas DI, Colquhoun WP, Colligan MJ (eds) Biological rhythms, sleep and shiftwork. Advances in sleep research, vol VII. Spectrum Publ. New York, pp 35-60

Wever RA (1982) Organization of the human circadian system: internal interactions. In: Wehr IA, Goodwin FK (eds) Circadian rhythms in psychiatry. Neuroscience Series. Boxwood Press, Los Angeles (in press)

Zulley J (1976) Schlaf und Temperatur unter freilaufenden Bedingungen. Berichte 30. Kongr Dtsch Ges Psychol. Hogrefe, Göttingen, pp 398-399

Zulley J (1979) Der Einfluss von Zeitgebern auf den Schlaf des Menschen. RG Fischer-Verlag, Frankfurt aM

Zulley J (1980) Duration and frequency of bedrest episodes in internally desynchronized rhythms. Sleep 2: 344

Zulley J, Schulz H (1980) Sleep and body temperature in freerunning sleep-wake cycles. In: Sleep 1978. 4th Eur Congr Sleep Res, Tirgu Mures, Karger, Basel, pp 341-344

Zulley J, Wever R, Aschoff J (1981) The dependence of onset and duration of sleep on the circadian rhythm of rectal temperature. Pfluegers Arch 391: 314-318

6.5 The Phase-Shift Model of Spontaneous Internal Desynchronization in Humans

C. Eastman[1]

A comprehensive and elegant multiple oscillator theory of human circadian rhythms, which I will call the "traditional" theory, has been developed primarily by Wever (Wever 1979, Aschoff and Wever 1976). The strongest evidence for this theory comes from humans in temporal isolation who show "spontaneous internal desynchronization". A typical case (Fig, 1, top) shows that the temperature and activity rhythms freeran mutually synchronized during the first section of the experiment (A), but became "desynchronized" during the second section (B). The periodograms for the second section showed two periods in each rhythm, one at 25.1 h which was most prominent in the temperature rhythm, and one at 33.4 h which was most prominent in the activity rhythm. In the traditional theory there are basically two oscillators, one primarily controls the temperature rhythm, and the other primarily controls the activity rhythm. During synchronized freeruns the oscillators are mutually synchronized, but during internal desynchronization they uncouple and freerun with their own natural frequencies.

This paper presents an alternative theory which developed from computer models of data produced by rats at the limits of entrainment (Eastman 1980). In one of these rat "phase-shift" models (Fig. 2), the "driving" oscillator freeran through a 22 h LD cycle until a certain phase relationship to the LD cycle was reached. At that point, the driving oscillator was abruptly shifted and then freeran again until the same phase relationship was reached, etc. The LD cycle will be called the "shift-inducing" oscillator. The periodogram shows a peak at 22 h corresponding to the shift-inducing oscillator, but there is no peak at 24 h corresponding to the driving oscillator. The peak at 25.2 h may be considered a mathematical artifact because it does not correspond to either oscillator.

The phase-shift model can be applied to cases of internal desynchronization in the human with intriguing consequences. In this interpretation of the case in Fig. 1, top, the driving oscillator free-ran with a period of 25.7 h during section A. During section B, the driving oscillator free-ran with the same period, but was shifted once every three days. By analogy with the rat phase-shift model, the periodogram peak at 25.1 h corresponds to a shift-inducing oscillator, the period of the driving oscillator (25.7 h) does not appear, and the 33.4-h period is considered an artifact. This interpretation prompts the question of what could act as a shift-inducing oscillator. There are no light-dark cycles or other known zeitgebers in temporal isolation. Therefore, the shift-inducing oscillator would appear to be endogenous to the subject. Another possibility stems from Wever's (1979)

1 University of Chicago Sleep Laboratory, 5741 S. Drexel Ave, Chicago, Ill. 60637/USA

Vertebrate Circadian Systems
Ed. by J. Aschoff, S. Daan, and G. Groos

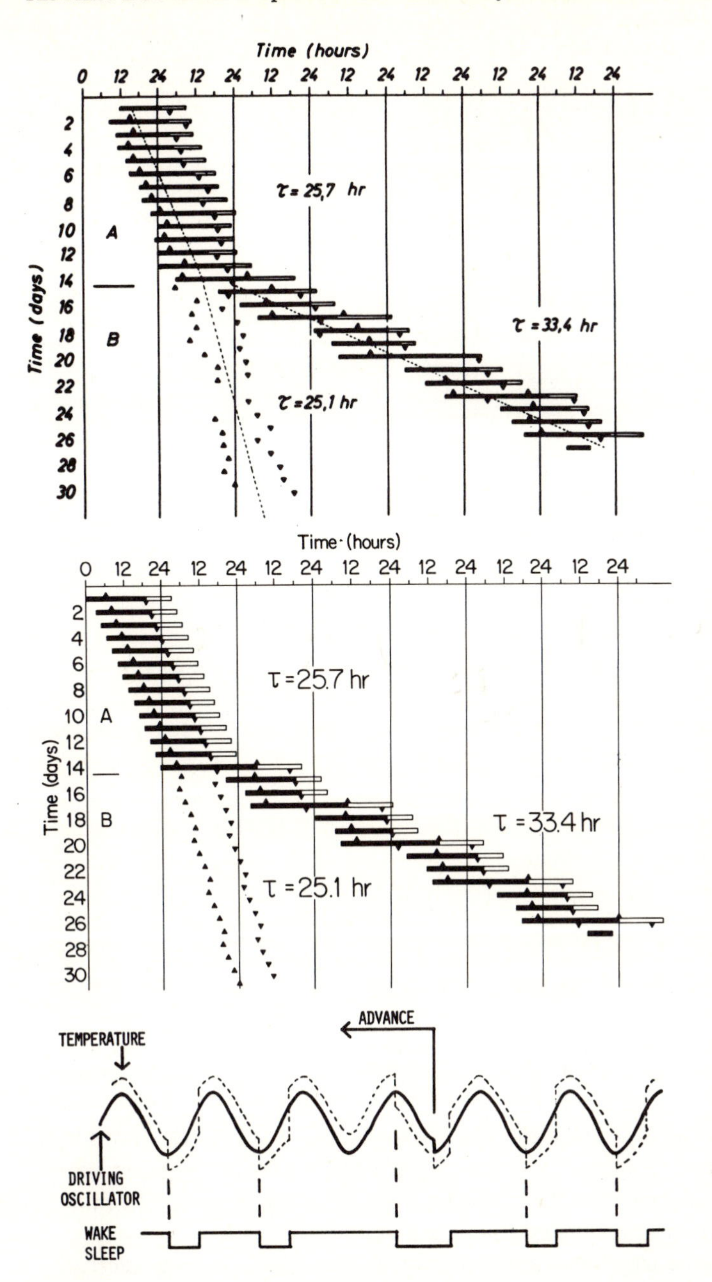

Fig. 1. *Top* Activity rhythm *(black bars* activity; *white bars* rest) and rectal temperature rhythm *(triangles* show daily maxima and minima) of a human in temporal isolation. *A* Synchronized free-running rhythms. *B* Spontaneous internal desynchronization. (Wever 1975). *Middle* Human phase shift model of the case shown in the top diagram which was made by the method shown in the bottom diagram. The driving oscillator had a period of 25.7 hrs. In section *A*, sleep episodes of 8 h started on the minima of the oscillation. In section *B*, sleep onset skipped every fourth minimum and occurred later (starting on the high of the oscillation). The misplaced sleep episodes lasted longer (14 h) and advanced the driving oscillator 2.34 h. *Bottom* Diagram of the method used to produce human phase-shift models. The driving oscillator was generated by a sine wave. Usually sleep episodes of normal length were started on the minimum of the oscillation. There were occasional misplaced sleep episodes which shifted the driving oscillator. In this diagram the third sleep episode was misplaced, lasted longer, and advanced the driving oscillator. "Internal desynchronization" was produced by a sequence of "normal" and misplaced sleep episodes. The position and length of the misplaced sleep episodes depended on the individual subject.

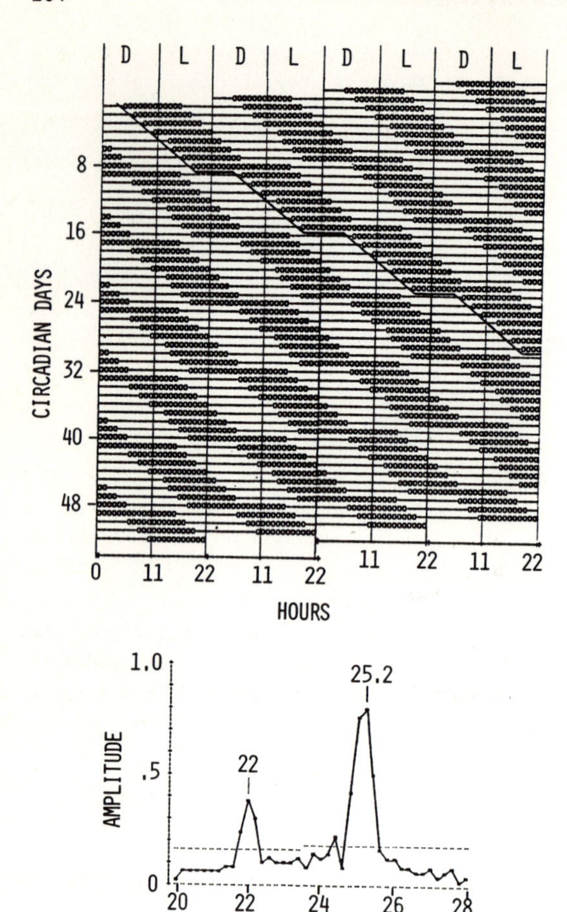

Fig. 2. *Top* Rat phase-shift model for an animal who was not entrained to a 22-h LD cycle and showed a pattern resembling relative coordination. The driving oscillator represented by a sine wave, free-ran with a period of 24 h until it hit a certain phase in the LD cycle, at which point it was shifted and then free-ran again, etc. In this quadruple plotted "wheel running style" graph, an "o" represents an hourly value above the mean for the circadian day. *Bottom* Periodogram for the data shown above. The *dashed horizontal line* represents the 95% confidence limit

finding that the distribution of all the near-25-h periods from the periodograms of internally desynchronized subjects is not significantly different from the lunar day cycle of 24.83 h. This suggests that if there is a shift-inducing oscillator, it could be the moon. A third possibility is that there is an endogenous shift-inducing oscillator which is entrained to the moon. Other considerations (see below) suggest that if may not be necessary to postulate a shift-inducing oscillator, and that the near-25-h periodogram peaks from internally desynchronized subjects may also be artifacts.

A disadvantage of the traditional model is that it is necessary to postulate that some people have "activity" oscillators with very short periods (between about 15 and 20 h) or very long periods (between about 30 and 40 h). There is considerable skepticism that circadian rhythms could be controlled by oscillators with such unusual periods (Aschoff 1965, Aschoff et al. 1967, Mills 1973). In the phase-shift model there are no oscillators with periods outside of the usual circadian range.

The phase-shift model was expanded to include the well-known masking effect of sleep and activity on temperature (Aschoff 1970) and recent descriptions of the relationship of sleep onset and sleep length to the phase of the temperature cycle during synchronized freeruns and during internal desynchronization (Czeisler 1978, Czeisler et al. 1980, Zulley et al. 1981). Human phase-shift models were made by computer as shown in Fig. 1, bottom. The driving oscillator controlled both the rhythms of temperature and activity (sleep-wake). Temperature followed the oscillator and included a masking component; temperature was raised by a constant amount during activity and was lowered by a constant amount during sleep. Usually sleep started on the minima of the oscillation, as it does during synchronized free-runs. Occasionally sleep occurred at an unusual phase of the oscillation and was often very long. There was feedback from the activity rhythm to the driving oscillator so that the oscillator was shifted (advanced) by the misplaced sleep episode. Thus, the phase-shift model includes two types of shifts, a shift of the activity rhythm caused by the misplaced sleep episode and a subsequent shift of the driving oscillator. Figure 1, middle, shows that the phase-shift model can mimic actual cases of internal desynchronization. The model also produced periodograms (not shown), which were similar to those produced by the empirical data.

Feedback from the activity rhythm to the driving oscillator may occur during normal entrainment, as well as during internal desynchronization. The direction and magnitude of this feedback shift may depend on when the sleep episodes occur relative to the phase of the driving oscillator. In order to entrain to the 24-h day, human circadian rhythms must be advanced by about an hour per day since the average free-running period is 25 h. In the 24-h day, sleep onset occurs about 6 h before the temperature minimum (Wever 1979, Aschoff and Wever 1976, Czeisler 1978). This placement of sleep may help advance the driving oscillator by the hour necessary for entrainment. During sleep deprivation, various circadian rhythms show a slight phase delay relative to the 24-h day (references in Wever 1979). This delay may result from the absence of advances produced by the usual sleep episodes.

In the case of internal desynchronization modeled in Fig. 1, and in many other cases in the literature, the misplaced sleep episodes and shifts occurred regularly. Regularity supports the idea that an oscillator was responsible for the shifts. Perhaps a shift-inducing oscillator prompted the subject to stay awake past the driving oscillator's minimum (see Fig. 1, bottom). However, in many other cases of internal desynchronization, the shifts do not occur regularly, so that the presence of a shift-inducing oscillator seems unlikely. In these cases, the subject may temporarily ignore the signal from the driving oscillator and have a misplaced sleep episode for any number of "random" reasons. For example, the subject may stay up late past his minimum to finish an interesting book. The human phase-shift model for a case of irregular shifting (Fig. 3) utilized only one periodic function, the sine wave representing the driving oscillator. There was no second periodicity or shift-inducing oscillator. The model produced a periodogram with two dominant peaks. In the traditional model the 24.6-h period is interpreted as the period of the "temperature" oscillator, and the 29.2-h period as the period of the "activity" oscillator. In terms of the phase-shift model both of these periodogram peaks are considered mathematical artifacts because they do not correspond to the period of the driving oscillator at 25.3 h.

The phase-shift model shows that spontaneous internal desynchronization can be produced while both the temperature and activity rhythms are controlled by the same driving oscillator. It is not necessary to postulate that there are separate "temperature"

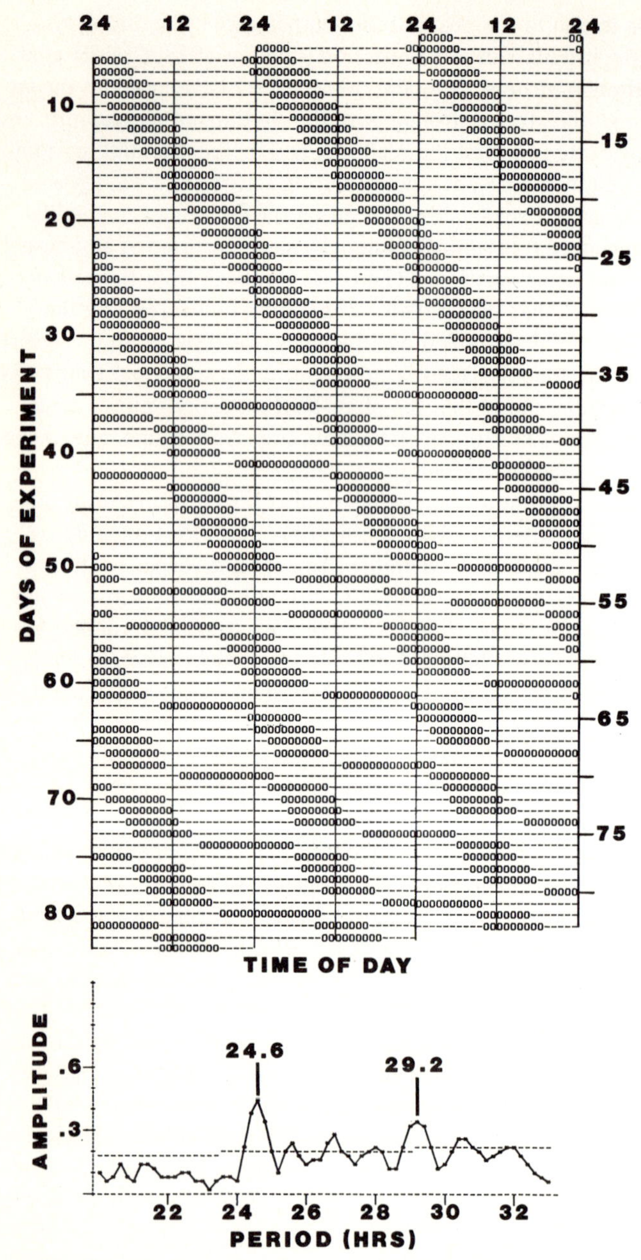

Fig. 3. Human phase-shift model of a case of spontaneous internal desynchronization reported by Czeisler (Czeisler 1978, Fig. 19; Czeisler et al. 1980, Fig. 1). *Top* Triple plotted sleep chart. An "o" represents an hour of sleep. The driving oscillator freeran with a period of 25.3 h. At first, sleep episodes of 8 h started on the minima of the oscillation (see Fig. 1, bottom). Starting on day 36, misplaced sleep episodes appeared in a "random" sequence. Each misplaced sleep episode skipped the minimum of the oscillation and occurred 8.43 h later. The misplaced sleeps were longer (14 h) and advanced the driving oscillator 4.5 h. *Bottom* Periodogram of the data above from days 35-74

and "activity" oscillators. Furthermore, cases of internal desynchronization with irregular shifting can be adequately explained with a single oscillator – the driving oscillator. Cases or regular shifting suggest there may be a second shift-inducing oscillator, but perhaps other reasons for regularly misplaced sleep periods could be found. In any case, it is not necessary to hypothesize that separate "temperature" and "activity" oscillators have uncoupled to free-run with different frequencies. Further work is necessary to determine which model, the phase-shift model, the traditional model, or some other variation, can best explain the patterns of human circadian rhythms.

References

Aschoff J (1965) Circadian rhythms in man. Science 148: 1427-1432

Aschoff J (1970) Circadian rhythm of activity and of body temperature. In: Hardy JD, Gagge AP, Stolwijk JAJ (eds) Physiological and behavioral temperature regulation. CC Thomas, Springfield, pp 905-919

Aschoff J, Wever R (1976) Human circadian rhythms: a multioscillatory system. Fed Proc 35: 2326-2332

Aschoff J, Gerecke U, Wever R (1967) Desynchronization of human circadian rhythms. Jpn J Physiol 17: 450-457

Czeisler CA (1978) Human circadian physiology: internal organization of temperature, sleep-wake and neuroendocrine rhythms monitored in an environment free of time cues. Ph D thesis, Stanford Univ

Czeisler CA, Weitzman ED, Moore-Ede MC, Zimmermann JC, Knauer RS (1980) Human sleep: its duration and organization depend on its circadian phase. Science 210: 1264-1267

Eastman CI (1980) Circadian rhythms of temperature, waking, and activity in the rat: dissociations, desynchronizations, and disintegrations. Ph D thesis, Univ Chicago

Mills JN (1973) Transmission processes between clock and manifestations. In: Mills JN (ed) Biological aspects of circadian rhythms. Plenum Press, New York, pp 27-84

Wever RA (1975) The circadian multi-oscillator system of man. Int J Chronobiol 3: 19-55

Wever RA (1979) The circadian system of man. Springer, Berlin Heidelberg New York

Zulley J, Wever R, Aschoff J (1981) The dependence of onset and duration of sleep on the circadian rhythm of rectal temperature. Pfluegers Arch 391: 314-318

7.1 Physiology of Photoperiodic Time-Measurement

B.K. Follett[1]

1 Introduction

Few seriously doubt that circadian rhythms are involved in photoperiodic time-measurement: the experimental evidence is difficult to interpret in any other way. Two sets of data may make the point. If sparrow are free-run in darkness there is a circadian rhythm of photoinducibility (Follett et al. 1974). Early in the subjective day a single block of 8 h light will not alter gonadotrophin secretion, but during the subjective night it increases the plasma concentration significantly. This rhythm of inductiveness recurs for at least five cycles with a periodicity close to 24 h. In hamsters, gonadal growth can be induced with only 1 h of light each day if the photoperiodic cycle differs slightly from 24 h (Elliott 1976), a finding which seems explicable only in terms of circadian entrainment theory. Given such results, it was inevitable that various models would be proposed as to how circadian rhythms might measure daylength and two classes of model have attracted the most attention. The first ("external coincidence") proposes that one of the many circadian oscillators is a rhythm of "photosensitivity". Should light coincidence with this daily peak of "photosensitivity" then induction occurs. The second ("internal coincidence") is based upon the belief that in a multioscillatory organism phase relationships between the oscillators alter as daylength changes, with the result that at one time of the year two of the oscillators may be in a state where they cause induction, whilst at another the phase relationships are non-inductive (Pittendrigh 1981). More complex variants of both models have been suggested (e.g. Beck 1980) but decisive evidence in favour of one or the other is lacking and, insofar as can be judged, all the experimental data will fit either class of model. Both seem equally plausible, or implausible, although for many years the external coincidence model has been in the ascendancy, probably because it is the older of the two and at first glance appears the simpler. More recently though advances in our understanding of entrainment has made the internal coincidence model the more attractive (Pittendrigh 1981). As a physiologist who often finds himself offering two rather different interpretations for his results, depending upon whether an external or an internal coincidence device might apply, I sometimes wonder if the models are an unmitigated blessing and suspect secretly that progress might be quicker if there were fewer preconceived notions. For this reason, amongst others, therefore, I will

1 ARC Research Group on Photoperiodism Reproduction, Department of Zoology, The University, Bristol 8, United Kingdom

Vertebrate Circadian Systems
Ed. by J. Aschoff, S. Daan, and G. Groos

use the terms "photoinducible phase (ϕ_i)" or "phase of photoperiodic sensitivity" to refer only to that period within the 24-h cycle which, when illuminated, leads to induction, and they are not to be seen as carrying any explicit meaning in terms of either coincidence model.

Progress in analysing photoperiodic time-measurement physiologically has been bedevilled by various problems, two of which might be mentioned. The first is that since the circadian oscillators involved in the photoperiodic clock lack any overt expression, it has proved impossible so far to monitor them directly and hence know their precise phase under a particular light treatment. Many workers have tried to deduce their state indirectly by tracking another rhythm (e.g. activity) and assuming that this accurately reflects those in the photoperiodic clock. In some organisms (e.g. hamsters) this seems successful but our experience with *Coturnix* quail is a reminder of the potential dangers. In this species the couplings between three circadian systems – activity, oviposition and photoperiodic time-measurement – are such that each shows different phase changes when exposed to various T-cycles (Simpson and Follett 1982) so that estimating phase of the photoperiodic oscillators from the activity rhythm is more of a hindrance than a help. The second problem is a more general one and involves the potential trap of assuming that the temporal output from the neural and/or endocrine organs ultimately responsible for the inductive processes (e.g. gonadal growth, diapause), will directly reflect the hour-by-hour state of the photoperiodic oscillators. As will be seen below, this is a problem encountered by those of us who have sought to measure patterns of hormone secretion.

In practice, various approaches have been attempted to unravel the photoperiodic oscillators and these are considered under four headings.

2 Location of the Photoperiodic Oscillators

Lesions within the suprachiasmatic nucleus (SCN) of the Syrian hamster *(Mesocricetus auratus)* disrupt the photoperiodic control of reproduction and the testes remain fully functional regardless of daylength. Since the SCN plays a central role in the circadian organization of rodents (see Section 3 and also Rusak and Zucker 1979) it seems reasonable to conclude that the blockade to the effects of short days is due to the hamsters no longer having a functional photoperiodic clock and so being unable to distinguish short from long daylengths, or to disruption in some other circadian event underlying seasonality, or to the SCN being a target site for the action of short days in terminating reproduction. There seems little doubt that the whole issue is bound up with the functioning of the pineal gland.

Although the story is incomplete, many findings suggest that the photoperiodic message is conveyed to the rodents's brain by melatonin secreted from the pineal gland (Reiter 1980, Hoffmann 1981). This statement carries with it the important implication that melatonin should not be seen as a specific "anti-gonadotrophin" but as a seasonal transducer. In species such as the Siberian hamster, short days not only suppress gonadotrophin secretion, but also alter pelage colour and body size. All of these effects are pineal-mediated and this argues that melatonin may act at some pivotal point in the brain above the level of the neuroendocrine circuits which individually regulate these three functions. The specific effect of the pineal's action may be different in other mammals (e.g. the ferret, Herbert 1981), but the unifying feature remains the organism's need for the gland in order to respond physiologically to changing photoperiods. The temporal pattern of melatonin secretion seems to be important in its transducing action (e.g. Goldman et al. 1979, Bittman et al. 1979). Single injections of melatonin to intact golden hamsters on long days will cause testicular regression if given in the afternoon but not in the morning, an effect which is explained by assuming that the exogenous melatonin late in the day somehow combines with the endogenous nighttime surge of melatonin to mimic the short day situation and switch off gonadotrophin secretion. This is supported by the finding that a daily treatment with three melatonin injections over a 6-h period causes testicular regression in pinealectomized hamsters. Such a conclusion fits well, of course, with the known secretion of melatonin in vivo where it is released only over a short period each night, this precise rhythmic pattern being under the control of a circadian rhythm emanating from the SCN. Such results led to a model in which it was thought that differences may exist between the patterns of melatonin secretion under short and long daylengths and that somehow this would lead to gonadal regression in the former case. The photoperiodic clock would then reduce to the question of how the SCN altered the phase and duration of melatonin output from the pineal gland. Unfortunately, this model has run into problems and the situation is even more confused. Estimates of pineal melatonin content in golden hamsters do not suggest striking differences between the amounts being produced in short and long daylengths (Tamarkin et al. 1979) and this has led on to the idea that if there is no great change in the circadian rhythm of melatonin secretion with photoperiod then perhaps there is a rhythm in the melatonin target site and this alters seasonally (e.g. Klein and Tamarkin 1980). At one time of the year it might coincide with high plasma melatonin levels and at another it would not, such a difference providing the basis for seasonal changes in response. This idea, however, also has its problems. Bittman et al. (1979) have shown that thrice-daily injections of melatonin cause testicular regression in SCN-lesioned hamsters. This fact, together with the knowledge that thrice-daily injections of melatonin are equally effective in pinealectomized hamsters during night or day (Goldman et al. 1979, Bittman et al. 1979), rather argues against a simple rhythm in target tissue sensitivity. Quite recently, Hoffmann et al. (1981) have shown that pineal NAT rhythms in Siberian hamsters

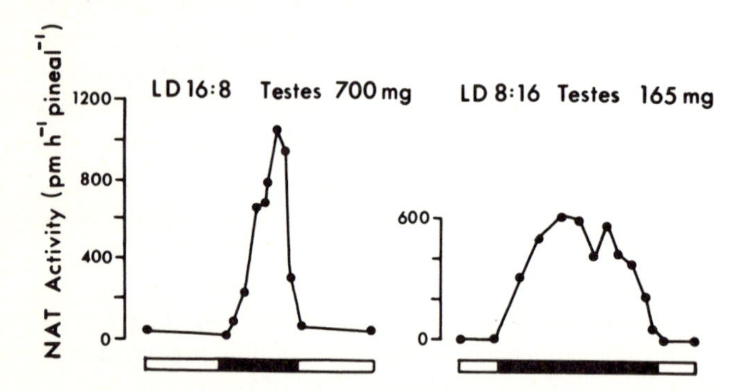

Fig. 1. Pineal N-acetyltransferase activity in pineal glands of Siberian hamsters maintained for 3-4 weeks in short or long photoperiods. (Adapted from Hoffmann et al. 1981)

are very different in LD 8 : 16 and LD 16 : 8 (Fig. 1). This seems to reopen the earlier possibility that the photoperiodic clock could be based on the characteristics of an SCN-driven rhythm in melatonin production (see also Arendt and Symons 1981), the indole itself having its transducing effect at a still unknown site in the brain. Distinguishing between these alternatives might be eased if it was possible to measure circulating melatonin and also know where it acts.

In birds, virtually all the evidence suggests the pineal (or melatonin) not to have a seasonal transducing role comparable with that in hamsters. The pineal gland may well be a component in the avian circadian system (see Chaps. 1.1, 4.4, 4.5, this Vol.) but photoperiodic time-measurement seems to proceed perfectly well in its absence. The SCN are also involved in the generation of circadian locomotor rhythms in birds (see Chaps. 3.5, 3.6, this Vol.) and it is particularly intriguing, therefore, that lesions in this general area in quail not only upset locomotor activity but also block photoinduced gonadal growth and ovulation (Davies 1980, Simpson and Follett 1981). All of these functions have a circadian component and it is not inconceivable that they could occur separately if the SCN were regionally differentiated. In species such as the quail distinct SCN are not visible, Whilst attractive it must be admitted that other explanations for the lesions are possible and no conclusions should be drawn without a more detailed analysis.

3 Patterns of Gonadotrophin Secretion Under Photostimulation

An initial approach here was to see if the reproductive hormones – LH, FSH and the sex steroids – show a circadian rhythm of secretion in photoperiodically stimulated animals and, if so, whether the rhythm's characteristics might directly reflect the time-measuring oscillators. The idea has not proceeded far, however, since once having risen under long days there is no strong daily rhythm visible in either quail, ducks or hamsters (Follett 1978). In other species, LH and FSH secretion is far more pulsatile and this is particularly true for sheep where the LH levels change strikingly every few minutes. Here, it is thought that photostimulation somehow increases the frequency of firing within the hypothalamic pulse generator responsible for surges in Gn-RH output (Lincoln and Short 1980, Goodman and Karsch 1981). However, the evidence is rather against this pulse generator having a strong daily (circadian) component to it and the number of pulses secreted each hour in photostimulated sheep changes only marginally (e.g. Lincoln et al. 1977).

These results are open to various interpretations in terms of the circadian oscillators underlying photoperiodic time-measurement. Perhaps they suggest that when light falls in the photoindicible phase the oscillators are switched into the "on" mode and the output to the neuroendocrine circuitry is effectively continuous. This is possible but it cannot be distinguished from the alternative in which the GnRH circuitry might be switched into a continuous "on" mode by a rhythmic input from the photoperiodic clock.

It might help if it were somehow possible to "see" the actual output of the clock immediately the animal has made its decision as to whether the day is short or long. So far this has only been possible with the secretion of LH during the first few days of photo-

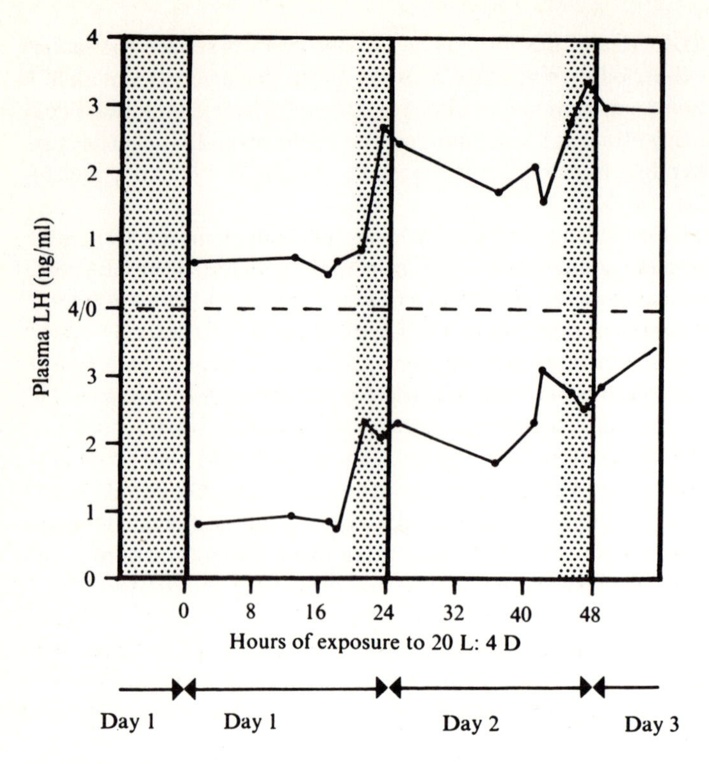

Fig. 2. Plasma LH levels in two *Coturnix* quail during the first 2 days of exposure to LD 20 : 4. Note the rapid changes in LH secretion occurring late in the photoperiod. Presumably, at or just prior to this time the day has been read as "long" and induction triggered. (Follett et al. 1977)

induction in quail. As Fig. 2 indicates, plasma levels of LH in quail remain at their low short-day concentrations throughout much of the first long day. However, at about hour 20 there is an abrupt rise in LH secretion and this continues through the short night (LD 20 : 4) and into the next day. During much of the second long day, plasma LH levels either remain constant or fall, but at about hour 18 they rise again. The third long day is sometimes like the second but subsequently rhythmicity breaks down and LH is maximal at all times of day and night. Other experiments (Follett et al. 1977) suggest that a photoperiodic "gate" (ϕ_i) exists, which on short days must begin about 12 h after dawn. If the quail is illuminated for even a short period during this gate, e.g. hours 12 to 16, then induction occurs and LH secretion rises a few hours later, beginning at hours 18-20. This rise must be one of the earliest events following photoperiodic induction. There is a considerable variation between quail in the precise pattern of LH secretion during the first few long days but we have the impression that induction is a progressive event which lasts longer on each successive day until eventually LH levels are always high. Because of the tight coupling between Gn-RH and LH secretion, it seems unlikely that any changes in the daily duration of hormone release can come about other than by a progressive change in the neural circuits. Once again, however, we do not know if the changes are occurring in the circadian clocks measuring the photoperiod, or in the neural circuits which control GnRH release. As in other areas of circadian research, it is difficult to disentangle the hands of the clock from the clock itself.

4 Corticosterone Rhythms

From the discovery that prolactin would only induce fattening in migratory white-throated sparrows if given at certain times of days, Meier and his colleagues have developed a sophisticated coincidence model which seeks to explain many photoperiodic phenomena in terms of phase-angle differences in the secretion of various hormones (Meier and Ferrell 1978). The central feature is the daily rhythm in corticosterone secretion which is seen as a master oscillator entraining other circadian rhythms. In one specific version, for example, the corticosterone rhythm is entrained by dusk, and peak secretion occurs 12 h later. This peak entrains ϕ_i in such a way that it is illuminated in LD 16 : 8 but not in LD 10 : 14. This is, of course, an external coincidence model with the responsibility for entraining ϕ_i removed one step from the light-dark cycle. An analogous arrangement is suggested for the rhythms in prolactin secretion/target tissue sensitivity. As the phase relationships between the corticosterone and prolactin rhythms change seasonally, different physiological events are triggered: the induction and breaking of refractoriness, migratory fat deposition, and Zugunruhe. Support for these ideas is twofold. Firstly, responses to hormone treatments (e.g. gonadal growth, fattening) can be altered if the times between the injections are changed, and secondly, there seem to be rhythms in corticosterone and prolactin output whose time relationships fit with the injection experiments. The strength of the evidence is under question, however, mainly because some workers have found it difficult to obtain dramatic effects on fattening etc. after treatment with corticosterone and prolactin, while the clarity of the hormonal rhythms leaves something to be desired. Perhaps the concept is more important than the specifics. Namely, that as rhythms change their relative phase angles seasonally – surely a near certainty in terms of Pittendrigh's entrainment model (1981) – there may be a major shift in their physiological effect.

5 Biochemistry within the Brain

Neurochemical approaches to analysing photoperiodism are surprisingly few when one considers the potential to study the biochemistry of discrete areas within the brain and the range of highly specific neuropharmacological probes available. The problem exists of course, that as with the hormonal measurements, it is difficult to distinguish between effects downstream from those on the clock itself. The early analyses on monoamine oxidase rhythms (Follett 1969) failed to throw up any obvious correlation with the circadian basis of photoperiodic time-measurement but perhaps the method was too crude and the chosen enzyme inappropriate. Certainly, there is an involvement of the adrenergic system in photoinduced gonadal growth in quail (e.g. El Halawani et al. 1980), but as what level is quite unclear.

In ferrets, there is a rhythm in serotonin (5HT) content within the hypothalamus which is different under short and long days (review, Herbert 1981), the differences disappearing if the ferrets are pinealectomized or treated with melatonin. As Herbert (1981) emphasizes, the serotonin content gives a superficial view only of underlying neuronal transmission and more dynamic measures of activity are needed. However, the results are encouraging and one might hope that similar kinds of studies might be undertaken in other *well-defined* photoperiodic systems: the potential seems untapped.

References

Arendt J, Symons C (1981) Comment in discussion at end of Herbert (1981)

Beck SD (1980) Insect photoperiodism. Academic Press, London New York

Bittman EL, Goldman BD, Zucker I (1979) Testicular responses to melatonin are altered by lesions of the suprachiasmatic nuclei in golden hamsters. Biol Reprod 21: 647-656

Davies DT (1980) The neuroendocrine control of gonadotrophin release in the Japanese quail. III. The role of the tuberal and anterior hypothalamus in the control of ovarian development and ovulation. Proc R Soc London Ser B 208: 421-437

El Halawani ME, Burke WH, Ogren LA (1980) Involvement of catecholaminergic mechanisms in the photoperiodically induced rise in serum luteinizing hormone of Japanese quail. Gen Comp Endocrinol 41: 14-21

Elliott JA (1976) Circadian rhythms and photoperiodic time measurement in mammals. Fed Proc 35: 2339-2346

Follett BK (1969) Diurnal rhythms of monoamine oxidase activity in the quail's hypothalamus during photoperiodic stimulation. Comp Biochem Physiol 29: 591-600

Follett BK (1978) Photoperiodism and seasonal breeding in birds and mammals. In: Crighton DB et al. (eds) Control of ovulation. Butterworths, London, pp 267-293

Follett BK, Mattocks PW, Farner DS (1974) Circadian function in the photoperiodic induction of gonadotrophin secretion in the white-crowned sparrow. Proc Natl Acad Sci USA 71: 1666-1669

Follett BK, Davies DT, Gledhill B (1977) Photoperiodic control of reproduction in Japanese quail: changes in gonadotrophin secretion on the first day of induction and their pharmacological blockade. J Endocrinol 74: 449-460

Goldman BD, Hall V, Hollister C, RoyChoudhury P, Tamarkin L, Westrom N (1979) Effects of melatonin on the reproductive system in intact and pinealectomized male hamsters maintained under various photoperiods. Endocrinology 104: 82-88

Goodman RL, Karsch FJ (1981) The hypothalamic pulse generator: A key determinant of reproductive cycles in sheep. In: Follett BK, DE (eds) Biological clocks in seasonal reproductive cycles. Wright, Bristol, pp 223-236

Herbert J (1981) The pineal gland and photoperiodic control of the ferret's reproductive cycle. In: Follett BK, Follett DE (eds) Biological clocks in seasonal reproductive cycles. Wright, Bristol p 261-276

Hoffmann K (1981) The role of the pineal gland in the photoperiodic control of seasonal cycles in hamsters. In: Follett BK and DE (eds) Biological clocks in seasonal reproductive cycles. Wright, Bristol pp 237-250

Hoffmann K, Illnerová H, Vanecek J (1981) Effect of photoperiod and of 1 minute light at right time on the pineal rhythm of NAT activity in the Djungarian hamster (*Rhodopus sungorus*). Biol Reprod 24: 551-556

Klein DC, Tamarkin L (1980) The role of the pineal gland in seasonal production: a theory of a coincidence of rhythms. Sixth Int Cong Endocr Melbourne. Abstract S-57 p. 154

Lincoln GA, Short RV (1980) Seasonal breeding: Nature's contraceptive. Rec Prog Horm Res 36: 1-52

Lincoln GA, Peet MJ, Cunningham RA (1977) Seasonal and circadian changes in the episodic release of follicle-stimulating hormone, luteinizing hormone and testosterone in rams exposed to artificial photoperiods. J Endocr 72: 337-349

Meier AH, Ferrell BR (1978) Avian endocrinology. In: Florkin M, Scheer B, Brush I (eds) Chemical zoology, vol X. Academic, London New York, pp 214-271

Pittendrigh CS (1981) Circadian organization and the photoperiodic phenomena. In: Follett BK, DE (eds) Biological clocks in seasonal reproductive cycles. Wright, Bristol, pp 1-135
Reiter RJ (1980) The pineal and its hormones in the control of reproduction in mammals. Endocrinol Rev 1: 109-131
Rusak B, Zucker I (1979) Neural regulation of circadian rhythms. Physiol Rev 59: 449-526
Simpson SM, Follett BK (1981) Pineal and hypothalamic pacemakers: Their role in regulating circadian rhythmicity in Japanese quail. J Comp Physiol 144: 381-389
Simpson SM, Follett BK (1982) Formal properties of the circadian system underlying photoperiodic time-measurement in Japanese quail. J Comp Physiol 145: 381-390
Tamarkin L, Reppert SM, Klein DC (1979) Regulation of pineal melatonin in the Syrian hamster. Endocrinology 104: 385-389

7.2 Pineal Influences on Circannual Cycles in European Starlings: Effects Through the Circadian System?

E. Gwinner and J. Dittami[1]

1 Introduction

The pineal organ plays a significant role in the control of circadian rhythms among passerine birds. In house sparrows *(Passer domesticus),* white-throated sparrows *(Zonotrichia albicollis)* and Java sparrows *(Padda oryzivora)* maintained in continuous darknness, pinealectomy has been shown to abolish the circadian rhythm of locomotor activity (Gaston and Menaker 1968, McMillan 1972, Ebihara and Kawamura 1980). In addition, the rhythm of body temperature disappeared in house sparrows following the same procedure (Binkley et al. 1971). In European starlings *(Sturnus vulgaris),* free-running locomotor activity rhythms became heavily disrupted after pineal removal (Gwinner 1978). Pinealectomy has affected circadian activity rhythms in the house sparrow even when entrained to a 24-h light-dark cycle (Gaston 1971, Laitman and Turek 1979). The results obtained from these and several other experiments are consistent with the hypothesis that the pineal organ contains a circadian pacemaker which drives overt circadian functions (Menaker and Zimmerman 1976, Takahashi and Menaker 1979).

Although the role of the pineal in the control of daily rhythms is well accepted, its significance in the regulation of avian annual cycles is still debatable. Pinealectomy has been shown to affect the photoperiodically controlled seasonal testicular cycles of the Indian weaverbird *(Ploceus phillipinus;* Balasubramanian and Saxena 1972) and the domestic duck *(Anas platyrhynchos*; Cardinali et al. 1971), but no effects in traditional photoperiodic experiments have been found in at least five other passerine species, including the house sparrow and the European starling (Gwinner and Dittami 1980, for review). This situation is unsatisfactory as it is known in the latter two species that some circadian rhythmicity is involved in photoperiodic time measurement (Menaker and Eskin 1967, Gwinner and Eriksson 1977). Hence, effects of the pineal on the circadian system should be reflected in the photoperiodic time-measuring system, and as a result an effect should be seen on the bird's annual cycles. Results of recent experiments on house sparrows which were specifically designed to test this general proposition have produced some positive evidence (Takahashi et al. 1978, Yokoyama 1981). However, it has not become clear in these experiments how pinealectomy might have altered the circadian system to produce the observed effects.

1 Max-Planck-Institut für Verhaltensphysiologie, Vogelwarte Radolfzell, Radolzell and Andechs, D-8138 Andechs

Vertebrate Circadian Systems
Ed. by J. Aschoff, S. Daan, and G. Groos

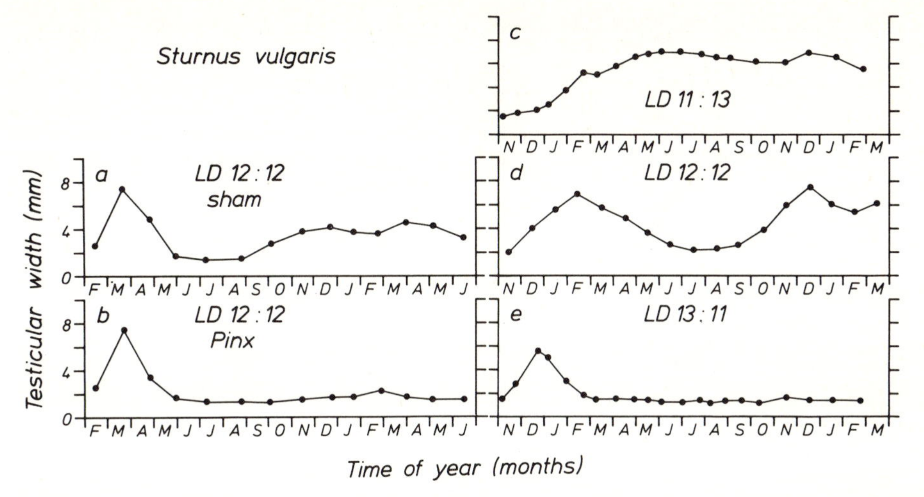

Fig. 1. Seasonal variations in testicular size in groups (7 to 18 birds each) of male European starlings held under various constant photoperiods. Groups *a* and *b* are from the same experiment and can be compared directly with each other; the same holds true for groups *c* and *e*. Other comparisons must be taken with reservation because experiments began at different times and experimental conditions differed in details (*a* and *b* after Gwinner and Dittami 1980; *c* and *e* after Hamner 1971; *d* after Schwab 1971)

In the European starling we found clear differences in the temporal course of testicular size between pinealectomized and sham-operated birds kept in LD 12 : 12 (Gwinner and Dittami 1980). The sham-operated birds went through a full testicular cycle following transfer to experimental conditions in February, and later 9 out of 11 birds spontaneous initiated a second cycle (Fig. 1a). This was consistent with other studies in that a circannual rhythmicity in testicular size persisted in a constant LD 12 : 12 (e. g. Schwab 1971, Gwinner 1981b, Fig. 1d). In contrast, only two out of ten individuals among the pinealectomized birds showed slight transient increases in testicular size at the time when the controls went through their second cycle (Fig. 1b). In this respect, our pinealectomized birds kept in the LD 12 : 12 behaved similarly to normal birds in LD 13 : 11 (Fig. 1e).

Other recent results have indicated that the failure of normal starlings kept under LD 13 : 11 to go through a second testicular cycle was due to these birds' failure to break photorefractoriness (Gwinner and Wozniak 1982). Therefore, it was tempting, to speculate the pinealectomy in starlings kept under LD 12:12 abolishes these birds' capacity to break photorefractoriness, perhaps by bringing them into a physiological state similar to that of intact birds held under LD 13 : 11. We proposed that changes in the circadian rhythmicity which are known to participate in photoperiodic time measurement are involved in this response (Gwinner and Dittami 1980). A specific version of this general model is shown in Fig. 2. It assumes that photorefractoriness in the starling is broken only if light fails to fall on a particular phase of a circadian oscillator (φ_r) located in the bird's late subjective day. In an intact bird this is the case only if photoperiod is 12 h or less. Pinealectomy leads to a phase advance of the circadian rhythm of photosensitivity relative to the entraining light-dark cycle. As a result, a pinealectomized bird should inter-

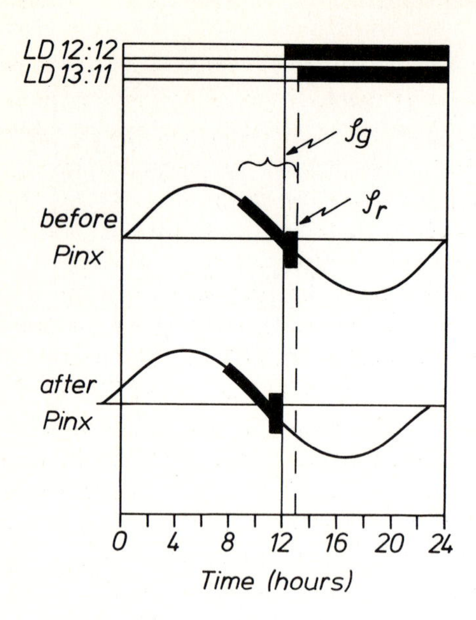

Fig. 2. Model to explain the effects of pinealectomy on the starlings' circannual rhythm in testicular function. The sine curve represents the postulated circadian rhythm in photosensitivity with φ_r and φ_g in normal *(upper curve)* and pinealectomized *(lower curve)* birds. *Black* and *white portions of the bars* in the upper part of the figure indicate light and dark fractions of LD 12 : 12 and LD 13 : 11 h

pret any photoperiod as being longer than it actually is. Thatis, LD 12 : 12 might act like LD 13 : 11 and photorefractoriness would not be broken.

This hypothesis is supported by two sets of empirical data:

1. It is well known that a circadian rhythm is indeed involved in the control of photoperiodic reactions of the starling as well as of other avian species. The available evidence is consistent with the assumption that testicular growth in early spring is initiated if light falls on a particular phase of a circadian rhythmicity ("external coincidence model of photoperiodic induction", see Pittendrigh 1972). At least for the white-crowned sparrow *(Zonotrichia leucophrys)* and the golden-crowned sparrow *(Zonotrichia coronata)* a similar mehcanism appears to be involved in controlling the termination of photorefractoriness; the data suggest that photorefractoriness is only broken if light fails to be coincident with the postulated circadian phase (Turek 1972). There is lastly strong evidence in the Japanese quail *(Coturnix coturnix)* that the circadian phase measuring light for testicular growth is indeed located in the birds' late subjective day (Nicholls and Follett 1974).
2. In both the house sparrow and the European starling, pinealectomy leads to a phase advance of circadian locomotor activity rhythms relative to the entraining light-dark cycle. In house sparrows kept under LD 3 : 21, 6 : 18, 0.5 : 23.5 and 6 : 20, the onset of activity was shifted forward several hours after the pineal had been removed (Gaston 1971). For the starling, no data from birds held under synchronizing light-dark cycles are available. However, in birds maintained in continuous dim light, pinealectomy almost invariably shortened the period of circadian activity rhythms (Gwinner 1978). Since a shortening of the natural period normally results in an increased phase-angle difference in the entrained state (Aschoff and Wever 1962), pinealectomy in the starling can be expected to have phase-advancing effects similar to those found in the house sparrow.

The present hypothesis not only accommodates the empirical data mentioned above; it also makes clear predictions, one of which we have tested in the present investigation. It concerns testicular growth rates during initial exposure to experimental photoperiods, i.e., the response for which a circadian involvement has initially been demonstrated (Gwinner 1975b for review). In many avian species the rate of testicular growth depends on photoperiod in such a way that growth rate is very small or zero below a particular critical photoperiod and maximal above a particular photoperiod with a graded response in between (e.g. Farner and Lewis 1971). If a logarithmic growth rate constant (k) is calculated (see Sect. 2) the dependence of k on photoperiod normally follows the dashed curve in Fig. 4a (Farner and Wilson 1957, Murton and Westwood 1977). The graded response between photoperiods of about 8 and 14 h can be easily interpreted in terms of the external coincidence model if one assumes the existence of an extended circadian phase φ_g. Testicular growth would depend on how much of φ_g was exposed, so the longer the portion of φ_g is illuminated the faster the testes will grow. Then, assuming that φ_g like φ_r occurs in the birds' late subjective day (and may even be partly or entirely identical with φ_r, Fig. 2) pinealectomy should shift the photoperiodic response curve to the left because of its postulated phase-advancing effect on the circadian system (Fig. 4, solid curve). Again the pinealectomized birds would interpret any photoperiod as being longer than it actually is, e.g., an 8-h photoperiod as a 9-h photoperiod (compare the temporal relationship between φ_g and the two vertical lines in Fig. 2).

To test this hypothesis we have examined testicular growth in pinealectomized and sham-operated starlings which were transferred to a number of different photoperiods in winter.

2 Methods

In autumn 1978 134 male European starlings were captured near Mannheim, Germany, and subsequently kept in large outdoor aviaries in Andechs. On January 2, 1979, 105 of these birds were moved to 50 x 45 x 40 cm wire mesh cages, housed in 120 x 120 x 55 cm constant conditions chambers. Five or six birds were in each cage. The remaining 30 birds were held in individual cages housed in three chambers. Temperature was constant at 18 ± 1 °C. Light was provided by 40 W fluorescent bulbs which produced a light intensity of about 400 lx during the light time. From January 1 to 10 all birds were exposed to a constant 8-h photoperiod.

Between January 3 and 7, 63 birds were pinealectomized and 72 birds were sham-operated following the procedure described in Gwinner (1978). On January 11 the birds were subdivided into six experimental groups which were subsequently held under six different constant photoperiods until the end of the experiment in July 1979.

Group I, consisting of 16 pinealectomized and 14 sham-operated birds was the group of 30 birds held in individual cages. They were exposed to LD 8 : 16, which was supplemented by 1 h twilight periods before lights went on and after lights went off.

Groups II to V, consisting of 11 or 12 pinealectomized and 11 or 12 sham-operated birds each, were exposed to LD 9 : 15, LD 10 : 14, LD 11 : 13 and LD 12 : 12, respectively. No twilight periods were used in these groups. Group VI, consisting of 12 sham-operated birds only ,was exposed to LD 13 : 11.

The birds of groups II to VI were housed, as mentioned, in subgroups of five to six birds per cage.

Prior to the transfer to experimental conditions and subsequently at 3- to 5-week intervals the birds were laparotomized and the width of the left testis was measured to the nearest tenth of a millimeter with a compass (Gwinner 1975a). To determine the initial growth rate constant (k) of the testes, the values of testicular widths were measured before and 20 days after transfer to experimental conditions and then converted into weight values according to the equation: weight = 3.292 + width 2.565, a relationship which has been established previously (Gwinner unpubl.). Then k-values were calculated from these data according to the formula: $k = (\log W_t - \log W_o)/t$, where W_o is the initial testicular weight, W_t the final testicular weight and t the time in days (Farner and Wilson 1957).

3 Results

Figure 3 summarizes the temporal variations in testicular size of all six experimental groups. It indicates that the pinealectomized and the sham-operated birds behaved rather similar. No differences were found between the pinealectomized and the sham-operated birds held under LD 11 : 13. In all other LD cycles tested, the testicular size of the pinealectomized birds lagged slightly behind that of the shams. This can also be seen on Fig. 4b where the mean growth rate constants (k) for the first 20 days of the experiment are plotted. Although none of the differences in k between the pinealectomized and the sham-operated birds of any experimental group were statistically significant, it is noteworthy that in four out of the five groups the pinealectomized birds had smaller k values. The values for the fifth group, the one held in LD 11 : 13, were indistinguishable. In two previous experiments (Gwinner unpubl.), in which photosensitive pinealectomized and sham-operated starlings were tested under LD 8 : 16 and LD 12 : 12, it was also found that the pinealectomized groups had slightly smaller testicular growth rates. Taken together, these results indicate a significant tendency of the pinealectomized birds to grow their testes more slowly than the controls.

4 Discussion

The results of the present experiment are clearly at odds with the predictions made by the hypothesis. The results obtained on initial testicular growth suggesting slightly slower growth-rate constants in the pinealectomized birds than in the controls tend to indicate pineal effects just opposite those predicted by the hypothesis. These discrepancies between predictions and results may be due to the following reasons:

1. The specific assumptions made about the temporal occurrence of φ_g are wrong. φ_r was assumed to occur late in the birds' subjective day to explain why pinealectomy due to its postulated phase-advancing effect would prevent the breaking of photorefractoriness in LD 12 : 12. The same assumption for φ_g would have predicted faster testicular growth rates in the pinealectomized birds compared with the controls. The fact, that rather the opposite was true can only be explained in terms of the present model if φ_r

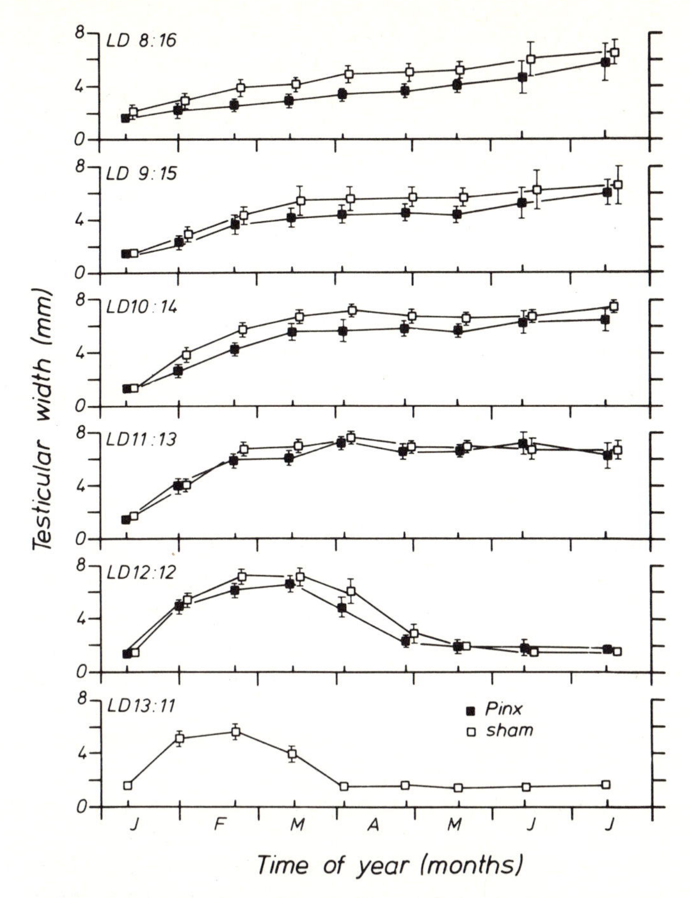

Fig. 3. Temporal variations in testicular width (means with standard errors) of starlings held under six different constant photoperiods. ▪ pinealectomized, ▫ sham-operated birds

and φ_g were separated by about 180°, for instance if one assumes that φ_r occurs late in the birds' subjective day and φ_g late in the birds' subjective night. The possibility of two widely separated photosensitive phases cannot be excluded even though there is also no positive evidence in support of it.

2. The specific assumptions made about the changes in the phase-relationship between the postulated circadian rhythmicity and the light-dark cycle are wrong. For our hypothesis we assumed that pinealectomy phase-advances the postulated circadian rhythmicity. With regard to the starling, this assumption was based on the observation that pinealectomy led to a shortening of the period of free-running circadian activity rhythms (Gwinner 1978). Generally such a decrease in the natural period length should result in an increasing phase-angle difference to the entraining LD-cycle (Aschoff and Wever 1962). However, it cannot be determined at present whether pinealectomy only affects the natural period of the circadian rhythm for perch-hopping activity and not that for measuring photoperiod. Moreover, it seems possible that pinealectomy may alter other parameters that affect the rhythm's response to the zeitgeber. As a result, the postulated

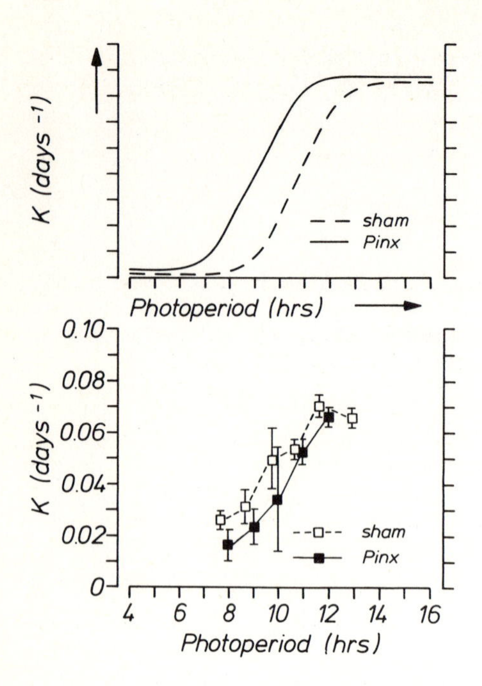

Fig. 4. *Upper diagram* predicted differences between the growth rate constants *(k)* of sham-operated and pinealectomized starlings held under various constant photoperiods. *Lower diagram* growth-rate constants measured in the experimental birds. Mean values with standard errors are given. For further explanations see text

phase-advancing effect of pinealectomy could be neutralized or even turned into phase-delays as a result of other changes in the circadian system. Finally, it must be mentioned that even in the house sparrow, the evidence for phase-advancing effects of pinealectomy has been qualified by recent results which indicate that activity onset apparently becomes advanced by pinealectomy only under short photoperiods and not under photoperiods of 5 h or more (Laitman and Turek 1979). The situation may indeed be much more complex than initially assumed.

3. The observed effects of pinealectomy on the starlings' circannual cycles are mediated by a circadian rhythmicity, but by a mechanism different from the one proposed here. In particular, it is possible that photoperiod is not measured by a mechanism of external coincidence but rather by a mechanism of internal coincidence (Pittendrigh 1972). Although such a hypothesis may eventually prove more successful than the one proposed here, we have not taken it into consideration because the internal-coincidence-hypothesis is not supported by any strong experimental results in birds (Gwinner 1981a). Moreover, we presently see no simple model based on this hypothesis that could explain the available results.

4. The observed effects of pinealectomy on the starling's circannual cycles are not mediated by the circadian system at all. The effects might rather be due to direct influences of pinealectomy on the hypothalamo-hypophyseal-gonadal axis, bypassing the circadian system. The parallel effects of pinealectomy on the circadian system would then have to be considered coincidental and not causally related to the effects on the circannual system. Since in all cases studied pinealectomy in the starling tended to reduce testicular size, a progonadal effect in a very general sense might be attributed to the starling's pineal organ. In most cases also the effects of pinealectomy on the starling's annual cy-

cles were relatively small, so they may have been in part due to a reduction of the birds' general well-being. In any case, it seems clear that the simple and therefore attractive mechanism proposed in this paper is not realized.

References

Aschoff J, Wever R (1982) Über Phasenbeziehungen zwischen biologischer Tagesperiodik und Zeitgeberperiodik. Z Vergl Physiol 46: 115-128

Balasubramanian KS, Saxena RN (1972) Effect of pinealectomy and photoperiodism in the reproduction of Indian weaver birds, *Ploceus phillippinus.* J Exp Zool 185: 333-340

Binkley S, Kluth E, Menaker M (1971) Pineal function in sparrows: circadian rhythms and body temperature. Science 174: 311-314

Cardinali DP, Cuello AE, Tramezzani JH, Rosner JM (1971) Effects of pinealectomy on the testicular function of the adult male duck. Endocrinology 89: 1082-1093

Ebihara S, Kawamura H (1980) Central mechanisms of circadian rhythms in birds. In: Tanabe W, Tanaka K, Ookawa T (eds) Biological rhythms in birds. Springer, Berlin Heidelberg New York, p 71

Farner DS, Lewis RA (1971) Photoperiodism and reproductive cycles in birds. In: Giese AC (ed) Photophysiology, vol VI. Academic Press, London New York, p 325

Farner DS, Wilson AC (1957) A quantitative examination of testicular growth in the white-crowned sparrow. Biol Bull 113: 254-267

Gaston S (1971) The influence of the pineal organ on the circadian activity rhythm in birds. In: Menaker M (ed) Biochronometry. Natl Acad Sci 1971, Washington DC, pp 451-548

Gaston S, Menaker M (1968) Pineal function: the biological clock in the sparrow? Science 160: 1125-1127

Gwinner E (1975a) Die circannuale Periodik der Fortpflanzungsaktivität beim Star *(Sturnus vulgaris)* unter dem Einfluß gleich- und andersgeschlechtiger Artgenossen. Z Tierpsychol 38: 34-43

Gwinner E (1975b) Circadian and circannual rhythms in birds. In: Farner DS, King JR (eds) Avian biology, vol V. Academic Press, London New York, pp 221-285

Gwinner E (1978) Effects of pinealectomy on circadian locomotor activity rhythms in European starlings, *Sturnus vulgaris.* J Comp Physiol 126: 123-129

Gwinner E (1981a) Relationship between circadian activity patterns and gonadal function: evidence for internal coincidence? Proc XVIIth Intern Ornithol Congr, Berlin 1978, pp 409-416

Gwinner E (1981b) Circannual rhythms: their dependence on the circadian system. In: Follett BK and DE (ed) Biological clocks in seasonal reproductive cycles. J Wright & Co Ltd, Bristol, pp 153-169

Gwinner E, Dittami J (1980) Pinealectomy affects the circannual testicular rhythm in European starlings, *Sturnus vulgaris.* J Comp Physiol 136: 345-348

Gwinner E, Eriksson LO (1977) Circadiane Rhythmik und photoperiodische Zeitmessung beim Star, *Sturnus vulgaris.* J Ornithol 118: 60-67

Gwinner E, Wozniak J (1982) Circannual rhythms in European starlings: Why do they stop under long photoperiods? J Comp Physiol 146: 419-421

Hamner WM (1971) On seeking an alternative to the endogenous reproductive rhythm hypothesis in birds. In: Menaker M (ed) Biochronometry. Natl Acad Sci, Washington DC, pp 448-462

Laitman RS, Turek FW (1979) The effect of pinealectomy on entrainment of the locomotor activity rhythm in sparrows maintained on various short days. J Comp Physiol 134: 339-343

McMillan JP (1972) Pinealectomy abolishes the circadian rhythm of migratory restlessness. J Comp Physiol 79: 105-112

Menaker M, Eskin A (1967) Circadian clock in photoperiodic time measurement. A test of the Bünning hypothesis. Science 157: 1182-1185

Menaker M, Zimmerman N (1976) Role of the pineal in the circadian system of birds. Am Zool 16: 45-55

Murton RK, Westwood NJ (1977) Avian breeding cycles. Clarendon Press, Oxford

Nicholls TJ, Follett BK (1974) The photoperiodic control of reproduction in Coturnix quail. The temporal pattern of LH secretion. J Comp Physiol 93: 301-313

Pittendrigh CS (1972) Circadian surfaces and the diversity of possible roles of circadian organization in photoperiodic induction. Proc Natl. Acad Sci USA 69: 2734-2737

Schwab RG (1971) Circannian testicular periodicity in the European starling in the absence of photoperiodic change. In: Menaker M (ed) Biochronometry. Natl Acad Sci, Washington DC, pp 428-447

Takahashi JS, Menaker M (1979) Brain mechanisms in avian circadian systems. In: Suda M, Hayaishi O, Nakagawa H (eds) Biological rhythms and their central mechanisms. The Naito Foundation. Elsevier/North Holland, Biomedical Press, pp 94-109

Takahashi JS, Norris C, Menaker M (1978) Circadian photoperiodic regulation of testis growth in the house sparrow: is the pineal gland involved? In: Gaillard PJ, Boer HH (eds) Comparative endocrinology. Elsevier/North Holland, Biomedical Press, Amsterdam, pp 153

Turek FW (1972) Circadian involvement in termination of the refractory period in two sparrows. Science 178: 1112-1113

Yokoyama K (1981) The possible role of the pineal for photoperiodic time measurement in two species of passerine birds. Proc XVIIth Int Ornithol Congr, Berlin 1978, pp 439-443

7.3 Complex Control of the Circadian Rhythm in N-Acetyltransferase Activity in the Rat Pineal Gland

H. Illnerová and J. Vaněček[1]

1 Introduction

The enzyme N-acetyltransferase (acetyl CoA: arylamine N-acetyltransferase, EC 2.3.1.5) (NAT) acetylates serotonin to N-acetylserotonin, precursor of melatonin (Weissbach et al. 1960). In the rat pineal gland, NAT activity exhibits a circadian rhythm, persisting in constant darkness, with night values 30-100 times higher than daytime values (Klein and Weller 1970). The rhythm is generated by daily changes in the turnover and presumably in the release of noradrenaline from sympathetic fibres innervating the pineal gland (Brownstein and Axelrod 1974). Noradrenaline released at night in darkness interacts with the beta-adrenergic receptors on the pineal cell membrane, enhancing the production of cAMP which mediates the induction of NAT (Klein and Berg 1970). Activated beta-adrenergic receptors are necessary not only for the evening NAT induction but also for the maintenance of the high night NAT activity (Deguchi and Axelrod 1972a). The NAT rhythm is abolished by denervation of the pineal gland through bilateral cervical ganglionectomy or decentralization of the ganglia (Klein et al. 1971), by destruction of the suprachiasmatic nuclei in the hypothalamus (Moore and Klein 1974) or by constant light (Klein and Weller 1970).

When rats are exposed to sudden light at night, NAT activity declines with the same halving time, 3.8 min, as after the administration of beta-blockers (Vaněček and Illnerová 1979). This observation indicates that light blocks immediately transmission of neural signals to the pineal gland and beta-adrenergic receptors cease to be activated. Even when rats are exposed at night to such a brief light as 1-min light pulse, NAT declines rapidly in the continuing darkness (Illnerová et al. 1979). Following a 1-min light pulse at 2200 h, NAT first decreases and after a lag period, begins to rise again; following a pulse at 0100 h, NAT declines and does not increase any more throughout the rest of night. The non-inducibility of NAT in darkness after the pulse at 0100 h is due to reduction or shut-off of the noradrenaline release from nerve-endings in the pineal gland (Illnerová and Vaněček 1979). A 1-min light pulse may be an effective phase-shifting signal of the pacemaker driving the daily rhythm in the noradrenaline release and hence driving the NAT rhythm. The different response of NAT to a 1-min light pulse at 2200 h and 0100 h may be caused by different phase-shifts of the pacemaker driving the NAT rhythm. To clarify this question and to learn more about the behaviour and organization

1 Institute of Physiology, Czechoslovak Academy of Sciences, 142 20 Prague 4 – Czechoslovakia

Vertebrate Circadian Systems
Ed. by J. Aschoff, S. Daan, and G. Groos

of the pacemaker, we followed phase-shifts of the NAT rhythm at different times after 1-min light pulses. In the following study we report on our results and on the ensuing conclusions about the organization of the pacemaker controlling the NAT rhythm.

2 Methods

We used 55-70 day old male Wistar rats from our own breeding colony. Unless otherwise stated rats were housed in a room lit by 40 W fluorescent tubes from 0600 h to 1800 h for at least three weeks prior to all experiments. During the exposure to 1-min light at night, the intensity of illumination at the cage level was around 100 lx. When rats were to be killed in darkness, they were exposed to a very faint red light for less than 1 min prior to decapitation. Other animals in the room were sheltered from the faint red light by a black curtain.

Pineal glands were removed rapidly and stored in Petri dishes on solid CO_2. Within 48 h after decapitation, NAT activity was determined by a modification (Parfitt et al. 1975) of the method of Deguchi and Axelrod (1972b). Units of NAT activity were defined as nmol N-acetyltryptamine formed in 1 h/1 mg of pineal tissue.

3 Effect of 1-Min Light Pulses

3.1 The NAT Rhythm One and Three Days After a Pulse at 2200 h and 0100 h

As a phase-reference point for the evening NAT rise, the time was chosen when NAT activity reached the value of 3 nmol mg^{-1} h^{-1}. Similarily, the time when NAT activity declined to 3 nmol mg^{-1} h^{-1} during its decrease was chosen as the phase-reference point for the morning NAT decline. A phase-shift in the NAT rise or decline in pulsed rats as compared to unpulsed controls is expressed as the time interval between the phase-reference point of pulsed rats and that of unpulsed controls.

Rats were released into constant darkness at the normal time of lights off, and received a 1-min light pulse at 2200 h, or at 0100 h, or no light. The NAT rhythm was determined after one and three days in DD (Fig. 1). Following the pulse at 2200 h the NAT rise was phase-delayed by about 1.6 h after 1 day and by about 1.9 h after 3 days; the NAT decline was phase-delayed by about 0.9 h after 1 day and by 1.4 h after 3 days. Following the pulse at 0100 h, the NAT rise was phase-shifted neither after 1 day nor after 3 days; however, the morning NAT decline was phase-advanced by about 1.1 h after 1 day as well as after 3 days.

The NAT rise and decline in this experimental arrangement showed phase-shifts of the pacemaker driving the NAT rhythm one and three days after the pulses. As the evening rise and the morning decline in NAT activity did not shift parallel after the pulses, they may possibly be driven by two coupled oscillators: the evening one, controlling the NAT rise and the morning one, controlling the NAT decline.

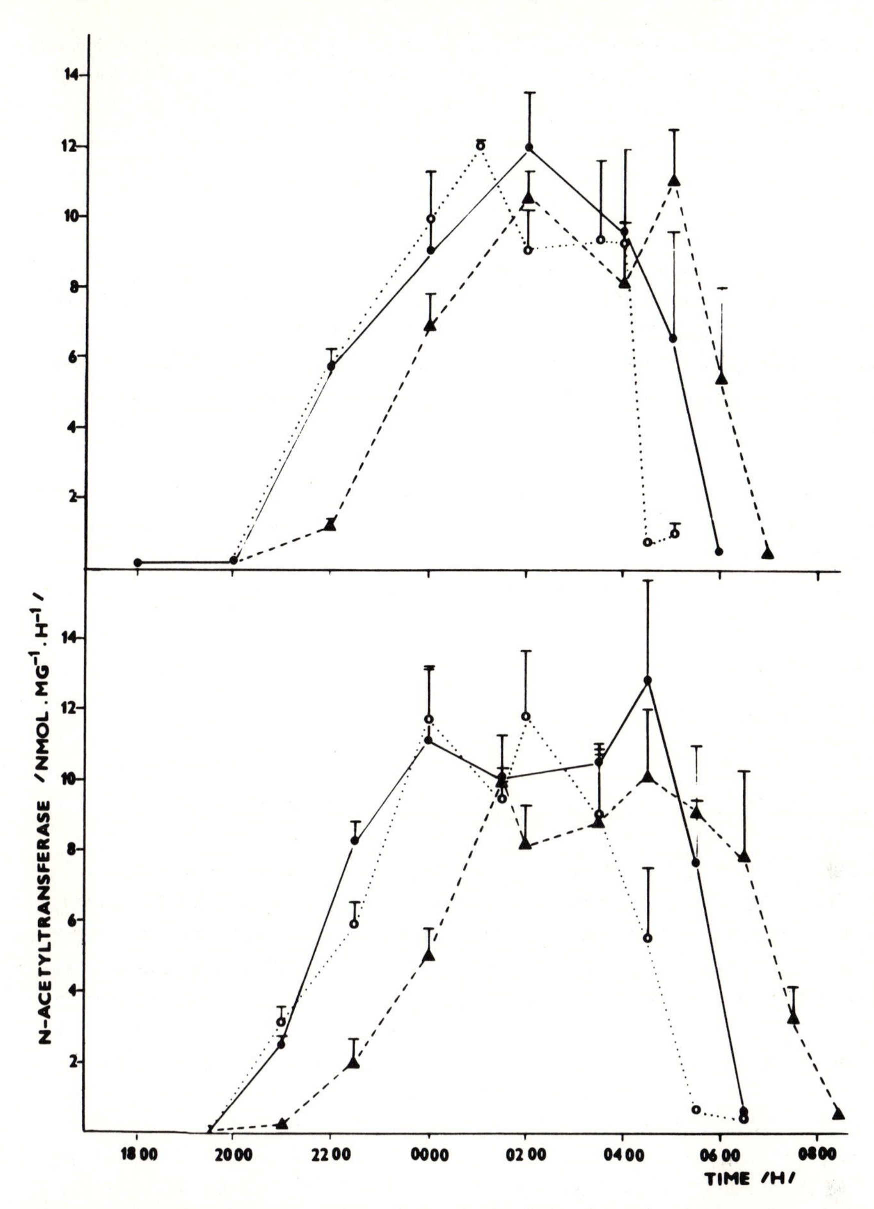

Fig. 1. Phase-shifts of the N-acetyltransferase rhythm 1 day *(upper)* and 3 days *(lower)* after 1 min light pulses. Rats were either pulsed at 2200 h *(filled triangles)*, or at 0100 h *(open circles)*, or they were not pulsed at all *(filled circles)* and then were released into constant darkness. Data are expressed as means + S.E.M. of three six animals

3.2 Phase-Response Curves (PRC's) for Evening and Morning Oscillator 1 day After the Pulses

If there are two oscillators driving the NAT rhythm, then their PRC's may not be identical. The existence of two oscillators would be proved best by finding two different PRC's, one for the oscillator controlling the evening NAT rise and the other for the oscillator controlling the morning NAT decline. However, it is not possible to find the exact PRC for each oscillator separately, as both oscillators are probably coupled into a complex pacemaker (Pittendrigh and Daan 1976). Phase-shifts of the evening or of the morning oscillators measured 1 or more days after the pulses may thus result from instantaneous phase-shifts of an individual oscillator and a subsequent interaction of both oscillators. Having this in mind, we nevertheless attempted to study PRC's of both supposed oscillators. We pulsed rats at different times of the night, then released them into darkness and the next night we determined the evening NAT rise and the morning NAT decline. As this approach needed many groups of animals, the evening rise was studied in two experiments and the morning decline in five experiments. Each experiment included new unpulsed control animals and phase-shifts were always related to proper unpulsed controls. Figure 2 shows, as an example of our procedure, one experi-

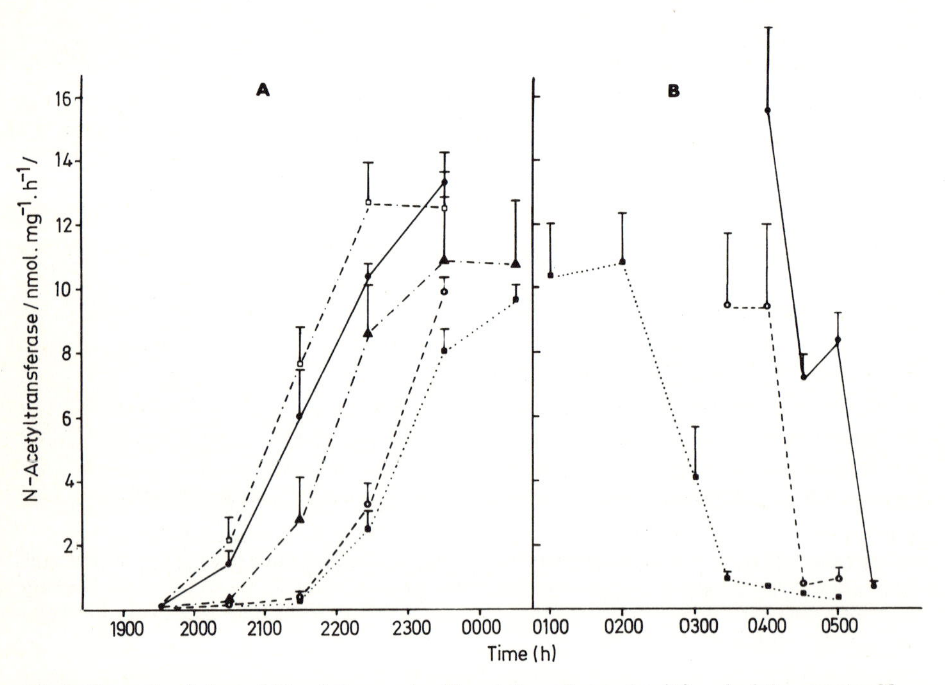

Fig. 2. Example of phase-shifts of the evening N-acetyltransferase rise **(A)** and of the morning N-acetyltransferase decline **(B)** 1 day after 1-min light pulses. Data are expressed as means + S.E.M. of four animals. **A** Rats were pulsed at 2100 h *(open circles)* or at 2200 h *(filled squares)* or at 2300 h *(filled triangles)* or at 0100 h *(open squares)* or not at all *(closed circles)* and then they were released into constant darkness. **B** Rats were pulsed at 0100 h *(open circles)* or at 0400 h *(filled squares)* or not at all *(filled circles)* and then they were released into darkness

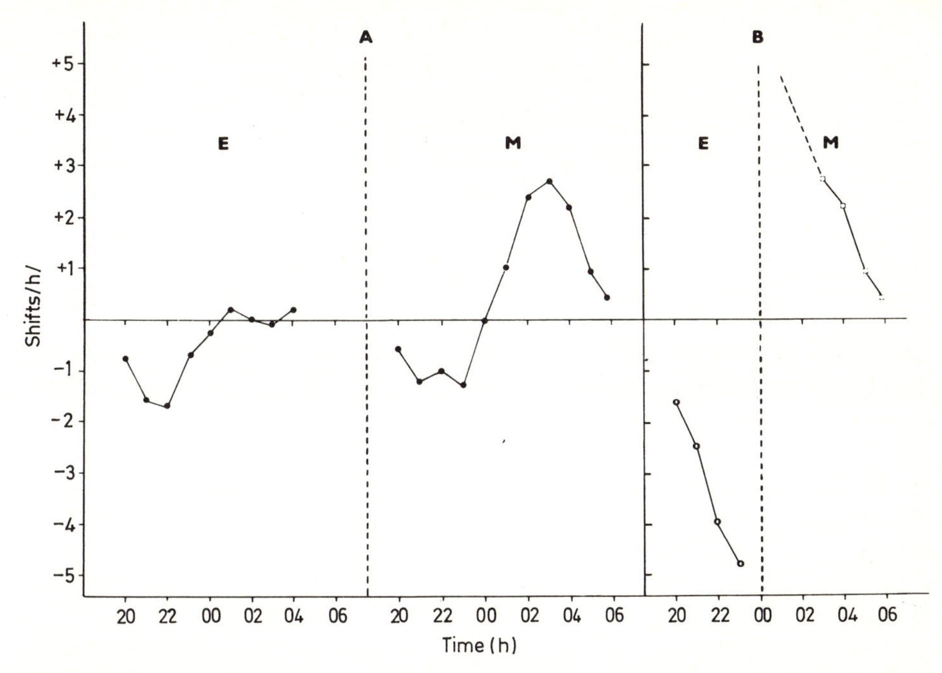

Fig. 3. Phase response curves of the evening oscillator *(E)* and of the morning oscillator *(M)* representing *A* phase-shifts one day after 1-min light pulses, *B* instantaneous ***separate*** phase shifts. *A* Values of phase-shifts *(filled circles)* were read from Fig. 2 and from five other similar experiments. *B* Values of instantaneous separate phase-shifts of the evening oscillator *(open circles)* were read from Fig. 4, values of phase-shifts of the morning oscillator *(open squares)* were taken from Fig. 3A, M

ment studying phase-shifts of the evening NAT rise (Fig. 2A) and one experiment studying phase-shifts of the morning NAT decline (Fig. 2B).

Figure 3A summarizes phase-shifts of the evening NAT rise into one PRC (Fig. 3A, E) and phase-shifts of the morning NAT decline into another PRC (Fig. 3A, M). The PRC demonstrating phase-shifts of the NAT rise 1 day after the pulses has only phase-delays, the PRC demonstrating phase-shifts of the NAT decline 1 day after the pulses has phase-delays as well as phase-advances, though phase-advances are more pronounced. From the existence of two different PRC's, we infer that these PRC's represent phase-shifts of two oscillators. We shall designate the evening oscillator E and the morning oscillator M according to Pittendrigh and Daan (1976) who proposed that E, coupled to dusk, controls the evening component of the locomotor activity in rodents, while M, coupled to dawn, controls the morning component.

3.3 The NAT Rhythm in the Course of the Night When the Pulses Are Applied

Phase-shifts one day after the pulses are results of instantaneous separate phase-shifts of E or M and of mutual interaction between E and M. We attempted to deduce the instantaneous separate phase-shifts of E and M from the behaviour of the NAT rhythm during

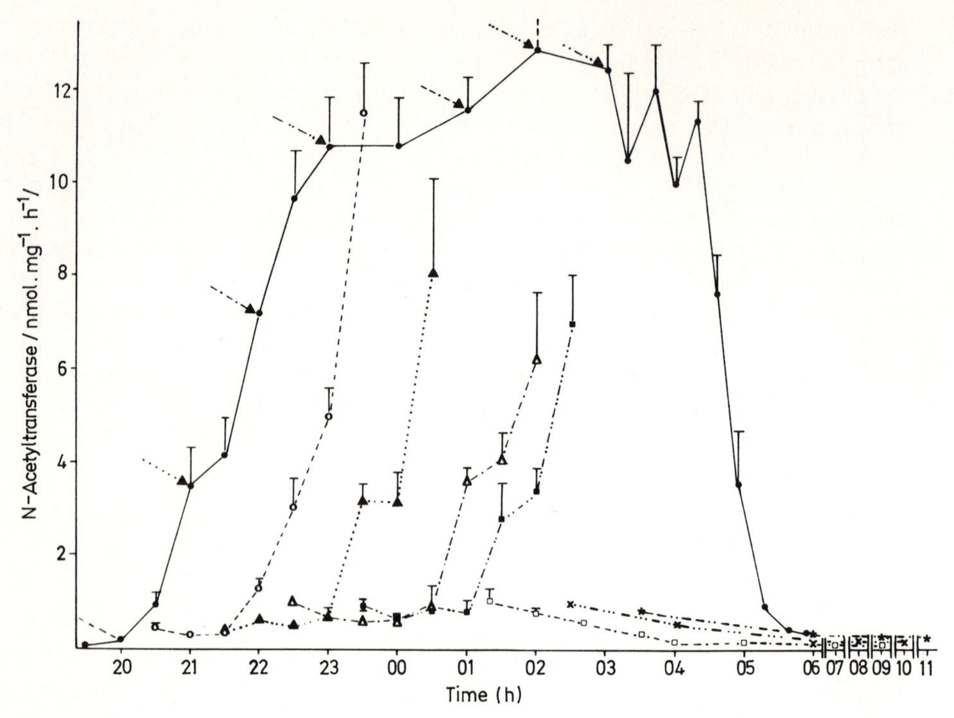

Fig. 4. The N-acetyltransferase rhythm in the course of night when 1 min light pulses were applied. Rats were either unpulsed *(filled circles)* or pulsed at 2000 h *(open circles)*, or at 2100 h *(filled triangles)* or at 2200 h *(open triangles)* or at 2300 h *(filled squares)* or at 0100 h *(open squares)* or at 0200 h *(crosses)* or at 0300 h *(asterisks)* and from that time on they were kept in darkness until killed. *Arrows* indicate time of the pulses and point to the N-acetyltransferase value in darkness at the moment of the pulse. Data are expressed as means + S.E.M. of four animals. N-acetyltransferase activity started to rise significantly 2 h after the pulse at 2000 h ($p < 0.01$), 2.5 h after the pulse at 2100 h ($p < 0.005$), 3 h after the pulse at 2200 h ($p < 0.005$) and between 2 and 3 h after the pulse at 2300 h ($p < 0.01$) (significance was evaluated by the Student's t-test)

the night when the pulses were applied. We pulsed rats at different night times and followed NAT activity after the pulse in the same night (Fig. 4). Following the pulse at 2000 h, 2100 h, 2200 h and 2300 h, NAT began to increase to high night values after a lag period; in rats pulsed at 2100 h, 2200 h and 2300 h, the rise occurred anew after the initial drop in NAT. Lag periods which elapsed between the pulses applied in the first half of night and the start of NAT rise may be due to different causes. First, after "cut-off" of NAT activity by the pulse, a certain time may be necessary for NAT to be resynthesized and reactivated anew. However, the application of beta-adrenergic agonist isoprotenerol 10 min after exposure of rats to light at night restores the initial high value within 1 h (Deguchi and Axelrod 1972a). Moreover, NAT is almost immediately partly reactivated, if the activation of beta-adrenergic receptors is resumed within a short time after exposure of rats to light at night (Vaněček and Illnerová 1980). Hence the necessary time for NAT resynthesis and reactivation is much shorter than the lag periods. Second, after the pulse, the noradrenaline release from nerve endings in the pineal gland may be for some time blocked. The impairment may be explained by the possibility that after "turning-off" of the noradrenaline release by the pulse, a certain time must elapse before noradrenaline is released again in an amount sufficient for NAT to be resynthesized and reactivated. This explanation seems also highly unlikely. Assuming that a sufficient amount of noradrenaline is being released when NAT starts to rise again, the periods between the pulse and the first significant rise of NAT above the previous value should be the same. However, NAT started to rise 2 h after the pulse at 2000 h, 2.5 h after the pulse at 2100 h and 3 h after the pulse at 2200 h (Fig. 4). Moreover, if this explanation were correct and noradrenaline would be released anew after a certain time following the pulse, what is the explanation for the fact that, following the pulses after midnight, NAT does not rise within the next 8 h in darkness (Fig. 4)? The plausible interpretation of our data is that the temporary impairment of the noradrenaline release in rats pulsed in the first night half may be caused by instantaneous phase-shifts of E to a phase when sufficient amount of noradrenaline is not yet bein released and hence NAT is low. Phase delays in the NAT rise, 1.6 h after the pulse at 2000 h, 2.5 h after the pulse at 2100 h, 4 h after the pulse at 2200 h and 4.8 h after the pulse at 2300 h, may thus represent instantaneous phase-shifts of E.

Following a 1-min light pulse at 0000 h, NAT activity, after the initial decline, began to increase in approximately half of the animals, in the other half it decreased further (data not shown). Following 1-min light pulse at 0100 h, 0200 h and 0300 h, NAT activity, after the initial rapid drop, declined slowly further and did not rise to high values within 8 h in darkness after the pulse (Fig. 4). The pulse after midnight may instantaneously phase-advance M to a phase when NAT declines or is already low. Even in the night following the pulses, phase-advances of M are large enough (Fig. 3A, M) to account for the non-inducibility of NAT by darkness after the pulse at 0200 h and 0300 h. The 2.5 h phase-advance of M observed the next day after the pulse at 0200 h can easily phase-shift M to a phase corresponding at least to 0430 h in the morning when NAT under LD 12 : 12 starts spontaneously to decline (Fig. 4). The almost 3-h phase-advance of M observed the next day after the pulse at 0300 h can phase-shift M to a phase corresponding to 0600 h when NAT activity is already low. Similarly, we may assume that the pulse at 0100 h may instantaneously phase-advance M to a phase when NAT is low even if the phase-advance of M measured the next day was much smaller.

Following the pulse at 2200 h, the morning NAT decline in the same night was delayed by only about 0.3 h (data not shown). Hence, when the pulse is applied in the first half of the night, M may be also somewhat phase-delayed in the same night.

3.4 PRC's for Instantaneous Separate Phase-Shifts of the E- and M-Oscillator

Assuming that phase-shifts in the NAT rise in the night when the pulses are applied represent instantaneous separate phase-delays of E, and further assuming that instantaneous phase-advances of M are at least as large as phase-shifts of M the next night after the pulses were applied, we propose the following PRC's for separate instantaneous phase-shifts of E and M (Fig. 3B). The PRC for E has only phase-delays (Fig. 3B, E). Even the next day following the pulses after the supposed interaction of E and M, the PRC for E lacks phase-advances (Fig. 3A, E). The PRC representing instantaneous separate phase-shifts of M has probably only phase advances (Fig. 3B, M). Yet the PRC for M in the night when pulses are applied may have also very small phase-delays; as has been already said, the pulse at 2200 h phase-delayed M in the same night by about 0.3 h. However, this small phase-delay of M may be due to interaction between E and M.

4 Two-Oscillator Structure of the Pacemaker Controlling the NAT Rhythm

4.1 General Discussion of the Two-Oscillator Structure

Our results indicate that the night high NAT activity is controlled by a complex pacemaker consisting of at least two oscillators, the evening one, E, controlling the NAT rise and the morning one, M, controlling the NAT decline. When the pulse is applied in the first half of night, E is instantaneously phase-delayed to a phase when NAT is still low; when the pulse is applied in the second half of night, M is instantaneously phase-advanced to a phase when NAT is already low. E and M are coupled in a mutual phase-relation, ψ EM. (ψ EM is the time-distance between the phase-reference point for E and for M. In this study, the time when NAT activity reaches the value of 3nmol mg^{-1} h^{-1} during its rise was chosen as the phase-reference point for E and the time when NAT activity declines to the same value during its decrease was chosen as the phase-ereference point for M). The strength of the mutual interaction between E and M depends on the phase-angle difference, ψ EM, between them (Pittendrigh and Daan 1976). The more ψ EM is compressed, the more the interaction between E and M may affect separate phase-shifts of E and M and thus influence the net phase-shifts of the coupled system. From the suggested PRC's for instantaneous separate phase-shifts of E and M, we may deduce that the compression of ψ EM is larger when the pulses are applied nearer to midnight and smaller when more distant. Consequently, differences between instantaneous separate phase-shifts and phase-shifts seen the next day after the pulses are larger when the time of the pulse is neare midnight. (cf. Fig. 3A, E and Fig. 3B, E). When the pulse is applied at midnight, the mutual interaction of the compressed E and M may be so strong, that the next night neither E nor M are phase-shifted (Fig. 3A). E may have greater impact on M than vice versa. The next night after the pulses, the PRC of M has phase-advances but also phase-delays, however the PRC of E has only phase-delays (Fig. 3A). The phase-delayed E might force M to phase-delay, but the phase-advanced M did not force E to phase-advance.

From the PRC's one day after the pulses (Fig. 3A), we may deduce that ψ EM, which represents simultaneously the period of the high night NAT activity, is compressed one day after the pulses as compared with the unpulsed controls, with the exception of ψ EM of rats pulsed at 2200 h, 2300 h and 0000 h. The compression of ψ EM persists for at least 3 days after the pulses, as may be deducted from Fig. 1.

4.2 Application of the Two-Oscillator Structure to the NAT Rhythm Under Different Artificial and Natural Photoperiods

When rats raised under LD 12 : 12 were adapted for 5 weeks to LD 8 : 16, the evening NAT rise was almost not phase-shifted as compared with rats under LD 12 : 12; how-

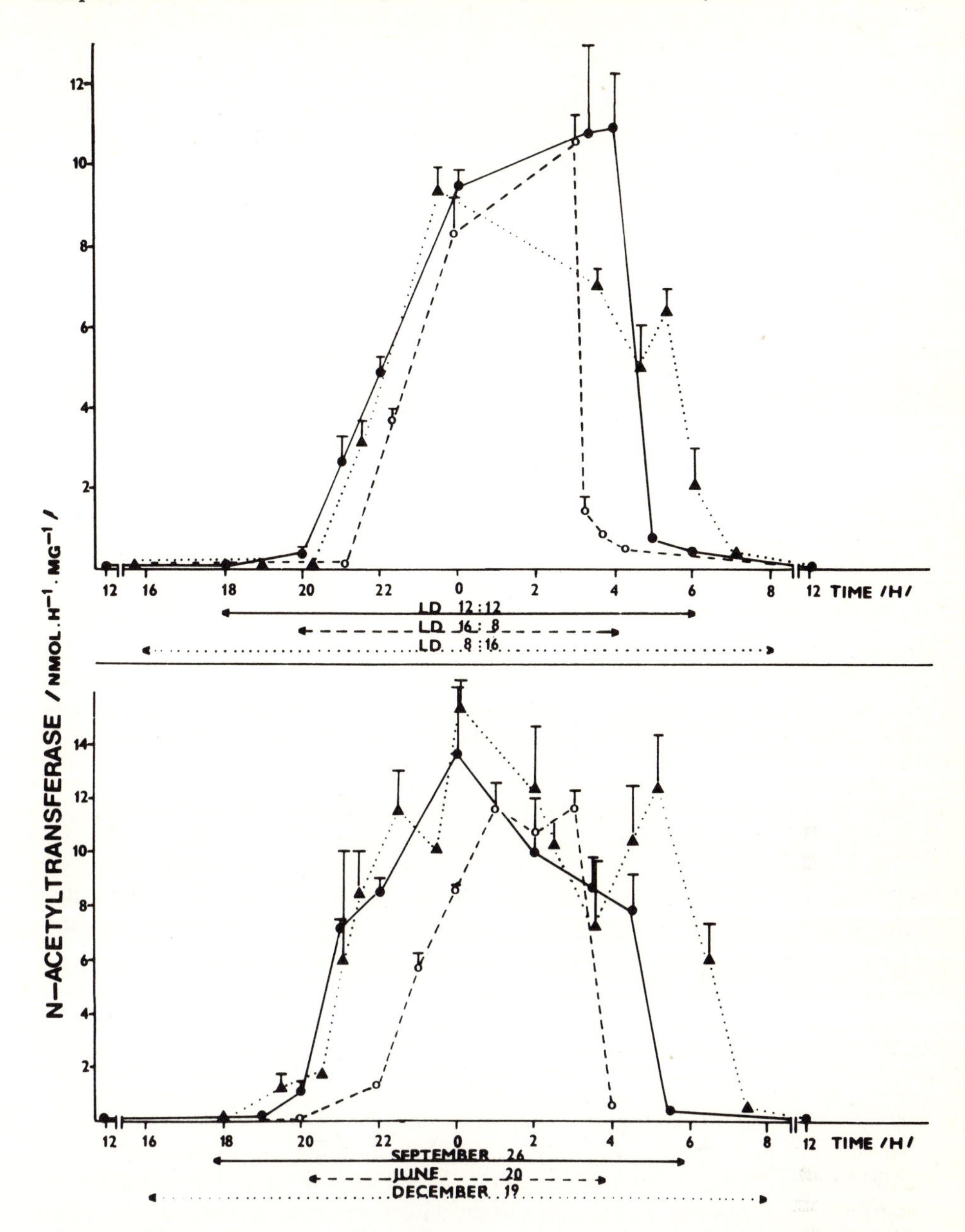

Fig. 5. The NAT rhythm under artificial light-dark regimes *(top)* and in natural daylight *(bottom)*. Data are expressed as means + S.E.M. of four *(top)* or three *(bottom)* animals. *Lines under abcissa* indicate periods of darkness or periods between sunsets and sunrises, respectively. *Top filled circles*, rats kept in LD 12 : 12; *open circles*, rats kept for 5 weeks in LD 16 : 8; *filled triangles*, rats kept for 5 weeks in LD 8 : 16. *Bottom* rats kept in natural daylight and killed on September 26 *(filled circles)*, on June 20 *(open circles)* and on December 19 *(filled triangles)*

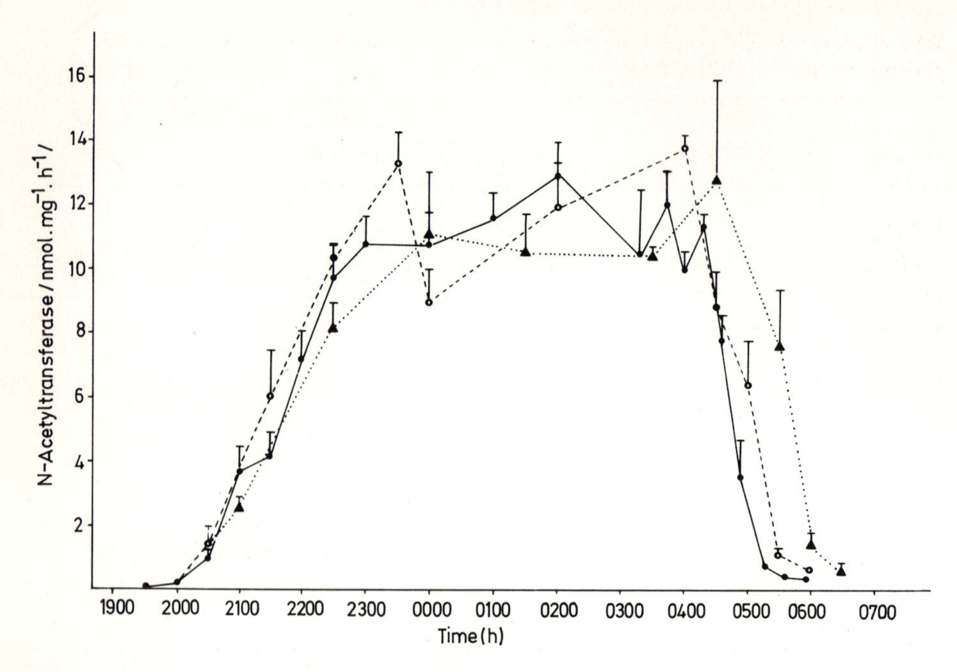

Fig. 6. The N-acetyltransferase rhythm 1 and 3 days after release into darkness. Rats were killed in LD 12 : 12 *(filled circles)* or they were released into darkness from LD 12 : 12 and killed 1 *(open circles)* and 3 days *(filled triangles)* later. Data are expressed as means + S.E.M. of four animals

ever, the morning NAT decline was phase-delayed by 1.2 h; conseuqnetly, ψ EM was decompressed (Fig. 5, top). Similarly, on December 19, in natural daylight comparable to LD 8 : 16, the evening NAT rise in rats was almost not phase-shifted as compared with rats on September 26, in natural daylight comparable to LD 12 : 12, but the morning decline was delayed by 2.0 h (Fig. 5, bottom) (Illnerová and Vaněček 1980). When rats were released into constant darkness from LD 12 : 12, the NAT rise after one and three days was again not phase-shifted as compared with rats under LD 12 : 12; however, the morning NAT decline was phase-delayed by 0.4 h after one day and by 0.9 h after three days (Fig. 6). It appears that light till 1800 h under LD 12 : 12 or on September 26 phase-delays E only a little, if at all, and hence after the release into short days or into continuous darkness, E does not phase-shift. On the contrary, light from 0600 h under LD 12 : 12 may phase-advance M and therefore, after the release into short days or in constant darkness, M phase-delays. Consequently, the decompression of ψ EM from LD 12 : 12 progresses more into morning hours than into evening hours. However, the reality may be more complex. The above conclusion is not consistent with out observation that the evening NAT rise in rats kept under LD 12 : 12 was advanced by half an hour when lights in the animal room were turned off 2 h earlier than usual (Illnerová and Vaněček 1980).

When rats were raised under LD 12 : 12 and then adapted for five weeks to LD 16 : 8, the evening NAT rise was phase-delayed by light till 2000 h, as compared with rats under LD 12 : 12, and the morning NAT decline was phase-advanced due to light

Table 1. Phase-angle difference between the evening and the morning oscillator under artificial photoperiods and in natural daylight

Artificial photoperiod	ΨEM (h)	Natural daylight	ΨEM (h)
LD 16 : 8	5.1	June 20	5.3
LD 12 : 12	7.6	September 26	8.7
LD 8 : 16	8.5	December 19	10.3

ΨEM indicates the interval between the time when N-acetyltransferase activity reaches the value of 3 nmol mg^{-1} h^{-1} and the time when the activity declines to the same value. ΨEM values were derived from Fig. 5

from 0400 h (Fig. 5, top). Similarly, on June 20, E was phase-delayed by the late sunset and M was phase-advanced by the early sunrise as compared with phases of E and M on September 26 (Fig. 5, bottom).

The phase-angle difference between E and M under different photoperiods is shown in Table 1. ψ EM is smaller under long photoperiods and larger under short photoperiods. The difference between ψ EM under LD 16 : 8 and under LD 12 : 12 is much larger than between ψ EM under LD 12 : 12 and under LD 8 : 16. From the PRC's for instantaneous separate phase-shifts extrapolated to zero phase-shifts, we may deduce that only light falling between 1800 h and 0645 h may phase-shift E and M and thus compress ψ EM. Consequently, ψ EM under all regimes with a dark period longer than 12.5-13.00 h may be decompressed and hence similar. ψ EM is larger in natural daylight than under corresponding artificial light-dark regimes. The difference indicates that ψ EM is dependent on the previous light history of animals as has been proposed by Pittendrigh and Daan (1976).

5 Conclusions

We report that the circadian rhythm in NAT activity in the rat pineal gland is driven by a complex pacemaker consisting of two coupled oscillators. The two-oscillator model of Pittendrigh and Daan (1976), proposed originally for the regulation of the locomotor activity rhythm, may be hence valid for other circadian rhythms as well.

Acknowledgments. We thank Mrs Marie Svobodová for her excellent technical assistance and Mrs Milada Sumová for typing the manuscript.

References

Brownstain M, Axelrod J (1974) Pineal gland. 24-hour rhythm in norepinephrine turnover. Science 184: 163-164

Deguchi T, Axelrod J (1972a) Control of circadian change of serotonin N-acetyltransferase activity in the pineal organ by the beta-adrenergic receptor. Proc Natl Acad Sci USA 69: 2547-2550

Deguchi,T, Axelrod J (1972b) Sensitive assay for serotonin N-acetaltransferase activity in rat pineal. Anal Biochem 50: 174-179

Illnerová H, Vaněček J (1979) Response of rat pineal serotonin N-acetyltransferase to one min light pulse at different night times. Brain Res 167: 431-434

Illnerová H, Vaněček J (1980) Pineal rhythm in N-acetyltransferase activity in rats under different artificial photoperiods and in natural daylight in the course of a year. Neuroendocrinology 31: 316-320

Illnerová H, Vaněček J, Křeček J, Wetterberg L, Sääf J (1979) Effect of one minute exposure to light at night on rat pineal serotonin N-acetyltransferase and melatonin. J Neurochem 32: 673-675

Klein DC, Berg GR (1970) Pineal gland: Stimulation of melatonin production by norepinephrine involves cyclic AMP-mediated stimulation of N-acetyltransferase. Adv Biochem Psychopharmacol 3: 241-263

Klein DC, Weller JE (1970) Indole metabolism in the pineal gland. A circadian rhythm in N-acetyl-transferase activity. Science 169: 1093-1095

Klein DC, Weller JE, Moore RJ (1971) Melatonin metabolism. Neural regulation of pineal serotonin: acetylcoenzyme A N-acetyltransferase activity. Proc Natl Acad Sci USA 68: 3108-3110

Moore RJ, Klein DC (1974) Visual pathways and the central neural control of a circadian rhythm in pineal serotonin N-acetyltransferase activity. Brain Res 71: 17-33

Parfitt A, Weller JL, Klein DC, Sakai KK, Marks BH (1975) Blockade by oubain or elevated potassium ion concentration of the adrenergic and adenosine cyclic 3', 5'-monophosphate induced stimulation of pineal serotonin N-acetyltransferase activity. Mol Pharmacol 11: 241-255

Pittendrigh CS, Daan S (1976) A functional analysis of circadian pacemakers in nocturnal rodents. V. Pacemaker structure: A clock for all seasons. J Comp Physiol A 106: 333-355

Vaněček J, Illnerová H (1979) Changes of a rhythm in rat pineal serotonin N-acetyltransferase following a one-minute light pulse at night. In: Arriens Kappers J, Pevet P (eds) The pineal gland of vertebrates including man. Progress in brain research, vol II. Elsevier/North Holland Biomedical Press, Amsterdam, p 245

Vaněček J, Illnerová H (1980) Some characteristics of the night N-acetyltransferase in the rat pineal gland. J Neurochem 35: 1455-1467

Weissbach H, Redfield BG, Axelrod J (1960) Biosynthesis of melatonin; enzymic conversion of serotonin to N-acetylserotonin. Biochem Biophys Acta 43: 352-353

7.4 The Critical Photoperiod in the Djungarian Hamster *Phodopus sungorus*

K. Hoffmann[1]

1 Introduction

In many vertebrates the annual cycle of reproductive and other functions is regulated by photoperiod (Turek and Campbell 1979, Hoffmann 1981a). In some bird species it has been shown that once the photoperiod exceeds a critical duration, gonadal growth is stimulated, and the rate of growth is proportional to day length (Farner 1975, Follett and Robinson 1980). In mammals, the critical photoperiod has been determined in the golden hamster (Gaston and Menaker 1967, Elliott 1976). Twelve and a half h of light or more maintained gonadal activity or stimulated recrudescence, photoperiods of 12 h or less induced regression. Above or below this critical photoperiod, no influence of the length of the light period could be detected. Since the golden hamster is the only mammalian species in which the critical photoperiod has been determined precisely, we examined the reaction in different photoperiods in a related species, the Djungarian hamster *Phodopus sungorus.* This hamster also shows strong photoperiodic reactions (Hoffmann 1972, 1978a, 1981b). The results are in agreement with, and lend support to, the suggestion that the pineal is involved, not only in the transduction of the inhibitory effects of short photoperiods, but also participates in the stimulatory effects of long photoperiods (Hoffmann 1977, 1979a, 1981b, Hoffmann and Küderling 1975, 1977).

2 Material and Methods

Phodopus differs from the golden hamster in one important aspect: not only adults react to photoperiod, but development to puberty is also photoperiodically controlled (Hoffmann 1978b). In long photoperiods, first spermatozoa are found in the epidydimal caudae at about 35 days, while in short photoperiods testicular development is arrested for a considerable time. In order to strictly standardize the experiments, 35-day-old male *Phodopus* were used in the experiments. Breeding pairs were maintained in long photoperiods. Within 48 h after parturition, they were moved, with their litter, into long (LD 16 : 8) or short (LD 8 : 16) photoperiods. Temperature was kept at 20 ± 1°C, light intensity above the cages was between 40 and 600 lx, depending on position of cage in the

1 Max-Planck-Gesellschaft, Clinical Research Unit for Reproductive Medicine, Steinfurter Str. 107, D - 4400 Münster

Vertebrate Circadian Systems
Ed. by J. Aschoff, S. Daan, and G. Groos

room. No differences in development, due to differences in light intensity, were observed within this range. The young were weaned at 21 days of age, and males were placed singly in plastic cages in the same room under the same conditions.

At 35 days, the hamsters were palpated for testis size. In LD 16 : 8, all animals had large testes. In LD 8 : 16, testes were not palpable in most cases, the few males with palpable testes were discarded (cf. Hoffmann 1978b). From each of the two light schedules, 16 animals were killed to serve as initial controls (IC). Other males were placed in one of the following light regiments: LD 1 : 23, 4 : 20, 8 : 16, 10 : 14, 11 : 13, 12 : 12, 13 : 11, 14 : 10, 15 : 9, 16 : 8, 20 : 4 or 24 : 0 (LL). Light-times were centered around local noon in all schedules. After 45 days in these conditions, the hamsters were killed. Body weight and fresh weight of testes and accessory glands (seminal vesicles, coagulating glands, ampullary glands) were determined. Testes were fixed in Bouin, embedded in Paraplast, sectioned at 10 μm and stained in Mayer's haemalum and eosin. Tubular diameter in the left testis was determined from 10 measurements in each animal. Experimental groups consisted of 15-19 animals in each light schedule, a total of 460 young male *Phodopus* was used. For statistical evaluation, the two-tailed U-test was employed throughout. In addition, older males (6-8 months) which were in winter condition due to exposure to natural daylight, were placed in LD 8 : 16, 16 : 8 or 24 : 0 (LL) for 47 days, starting in early December. Otherwise, treatment was identical to that of 35 days old young males.

3 Results

Figure 1 shows testis weight of the young hamsters after 45 days in the different light schedules. In males coming from LD 16 : 8 (Fig. 1A), testes had significantly regressed in all groups that were exposed to photoperiods with 12 h light or less (in all groups $p < 0.002$ versus IC). There was no significant difference between these groups. In some animals, testes did not regress in the short photoperiod. This also occurred in previous experiments with short days and seems to be a regular phenomenon (Figala et al. 1973, Hoffmann 1978b). In LD 13 : 11, there was regression in some animals only ($p < 0.01$ versus all groups with shorter photoperiods; $p < 0.02$ versus IC; $p < 0.002$ versus all groups with longer photoperiods). In longer photoperiods, there was some further development compared to initial controls ($p < 0.01$ for photoperiods with 15 h or more light per day). No significant difference between hamsters maintained in LD 14 : 10, 15 : 9 and 16 : 8 could be detected. In even longer light periods, however, testicular weight was slightly higher ($p < 0.02$ for LD 20 : 4 versus 14 : 10; $p < 0.02$ to < 0.002 for LD 24 : 0 versus LD 14 : 10, 15 : 9 or 16 : 8). In general, the data in animals coming from long photoperiods show that there is a marked critical photoperiod at around 13 h light per day. In addition, they suggest that very long photoperiods and constant light are somewhat more stimultory (cf. also Fig. 2a).

In hamsters with initially small testes, coming from LD 8 : 16, the results are similar for short photoperiods (Fig. 1B). There was even some further regression compared to initial controls ($p < 0.05$ for groups with 1 to 10 h light per day). In LD 11 : 13 and 12 : 12, slight development could be discerned in some animals ($p < 0.05$ to < 0.002 versus any of the shorter photoperiods), and development in LD 12 : 12 was slightly increased versus LD 11 : 13 ($p < 0.05$). In LD 13 : 9, testis size was significantly higher than in

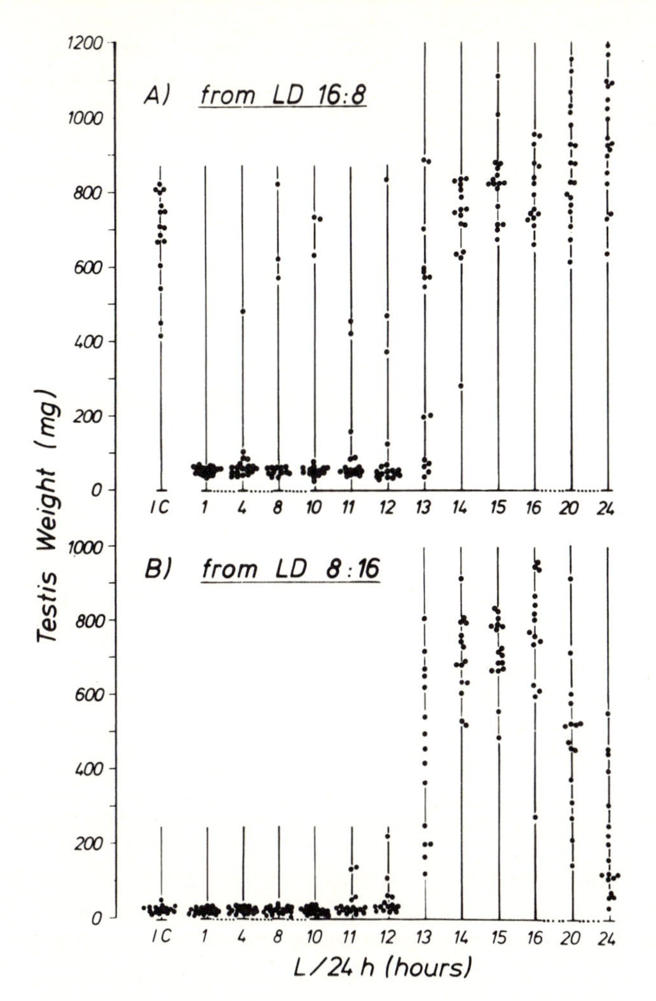

Fig. 1. Testis weight of young *Phodopus* that had lived from birth to 35 days in long **(A)** or short **(B)** photoperiods and were then exposed for another 45 days to the photoperiods indicated. *IC* initial controls at age 35; *1, 4* etc. *L/24 h* refers to LD 1 : 23, 4 : 20 etc.; each *point* gives the weight of both testes for one male

any of the shorter photoperiods or in IC ($p < 0.002$), but less than in LD 14 : 10, 15 : 11 and 16 : 8 ($p < 0.002$). In the latter groups, the values did not differ significantly.

Unexpectedly, in very long photoperiods there was again less testicular development. In LD 20 : 4 testis weight was significantly less than in LD 14 : 10, 15 : 9 or 16 : 8 ($p < 0.002$). This was confirmed in an additional experiment in which litter mates in LD 16 : 18 and LD 20 : 4 were compared (see Fig. 2b). In LL there was even less development, the values differ significantly, not only from LD 14 : 10, 15 : 11 and 16 : 8 but also from LD 20 : 4 and LD 13 : 11 ($p < 0.002$). In general, thus, while the results obtained in short photoperiods agree well with those observed in animals coming from LD 16 : 8, in very long photoperiods or in constant light there are drastic differences. However, even in LL there was significant testicular growth as compared to initial controls or to any of the short photoperiods with up to 12 h light per day ($p < 0.002$).

Essentially the same picture emerged from the other parameters measured. Figure 2 gives the mean values for testis weight, tubular diameter and weight of accessory glands. This figure includes additional experiments; i.e., in hamsters coming from LD 16 : 8 one

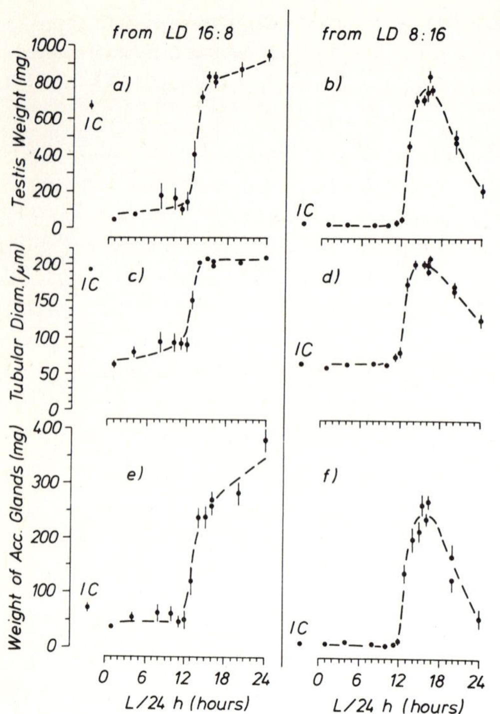

Fig. 2. Weight of testes **(a, b)**, tubular diameter **(c, d)** and accessory gland weight **(e, f)** of young *Phodopus* that lived for the first 35 days in LD 16 : 8 **(a, c, e)** or LD 8 : 16 **(b, d, f)** and for another 45 days in the photoperiods indicated (cf. Fig. 1). Each *point* represents the mean ± SE from 15-19 animals. *IC* initial control at age 35

further experiment with males maintained in this photoperiod, using litter-mates of males exposed to LD 1 : 23 or 24 : 0. In hamsters coming from LD 8 : 16, two further groups in LD 16 : 8 are included, one running parallel with the groups in LD 1 : 23 and 24 : 0, the other as a control to the repetition of the LD 20 : 4 experiment; here again the males in the different conditions were often litter-mates.

Tubular diameter in the testes of hamsters coming from LD 16 : 8 (Fig. 2c) was significantly reduced in all groups exposed to 12 h of light or less per day ($p < 0.002$ versus IC or longer light-times). In LD 1 : 23, the diameter was slightly smaller than in the other short photoperiods ($p < 0.01$). In LD 13 : 11, mean tubular diameter was higher than in any of the shorter photoperiods ($p < 0.002$), but lower than in initial controls ($p < 0.05$) or in any of the longer light periods ($p < 0.01$ or less). In the longer photoperiods, tubular diameter was significantly higher than in IC only in LD 24 : 0 ($p < 0.002$); there were no significant differences between the groups with long light periods. In males coming from LD 8 : 16 (Fig. 2d), tubular diameter did not differ significantly between groups with 11 h light or less per day, and corresponded to that of initial controls. In LD 12 : 12, a slight increase was observed ($p < 0.01$ versus IC or shorter photoperiods). LD 13 : 11 resulted in markedly higher values than in shorter photoperiods ($p < 0.002$) but in lower values as compared to LD 14 : 10, 15 : 9 and 16 : 8 ($p < 0.01$). In LD 20 : 4, tubular diameter was smaller than in LD 14 : 10, 15 : 9 or 16 : 8 ($p < 0.002$), and in LL tubular diameter was even less than in LD 20 : 4 ($p < 0.002$) though it was higher than in initial controls or in any of the photoperiods with 12 h or less of light per day ($p < 0.002$).

Weight of the accessory glands showed the same dependence on length of photoperiod as did testis weight. In animals from LD 16 : 8 (Fig. 2e), values for initial controls were quite low, since their development lags behind that of the testes (Hoffmann 1978b). Nevertheless, after 45 days in LD 12 : 12 or shorter photoperiods, the accessory gland weight had decreased ($p < 0.05$ to < 0.002), suggesting regression. In photoperiods with 14 h light or more, the glands had grown significantly ($p < 0.002$ versus IC and all shorter photoperiods). In LD 13 : 11, accessory gland weight was higher than in the short photoperiods ($p < 0.05$ to < 0.002). In constant light the value was higher than in all other photoperiods ($p < 0.002$). In animals coming from LD 8 : 16 (Fig. 2f) weight of accessory glands in photoperiods with 12 h of less light did not differ from the initial controls. In LD 12 : 12, the weight was slightly higher ($p < 0.05$ to < 0.002, except versus LD 11 : 13). In LD 13 : 11, gland weight was still higher ($p < 0.002$ versus all shorter photoperiods), and in photoperiods with 14 to 16 h of light, values were again higher ($p < 0.01$ versus LD 13 : 11, $p < 0.002$ versus all shorter photoperiods). In LD 20 : 4, weight of accessory glands was significantly less than in LD 14 : 10, 15 : 11 and 16 : 8 ($p < 0.002$), but higher than in LD 12 : 12 or shorter photoperiods. In LL, an even smaller value was found ($p < 0.002$ versus photoperiods with 14 to 20 h of light), but it was significantly higher than in LD 12 : 12 or shorter photoperiods including IC ($p < 0.01$ to < 0.002).

In *Phodopus*, body weight is also strongly influenced by photoperiod, and runs closely parallel to weight of gonads and accessory glands (Hoffmann 1978a, b, 1979a, b). Accordingly, for body weight a curve corresponding to the results mentioned above was observed (Fig. 3). In hamsters coming from LD 16 : 8 after 45 days in photoperiods with up to 12 h light, body weight was only slightly higher than before ($p < 0.05$ for LD 8 : 16, 10 : 14 and 11 : 13 versus IC) and there was no significant difference between these groups. In LD 13 : 11, body weight was higher than in any of the shorter photoperiods ($p < 0.01$ to < 0.002), but lower than in longer photoperiods ($p < 0.01$ to < 0.002). There was no significant difference between the groups in longer photoperiods, only in LL was body weight significantly higher than in any other group ($p < 0.05$ to < 0.002). In hamsters that had spent their first 35 days in LD 8 : 16 (Fig. 3b), body weight was lower than in those that had lived in LD 16 : 8 during that time (cf. IC in Fig. 3a and b). During the next 45 days, in photoperiods of 12 h or less light it increased to about the same value as in animals coming from LD 16 : 8 ($p < 0.002$ versus IC). There was no significant difference between these groups. In LD 13 : 11 to 16 : 8, body weight was higher than in any of the shorter photoperiods ($p < 0.002$), in LD 20 : 4 and in LD 24 : 0 body weight was again lower (significant only for LD 24 : 0 versus LD 13 : 11, 14 : 10, 15 : 9 and 16 : 8, $p < 0.01$ to < 0.002). In general, the curves resemble those found for the reproductive parameters.

The observation that in *Phodopus* with undeveloped gonads, coming from short photoperiods, very long photoperiods or constant light were less stimulatory than normal long photoperiods was unexpected. In the golden hamster no such effect has been reported (Elliott 1974, 1976). Since in our experiments young males were used that had never experienced full testicular development, the question arose whether the same rule would apply to older animals, in which long photoperiods stimulate gonadal recrudescence. We therefore repeated some such experiments with older males. Figure 4 shows testicular weight of adult hamsters whose testes had regressed during exposure to natural

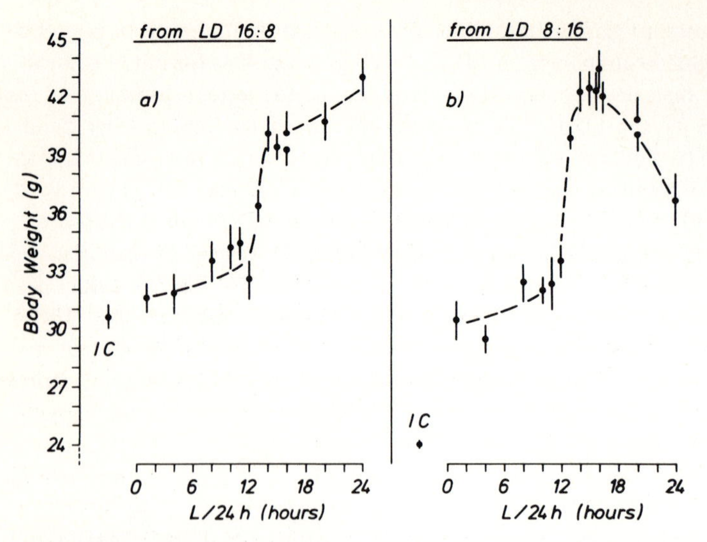

Fig. 3. Body weight of male *Phodopus* after 45 days in the photoperiods indicated. Compare with Figs. 1 and 2 for further explanation

illumination (IC) and which were then maintained in different photoperiods for 47 days. Gonadal development in LL was significantly less than in LD 16 : 8 ($p < 0.002$). There was some development even in LD 8 : 16 due to spontaneous recrudescence (see Hoffmann 1981b). In LL development was slightly advanced compared to LD 8 : 16 ($p < 0.05$). The data for weight of accessory glands and for body weight corroborate these findings; for both parameters the values in LL were significantly lower than those in LD 16 : 8 ($P < 0.002$). In an additional experiment, adult males in winter condition due to short natural photoperiods were exposed to LD 20 : 4 and LD 16 : 8. Again, development in LD 20 : 4 was less than in LD 16 : 8 (Hoffmann unpubl.).

In general, the results demonstrate that for stimulation of gonadal development very long photoperiods are less effective than medium long photoperiods. This holds for

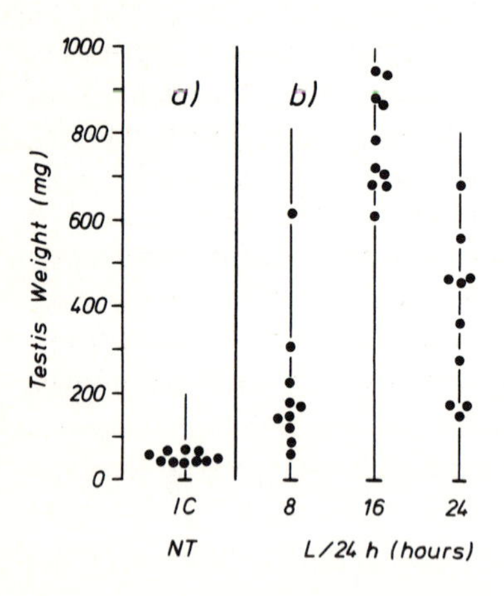

Fig. 4. Testis weight of adult male *Phodopus* that had regressed testes due to natural short photoperiods *(a)*, and were then exposed, for 47 days, to one of the photoperiods indicated (24 = LL). Each *point* gives the combined weight of both testes for one hamster

young males with undeveloped gonads, as well as for older males in which recrudescence is initiated by the long photoperiods.

4 Discussion

The results demonstrate a marked critical photoperiod at around 13 h light per day, which, in animals coming from long photoperiods, divides light periods which maintain or further stimulate gonadal activity from those that induce involution. This is slightly longer than the critical photoperiod reported for the golden hamster (Elliott 1974, 1976) which corresponds with the more northern distribution of the Djungarian hamster (Figala et al. 1973). In *Phodopus* with small gonads, coming from short photoperiods, the same value for the critical photoperiod was found, though there is some indication that 11 or 12 h of light per day can be slightly stimulatory in some animals.

The most noteworthy finding was that in males with undeveloped testes, very long photoperiods or constant light were less stimulatory than medium long photoperiods. In the golden hamster no such effect was reported (Elliott 1974). The observation can, however, be explained by, and renders support to, a concept of pineal involvement in the transduction of photoperiodic effects. Several findings suggest that in mammals melatonin from the pineal is involved in conveying the photoperiodic message, and that the temporal pattern of melatonin release is an important feature in this process (for examples and discussion see Hoffmann 1979a, 1981a, b, c).

In *Phodopus*, as well as in some other mammals, there is evidence that the pineal participates, not only in transducing the effects of short photoperiods which induce regression, but also in conveying the effects of long photoperiods which stimulate development, and that the latter is an active process rather than the abolishment of an inhibition (Hoffmann 1977, 1979a, 1981b, Hoffmann and Küderling 1975, 1977, Brackmann and Hoffmann 1977). In a recent brief note, Carter and Goldman (1981) report that in young pinealectomized *Phodopus*, daily infusion of melatonin for 12 h inhibited gonadal development, while daily infusion for only 4 h stimulated gonadal development versus untreated controls. These results strongly support the concept that the temporal pattern of melatonin is the main factor in conveying the photoperiodic message, and that long times of elevated melatonin, as can be expected in short photoperiods, are inhibitory, while short periods of elevated levels, as can be observed in medium long photoperiods, are stimulatory (cf. Hoffmann et al. 1981). In practically all mammals, constant light suppresses the nocturnal rise in melatonin. In the golden hamster in LD 20 : 4 no consistent and reproducible nocturnal rise in pineal melatonin was found (Tamarkin et al. 1980). If similar effects can be expected in *Phodopus*, the fact that very long photoperiods or LL are less stimulatory than medium long photoperiods can easily be explained. They diminish or completely suppress the brief nocturnal peak in pineal melatonin synthesis and release which is necessary to obtain maximal stimulation of gonadal development. That there is some development versus animals in short photoperiods is due to the lack of the inhibitory effect of the long times of elevated melatonin in short photoperiods.

References

Brackmann M, Hoffmann K (1977) Pinealectomy and photoperiod influence testicular development in the Djungarian hamster. Naturwissenschaften 64: 341-342

Carter DS, Goldman BD (1981) Antigonadal and progonadal effects of programmed melatonin infusion into juvenile Djungarian hamsters *(Phodopus sungorus)*. Biol Reprod 24: (Suppl 1) 23A

Elliott JA (1974) Photoperiodic regulation of testis function in the golden hamster: Relation to the circadian system. Ph D thesis, Univ Texas, Austin

Elliott JA (1976) Circadian rhythms and photoperiodic time measurement in mammals. Fed Proc 35: 2339-2346

Farner DS (1975) Photoperiodic controls in the secretion of gonadotropins in birds. Am Zool 15: (Suppl 1) 117-135

Figala J, Hoffmann K, Goldau G (1973) Zur Jahresperiodik beim Dsungarischen Zwerghamster *Phodopus sungorus* Pallas. Oecologia 12: 89-118

Follett BK, Robinson JE (1980) Photoperiod and gonadotrophin secretion in birds. In: Reiter RJ, Follett BK (eds) Progress in reproductive biology, vol V. Karger, Basel, pp 39-61

Gaston S, Menaker M (1967) Photoperiodic control of hamster testis. Science 158: 925-928

Hoffmann K (1972) The influence of photoperiod and melatonin on testis size, body weight, and pelage colour in the Djungarian hamster *(Phodopus sungorus)*. J Comp Physiol 95: 267-282

Hoffmann K (1977) Die Funktion des Pineals bei der Jahresperiodik der Säuger. Nova Acta Leopold NF 46: 217-229

Hoffmann K (1978a) Photoperiodic mechanism in hamsters: The participation of the pineal gland. In: Assenmacher I, Farner DS (eds) Environmental endocrinology. Springer, Berlin Heidelberg New York, pp 94-102

Hoffmann K (1978b) Effect of short photoperiods on puberty, growth and moult in the Djungarian hamster *(Phodopus sungorus)*. J Reprod Fertil 54: 29-35

Hoffmann K (1979a) Photoperiod, pineal, melatonin and reproduction in hamsters. Progr Brain Res 52: 397-415

Hoffmann K (1979b) Photoperiodic effects in the Djungarian hamster: One minute of light during darktime mimics influence of long photoperiods on testicular recrudescence, body weight and pelage colour. Experientia 35: 1529-1530

Hoffmann K (1981a) Photoperiodism in vertebrates. In: Aschoff J (ed) Handbook of behavioral neurobiology, vol IV. Biological rhythms. Plenum Press, New York, pp 449-473

Hoffmann K (1981b) Pineal involvement in the photoperiodic control of reproduction and other functions in the Djungarian hamster *Phodopus sungorus*. In: Reiter RJ (ed) The pineal gland, vol II. Reproductive effects. CRC Press, Boca Raton, Fla pp 83-102

Hoffmann K (1981c) Photoperiodic function of the mammalian pineal organ. In: Oksche A, Pévet P (eds) The Pineal Organ. Photobiology – Biochronometry – Endocrinology. Developments in Endocrinology Vol. 14. Amsterdam Elsevier/North-Holland Biomedical Press, pp 123-138

Hoffmann K, Küderling I (1975) Pinealectomy inhibits stimulation of testicular development by long photoperiods in a hamster. Experientia 31: 122-123

Hoffmann K, Küderling I (1977) Antigonadal effects of melatonin in pinealectomized Djungarian hamsters. Naturwissenschaften 67: 408-409

Hoffmann K, Illnerová H, Vaněček J (1981) Effect of photoperiod and of one minute light at nighttime on the pineal rhythm on N-acetyltransferase activity in the Djungarian hamster *Phodopus sungorus*. Biol Reprod 24: 551-556

Tamarkin L, Reppert SM, Klein DC, Pratt B, Goldman BD (1980) Studies on the daily pattern of pineal melatonin in the Syrian hamster. Endocrinology 107: 1525-1529

Turek FW, Campbell CS (1979) Photoperiodic regulation of neuroendocrine-gonadal activity. Biol Reprod 20: 32-50

8.1 Circadian Contributions to Survival

S. Daan[1] and J. Aschoff[2]

1 Introduction

The significance of biological rhythms can be discussed under at least two aspects. They serve, on the one hand, to attain an optimal temporal arrangement of animal behaviour within the cycles of the environment, as in the four "circa-clocks" (Aschoff 1981). On the other hand, this *external* adaptation results in *internal* temporal order which in itself may have selective value. In addition, there are many rhythmic processes within the organism, not related to any environmental periodicity, which in various ways contribute to the maintenance of functional integrity of the internal milieu (Aschoff and Wever 1961). In focussing on how circadian rhythms contribute to survival, we do well to consider them, first, as part of a spectrum of rhythms and to evaluate their possible intrinsic function regardless of the environmental day-night cycle. We then will proceed to a discussion of possible benefits to be derived from the adjustment to the periodic environment.

2 Rhythms and Temporal Scaling

Animals' lives are controlled by a multitude of periodic processes which are not correlated with any environmental cycle. Foremost among these are processes involved in energy transfer, such as lung ventilation and heart beat (renewal of oxygen supply to body and tissues), meal-digestion cycles and periodic gut contractions (renewal of fuel supply). Among homeotherms the frequencies of these rhythms are linearly related to average tissue metabolic rate. In the comparison *across* mammalian species (Kleiber 1971) metabolic rate per gram body mass (M) is proportional to $M^{-0.27}$. (*Within* a species, the exponent is -0.33; cf. the recent reviews by Wilkie 1977 and Heusner 1982). Pulse time, breath time and gut beat time all increase with about $M^{0.27}$ (Fig. 1a, b, c; cf. also Adolph 1949). A similar relationship between maximal life span and body size (Fig. 1g) led Hill (1950) to introduce the concept of "physiological time". Generation time (= age at sexual maturity + gestation time), reflects energy transfer on a population level, and again follows the same relationships with a body weight exponent of 0.26 (Fig. 1f). Final-

1 Zoological Laboratory, Groningen University, Haren, Netherlands
2 Max-Planck-Institut für Verhaltensphysiologie, D-8138 Andechs, FRG

Vertebrate Circadian Systems
Ed. by J. Aschoff, S. Daan, and G. Groos

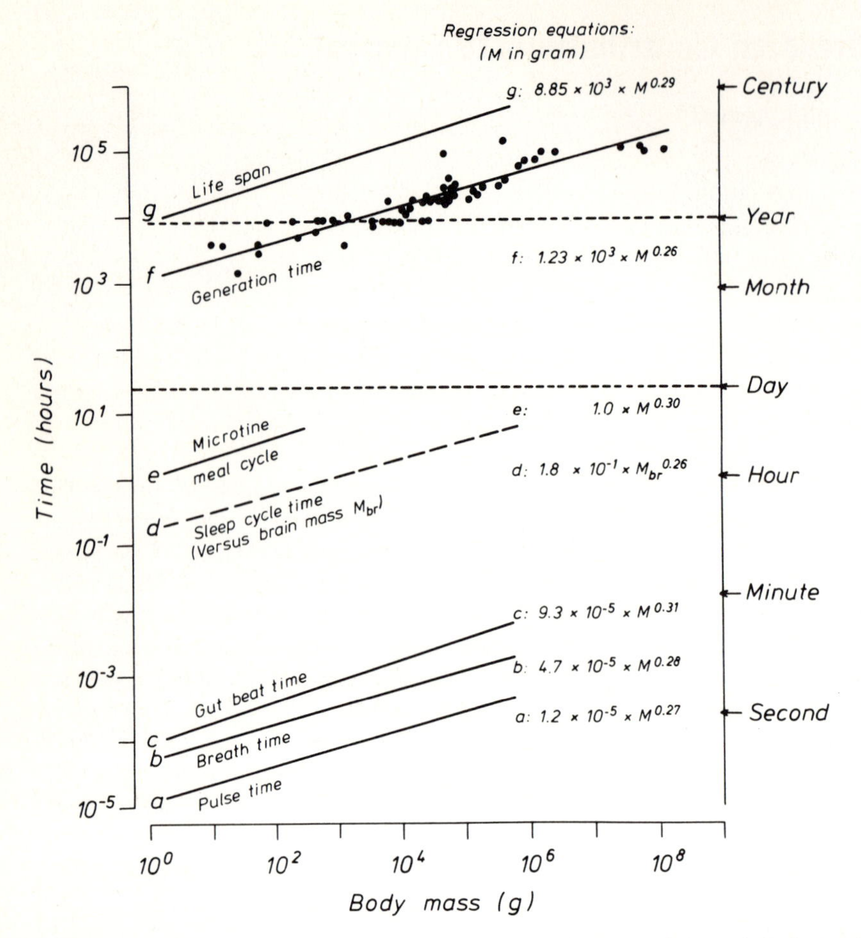

Fig. 1. Exponential relationships between body size and life span and the period of several biological cycles in different species of mammals. Regression equations express the period lengths in hours as a function of body mass (or brain mass, d).
a Heart beat duration. (Stahl 1962); *b* breath time. (Stahl 1962); *c* gut beat time. (Stahl 1962); *d* sleep cycle time as a funtion of brain mass according to Allison et al. (1977); *e* meal cycles in Microtine rodents. (Daan and Slopsema 1978); *f* generation time derived from data on gestation time + age at sexual maturity presented by Corbet and Southern (1977) and Western (1979); *g* life span. (Stahl 1962)

ly, the allometric relationship between microtine meal cycle time and body size is tentatively included in Fig. 1e as well as the relationship between sleep cycle time and brain weight (d) published by Allison and coworkers (1977). Sleep cycle time is usually measured as the average time from the end of one REM-sleep sequence till the end of the next. The exponent of mass (in this case brain mass, which is a better predictor of sleep cycle time than body weight: Zepelin and Rechtschaffen 1974) is 0.26, suggesting a metabolic temporal scaling of this periodic process also. Other similar relationships, including time constants of non-cyclic biological processes were recently presented by Lindstedt and Calder (1981) whose review covers the subject more broadly. The point we wish to emphasize here is that a homeotherm's life runs down along a time scale primarily dictated by its size. In contrast, the main environmental periodicities – tides, day and night, phases of the moon, and season – are the same for all creatures. A year is close to the maximum life span for the smallest mammals, it is more likely one-hundreth of that for the largest. The regression for generation time against body mass (Fig. 1f) reaches a value of one year at a mass of about 1 kg. In strongly seasonal environments, smaller animals may synchronize their generation cycles to the year by delay mechanisms such as deferred maturity, fertilization or implementation; larger animals may show generation cycles accelerated to one year, or have them set at whole multiples of a year. The adaptive value of this synchronization with the seasons has often been described. On the other hand, synchronous reproduction within a population may have its *intrinsic* merits irrespective of environmental conditions. These may be related to such benefits conferred by colonial breeders onto each other as increased chances for mating, facilitation of the detection of variable food sources (Emlen and DeMong 1975) and relative safety from predators (Darling 1938). It is hence not surprising that in environments with less pronounced seasonal variation there still may be conspicuous breeding cycles not related to an external annual periodicity. The classic example is the breeding of sooty terns *(Sterna fuscata)* every 10 months on the island of Ascension (Chapin and Wing 1959).

In the time domain of the *daily* cycle, the size-dependent periodicities discussed above do not seem to be represented. Size-predicted sleep cycle time would reach a value of 24 h at a brain weight of about 149,000 kg, which is of course realized nowhere among mammals. If a microtine rodent of 30 kg existed, we could expect it to have its ultradian feeding cycle lengthened to one meal per day. However, the microtine feeding cycle is probably a phenomenon related to the particular dietary habits of this group of small rodents which primarily use intestinal fermentation of bulky plant material (Daan and Slopsema 1978). The regression line e in Fig. 1 cannot be extrapolated to larger animals with quite different food habits, and less comparable meals. Thus, while annual rhythms often can be regarded as reproductive cycles modified by seasonal environments but present also in aseasonal conditions, the intrinsic value of daily rhythms is at least much less obvious.

3 The Circadian System

3.1 Intrinsic Significance of Circadian Order?

There are daily rhythms in almost any aspect of animal behaviour and physiology; a sample of rat rhythms is compiled in Fig. 2. In principle their function may be partly in preparing the organism for different activities which have to be performed in periodic alternation irrespective of the time of day and in ordering the sequence of mutually exclusive processes within the tissues (" temporal compartmentalization"). Which arguments could be put forward in support to the hypothesis that circadian temporal order is of value to the organism apart from the external cycle? In animals adapted to constant, non-daily environments, such as deep subterranean caves, there is little evidence of periodicities in the 24-h-domain (e.g. the blind fish *Anoptichthys antrobius:* Thinès et al. 1965), although in other species remnants of rhythmicity can be detected by spec-

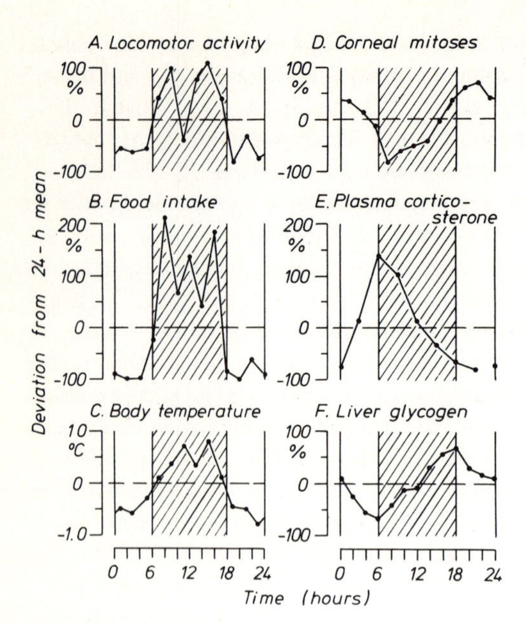

Fig. 2. Daily rhythms in behaviour and physiology of the laboratory rat. **A** Locomotor activity. (Aschoff and Meyer-Lohmann 1954). **B** Food intake. (Nishio et al. 1979). **C** Body temperature. (Honma and Hiroshige 1978). **D** Corneal mitoses. (Scheving and Pauly 1967). **E** Plasma corticosterone. (Hilfenhaus 1976). **F** Liver glycogen. (Peret et al. 1973)

tral analysis at least in some specimens (crayfish *Orconectes pellucidus:* Jegla and Poulson 1968).

In a temporary aperiodic environment, under cover of ice and snow, beavers may loose external synchrony and still show free-running circadian rhythms (Bovet and Oertli 1974). This, however, presents no argument for a role of circadian rhythmicity in the absence of external cycles because these rhythms may be the consequence of a mechanism which derives its function from other (non-constant) conditions. It further has been shown that, in man, the internal temporal order of a free-running system differs drastically from that of the entrained system. If the same is true for animals, one might expect detrimental effects from long-lasting exposures to constant conditions and to an "abnormal" circadian order. There are no data supporting this hypothesis. On the contrary, mice raised for six generations in continuous illumination showed free-running activity rhythms throughout without any symptoms of impairment and had an unchanged reproductivity (Aschoff 1960).

Another line of evidence could be derived from animals in which rhythmicity is disturbed by experimental manipulation. We know of no study where the effect of arrhythmicity as induced by constant light on individual well-being or life expectancy has been tested. Rodents made arrhythmic by bilateral lesions of the nucleus suprachiasmaticus are generally quite healthy and survive – in admittedly permissive laboratory conditions – no less than their intact conspecifics (Rusak 1982). Their sexual performance is impaired, but only when competing with a rhythmic rival for a rhythmic mate, as was elegantly demonstrated by Eskes (cf. Chap. 8.4 this Vol.).

Disturbances of the circadian system by repeated phase shifting have been shown to reduce the life expectation of an invertebrate (Aschoff et al. 1971), but the results of such experiments do not necessarily indicate the intrinsic value of circadian order, but the importance of a stable environmental cycle for the circadian system.

Internal integrity maintained by various circadian rhythms keeping pace with each other thus remains an elusive concept. It may be that the long-term effects of different feeding schedules on animal well-being and perhaps life expectation would present a valuable tool to study the importance of appropriate internal phasing of physiological rhythms to each other. However, no such studies exist so far and no solid data support the idea that circadian rhythmicity is of functional meaning in itself regardless of the environmental periodicity.

3.2 Benefits from Adjustment to the Periodic Environment

With respect to external optimalization of behaviour within the 24-h cycle, there is likewise no good experimental approach available, However, one may try to quantify circadian contributions to survival by comparing the chances for individuals in a population. We shall first discuss how single behavioural events may be timed optimally with respect to time of day and then attempt a quantitative estimation of the contribution of circadian organization of everyday life.

3.2.1 Daily Timing of Single Events

There are events which occur only once in animal's lifetime but are of crucial importance for its survival. They reflect a stage in ontogeny such as birth or egg hatching, metamorphosis and emergence of insects, fledging and in some species, copulation. For such events there may be an optimal time of day. The insect literature abounds with examples of daily rhythms of pupal eclosion and imaginal emergence (Remmert 1962). The functional significance of such rhythms has seldom been analysed. Where emergence is nearly immediately follwed by mating, circadian rhythmicity contributes to the chance of reproduction. In intertidal chironomid midges, for instance, the short exposure of the substrate necessary for oviposition demands a sharp synchronization of the emergence and adult sexual behaviour with respect to time of day (review in Neumann 1976). In other species, the success of the emergence act itself depends on the time of day. In *Drosophila pseudoobscura,* the eclosion takes place in a short temporal window just after sunrise, and it has been suggested that usually higher evaporative power of the environment at other times of day would interfere with normal eclosion and thus select for appropriate circadian timing of the act (Pittendrigh 1958). However, the different daily distributions of eclosion in other *Drosophila* species remain to be clarified.

A central theme in the daily timing of single events is probably the synchronization among conspecifics in a population. In vertebrates one sample case, that of fledging in Brünnich's guillemot *(Uria lomvia)*, has been analyzed in some detail. Guillemots are seabirds breeding in dense colonies on cliff ledges in the arctic and subarctic. Mature females produce a single egg per year, and the adults feed their single young initially in the nest. In *Uria* species, the young leave the nest at three weeks of age, long before attaining adult size and shape. Their wings are still undeveloped and when jumping from their ledge and gliding down to the sea they are easy prey for several predators. Fledging of young from the colony is synchronized with respect to time of day. At lower latitudes,

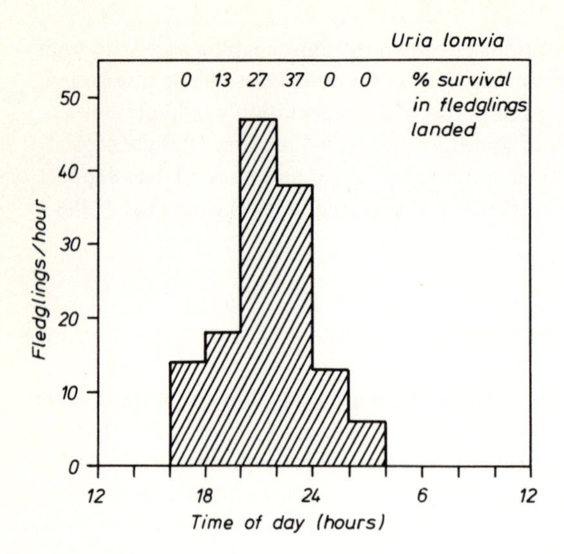

Fig. 3. Daily rhythm of juvenile fledgling in Brünnich's guillemot *(Uria lomvia)* at a colony in Spitsbergen, and in the relative frequencies of safely reaching the sea in fledglings dropped on the ground. (Daan and Tinbergen 1980)

in *Uria aalge,* fledging occurs shortly after sunset (Greenwood 1964). In the high arctic, *Uria lomvia* young similarly fledge in the evening hours, although daylight is continuously available (Daan and Tinbergen 1980). Some fledglings do not reach the sea immediately. Predation by glaucous gulls *(Larus hyperboreus)* on this part of the population was quantified by direct observation in a colony on Spitsbergen. Predation pressure turned out to be especially severe in early evening and early morning, and was relaxed at the daily peak of fledging between 2000 and 2400 h (Fig. 3). Satiation of the predators probably led to increased chances of survival for those individuals fledging simultaneously with the majority of the population. Thus, daily timing of fledging of the single young may be of crucial importance for a whole year's reproductive output of the parents. In the case of the guillemot, it remains to be established whether the timing of fledging is controlled by an endogenous circadian rhythm. The nestlings may well be using direct information, e.g., from abundant acoustical stimuli, instead of endogenous temporal cues. Only in insects has the endogenous circadian gating system, suppressing eclosion at maladaptive times of day, been thoroughly established (Pittendrigh and Skopik 1970).

3.2.2 Energy Balance in Daily Behaviour

More complex than the once-in-a-lifetime events, but also much more general, is the circadian organization of everyday life. Day after day, most animals have to perform activities which are related to the maintenance of body condition as well as of their food resources, territorial advertisement and defence, to mating, reproduction and parental care, and to safeguarding from predation. All of these may be optimally arranged within a species' day, and the host of possible solutions to the problem is bewildering. As the effects of such temporal arrangements will be on individual condition and wellbeing and only distantly on future reproductive success, there seems to be little hope of empirically establishing survival value directly. However, at least in the case of for-

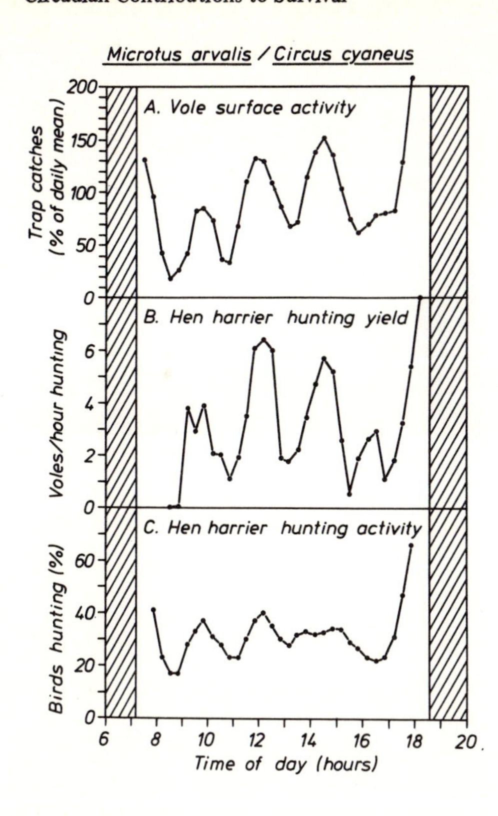

Fig. 4. Synchronous short-term variations in **A** common voles *(Microtus arvalis)* trapped and **B** yield and **C** activity of hen harriers *(Circus cyaneus)* hunting for voles. Based on data collected simultaneously during three days of observation. (Unpubl. data Zoology Dept. Groningen)

aging behaviour we may attempt to estimate effects of temporal organization on the daily balance of energy as an intermediate step. This section is an attempt to work this out for one particular case.

The daily distribution of foraging activity of animals can be adjusted to the daily pattern of food availability. As an example, Fig. 4 shows the daily changes of hunting activity in hen harriers *(Circus cyaneus)* over a large lucerne field in relation to the surface activity patterns of their prey, the common vole *(Microtus arvalis)* measured simultaneously in the same field. Common voles have an ultradian, ca. 2-h rhythm in feeding, and such meal cycles run more or less in synchrony in the field population (Daan and Slopsema 1978). Since the voles retreat in burrows between meals, peaks and troughs in prey availability for the harriers alternate periodically (Fig. 4a). Their hunting yield, expressed in the numbers of voles caught per hour of hunting, follows this periodicity closely (Fig. 4B). With randomly distributed hunting sessions, we would expect more hunting during vole trough times, when longer sessions are needed per catch. In contrast, we observed that slightly more birds were hunting during the vole peaks (Fig. 4C), suggesting that the tendency to hunt was actively increased at times of high expected yield.

Such temporal adjustment of hunting behaviour is presumed to have a circadian basis. In an other predator, the European kestrel *(Falco tinnunculus)* which hunts for the same prey, we have recently obtained field evidence (Rijnsdorp et al. 1981) supporting that 24 h after prey capture there is an increased tendency to flight-hunt and also to

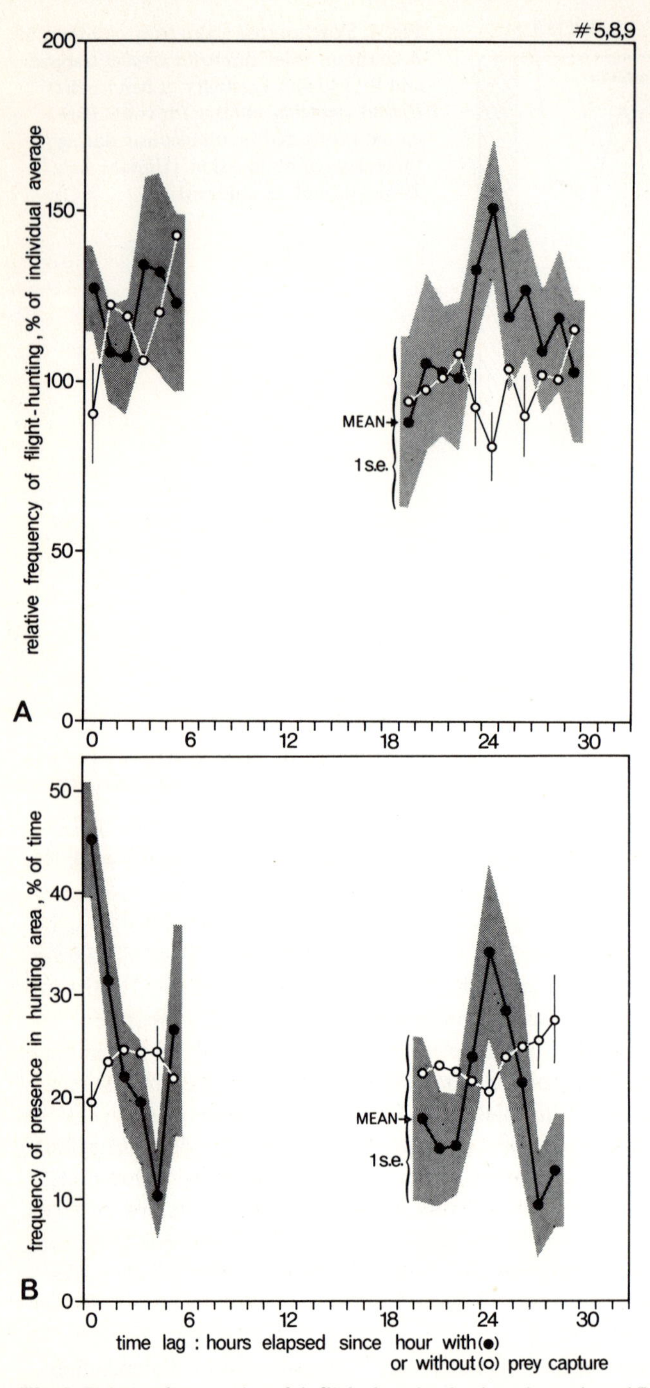

Fig. 5. Relative frequencies of **A** flight-hunting in three kestrels and **B** presence in a specific area of the home range in one kestrel *(Falco tinnunculus)* for observation hours with *(filled circles)* and without *(open circles)* prey capture t hours ago, plotted as a function of time lag t. (Rijnsdorp et al. 1981). Note the increased frequencies of flight-hunting **A** and of presence in the area of prey capture **B** 24 h after prey capture. (Rijnsdorp et al. 1981)

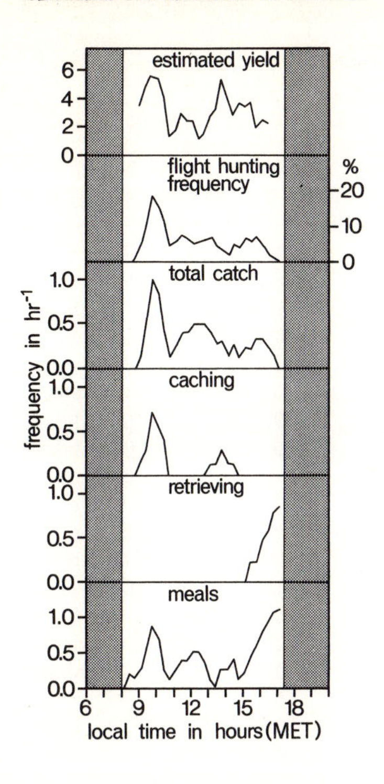

Fig. 6. Summary data for the daily variation in aspects of foraging behaviour in a kestrel *(Falco tinnunculus)*. Curves connect 3-pt running means of 20 min values and are commented in text. (Rijnsdorp et al. 1981)

return to the place capture (Fig. 5). The birds are most probably making use of mechanisms much as the time memory of honeybees, which is based on an endogenous circadian clock (Beier and Lindauer 1970), and like the SCN-mediated 24-h memory retention of rats (Stephan and Kovacevic 1978). For the kestrel, it is possible to obtain a crude estimate of the energetic savings incurred by thus adjusting their temporal distribution of hunting to times of day when hunting yield is expected to be high.

The daily behaviour patterns associated with food intake are summarized for one individual kestrel in Fig. 6. Estimated hunting yield fluctuated in a 2-h rhythm as in the hen harriers (Fig. 4B), and by adjusting her flight-hunting frequency, this kestrel exploited the first morning peak of vole surface activity around 1000 h. At this time of day most catches were made, but some of the prey were cached away and stored for retrieval in the evening. So that the pattern of food intake had a peak around dusk. Apparently the caching-retrieving behaviour enabled the bird to optimize the daily times of hunting and eating separately. In Table 1, we attempt to express the benefits derived from this organization in a common unit of energy saved, on the basis of field data in two individuals presented by Rijnsdorp et al. (1981). The flight-hunting time they required to catch one vole was 0.33 and 0.29 hours in these two birds respectively. In both, average hunting time per vole obtained by weighing data from each 20'-interval of the day equally, were slightly higher: 0.44 and 0.37 h/vole. From these figures it can be deduced how much time the birds would have had to spend in randomly timed flight-hunt to obtain the same amount of food per day. The time thus saved was 0.39 and 0.16 h. Such savings seem very small, yet amount to 25% and 22% of the total flight-hunting time. Since flying is the most expensive item on the birds' daily energy budget, the ener-

Table 1. Estimation of energetic savings by daily adjustment of hunting and eating times in two freeliving kestrels, based on observations by Rijnsdorp et al. (1981)

	Kestrel sex and registration Observation dates	♂ No 9 Dec. 7-23, 1977	♀ No 5 Jan. 17-24, 1978	
a	Total time observed:	115.10	67.17	h
b	Length of active day:	8.28	9.12	h
c	Flight hunt per day:	1.16	0.57	h
d	Other flight per day:	0.74	0.28	h
e	Catch per day (including bird prey in vole equivalents);	6.69	3.73	voles
f	Catch per day from flight hunt:	3.53	1.97	voles
g	Flight hunting time per vole (= c : f)	0.33	0.29	h/vole
h	Estimated 20' average flight hunting time per vole:	0.44	0.37	h/vole
i	Random hunting: time to catch f voles:	1.55	0.73	h
j	Hunting time saved (= i-c):	0.39	0.16	h
k	Body mass, mean for sex and month:	217	242	g
l	Wingspan, mean for sex and month:	74.5	77.6	cm
m	Minimum flight cost, calculated from k, l (Pennycuick 1975):	41.4	46.4	kJ/h
n	Saved in flight-hunt energy (= m.j):	16.1	7.4	kJ
o	Prey eaten, not caught in last 2 h of day:	12	9	voles
p	Meals shifted to evening per day (= o.b/a):	0.86	1.22	voles
q	Estimated weight reduction by shifted meals (= 7.5 p):	6.4	9.2	g
r	Flight cost reduction by shifted meals:	2.4	2.4	kJ/h
s	Saved in flight per day (= r. (c + d)):	4.6	2.0	kJ
t	Weight of meals shifted to evening per day (= p.15):	12.9	18.3	g
u	Specific dynamic action of meals shifted (= 0.741.t):	9.6	13.6	kJ
v	Total daily savings: minimum (n + s), maximum (n + s + u):	21-30	9-23	kJ

getic savings are not negligible. Their value was expressed in kilojoules, making use of Pennycuick's (1975) flight-cost model and of average measurements of body mass and wingspan, obtained in the months (December, January) when birds were observed (Daan et al. 1982). The values obtained (16.1 and 7.4 kJ saved per day by reducing hunting time) were not heavily dependent on the particular model used, since calculations using two other models gave similar results.

The daily organization of food intake, by caching and retrieval, contributes further to the energy balance in two ways (Rijnsdorp et al. 1981): body weight is kept lower during the day then during the night, reducing the cost of flying, and the extra heat dissipated in processing and digestion of the meal (specific dynamic action (SDA), may be partly lost in daytime meals but exploitable for thermoregulation below the thermoneutral zone at night. The cost reduction of daytime flight (flight hunting + directional flight) was estimated at 4.6 and 2.0 kJ/day, again using Pennycuick's model. The specific dynamic action of the meals shifted to the last 2 h of the day was calculated on the basis of SDA measurements in captive kestrels obtained by J. de Vries (unpublished). These figures (9.6 and 13.6 kJ) present a maximum estimate for the energy saved by meal-shifting, since the conditions in which this energy may or may not be exploited for thermoregulation have not been precisely defined. Remaining on the safe side, we

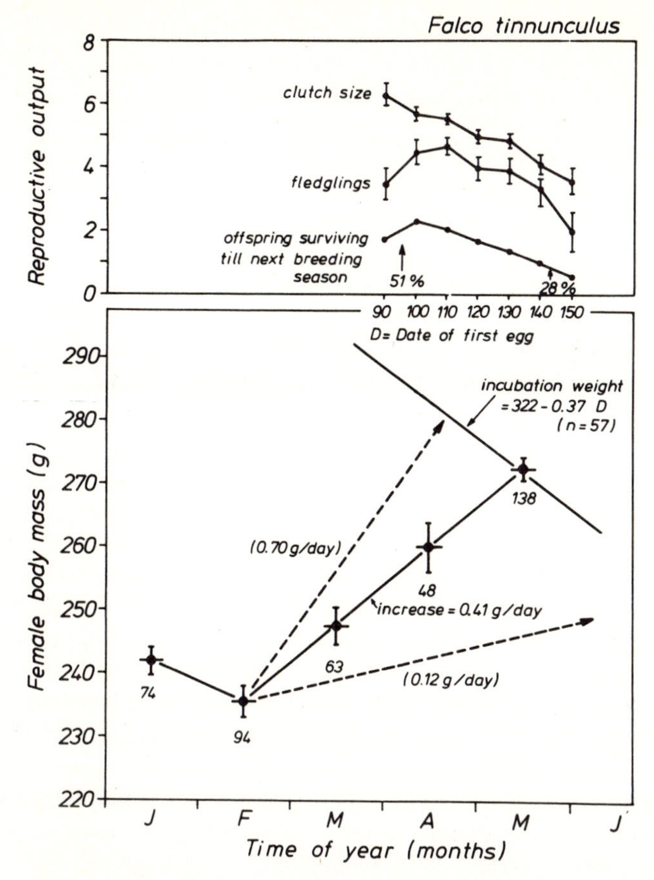

Fig. 7. Scheme to illustrate the relation between female body mass increase in kestrels **(Falco tinnunculus)** in spring and their production of offspring surviving till breeding.
Lower panel Average monthly body mass of wild caught female kestrels (± 1 S.E.M., numbers of birds shown), and linear regression of body weights during incubation on laying date of the first egg *(D* days since December 31). *Dashed lines* show mass increases 0.29 g/day steeper or less steep than average increase. (Data from Daan et al. 1982). *Upper panel* Average numbers (± 1 S.E.M.) of clutches laid and of young fledged by females starting to lay on different days *(D)* of the year and estimated number of offspring surviving till next breeding season based on ring recoveries (unpubl.)

conclude that altogether the daily organization of hunting and eating, as far as deviating from randomness, would give the birds a premium in the order of at least about 10 kJ per day on their daily balance of energy, although the actucal figure may have been more likely in the order of 20 kJ.

This is certainly a small item, since the total daily metabolizable energy is in the order of 300 kJ, according to preliminary calculations based on food intake in the wild, estimations of energetic equivalents of activity and gas exchange measurements (J. de Vries unpubl.). Yet it is possible to show that similar small energetic savings may contribute significantly to the number of offspring eventually propagated.

During springtime female kestrels increase considerably in body mass, as do other species of raptors (Newton 1979). Fig. 7 (lower panel) shows that the average increase from February till May is 0.41 grams per day, based on body weights taken from wild caught birds (Daan et al. 1982). Some birds increase rapidly in weight, others more slowly, and such variations result in variations in their date of laying. Heavy females lay earlier, light-weights later in the year, as shown by a significant negative slope in the regression of body weights during incubation and laying date (Fig. 7). This variation has considerable impact on the number of offsping produced. Early breeders produce large, late breeders small clutches of eggs (Cavé 1968). Survival in the nest may be somewhat less in the extremely early birds, but on the whole there is a trend towards fewer nestlings fledged with proceeding laying date. Of crucial importance is not how many young are produced but how many survive and enter the next breeding generation. This fraction was determined for different laying dates from ring recoveries, as the percentage of recovered birds which died before reaching 1 year of age. More than half of the early young survive to breed, against only 28% of the late born young. By breeding a month earlier, a female's expectation of suviving offspring may roughly double (Fig. 7, upper panel), and we have indications that his figure underestimates the real effect by bias involved in using ring recoveries. That the daily balance of energy is involved in affecting laying date and hence production of surviving young is supported by field experiments with surplus food (preliminary data reported in Drent and Daan 1980). Extra food given to females in spring led to a reduction in hunting effort, a steeper increase in body weight, advanced laying dates and larger clutch size. These experiments (to be published extensively elsewhere) suggest that the rate of energy storage in spring may be crucial in determining onset of breeding. The average increase in body weight accomplished in the population was 0.41 g/day. The energetic cost involved in the storage of body reserves was recently measured accurately in the kestrel as being 34.7 kJ/g (Kirkwood 1981). Thus the population weight data would point to an average of 14 kJ per day available for storage. This is quite close to our minimum of 10 kJ saved by circadian organized of foraging. In fact, with an increase of only 0.12 gram body weight (= 14-10/34.7) from February onwards they would not reach breeding condition until July (according to Fig. 7), when the breeding season is over. A 10 kJ or 0.29 g/day steeper increase than the population weights show, would correspond to almost a month earlier breeding and doubling of the reproductive output (cf. the 0.70 g/d – line in Fig. 7).

This exercise was done to demonstrate, not that circadian organization precisely doubles a bird's reproductive output, but generally that even small energetic savings, hardly detectable on a day's balance, may eventually have strong repercussions for the chances of individual survival. We made this calculation only for one sex over three months of the year, and for one aspect – foraging – of a whole daily behavioral repertoire. The point emerging is that by the recurrence day after day of small circadian contributions to energy retention the probability of producing viable offspring may be strongly affected in the long run. There are many other circadian aspects to behaviour and physiology, here neglected and perhaps less easy to quantify, such as variations in metabolic rate, heat conductance, preening behaviour etcetera. Together they may indeed have effects of such significance on an animal's performance and condition that we may expect that absence of circadian organization should be wiped out by selection in a single generation. It is left to field experiments with artificially clockless animals to demonstrate the strength of such selective pressures.

3.3 A Functional View of Circadian Phenomena

If there is a high selective premium on circadian patterns of behaviour, as suggested by two avian examples, it is appropriate to search for functional meaning in the properties of endogenous oscillators which underlie these patterns. There are several general features in vertebrate circadian systems, which probably ask for understanding in terms of survival value as well as for physiological insight.

Most troublesome to a functional approach is the considerable precision displayed by circadian oscillations in constant conditions, especially in the onset of spontaneous activity of higher vertebrates (Pittendrigh and Daan 1976a). Such precision is hardly present in the daily environmental variations which have been identified as selective factors. Although some of the pacemaker precision may be lost in the entrainment process, the accuracy of phase-angle differences with natural daylight (Daan and Aschoff 1975) is much greater than conceivable in daily patterns in predation pressure or food availability. The animal's need for maintaining a very stable and precise phase with respect to the light-dark cycle was axiomatic in Pittendrigh and Daan's (1976b) functional analysis of circadian properties. Why this should be so remains, however, unclear.

A major impediment is that our knowledge of circadian systems is largely based on spontaneous locomotor activity, especially wheel-running, which in itself is poorly understood. Functional hypotheses on circadian organization have been concerned mainly with foraging behaviour, as in the case of the kestrel discussed above. Evidence is now accumulating that circadian oscillators involved in feeding are somewhat independent from those in locomotor activity rhythms (see Chap. 1.2, this Vol.). In rodents they can persist after ablation of the suprachiasmatic nuclei, responsible for spontaneous activity rhythms (see Chap. 3.7, this Vol.). In starlings a circadian feeding rhythm persists in LL of intensities high enough to abolish the activity rhythm (Gänshirt, unpubl.). In restricted food schedules, locomotor activity may be associated with the anticipatory expectation of food (cf. Chap. 1.2, this Vol.), and indeed sometimes be oriented towards the site where food will come (Daan and Koene 1981). In ad lib feeding conditions, however, spontaneous locomotor activity does not usually seem to reflect drives as involved in natural feeding behaviour.

The distribution of spontaneous locomotor activity in small rodents and songbirds – species on which most of the circadian analyses were based – suggests that some function is served especially early in the active phase. Perch-hopping activity in caged songbirds is concentrated early in the day, often long before food intake starts. Wheel-running in nocturnal rodents is concentrated early in the night, while feeding is more regularly spread through the night or even throughout the entire 24 h (as in voles: Lehmann 1976, Daan and Slopsema 1978). Songbirds in nature begin their day quite commonly with territorial advertisement. By removal of great tits *(Parus major)* from their nest-boxes, Kacelnik and Krebs (1982) have elegantly shown that the risk of losing unadvertised territories was greatest early in the morning. If perch-hopping activity in caged birds somehow reflects motivational states as involved in the "dawn chorus", its circa-

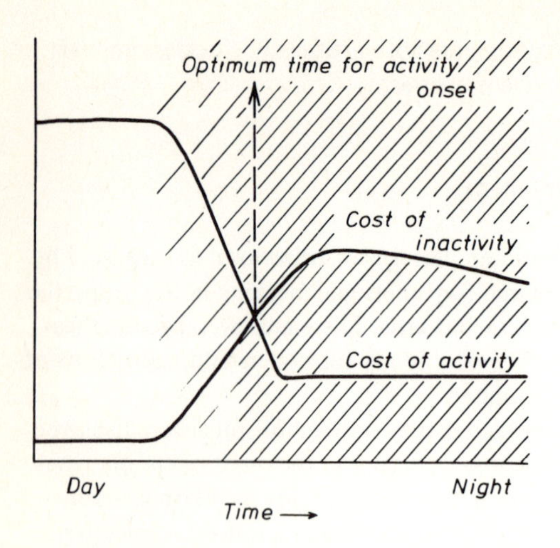

Fig. 8. Hypothetical scheme for the evolutionary costs of a small nocturnal mammal of being active or inactive as a function of time around sunset. The hypothesis assumes decreasing costs of activity due to reduced predatory risk with darkness and increasing costs of inactivity related to endangered territorial status. Optimal activity onset would be at the cross-over time for the two curves

dian organization may become understood. Large positive phase-angle differences, i.e., early activity onsets, are found early especially in the breeding season, both in natural conditions (Nice 1943) and in captivity (e.g. Aschoff 1969). The advanced onset of spontaneous perch-hopping activity may be directly involved in guaranteeing early advertisement of territorial occupancy at times of year and of day when the risk of losing territories to invaders and perhaps the chance of attracting an unmated partner are both large. The selective value of dawn singing in birds has also been associated with the fact that the broadcast area of sounds is greater in the morning than at other times of day, due to atmospheric attenuation mechanisms (Henwood and Fabrick 1979).

Small rodents are organized in more complex social structures than territorial songbirds. Yet they use generally restricted home ranges where others than the near kin are excluded. Intraspecific and intersexual communication makes use of olfactory rather than acoustical signals, but parallel temporal strategies may apply. Wheel-running activity in cages at least in some microtine species is associated with scent marking (De Kock and Rohn 1971), and may indeed reflect natural running along territorial pathways refreshing chemical sign posts in the early night. Dusk is the peak time for agonistic encounters as well as reproductive interactions in free-living rats (Robitaille and Bovet 1976) and similar daily rhythms in agonistic behaviour have been found in various small rodents in laboratory conditions (e.g. Landau 1975). It is then likely that a late start in the night (for nocturnal small rodents) or in daytime (for diurnal songbirds) would increase the risk of losing a territorial position or a chance to mate.

If the evolutionary costs associated with spontaneous activity for a nocturnal rodent could be measured, we would expect a rapid decline in these around dusk, when diurnal predation (Fig. 4) fades out. Simultaneously, a rapid rise in the costs of being inactive would be the consequence of intraspecific competition for space and mates. A well-defined intersection point (Fig. 8) would represent the optimum time for the onset of activity. There is, unfortunately, too little knowledge about detailed behaviour patterns of small mammals in nature and about the costs and benefits associated with these.

Also for the animal, these costs and benefits involved in the timing of activity onset are probably not easily maesurable, in contrast with the possibility it has to assess continuously the quality and availability of its food resources. This may be why spontaneous locomotor activity is often programmed rather precisely and rigidly, while circadian feeding systems tend to be flexible and adjustable to experience.

If indeed the evolutionary urge to start activity early is largely related to intraspecific competition, then conditions generally enhancing activity should shift the optimal phase to an earlier time. It is then from a functional point of view interesting to note that the free-running period (τ) of circadian activity rhythms is virtually always negatively correlated with the amount of activity and in fact with metabolic rate. (Spontaneous changes in α and τ may represent an exception, see Chap. 4.1, this Vol.). This is of course well known in the case of the circadian rule (Aschoff 1960, 1979): both diurnal and nocturnal animals have longer τ when α is shorter and activity is less due to changes in light intensity. Several hormones affect τ and α with opposite sign (see Chap. 5.1, this Vol.). Reductions in metabolic state are likewise associated with reduced activity and lengthened τ (as suggested in free-living beavers in winter: Potvin and Bovet 1975, and hibernators: Daan 1973). Apart from its special interpretation in the case of light effects on τ (Pittendrigh and Daan 1976b), it is possible that conditions enhancing activity generally would shift the optimum time for activity onset slightly forward, and that a shorter free-running τ reflects the mechanism by which this is accomplished.

This is a highly speculative set of suggestions. They cannot replace empirical fact, but should serve to urge for a deeper comparative analysis in species where both the physiological basis of temporal organization and the evolutionary forces shaping this can be studied.

References

Adolph EF (1949) Quantitative relations in the physiological constituents of mammals. Science 109: 579-585

Allison T, Gerber SD, Breelove SM, Dryden GL (1977) A behavioral and polygraphic study of sleep in the shrews *Suncus muriuns, Blarina brevicauda* and *Cryptotis parva.* Behav Biol 20: 354-366

Aschoff J (1960) Exogenous and endogenous components in circadian rhythms. Cold Spring Harbor Symp Quant Biol 25: 11-28

Aschoff J (1969) Phasenlage der Tagesperiodik in Abhängigkeit von Jahreszeit und Breitengrad. Oecologia 3: 125-165

Aschoff J (1979) Circadian rhythms: Influence of internal and external factors on the period measured in constant conditions. Z Tierpsychol 49: 225-249

Aschoff J (ed) (1981) Biological rhythms. Handbook of behavioral neurobiology, vol IV. Plenum Press, New York

Aschoff J, Meyer-Lohmann J (1954) Die 24-Stunden-Periodik von Nagern im natürlichen und künstlichen Belichtungswechsel. Z Tierpsychol 11: 476-484

Aschoff J, Wever R (1961) Biologische Rhythmen und Regelung. Bad Oeynhauser Gespr 5: 1-15

Aschoff J, Saint Paul von U, Wever R (1971) Die Lebensdauer von Fliegen unter Einfluß von Zeitverschiebungen. Naturwissenschaften 58: 574-575

Beier W, Lindauer M (1970) Der Sonnenstand als Zeitgeber für die Biene. Apidologie 1: 5-28

Bovet J, Oertli EF (1974) Free-running circadian activity rhythms in freeliving beaver (Castor canadensis). J Comp Physiol 92: 1-10

Cavé AJ (1968) The breeding of the kestrel, *Falco tinnunculus,* in the reclaimed area of Oostelijk Flevoland. Neth J Zool 18: 313-407

Chapin JP, Wing LW (1959) The wideawake calender, 1953 to 1958. Auk 76: 152-158

Corbet GB, Southern HN (eds) (1977) The handbook of British mammals, 2nd edn. Blackwell, Oxford

Daan S (1973) Periodicity of heterothermy in the Garden Dormouse, *Eliomys quercinus* (L.). Neth J Zool 23: 237-265

Daan S. Aschoff J (1975) Circadian rhythms of locomotor activity in captive birds and mammals: Their variations with season and latitude. Oecologia 18: 269-316

Daan S, Koene P (1981) On the timing of foraging flights of oystercatchers *Haematopus ostralegus* on tidal mudflats. Neth J Sea Res 15: 1-22

Daan S, Pittendrigh CS (1976) A functional analysis of circadian pacemakers in nocturnal rodents. II. The variability of phase response curves. J Comp Physiol 106: 252-266

Daan S, Slopsema S (1978) Short-term rhythms in foraging behaviour of the common vole, *(Microtus arvalis)*. J Comp Physiol 127: 215-227

Daan S, Tinbergen JM (1980) Young guillemots *(Uria lomvia)* leaving their arctic breeding cliffs. A daily rhythm in numbers and risk. Ardea 67: 96-100

Daan S, Cavé AJ, Dijkstra C (1982) Body mass and size in relation to breeding behaviour in the kestrel *(Falco tinnunculus (L.))*. Ardea (in press)

Darling FF (1938) Bird flocks and the breeding cycle: a contribution to the study of avian sociality. Cambridge Univ Press, Cambridge

Drent RH, Daan S (1981) The prudent parent: Energetic adjustments in avian breeding. Ardea 68: 225-252

Emlen ST, DeMong NJ (1975) Adaptive significance of synchronized breeding in a colonial bird: a new hypothesis. Science 188: 1029-1031

Greenwood J (1964) The fledging of the guillemot *Uria aalge* with notes on the razor bill *Alca torda*. Ibis 106: 469-481

Henwood K, Fabrick A (1969) A quantitative analysis of dawn chorus: temporal selection for communicatory optimization. Am Nat 114: 260-274

Heusner AA (1982) Energy metabolism and body size. I: Is the 0.75 mass exponent of Kleibers equation a statistical artifact? Respir Physiol 48: 1-12

Hilfenhaus M (1976) Circadian rhythm of plasma renin activity, plasma aldosterone and plasma corticosterone in rats. Int J Chronobiol 3: 213-219

Hill AV (1950) The dimensions of animals and their muscular dynamics. Sci Prog 38: 209-230

Honma K, Hiroshige T (1978) Simultaneous determination of circadian rhythms of locomotor activity and body temperature in the rat. Jpn J Physiol 28: 159-169

Jegla TC, Poulson TL (1968) Evidence of circadian rhythms in crayfish. J Exp Zool 168: 273-282

Kacelnik A, Krebs JR (1982) The dawn chorus in the great tit *(Parus major)*: a causal and functional analysis Behaviour (in press)

Kirkwood JK (1981) Bioenergetics and growth in the kestrel, *(Falco tinnunculus)*. Ph D thesis, Univ Bristol

Kleiber M (1961) The fire of life. John Wiley and Sons, New York

Kock De LL, Rohn I (1971) Observations on the use of the exercise-wheel in relation to the social rank and hormonal conditions in the bank vole *(Clethrionomys glareolus)*, and the Norway lemming *(Lemmus lemmus)*. Z Tierpsychol 29: 180-195

Landau IT (1975) Light dark rhythms in aggressive behavior of the male golden hamster. Physiol Behav 14: 767-774

Lehmann U (1976) Short-term and circadian rhythms in the behaviour of the vole, *(Microtus agrestis (L.))*. Oecologia 23: 185-199

Lindstedt SL, Calder WA (1981) Body size, physiological time, and longevity of homeothermic animals. Q Rev Biol 56: 1-16

Neumann D (1976) Adaptations of chironomids to intertidal environments. Annu Rev Entomol 21: 387-414

Newton I (1979) Population ecology of raptors. Poyser, Berkhamsted

Nice M (1943) Studies in the life history of the song sparrow II *(Melospiza melodia)*. Trans Linn Soc N Y 6: 1-328

Nishio T, Shiosaka S, Nakagawa H, Sakumoto T, Sato K (1979) Circadian feeding rhythm after hypothalamic knive-cut isolating suprachiasmatic nucleus. Physiol Behav 23: 763-769

Pennycuick CJ (1975) Mechanics of flight. In: Farner DS, King JR (eds) Avian biology, vol V. Academic Press, London New York, pp 1-75

Peret J, Macaire I, Chanez M (1973) Schedule of protein ingestion, nitrogen and energy utilization and circadian rhythm of hepatic glycogen, plasma corticosterone and insulin in rats. J Nutr 103: 866-874

Pittendrigh CS (1958) Perspectives in the study of biological clocks. In: Buzzati–Traverso AA (ed) Perspectives in marine biology. Univ Calif Press, Berkeley, pp 239-268

Pittendrigh CS, Daan S (1976a) A functional analysis of circadian pacemakers in nocturnal rodents. I The stability and lability of spontaneous frequency. J Comp Physiol 106: 223-252

Pittendrigh CS, Daan S (1976b) A functional analysis of circadian pacemakers in nocturnal rodents. IV; Entrainment: pacemaker as clock. J Comp Physiol 106: 291-331

Pittendrigh CS, Skopik SD (1970) Circadian systems V. The driving oscillation and the temporal sequence of development. Proc Natl. Acad Sci USA 65: 500-507

Potvin CL, Bovet J (1975) Annual cycle of patterns of activity rhythms in beaver colonies *(Castor canadensis)* J Comp Physiol 98: 243-256

Remmert H (1962) Der Schlüpfrhythmus der Insekten. Steiner, Wiesbaden

Rijnsdorp A, Daan S, Dijkstra C (1981) Hunting in the kestrel, *(Falco tinnunculus)*, and the adaptive significance of daily habits. Oecologia 50: 391-406

Robitaille JA, Bovet J (1976) Field observations on the social behaviour of the Norway rat, *Rattus norvegicus* (Berkenhout). Biol Behav 1: 289-308

Rusak B (1982) Neural control and functional aspects of Mammalian behavioral rhythms. In: Loher W (ed) Behavioral expressions of biological rhythms. Garland, New York (in press)

Scheving LE, Pauly JE (1967) Effect of adrenalectomy, adrenal medullectomy and hypophysectomy on the daily mitotic rhythm in the corenal epithelium of the rat. In: Mayersbach H von (ed) The cellular aspects of biorhythms. Springer, Berlin Heidelberg New York, pp 167-174

Stahl WR (1962) Similarity and dimensional methods in biology. Science 137: 205-212

Stephan FK, Kovacevic NS (1978) Multiple retention deficits in passive avoidance in rats is eliminated by SCN lesions. Behav Biol 22: 456-462

Thinès G, Wolff F, Boucqney C, Soffie M (1965) Etude comparative de l'activité du poisson cavernicole *Anoptichthys antrobius* Alvarez, et de son ancêtre epigé *Astyanax mexicanus* (Fillippi). Ann Soc R Zool Belg 96: 61-116

Western D (1979) Size, life history and ecology in mammals. Afr J Ecol 17: 185-204

Wilkie DR (1977) Metabolism and body size. In: Pedley TJ (ed) Scale effects in animal locomotion. Academic Press, London New York, pp 23-36

Zepelin H, Rechtschaffen A (1974) Mammalian sleep, longevity and energy metabolism. Brain Behav Evol 10: 425-470

8.2 Daily Temporal Organization of Metabolism in Small Mammals: Adaptation and Diversity

G.J. Kenagy and D. Vleck[1]

1 Introduction

An animal's daily schedule of social interaction, acquisition and processing of food, and exposure to the environment comprises a complex suite of cyclic behavioral events. Many of these events are coordinated by an endogenous "circadian clock". Clearly there are ecological advantages to performing particular functions at particular times of day, and it is adaptive for an organism to anticipate its daily schedule by relying on an endogenous temporal program. In addition, it has been pointed out that temporal organization may also have intrinsic merits besides those of ecological adaptation (Pittendrigh 1961, Aschoff 1964, Enright 1970).

The daily course of resting metabolic rate (RMR) in birds is about 25% higher during the normally active phase (α) of the daily cycle (daytime for most species) than during the normal resting phase (ρ) of the daily cycle (Aschoff and Pohl 1970). Furthermore these patterns of metabolism are endogenous because they persist under constant environmental conditions. Birds prepare for the normally active phase of their daily cycle by increasing RMR even when held in constant darkness, which precludes activity in most species. Daily rhythmicity of RMR has not previously been treated systematically in any major group other than birds (Aschoff and Pohl 1970).

The diversity of life styles (diets, activity patterns, environments) among small mammals makes them particularly interesting subjects for an investigation of the adaptive diversity of circadian organization. Measurements of whole-animal energy metabolism sum the energy transactions that occur throughout the body, and consequently the daily course of RMR of any species (measurements on fasting animals at rest in a thermally neutral environment) should reflect a pattern of temporal organization that has evolved along with the life style of that species.

We sought to determine (1) whether divergent patterns of circadian rhythmicity in RMR can be recognized and correlated with the behavior and ecology of species, and (2) to what extent circadian organization of RMR is subject to adaptive modification. An analysis of RMR as a function of body mass as well as time of day is fundamental to answering these questions, because adaptive modification can thereby be identified by comparison to allometric norms. We have undertaken such a comparative investigation in 18 species of small mammals (body mass 8-270 g) of the orders Insectivora and Roden-

1 Department of Zoology, University of Washington, Seattle, Washington 98195 USA

Vertebrate Circadian Systems
Ed. by J. Aschoff, S. Daan, and G. Groos

tia. In order to maximize the comparative power of our analysis, we limit the analysis to data that we have obtained ourselves by means of a single, uniform protocol.

2 Materials and Methods

All animals were collected in the field during 1977-1979, and held in an animal room at 23 ± 2°C with a photocycle of LD 12 : 12 (light on 0600-1800 PST). We recorded rate of oxygen consumption ($\dot{V}_{O_2}$) continuously for at least 30 h on each animal, using an open-flow system and glass metabolism chambers of either 2 or 4 l. Animals were placed on wire mesh over mineral oil in order to isolate feces and urine. During all measurements chambers were continuously dark (DD) and temperatures were within the animals' thermal neutral zones. Chamber temperature ranged from 28° to 32°C depending on the species being tested, but did not vary more than ± 0.2°C during an experiment. Animals had access to water but not food during measurements. We report body mass as the mean of measurements at the start and end of an experiment.

Air was circulated through the metabolism chambers at constant flow rates (measured with Matheson rotameter flowmeters upstream from chambers) that ranged between 0.4 and 1.0 l/min depending on size of animal. Percent oxygen concentration was measured with an Applied Electrochemistry S-3A oxygen analyzer and recorded on a Leeds and Northrup Speedomax chart recorder. Water vapor was removed from the air stream with silica gel and Drierite (anhydrous $CaSO_4$) prior to measurements of air flow and oxygen concentration. Carbon dioxide was absorbed with soda lime and Ascarite (sodium hydroxide-coated asbestos) prior to oxygen analysis. Rates of oxygen consumption were calculated using equation 2 of Hill (1972). Gas volumes are reported at standard temperature (0°C) and pressure (101.3 kPascals). We did not remove CO_2 before measuring flow rates; this results in an overestimate of $\dot{V}_{O_2}$ (Hill 1972) but the error is less than 0.03%.

We estimated hourly minimum resting metabolic rate (RMR) as the minimum $\dot{V}_{O_2}$ maintained continuously for at least 5 min during each h. For each run we analyzed a continuous 24-h segment of data in this fashion, beginning after the animal had been in the respirometer for at least 2 h. We established minimal RMR for the activity phase (α) and for the rest phase (ρ) of the circadian cycle in two ways:

1. *12-h absolute minimum RMR ("extremes" method).* We took the single lowest of the hourly minimal RMR's that occurred during the 12 h of the usual light phase and the lowest within the 12 h of the usual dark phase. These values were used to calculate the $\alpha : \rho$ ratios of RMR in column (a) of Table 1 and are plotted in Fig. 1. Measurements were actually made in DD, so that the division in terms of light and dark is based on the LD 12 : 12 light-dark cycle to which the animals were exposed prior to measurement of $\dot{V}_{O_2}$. The light phase is the normal α of diurnal animals and ρ of nocturnal animals, and the dark phase is the opposite. Because it was not always certain whether observations within 1 h of the normal L-D or D-L transitions represent metabolism in α or in ρ, some of these borderline cases were rejected as 12-h minima.
2. *Twelve-hour average hourly minimum RMR ("averages" method).* We took the average of all 12-hourly RMR's in the light phase and the average of all 12-hourly RMR's in the dark phase. These values correspond to the $\alpha : \rho$ ratios of RMR of column (b) in

Table 1. Day and night minima of resting metabolism in thermal neutrality and their ratios in 18 species of small mammals

Taxa	α	n	Mass (g)	$\dot{V}_{O_2}$ (cm³/h) (12-h extremes) α	ρ	α : ρ Ratio (a) 12-h extremes	(b) 12-h averages
INSECTIVORA Soricidae							
Sorex cinereus	N*	1	7.9	58.4	54.8	1.07	1.07
Talpidae							
Scapanus orarius	N*	1	61.2	65.8	62.4	1.05	1.08
Scapanus townsendii	N*	3	130.1 (34.3)	111.3 (28.9)	106.4 (17.8)	1.05	1.03
RODENTIA Cricetidae-Cricetinae							
Peromyscus crinitus	N	3	13.6 (1.9)	28.7 (2.6)	18.1 (3.7)	1.59	1.49
Peromyscus maniculatus	N	3	16.7 (0.9)	49.9 (7.5)	29.1 (4.0)	1.71	1.51
Neotoma cinerea	N	1	158.1	158.8	120.4	1.32	1.29
Cricetidae-Microtinae							
Microtus longicaudus	N	1	41.4	77.3	70.2	1.10	1.04
Microtus townsendii	N	2	52.2 (0.8)	95.2 (13.4)	85.6 (2.6)	1.11	1.08
Heteromyidae							
Perognathus longimembris	N	2	8.0 (0.6)	17.7 (6.9)	7.1 (2.4)	2.49	1.14
Perognathus parvus	N	3	19.2 (2.7)	37.2 (6.6)	28.8 (6.7)	1.29	1.20
Dipodomys merriami	N	3	43.4 (2.6)	65.5 (26.0)	40.1 (9.0)	1.63	1.47
Dipodomys ordii	N	3	48.8 (9.7)	70.5 (18.3)	57.4 (14.2)	1.23	1.19
Dipodomys deserti	N	4	107.5 (11.2)	108.8 (22.8)	86.6 (17.6)	1.26	1.28
Geomyidae							
Thomomys talpoides	N*	3	82.6 (11.3)	107.3 (24.6)	97.9 (19.5)	1.09	0.96
Sciuridae							
Eutamias minimus	D	3	33.5 (1.4)	90.5 (20.6)	51.3 (10.6)	1.76	1.61
Eutamias amoenus	D	3	52.7 (3.2)	93.2 (14.7)	76.6 (10.4)	1.22	1.21
Ammospermophilus leucurus	D	3	112.8 (2.7)	117.8 (20.4)	98.0 (22.3)	1.20	1.33
Spermophilus saturatus	D						
Spring		3	252.2 (6.2)	178.6 (20.6)	163.0 (26.4)	1.10	1.14
Winter		3	270.1 (51.5)	116.3 (5.7)	122.1 (13.5)	0.95	0.99
Mean						1.33	1.22
SD						0.37	0.19

Under α in second column are indicated species with diurnal (D) or nocturnal (N) activity patterns; * indicates arbitrary assignment to (N) because a distinctive difference between activity (α) and rest (ρ) phases of the day was not apparent. n = number of individuals of each species tested. Parentheses contain standard deviations. The values of resting oxygen consumption ($\dot{V}_{O_2}$) shown are the absolute minimum hourly values during α and ρ. The ratio of these two values is given in column (a) (12-h extremes) under α : ρ ratio. The ratio in column (b) represents the average of all 12 hourly minima of $\dot{V}_{O_2}$ in α and ρ respectively. All data are for normothermic, fasting animals, measured at seasons of homeothermy with the following three exceptions: *Sorex cinereus* received ad lib food while in the metabolic chamber; *Perognathus longimembris* became torpid briefly in the daytime; and the same three *Spermophilus saturatus (= lateralis)* were measured initially during the end of the hibernation season ("winter") and later during the season of strict normothermy ("spring").

Table 1. This analysis produced values of minimal RMR equivalent to the areas under the curves during the light phase and during the dark phase, as seen in Figs. 2-5.

3 Results and Analysis

3.1 General Relationships and Allometry of RMR in α and ρ

We examined 24-h endogenous patterns of metabolic rate in 18 species of small mammals and compared minimal resting rates during the times of normal light phase (12 h) and dark phase (12 h) (Table 1). Hourly minima during the usual 12-h phase of activity (α) were higher than hourly minima during the usually inactive 12-h phase (ρ). We calculated ratios of (1) the minimum of all 12-hourly minima of $\dot{V}O_2$ during α to that during ρ and (2) the average of all 12-hourly minima during α to an average during ρ. The $\alpha : \rho$ ratios of RMR calculated by both methods were remarkably similar in nearly all individual cases, and the mean ratios did not differ significantly (paired t-test; $p > 0.1$). The $\alpha : \rho$ metabolic ratio for the absolute minima of *Perognathus longimembris* was more than twice the ratio of average hourly minima; but as previously pointed out (Bartholomew and Cade 1957), this is due to the occurrence of torpor during ρ in this small hibernating rodent and should therefore not be compared with ratios for normothermic animals.

The resting metabolic rate of all 16 species in Fig. 1 taken together averages 28% greater during the α phase than in ρ, and the $\alpha : \rho$ ratio of RMR does not vary significantly with body mass. Regressions of metabolic rate on body mass are represented by the following equations:

$$\text{for } \alpha: \quad \dot{V}O_2 = 480\ M^{0.61} \tag{1}$$

$$\text{for } \rho: \quad \dot{V}O_2 = 376\ M^{0.61} \tag{2}$$

where $\dot{V}O_2$ = metabolic rate as oxygen consumption in cm^3/h and M = body mass in kg. We originally determined that the slope of the α regression was 0.55 and the slope of the ρ regression was 0.68. Using analysis of covariance (Dunn and Clark 1974) we found, however, that these slopes were not significantly different ($F = 2.404$, $p > 0.05$), and therefore we combined all of the values of α and ρ to obtain the best estimate of a common slope, 0.61 which is shown in equations (1) and (2) and Fig. 1. Because the slopes of α and ρ regressions of RMR on body mass are not significantly different, we conclude that the $\alpha : \rho$ ratio of RMR does not vary significantly with body mass among small mammals weighing less then 300 g. In contrast to the indistinguishable slopes of the α and ρ regressions, the intercepts differ. The α intercept is significantly higher, by 28%, than that of the ρ regression ($F = 12.415$, $p < 0.01$). This 28% average greater RMR in α than in ρ is merely a general "norm" that results from analysis of a diversity of species (Fig. 1). In light of the following Section 3.2, Diversity of Metabolic Rhythmicity, where we account for the $\alpha : \rho$ relationship of RMR in different species in terms of their behavior and ecology, one should be cautioned against using the average $\alpha : \rho$ value for RMR of 1.28 as a prediction of day-night rhythmicity of RMR for single species.

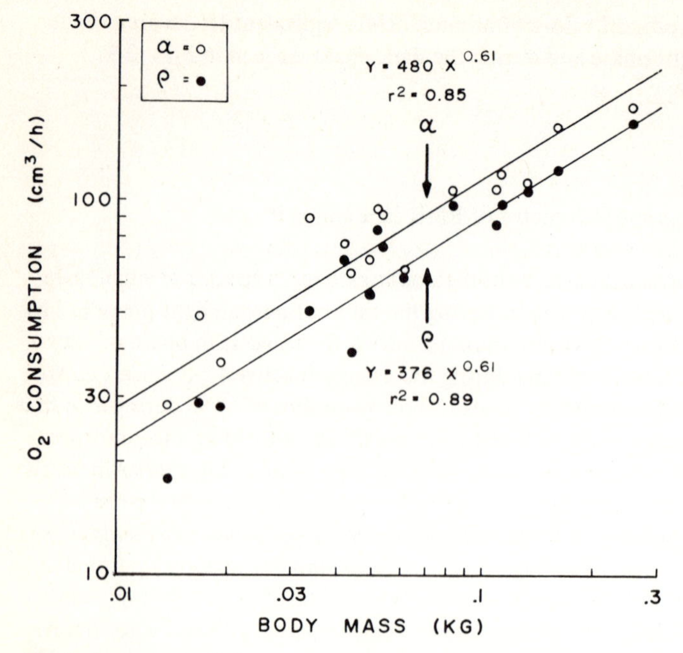

Fig. 1. Resting metabolic rate of small mammals in relation to body mass and time of day. *Open circles* represent minimum $\dot{V}_{O_2}$ maintained for at least 5 min during the α phase of the circadian cycle and closed circles represent minimum $\dot{V}_{O_2}$ during the ρ phase for 16 of the 18 species in Table 1. (Cases are excluded in which animals were not fasting and normothermic during measurement; see text and Table 1)

For convenience and comparison, Eq. (1) and (2) respectively can also be expressed in SI units as follows, assuming the energy equivalence of oxygen consumption to be 20.1 J/cm^3 O_2:

$$\text{For } \alpha: \quad P = 2.67\, M^{0.61} \tag{3}$$

$$\text{For } \rho: \quad P = 2.10\, M^{0.61} \tag{4}$$

where P = metabolic rate (power) in watts and M = body mass in kg. Metabolic rate can be expressed in kcal/24 h by multiplying the coefficients of equations 3 and 4 by 20.64.

3.2. Diversity of Metabolic Rhythmicity

The daily temporal patterns of resting metabolic rate that we observed in small mammals are diverse and can be correlated with species-specific patterns of behavior, including activity and feeding. The species we studied fall into a series of behavioral and ecological categories, each of which is typified by a different pattern of circadian metabolic organization.

Nocturnal Activity. Eight species of nocturnally active rodents of the families Cricetidae (subfamily Cricetinae) and Heteromyidae had levels of resting $\dot{V}_{O_2}$ at night (α) that were substantially greater than those of daytime (ρ), mostly with α about 20%-70% greater than ρ (Table 1). The form of the daily metabolic cycle of these nocturnally active animals indicates a strong temporal segregation of resting metabolic level between night and day (Fig. 2). This pattern reflects an increased level of basal metabolic support for the activities of night, when the animals are intermittently foraging and exposed to thermal conditions of the macroenvironment, and when they may interact socially; the pattern also reflects the stillness of these animals during their daytime of refuge in burrows or nests.

Diurnal Activity. Four species of day-active rodents of the squirrel family all showed a strong daily segregation of metabolic level that corresponds to that of the above nocturnal rodents in that resting metabolic rate in α was about 10%-75% greater than in ρ (Table 1). Within this ecologically and behaviorally similar series of four species, there is a noteworthy trend that the differences between α and ρ level of RMR are greatest in the smallest species and least in the largest. (Recall that this trend was not significant for small mammals all taken together, as shown in Sect. 3.1).

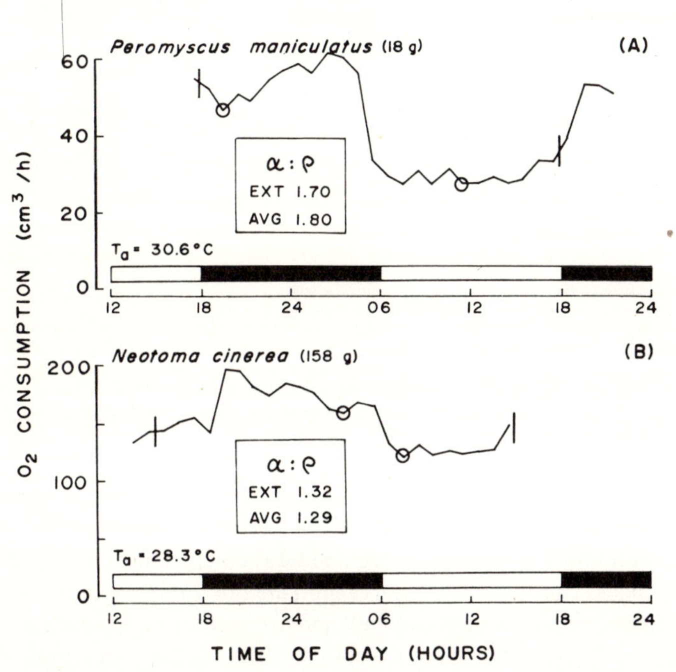

Fig. 2. Daily rhythm of resting metabolic rate in two nocturnal rodents: **A** *Peromyscus maniculatus* (deer mouse) and **B** *Neotoma cinerea* (bushy-tailed wood rat). Data obtained on fasting animals provided with water and held in continuous darkness and at thermal neutrality. Prior to days when oxygen consumption was being measured, the animals were held in a 12-h photoperiod; this photocycle is indicated by *light* and *dark bars* above the time scale. *Lines* connect hourly values of minimal metabolism and the lowest minima in α and ρ are circled. The *two vertical bars* on the curve delimit the 24-h period over which the analysis was performed. The two values in the $\alpha : \rho$ box represent respectively the ratios of the absolute minima ("extreme") and the average of hourly minima ("average") of RMR

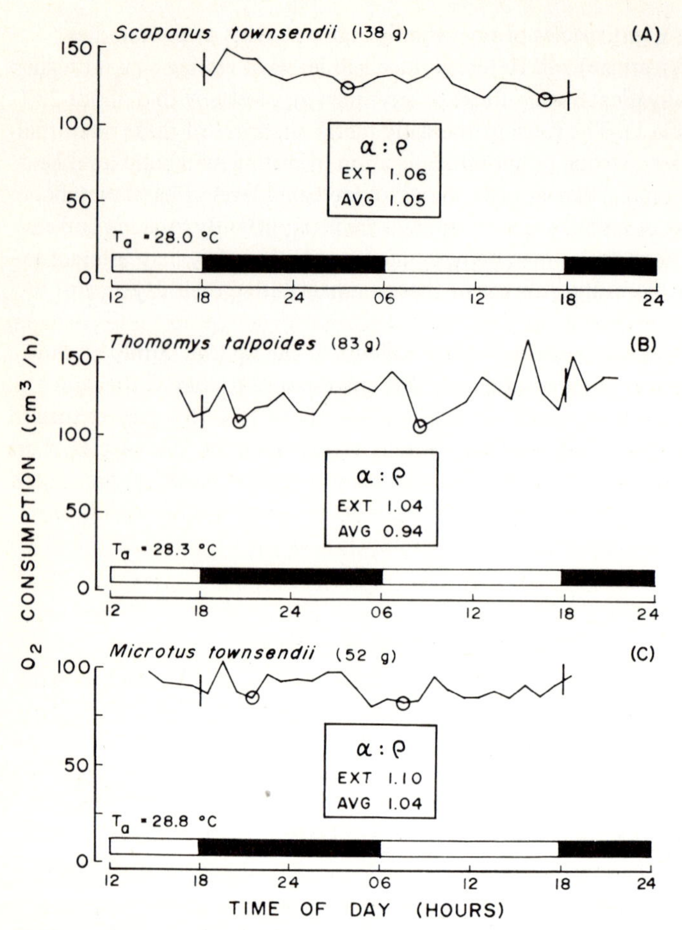

Fig. 3. Daily rhythm of resting metabolic rate in two fossorial species: **A** *Scapanus townsendii* (Townsend's mole) and **B** *Thomomys talpoides* (Northern pocket gopher), and in the herbivorous rodent **C** *Microtus townsendii* (Townsend's vole). Details as in Fig. 2

Fossorial Activity. Continuously fossorial (underground-dwelling) mammals such as moles *(Scapanus* spp.) and pocket gophers *(Thomomys* spp.) are not directly exposed to the daily macroenvironmental cycle of light and dark and other physical parameters such as temperature. They live in an environment that places little constraint upon the phase and timing of their activity. In contrast to the above cases of strong daily temporal segregation of resting metabolic rate in strictly nocturnal or diurnal animals, moles and pocket gophers showed essentially the same level of metabolic rate both day and night (Fig. 3A, B). For two species of moles and one pocket gopher, day and night minimal metabolism generally differed by no more than 5% (Table 1). (Our designation of 12 h night as the α phase for these species was arbitrary). Despite the biological dissimilarities of gophers (herbivores of the order Rodentia) and of moles (members of the order Insectivora that feed on small invertebrates) the fossorial mode of life has apparently led in both cases to a relaxation of the constraints for temporal organization of daily metabolic physiology and given rise to relative temporal uniformity of resting metabolic rate.

Herbivorous, High-Bulk Diet. Microtine rodents (including *Microtus* spp.) are a conspicuously successful group of rodents that consume large volumes of plant foods of high bulk and low energy content; these animals are at times active both day and night in order to maintain this dietary habit (Pearson 1960, review by Erkinaro 1969). Our metabolic measurements show relative uniformity of metabolic level between night and day for *Microtus* spp. (Fig. 3C), with values of minimal metabolism that do not generally differ by more than 10% from day to night (Table 1).

Lower Limit of Mammalian Body Size. Intense and sustained activity is characteristic of shrews (*Sorex* spp., order Insectivora), which are among the smallest of homeothermic animals and prey on small invertebrate animals. They cannot survive for more than a few hours without food and thus – as an exception to our usual experimental protocol – we measured oxygen consumption of a *Sorex cinereus* (masked shrew) continuously for 32 h by providing the animal with food in the metabolic chamber. Minimal nighttime metabolic rate was only greater than that of daytime by 7%; this comparison was identical both for the two single minima of α and ρ respectively and for the average hourly minima of α and ρ over 12 h (Table 1). The intense metabolic requirements and activity patterns of small shrews impose a pattern of temporal organization that has a time component of much shorter duration than the 24-h day-night cycle.

Seasonal Effect on Circadian Organization in a Hibernator. The daily normothermic metabolism of golden-mantled ground squirrels in thermal neutrality ($T_{air} \approx 30°C$) differed in two ways during the season of hibernation as compared to the daily pattern during the active season of the year: (1) overall level of metabolism was lower in the hibernation season, and (2) the amplitude of day-night change was much reduced during winter (Fig. 4). During the season of hibernation the mean ratio of average hourly daytime metabolism to nighttime metabolism was 0.99, indicating a uniformity of "daytime" and "nighttime" levels of metabolism (Table 1). The lack of circadian segregation of metabolic level during the hibernation season correlates with both the dark and thermally stable environment of the winter hibernaculum and with the behavior of the hibernating animals, i.e., there is no functional requirement that the animal show a daily rhythmic alternation between higher and lower levels of metabolism. By contrast, in the active season the daytime minimal metabolic rate was at least 10% greater than that of the nighttime (Table 1). Ratios of $\alpha : \rho$ RMR (based on average hourly minima) for the three animals respectively were 0.96, 0.87 and 1.13 during hibernation season and 1.10, 1.09, and 1.23 during the normothermic season. Thus each individual increased its daytime resting $\dot{V}_{O_2}$ with respect to nighttime $\dot{V}_{O_2}$ as it entered the active season.

Bimodal Peaks of Circadian RMR in Squirrels. In contrast to the form of the circadian rhythm of RMR in other small mammals (Figs. 2, 3, 4), the pattern in some of the squirrels we examined showed two substantial peaks in hourly metabolic rate, one associated with beginning of activity and the other occurring near the end of daily activity (Fig. 5). These peaks are sufficiently high and sustained in *A. leucurus* for the $\alpha : \rho$ metabolic ratio of average hourly minima to be higher than that of single daily minimum of α and ρ (Fig. 5; Table 1). These peaks, for example in Fig. 5, are as great as 2.5 times the RMR at other times of day and night. Probably these animals were never truly resting during these periods; nonetheless each point represents the 5-min value of minimal MR

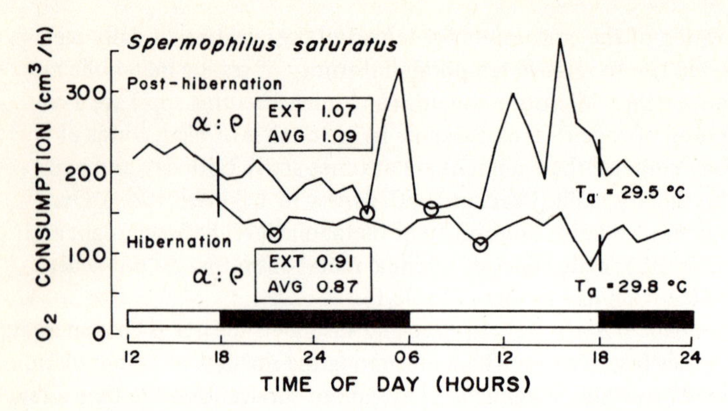

Fig. 4. Daily rhythm of resting metabolic rate in the same individual, normothermic *Spermophilus saturatus (= lateralis)* (Cascades golden-mantled ground squirrel) at two seasons: near end of hibernation season (255 g) and 3 months later, after hibernation season (259 g). Details as in Fig. 2

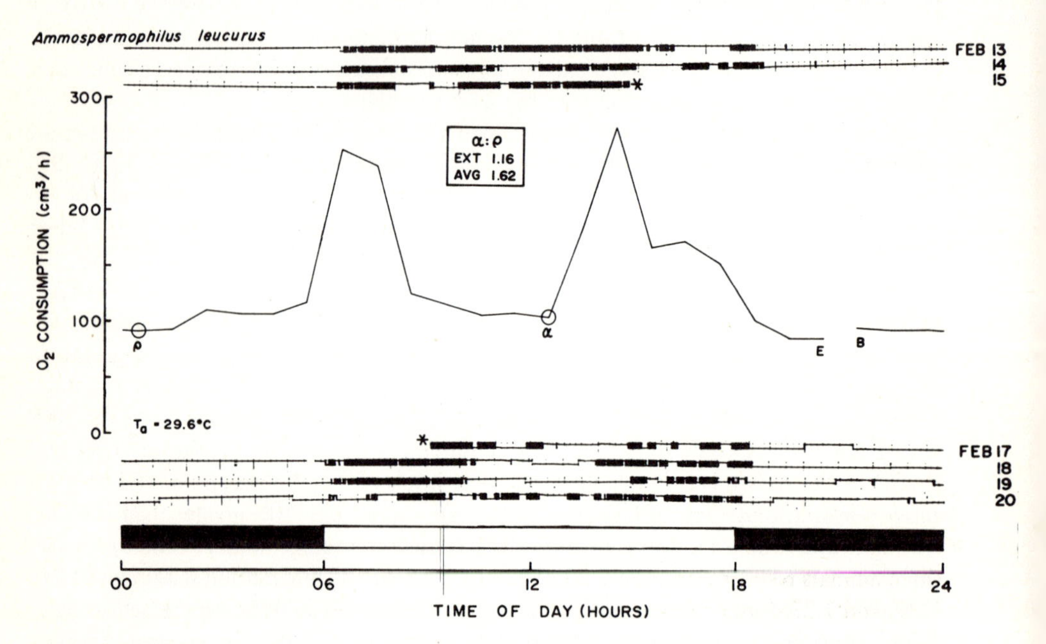

Fig. 5. Daily rhythm of resting metabolic rate for *Ammospermophilus leucurus* (antelope ground squirrel) in continuous darkness. Details as in Fig. 2. Also shown are 3 days of running-wheel activity in LD 12 : 12 recorded by event recorder from the animal before the metabolic measurements and 4 days of activity recorded afterward in LD 12 : 12; the *asterisks* indicate the time of removal of the animal from its running wheel in LD 12 : 12 on 15 February and the return of the animal to its running wheel on 17 February. Metabolic rate is shown beginning 2100 h on 15 February *(B)* and ending 2000 h 16 February *(E)*

within each hour in which oxygen consumption was measured. These peaks are also apparent in the daily metabolism of *S. saturatus* after but not during the season of hibernation (Fig. 4).

These peaks in "RMR" are probably associated with the kind of bimodal peaks in arousal and activity that have been described previously and for which an adaptive basis has been proposed (Aschoff 1966). However, it is important to notice the occurrence of this phenomenon as a possible undesirable effect in laboratory studies of metabolism in which measurements of RMR are sought. *A. leucurus* were maintained under the same conditions as all other animals, except that *A. leucurus* had access to running wheels. Though no running wheels were available during measurements of metabolic rate, peaks in resting metabolic rate occurred at the beginning and end of the usual times of activity as measured in the running wheel (Fig. 5). In such cases of dawn and dusk peaks in $\dot{V}O_2$ the average hourly minimal metabolic rate would give an overestimate of the $\alpha : \rho$ ratio that one would get on the basis of single daily α and ρ minima of hourly RMR.

4 Discussion

4.1 Adaptive Nature of Endogenous Circadian Rhythms of Metabolism

Small Mammals. Twelve species (8 strictly nocturnal and 4 strictly diurnal) out of the 18 we investigated showed a marked segregation of resting metabolic rate (RMR) between the daily phase of normal activity time (α), in which RMR is higher, and the daily phase of normal resting time (ρ), in which RMR is lower (Table 1). The elevated RMR of these animals during α presumably represents a cost of alertness and higher idling level of basal metabolic functions than one observes during ρ, when the animals are not primed to carry out peak behavioral and physiological performances.

Because all our measurements in both α and ρ were for animals in thermal neutrality and during time segments of inactivity, we conclude that the observed elevations of RMR during α for these 12 species are not due to metabolic demands of either thermoregulation or locomotor exertion. Furthermore, because the measurements were made in continuous darkness, the rhythmic patterns were not stimulated by light-dark transitions and can thus be considered to be endogenous, as is the case in birds (Aschoff and Pohl 1970).

Although the segregation of daily metabolism into distinct α and ρ phases is a common pattern in small mammals, there are situations in which ecological life styles either do not allow or do not demand the segregation of behavioral and physiological performances into two temporally distinct daily phases and where $\alpha : \rho$ of RMR approximates 1.0 (Table 1; Figs. 3 and 4). Each of these situations involves a special adaptive suite of behavioral, physiological, or morphological characteristics which predisposes the animal for patterns of activity that are at least somewhat independent of the constraints of the day-night environmental cycle. Both the proximate and the ultimate roles of the daily environmental cycle in the behavior and physiology of these animals are worthy of further examination.

Our survey shows four situations in which animals have relative day-to-night uniformity of resting metabolic level:

1. continuously fossorial species,
2. herbivorous species that consume large volumes of low-quality food, and must forage day and night,
3. species at the lower limit of mammalian body size, and
4. hibernating species during their season of dormancy.

With one exception, these situations pertain to mammalian life styles that are represented only by relatively small species. High-bulk, herbivorous diet also characterizes many large mammals, particularly the ungulates, many of which must, like small microtine rodents, feed both day and night (review by Eriksson et al. 1981). These large mammals should also be expected to have low $\alpha : \rho$ ratios of RMR.

Our observations on animals isolated from nature and held in darkness and at thermal neutrality presumably reflect the capacity of the central nervous system to generate circadian programs of basal metabolism. The level and form of the daily metabolic rhythm of these animals in nature should not necessarily be expected to coincide directly with our laboratory observations, particularly due to complexities introduced by daily cyclic energy expenditures for thermoregulation and activity. The lack of metabolic rhythmicity in a hibernating species in thermal neutrality (*S. saturatus,* Fig. 4), is matched by a similar lack of rhythmicity at cold ambient temperature when the animal is dormant. The damping of the daily metabolic rhythm in dormant rodents and bats at 5°C has previously been demonstrated, but indirect evidence suggests that the underlying circadian rhythm in the central nervous system continues to function (Pohl 1961, 1969).

Daily rhythms of metabolism measured under a variety of laboratory and seminatural conditions that differ from those of the present study (e.g., temperature, photoperiod, food regime, etc.) nonetheless show $\alpha : \rho$ metabolic ratios that resemble some of the results of the present study (Table 2). Although differences in both observational protocol and analysis of metabolic data make it difficult to draw precise comparisons among the examples of Table 2 or between the data of Table 2 and those of the present study (Table 1), it nonetheless seems justified to draw some initial conclusions.

Metabolic costs of thermoregulation, locomotion, and food processing are a substantial component of natural daily energy budgets, and a greater part of these added-on costs is probably expended during α than during ρ. However, the general resemblance of $\alpha : \rho$ metabolic ratios under conditions that involve many of the above natural costs of living (Table 2) to the $\alpha : \rho$ metabolic ratios of animals isolated from environmental factors in DD and at thermal neutrality (Table 1) is interesting and significant. It seems especially important that $\alpha : \rho$ ratios of metabolism are not substantially greater in animals experiencing normal costs of living than in animals in environmental neutrality. For example, note the similarity of metabolic $\alpha : \rho$ in *Peromyscus leucopus* under field conditions (Table 2) to *P. maniculatus* in DD at thermal neutrality (Table 1). Many of the above factors could be responsible for differences in α, ρ or both under these different conditions, but present data do not permit the pinpointing of factors involved. The main point, for now, is that $\alpha : \rho$ ratios of metabolism in individual species of small mammals seem to be of about the same magnitude under a variety of environmental conditions and under different methods of analysis.

P. leucopus and *Blarina,* both of which are non-hibernators, showed greater $\alpha : \rho$ of metabolism in winter under semi-natural conditions than at any other season, which may reflect increased costs of thermoregulation and locomotion during α as compared to ρ. On the other hand, the hibernating ground squirrel *Spermophilus tridecemlineatus* showed a reduction in $\alpha : \rho$ of metabolism to 1.0 during the season of dormancy and cold environmental temperatures (Table 2), just as we found in *S. saturatus* during the season of dormancy in the laboratory at thermal neutrality. The fossorial rodent *Thomomys bottae,* measured in a light-dark cycle, shows $\alpha : \rho$ for metabolism close to 1 (Table 2) as in other fossorial species that we studied in DD (Table 1).

The cost of locomotion has been measured both night and day in the gerbil *Meriones unguiculatus;* metabolic rate during running at night (α) was greater than that during day (ρ) by about 10% (Table 2). However, it seems likely that this difference between metabolism in α and ρ is due to a rhythm of RMR and not to a rhythm of metabolic rate for running.

Birds. The $\alpha : \rho$ ratio of resting metabolic rate is more similar among avian species than among mammalian species, and indeed birds as a group show greater uniformity in their morphology, behavior, and ecology than do the mammals. It is noteworthy that although the metabolic rates of small birds (represented by many species of the order Passeriformes in particular) are substantially higher than those of small mammals of the same body size, the average amplitude of daily metabolic rhythmicity is nonetheless similar in birds and mammals (Fig. 6). The mean $\alpha : \rho$ of RMR for the 18 species of birds investigated (Aschoff and Pohl 1970) is 1.28, which is identical to the ratio of α and ρ intercepts for mammalian RMR [Eqs. (1) and (2)] and closely approximates the mean $\alpha : \rho$ of RMR for the mammalian species we investigated (Table 1); however, the standard deviation (0.14) and range (1.08-1.62) for $\alpha : \rho$ of RMR in birds are considerably smaller than those of mammals (cf. Table 1). The sample of 18 bird species selected by Aschoff and Pohl (1970) also appears uniform in that 16 of 18 species are diurnal and 11 belong to the single order Passeriformes. None of the birds studied by Aschoff and Pohl lives in environments with little or no daily physical fluctuation like those of fossorial mammals, and none employs torpor or possesses other ecological or behavioral characteristics of the sort that represents the basis of adaptive modification in $\alpha : \rho$ of metabolism that we found in small mammals.

Lizards. Ectothermic vertebrates such as lizards show daily cycles of metabolism in nature that reflect cycles of environmental temperature and the temperature selection behavior of the animals. Mautz (1979) has shown that apart from these environmental and behavioral influences on metabolic rate, there is also a daily rhythm of metabolism under conditions of constant temperature in lizards of the family Xantusiidae. One cave-dwelling species showed uniformity of day and night RMR – as found in fossorial mammals in the present study – which demonstrates adaptive flexibility of daily metabolic rhythmicity among lizard species.

4.2 Allometry of Resting $\dot{V}_{O_2}$ in α and ρ

We have shown that the $\alpha : \rho$ ratio of resting $\dot{V}_{O_2}$ has an overall average of 1.28 and does not vary significantly with body mass among small mammals. There remains a trend for decrease in $\alpha : \rho$ ratio of RMR in larger mammals. This trend is most apparent within uniform taxonomic groups, e.g., the low circadian metabolic amplitude in the

Table 2. α and ρ ratios of daily metabolic rhythmicity for mammals studied under a variety of different conditions and analyzed in a variety of different ways

	Conditions	Lighting	T_a	Body mass (g)	Metabolism Measured	Metabolism Analyzed	$\alpha : \rho$	Source
Spermophilus tridecemlineatus								
Thirteen-lined ground squirrel	L, Fed, Sp-Su	NLD	22-28 (c)	109	A.T.	Avg.	1.6	Scheck and Fleharty 1979
	Fa	NLD	20-25 (c)	146	A.T.	Avg.	1.0	Scheck and Fleharty 1979
	Wi	NLD	0 0	117	A.T.	Avg.	1.0	Scheck and Fleharty 1979
Tamias striatus								
Eastern chipmunk	F, Fed, Su	NLD	18-23 (c)	92	A.T.	Avg.	1.1	Randolph 1980a
	Fa	NLD	12-24 (c)	100	A.T.	Avg.	1.2	Randolph 1980a
	Wi	NLD	2-14 (c)	103	A.T.	Avg.	1.1	Randolph 1980a
Peromyscus leucopus								
White-footed mouse	F, Fed, Su	NLD	18-23 (c)	20	A.T.	Avg.	1.4	Randolph 1980a
	Fa	NLD	12-24 (c)	24	A.T.	Avg.	1.4	Randolph 1980a
	Wi	NLD	2-14 (c)	19	A.T.	Avg.	1.8	Randolph 1980a
Thomomys bottae								
Botta pocket gopher	L, Fst, NS	LD 12:12	15	232	R.M.	Min.	0.95	Vleck 1979
Meriones unguiculatus								
Gerbil	L, Fst, NS	LD 12:12	20-22	72	Run	Min.	1.1	Raab and Brady 1976
Blarina sp.	F, Fed, Su	NLD	18-25 (c)	20	A.T.	Avg.	1.4	Randolph 1980b
Short-tailed shrew	Fa	NLD	6-19 (c)	21	A.T.	Avg.	1.3	Randolph 1980b
	Wi	NLD	3-23 (c)	22	A.T.	Avg.	1.6	Randolph 1980b
Planigale gilesi								
Marsupial "shrew"	L, Fed, NS	LD 12:12	14	10	R.M.	Min.	1.4	Dawson and Wolfers 1978
Herpestes sanguineus								
Slender mongoose	L, Fst, NS	LD 12:12	26	500	A.T.	Avg.	1.1	Kamau et al. 1979

larger squirrel species and the high amplitude in the smaller species (Table 1). However, because the α and ρ slopes of whole-animal RMR as a function of body mass did not differ significantly, we conclude that the $\alpha : \rho$ ratio of RMR does not change systematically and significantly with body mass in small mammals as a whole.

Although the $\alpha : \rho$ ratio of RMR does not change as body size increases, the difference between α and ρ values does. Aschoff (1982) reported circadian ranges of oscillation in birds and mammals as the *difference* between α and ρ values. He found a negative correlation between body mass and the $\alpha : \rho$ difference in both body temperature and *mass-specific* RMR. Aschoff's results are not strictly comparable to ours because he used data that spanned a much wider range of body mass and because he included data for animals in LD cycles, whereas our measurements were in DD. (Amplitude of circadian rhythms is known to be greater in LD cycles than in DD; Aschoff 1982). However, the trend in α-ρ difference reported by Aschoff can be seen in our data. Substracting Eq. (2) for metabolic rate during ρ from Eq. (1) for metabolic rate during α and dividing this difference by body mass ($M^{1.0}$), one obtains the α-ρ difference in mass-specific RMR:

$$\dot{V}O_2 = 104\ M^{-0.39} \tag{5}$$

where $\dot{V}O_2$ = mass-specific oxygen consumption in $cm^3/kg \cdot h$ and M = body mass in kg.

It is somewhat surprising to find that the ratio of the lowest 5-min period of metabolism during all of α to the lowest 5-min period during ρ usually provides a reliable estimate of the ratios of total resting metabolism during α and ρ (Table 1). The practicality of this finding is that such minimal values are more easily extracted from either original data or published data than are more complex functions of metabolism. The minimal points of α and ρ metabolism would also be unambiguous if one needed to select metabolic data from experiments in which photoperiod was other than LD 12 : 12. Selection of single daily minimal points of α and ρ also avoids introducing bias by including "minimal" values during intervals when an animal is not truly resting (as in Fig. 5). On the other hand we observed that a brief bout of torpor produces a single daily minimal point that is not representative of normothermic metabolism; and in this sort of case average hourly metabolism would avoid the bias introduced by a single daily minimal value of RMR that occurred during a trough associated with a brief bout of diurnal torpor.

Artifacts in Allometric Analyses of RMR: Possible Biases in Species Selected, Their Body Size and Time of Activity, and the Time of Measurement by Investigators. The slope and intercept of our initial regressions of $\dot{V}O_2$ during ρ on body mass do not differ significantly from those of the equation of Kleiber (1961) for RMR of mammals based on 17 species over a wide range of body size (Fig. 6; F = 2.592, P >0.05 for slope; F = 0.47, p >0.05 for intercept). On the other hand our regression of $\dot{V}O_2$ during α has a

←

Conditions: L = Laboratory or F = field under semi-natural conditions in metabolism chamber; Fed = ad lib food provided or Fst = fasted: Seasons: Wi, Sp, Su, Fa, or NS = no season stated. Lighting: LD = light-dark ratio of fixed photoperiod, NLD = natural light-dark cycle. T_a = air temperature in °C, (c) = cyclic temperature over indicated range. Metabolism measured: A.T. = average total per h, R.M. = resting minimum, Run = running at 0.5 and 1.5 km/h. Metabolism analysis: Avg. = average of all α and ρ values; Min. = single minimal values for α and ρ.

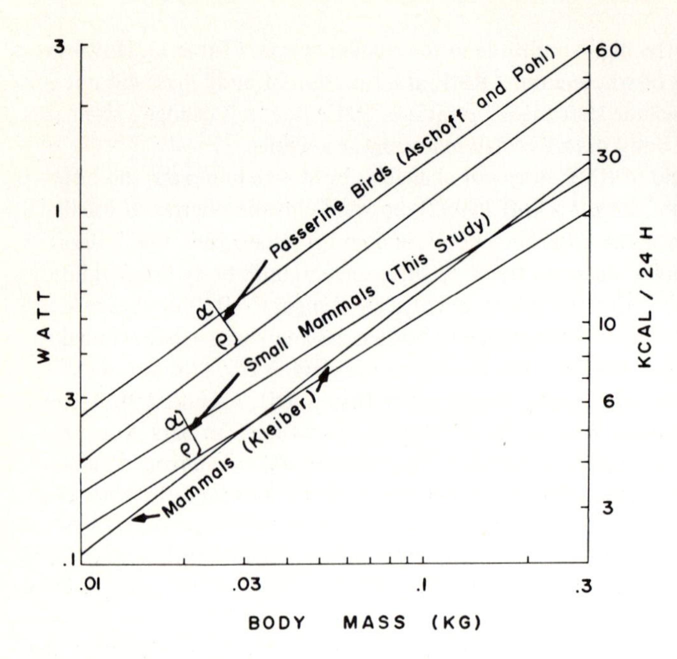

Fig. 6. Relationship between body mass and metabolism as expressed in different studies: present study of small mammals, Eqs. (3) and (4); Kleiber's (1961) data for mammals in general; Aschoff and Pohl's (1970) data for passerine birds. Measurements of resting metabolism as a function of time of day are only available in the present study and in that of Aschoff and Pohl

significantly lesser slope than that of Kleiber ($F = 20.091, p < 0.01$). The similarity of Kleiber's curve to our data for RMR of small mammals during ρ may be due to the preponderance of nocturnal species among the small mammals surveyed by Kleiber. It is probable that these nocturnal forms were measured during their ρ phase, in the daytime. Furthermore, the larger mammals surveyed by Kleiber were mostly day active and were probably measured during α, which would further contribute to the significantly steeper slope of Kleiber. Our slope of 0.61 for 16 species of small mammals is also matched by a slope of 0.62 in a study of 6 small-mammal species by Withers et al (1979), whereas other surveys that include larger mammals have yielded greater slopes (Kleiber 1961, slope = 0.75; Lechner 1978, slope = 0.73; Stahl 1967, slope = 0.76). An increased slope could be an artifact of conducting laboratory work during daytime and thereby measuring small (mostly nocturnal) species during ρ and large (mostly diurnal) species during α (see below). Such an effect has also been indicated by Aschoff (1981) in a reanalysis of data on minimal thermal conductance and body size.

The daily work-shift rhythms of scientists may have had an important impact on the data available on allometry of metabolism and body size in mammals: it is likely that where not stated in the literature, the time of measurement of metabolism is day time, which corresponds to α of day-active animals and ρ of night-active animals. This problem is compounded in mammals because the small mammals included in allometric analyses of metabolism are mostly nocturnal, whereas the larger mammals are mostly diurnal (Kleiber 1961, Stahl 1967). The steeper slope of Kleiber's analysis for mammals as compared to that of the present study (Fig. 6) may be a result of the artifactual biases stated above. Care should be taken in future analyses to use a balance of data for both diurnal and nocturnal mammals throughout the full size range of mammals, and to present data from both α and ρ.

5 Conclusions

We measured the 24-h course of resting metabolic rate (RMR) in 18 species of small (8-270 g) mammals in continuous darkness and thermal neutrality. Most species showed an endogenous elevation of RMR during the normally active phase (α) of the daily cycle and a lower level of RMR during the normally inactive, or rest phase (ρ). The data for 16 species measured under fasting, normothermic conditions give the following equations for RMR – in α: $P = 2.67\ M^{0.61}$ and in ρ: $P = 2.10\ M^{0.61}$, where P = metabolic rate (power) in W and M = body mass in kg. The ratio of RMR in α to that in ρ for the two equations is 1.28; however, the $\alpha : \rho$ ratios of RMR for individual species are as great as 1.76 or so small as to approximate 1.0. This interspecific variability in $\alpha : \rho$ of RMR is correlated with adaptive patterns of temporal organization that have evolved in response to the environment and in concert with the ecology and behavior of each species. Strictly nocturnal or diurnal species show conspicuous elevation of RMR in α as compared to RMR in ρ. On the other hand several other categories of small mammals show little day-night segregation of RMR; these include the continuously fossorial moles and pocket gophers, small herbivorous rodents that feed both day and night, the smallest mammalian homeotherms – shrews, and hibernating rodents during their season of hibernation. Although birds show slightly higher RMR's than mammals of the same size, the overall $\alpha : \rho$ ratios of RMR are about 1.28 both for birds and small mammals. Our equations for RMR of small mammals during α and ρ as a function of body mass have a shallower slope (0.61) than that of the equation of Kleiber (0.75), which represents data for mammals in general at unspecified times of day. It is possible that the slope of Kleiber's equation is biased by a preponderance of RMR values taken on small, mostly nocturnal species during ρ (daytime) and on large, mostly diurnal species during α (also daytime).

Acknowledgments. This study was supported by National Science Foundation grants PCM 77-05397 and DEB 77-06706 and by a grant from the University of Washington Graduate School Research Fund. We thank J. Aschoff and G.A. Bartholomew for critical comments on the manuscript and D. Koch for technical assistance.

References

Aschoff J (1964) Survival value of diurnal rhythms. Zool Soc London Symp 13: 79-98

Aschoff J (1966) Circadian activity pattern with two peaks. Ecology 47: 657-662

Aschoff J (1981) Thermal conductance in mammals and birds: its dependence on body size and circadian phase. Comp Biochem Physiol 69A: 611-619

Aschoff J (1982) The circadian rhythm of body temperature as a function of body size. In: Taylor CR, Johansen K, Bolis LA (eds) Companion to animal physiology. Cambridge Univ Press, New York (in press)

Aschoff J, Pohl H (1970) Der Ruheumsatz von Vögeln als Funktion der Tageszeit und der Körpergröße. J Ornithol 111: 38-47

Bartholomew GA, Cade TJ (1957) Temperature regulation, hibernation, and estivation in the little pocket mouse, *Perognathus longimembris.* J Mammal 38: 60-72

Dawson TJ, Wolfers JM (1978) Metabolism, thermoregulation and torpor in shrew sized marsupials of the genus *Planigale.* Comp Biochem Physiol 59: 305-309

Dunn OJ, Clark VA (1974) Applied Statistics: Analysis of Variance and Regression. John Wiley and Sons New York

Enright JT (1970) Ecological aspects of endogenous rhythmicity. Annu Rev Ecol Syst 1: 221-238

Eriksson LO, Källqvist ML, Mossing T (1981) Seasonal development of circadian and short-term activity in captive reindeer, *Rangifer tarandus* L. Oecologia 48: 64-70

Erkinaro E (1969) Der Phasenwechsel der lokomotorischen Aktivität bei *Microtus agrestis* (L.), *M. arvalis* (Pall.) und *M. oeconomus* (Pall.). Aquilo Ser Zool 8: 3-31

Hill RW (1972) Determination of oxygen consumption using the paramagnetic oxygen analyzer. J Appl Physiol 33: 261-263

Kamau JMZ, Johansen K, Maloiy GMO (1979) Thermoregulation and standard metabolism of the slender mongoose *(Herpestes sanguineus)*. Physiol Zool 52: 594-602

Kleiber M (1961) The fire of life. John Wiley and Sons, N.Y. 454 p.

Lechner AJ (1978) The scaling of maximal oxygen consumption and pulmonary dimensions in small mammals. Respir Physiol 34: 29-44

Mautz WJ (1979) The metabolism of reclusive lizards, the *Xantusiidae*. Copeia 1979: 577-584

Pearson OP (1960) Habits of *Microtus californicus* revealed by automatic photographic records. Ecol Monogr 30: 231-249

Pittendrigh CS (1961) On temporal organization in living systems. Harvey Lect 56: 93-125

Pohl H (1961) Temperaturregulation und Tagesperiodik des Stoffwechsels bei Winterschläfern. Z Vergl Physiol 45: 109-153

Pohl H (1969) Circadiane Periodik im Winterschlaf. Naturwissenschaften 52: 269

Raab JL, Brady MS (1976) Do nocturnal rodents run more efficiently at night? Nature (London) 260: 38-39

Randolph JC (1980a) Daily energy metabolism of two rodents *(Peromyscus leucopus* and *Tamias striatus)* in their natural environment. Physiol Zool 53: 70-81

Randolph JC (1980b) Daily metabolic patterns of short-tailed shrews *(Blarina)* in three natural seasonal temperature regimes. J Mammal 61: 628-638

Scheck SH, Fleharty ED (1979) Daily energy budgets and patterns of activity of the adult thirteen-lined ground squirrel, *Spermophilus tridecemlineatus*. Physiol Zool 52: 390-397

Stahl WR (1967) Scaling respiratory variables in mammals. J Appl Physiol 22: 453-460

Vleck D (1979) The energy cost of burrowing by the pocket gopher *Thomomys bottae*. Physiol Zool 52: 122-136

Withers PC, Casey TM, Casey KK (1979) Allometry of respiratory and haemotological parameters of arctic mammals. Comp Biochem Physiol 64A: 343-350

8.3 Characteristics and Variability in Entrainment of Circadian Rhythms to Light in Diurnal Rodents

H. Pohl[1]

Introduction

Entrainment of circadian systems of vertebrates to light cycles has been studied most thoroughly in rodents. A model of non-parametric entrainment, proposed by Pittendrigh (1966) for the pupal eclosion rhythm of *Drosophila,* has been found adequate to explain to a certain extent entrainment of circadian pacemakers in night-active (nocturnal) rodents (Daan and Pittendrigh 1976a, b, Pittendrigh and Daan 1976a, b). In a comparative analysis of functional aspects of circadian pacemakers in four species of nocturnal rodents, Pittendrigh and Daan have shown that various pacemaker properties, such as the stability of the period (τ), the shape of the phase response curve (PRC) for light pulses, the light-dependence of τ, and the ranges of entrainment to light cycles, vary interdependently. Empirical regularities of these characteristics on an interspecific, inter- and intra-individual basis indirectly support the usefulness of the non-parametric entrainment model.

Two main questions are of interest: (1) Do differential (non-parametric) properties of light cycles also contribute substantially to the entrainment mechanism of circadian pacemakers in day-active (diurnal) rodents? (2) Do diurnal rodents show similar interrelations between certain properties of their circadian systems as those found in nocturnal rodents? This presentation summarizes new findings from a comparative study related to functional aspects of circadian systems in three species of diurnal rodents (Pohl, in prep.). On the basis of our recent knowledge, differences in the light responsiveness and in entrainment properties of circadian systems between diurnal and nocturnal rodents are briefly discussed.

1 Methods

The circadian rhythm of wheel-running activity was measured by conventional methods (cf. Pohl 1972) in *Eutamias sibiricus* (Siberian chipmunk), *Ammospermophilus leucurus* (antelope ground squirrel) and *Tamiasciurus hudsonicus* (red squirrel). All animals were held individually in small cages with access to a running-wheel. Several cage units were

1 Max-Planck-Institut für Verhaltensphysiologie, D-8138 Andechs

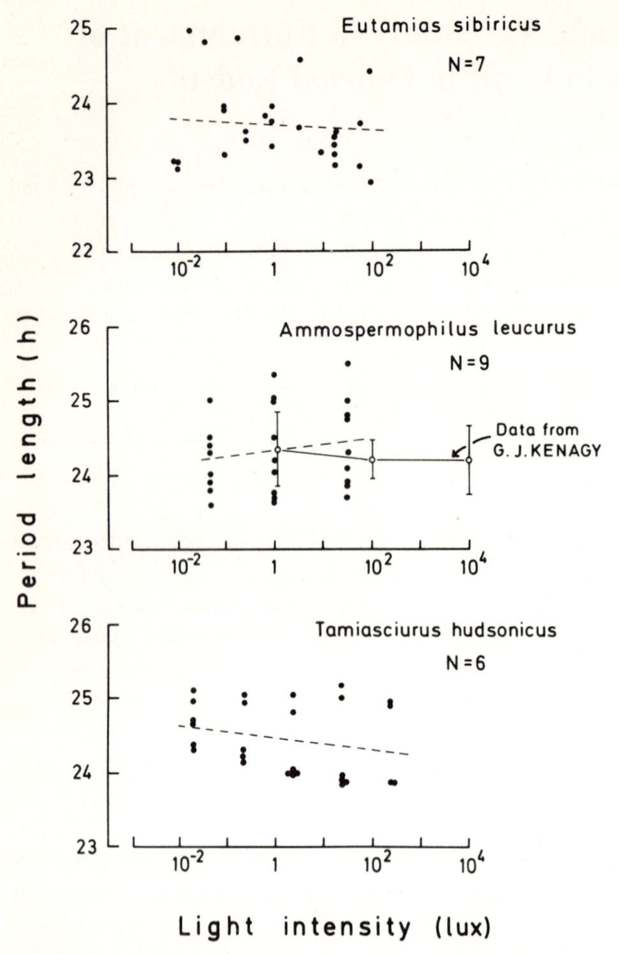

Fig. 1. Circadian period as function of light intensity in three species of diurnal rodents. *Points* refer to single measurements

placed in a temperature-controlled room (20 ± 2°C). Animals were subjected (1) to continuous light of various intensity levels (all three species); (2) to light-dark (LD) cycles (100 lx vs. total darkness) with different periods (T) (all three species); (3) to low-amplitude light cycles with short (ca. 20 min) and long (ca. 4 h) transition intervals between maximal and minimal intensity (*E. sibiricus* and *T. hudsonicus*); (4) to periodic light pulses with one or two 1-h (500 lx) pulses per cycle and different periods (T 23.75 and 24.25 h) (*A. leucurus*); and (5) to single 1-h and 6-h (500 lx) pulses (*A. leucurus*). Background light intensity was 0.05 lx in (4) and 1.0 lx in (5). The average time of exposure to each condition was between 3 and 4 weeks. Further details on the origin of the animals, the experimental techniques as well as original data will be documented elsewhere (Pohl, in prep.).

2 Results and Discussion

2.1 Circadian Period as Function of Light Intensity

The period (τ) of the free-running rhythm of locomotor activity does not vary systematically with changing light intensity over a relatively wide range of intensities in all three species (Fig. 1). This result contrasts with data reported from a large number of nocturnal rodent species (cf. Aschoff 1979). In individuals of all three species, τ remained relatively stable (maximal variation ca. 1%) in constant light of 1 lx over several

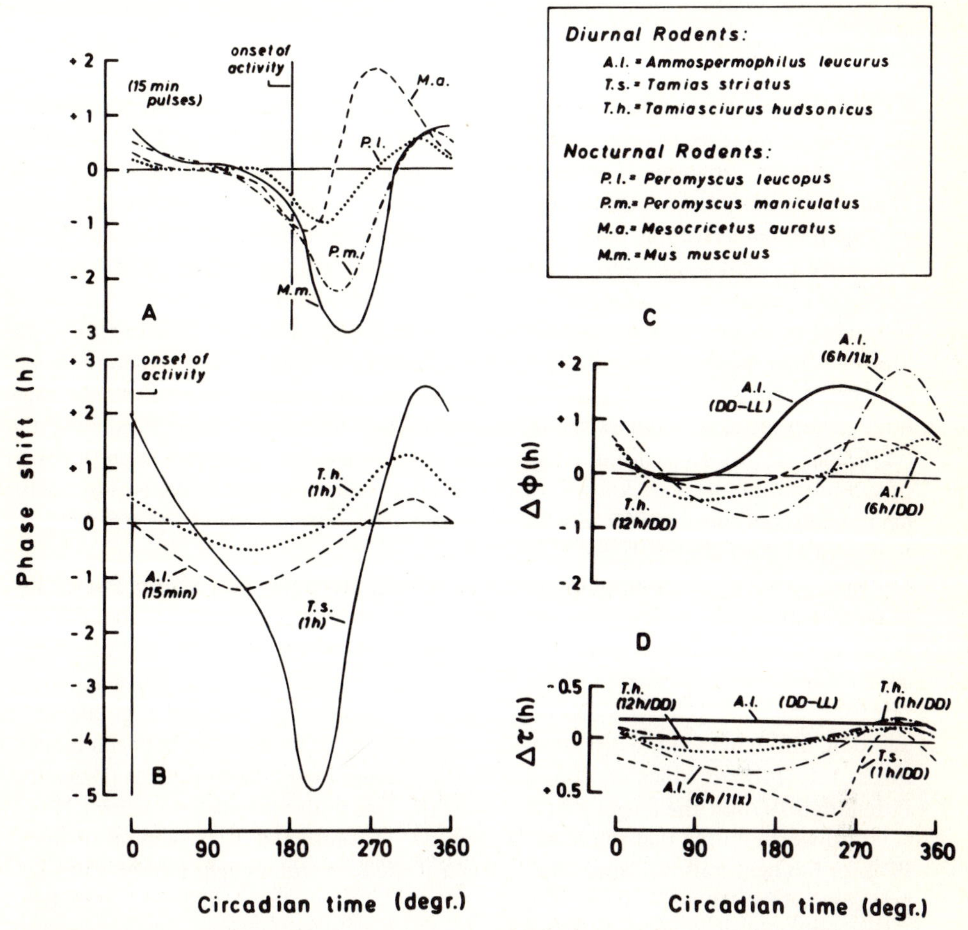

Fig. 2. Phase-responses ($\Delta\varphi$) to short (15-min or 1-h) light pulses in **A** nocturnal and **B** diurnal rodents, and **C** for long (6- or 12-h) light pulses or step transitions (DD-LL) in diurnal rodents. **D** Period-responses ($\Delta\tau$) associated with light pulses or step transitions in diurnal rodents. Curves for nocturnal rodents are fitted to data points measured by Daan and Pittendrigh (1976a); curves for diurnal rodents fitted to data measured by Kramm (1974, 1975b, 1976), Kramm and Kramm (1980) and Pohl (manuscript in prep.)

months during which single light pulses were given at various phases of the rhythm. This finding reflects a high intra-individual stability of τ despite a relatively large inter-individual variation (e.g. *A. leucurus* in Fig. 1). In total darkness (DD), the mean τ is close to 24 h in *E. sibiricus* and *A. leucurus* but significantly longer than 24 h in *T. hudsonicus.* A significant lengthening of τ_{DD} with time of exposure has been reported for *A. leucurus* and *T. hudsonicus* (Kramm 1976, 1975a). In diurnal species, $\tau_{DD} > 24$ h has functional significance for stabilization of the phase and the period (Pittendrigh and Daan 1976b). It is interesting to note, that in *E. sibiricus* and *A. leucurus* τ_{DD} resembles that found in *Mesocricetus auratus* and *Peromyscus leucopus,* two nocturnal rodents which show the highest stabilization of τ and the smallest dependence of τ on light intensity (Pittendrigh and Daan 1976a, Daan and Pittendrigh 1976b).

2.2 Phase-Response Curves

Phase-response curves (PRC) for short (15 min or 1 h) light pulses applied during DD are known from three species of diurnal rodents (Fig. 2B): *A.leucurus, T. hudsonicus,* and the eastern chipmunk (*Tamias striatus*) (Kramm 1976, Kramm and Kramm 1980). Their shapes show large species-specific differences comparable to those found among nocturnal rodents (see Fig. 2A). In contrast to nocturnal rodents, which are mostly irresponsive to light pulses during rest time (subjective day), the day-active rodents do not show a particular "dead zone" of the PRC. The insensitivity of the circadian system to sudden brief light-dark transitions may be of important adaptive value in ground-dwelling nocturnal species. It may protect them against adverse light effects when these animals occasionally leave their burrows during the day. In diurnal ground-dwelling rodents, intermittent retreate into dark shelter may be far less effective to their circadian systems (cf. Kenagy 1978). Evidence of a high sensitivity to very short light pulses (in the order of seconds or minutes) are known for the circadian and reproductive systems of nocturnal rodents (DeCoursey 1959, Hoffmann et al. 1980), but no comparable effects have been seen in diurnal species.

In *A. leucurus*, phase-response curves have been determined for light pulses of different durations and background intensities as well as for step transitions from DD to bright light of 100-1200 lx (Fig. 2C). For longer light pulses and step transitions the area under the advance section of the PRC is larger than the area under the delay section; only for short pulses do delay shifts prevail over advance shifts (cf. Fig. 2B). In addition, differences in the PRC shape as a function of τ are indicated between individuals of *A. leucurus* which are consistent with findings in nocturnal rodents (Daan and Pittendrigh 1976a): the longer τ is, the larger the area under the advance section (Fig. 3). It is also worth noting that the total "amplitude" (advance and delay section) of the PRC for 6-h light pulses is considerably larger if the background light intensity is 1 lx as compared to DD.

2.3 Period-Response Curves

Relatively small but significant changes in τ associated with light pulses or step transitions from DD to bright light or vice versa occur in *A. leucurus, T. hudsonicus,* and *T.*

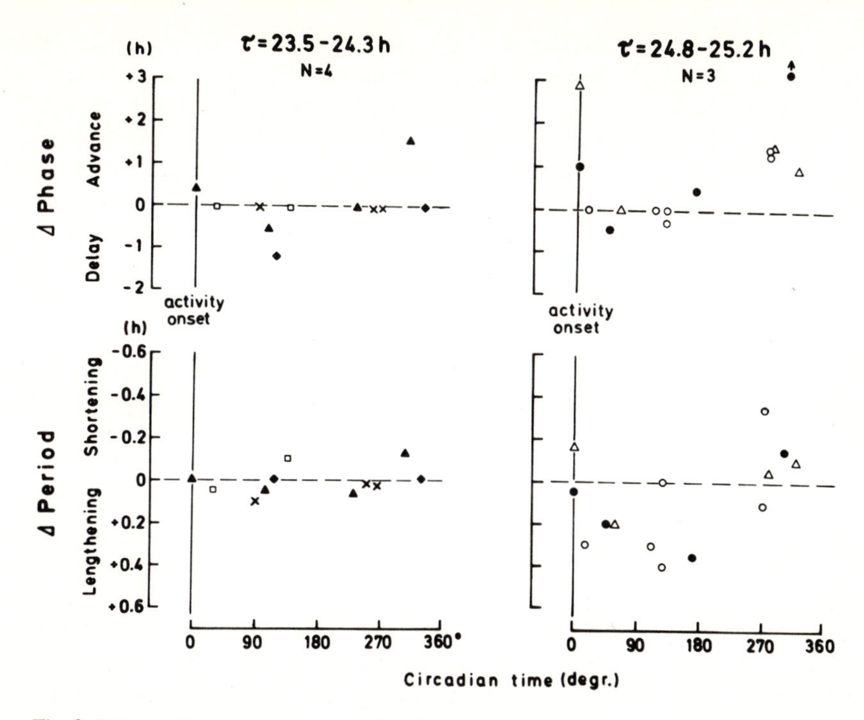

Fig. 3. Phase and period responses as function of circadian time (in degrees of arc) in *A. leucurus*. *Left* animals with relatively short τ; *right* animals with relatively long τ. *Different symbols* refer to different individuals. Circadian time zero = activity onset

striatus (Fig. 2D). These changes, which may also be considered as "after-effects" (Pittendrigh 1960), depend on the direction and/or amount of the phase shift, but they can also occur without any phase shift. In *A. leucurus*, the relationship between changes in τ and phase shifts depends on the duration of the pulse and on background light intensity (see also Kramm 1976). "Period-response curves" have been found in *T. hudsonicus* for 1-h and 12-h pulses (at DD), in *T. striatus* for 1-h pulses (at DD), and in *A. leucurus* for 6-h pulses (at 1 lx). Generally, maximal decreases (shortening) of τ are associated with maximal phase advances, whereas maximal increases (lenghtening) of τ are associated with maximal phase delays (cf. Fig. 2B, C, D). The coincidence of maximal increases of τ with maximal phase delays and the previously documented finding that the amount of phase advance is positively related with τ suggests that both phenomena are compensatory and may be part of a mechanism adjusting τ to T. For example, if a light pulse causes a phase delay and simultaneously increases τ, the next pulse may consequently lead to a smaller delay or greater advance shift, depending on which part of the PRC the light signal occurs.

2.4 Ranges of Entrainment

The ranges of entrainment of the activity rhythm to complete LD cycles (100 lx vs. DD) with T ranging from 23.5 to 25.0 h differ significantly between the three species (Table

Table 1. Number and percentage of animals entrained to light-dark cycles (LD 100 lx vs. total darkness) with different periods (T). N = number of animals tested at each T

T (h)	23.50	24.25	24.60	25.00
A. *E. sibiricus*				
N	6	7	6	7
Entrained (%)	6 (100)	7 (100)	6 (100)	7 (100)
Not entrained	-	-	-	-
Uncertain	-	-	-	-
B. *A. leucurus*				
N	13	14	13	13
Entrained (%)	5 (38)	13 (93)	9 (69)	3 (23)
Not entrained	7	1	4	8
Uncertain	1	-	-	2
C. *T. hudsonicus*				
N	6	6	6	6
Entrained (%)	1 (17)	4 (67)	2 (33)	0 (0)
Not entrained	5	1	3	6
Uncertain	-	1	1	-

1): *E. sibiricus* has the largest range which exceeds that tested experimentally (⩾1.5 h); *A. leucurus* has an intermediate range of about 1 h (from ca. 23.75 to 24.75 h), and *T. hudsonicus* has the smallest range of about 0.5 h (from ca. 23.9 to 24.4 h). Increasing the light intensity level from 100 to 600 lx does not significantly increase the limits of entrainability in all three species.

The differences in the ranges of entrainment between the three diurnal rodents partly correspond with the differences in the PRC amplitudes for either short or long light pulses. *T. hudsonicus* shows both the smallest range of entrainment and the smallest PRC-amplitude (advance and delay section). The largest range of entrainment of *E. sibiricus* from eastern Asia corresponds with the largest PRC-amplitude of *T. striatus*, a closely related species from the eastern part of North America.

2.5 Non-Parametric Versus Parametric Entrainment

One experiment to investigate the model of non-parametric entrainment is the simulation of complete light fractions ("photoperiods") of LD cycles by two short light pulses per cycle. The majority (ca. 70%) of *A. leucurus* exposed to "skeleton photoperiods" simulating a ca. 12 : 12 h LD cycle by two 1-h pulses (at T 24.25 h) entrained their activity rhythms to this zeitgeber. More than 50% even entrained to one 1-h pulse per cycle at both T 23.75 and 24.25 h. Evidence for differential effects of light during entrainment has also been shown in *T. hudsonicus* and *T. striatus* (Kramm and Kramm 1980). The animals have been tested in both complete and skeleton photoperiods of 1(3) to 18 h per cycle. The close correspondence between complete and skeleton photoperiods with regard to the phase-angle difference (ψ) between activity onset and lights-on or the "morning" pulse over a wide range of "photoperiods" in *T. hudsonicus* indicates than non-parametric entrainment by short light signals (pulses or steps) contributes substantially to the entrainment mechanism in this species (cf. Pittendrigh and Daan 1976b).

To test whether proportional properties of light cycles are effective in entraining circadian pacemakers of diurnal rodents, low-amplitude light cycles with long transition intervals were used as zeitgeber. (It should, however, be emphasized that, although differential or proportional properties of a light cycle can be minimized by experimental manipulation, neither of the two can be totally excluded.) About 50% of *E. sibiricus* successfully entrained their rhythms to a light cycle with a LD intensity ratio of 12 : 3 lx and transition intervals of 4 h. Similar small intensity ratios at even higher intensity levels and with different light fractions ("photoperiods") have been found effective in entraining the rhythms of ground squirrels (*Spermophilus columbianus, S. undulatus* and *A. leucurus*), as well as of nocturnal species (e.g. *M. auratus*) (Swade and Pittendrigh 1967). The significance of proportional (parametric) effects of light as a supplemental element in the entrainment mechanism has been postulated for nocturnal rodents (Pittendrigh and Daan 1976b). The previous findings suggest that parametric effects have a major share in the entrainment to light cycles in diurnal rodents. The relative importance of non-parametric versus parametric elements in entrainment mechanism can, however, not be assessed until we know more about their functional significance in different entrainment strategies.

In summary, it can be stated that diurnal and nocturnal mechanisms of entrainment of circadian systems to light cycles are not basically different. Both groups of organisms, however, show a high degree of variability probably related to special ecological requirements. This view is supported by species differences in the responsiveness (sensitivity) of the circadian systems to light changes, in the phase-response curves to light pulses, and by interrelated functions of properties of the circadian system which, to a certain extent, resemble those found in nocturnal rodents. It remains a matter of further studies to analyze the specific adjustments of entrainment mechanisms in closely related species experiencing different light conditions in their natural habitats.

References

Aschoff J (1979) Circadian rhythms: Influences of internal and external factors on the period measured in constant conditions. Z Tierpsychol 49: 225-249

Daan S, Pittendrigh CS (1976a) A functional analysis of circadian pacemakers in nocturnal rodents. II. The variability of phase response curves. J Comp Physiol 106: 253-266

Daan S, Pittendrigh CS (1976b) A functional analysis of circadian pacemakers in nocturnal rodents. III. Heavy water and constant light: Homeostasis of frequency? J Comp Physiol 106: 267-290

DeCoursey P (1959) Daily activity rhythms in the flying squirrel, *Glaucomys colans.* Ph D thesis, Univ Wisconsin

Hoffmann K, Illnerová H, Vaněček J (1980) Pineal N-acetyltransferase activity in the Djungarian hamster. Naturwissenschaften 67: 408

Kenagy GJ (1978) Seasonality of endogenous circadian rhythms in a diurnal rodent *Ammospermophilus leucurus* and a nocturnal rodent *Dipodomys merriami.* J Comp Physiol 128: 21-36

Kramm KR (1974) Phase control of circadian activity rhythms in ground squirrels. Naturwissenschaften 61: 34

Kramm KR (1975a) Circadian activity of the red squirrel, *Tamiasciurus hudsonicus*, in continuous darkness and continuous illumination. Int J Biometeorol 19: 232-245

Kramm KR (1975b) Entrainment of circadian activity rhythms in squirrels. Am Nat 109: 379-389

Kramm KR (1976) Phase control of circadian activity in the antelope ground squirrel. J Interdiscip Cycle Res 7: 127-138

Kramm KR, Kramm DA (1980) Photoperiodic control of circadian rhythms in diurnal rodents. Int J Biometeorol 24: 65-76

Pittendrigh CS (1960) Circadian rhythms and circadian organization of living systems. Cold Spring Harbor Symp Quant Biol 25: 159-184

Pittendrigh CS (1966) The circadian oscillation in *Drosophila pseudoobscura* pupae: A model for the photoperiodic clock. Z Pflanzenphysiol 54: 275-307

Pittendrigh CS, Daan S (1976a) A functional analysis of circadian pacemakers in nocturnal rodents. I. The stability and lability of spontaneous frequency. J Comp Physiol 106: 223-252

Pittendrigh CS, Daan S (1976b) A functional analysis of circadian pacemakers in nocturnal rodents. IV. Entrainment: Pacemaker as clock. J Com Physiol 106: 291-331

Pohl H (1972) Die Aktivitätsperiodik von zwei tagaktiven Nagern, *Funambulus palmarum* und *Eutamias sibiricus*, unter Dauerlichtbedingungen. J Comp Physiol 78: 60-74

Swade RH, Pittendrigh CS (1967) Circadian locomotor rhythms of rodents in the arctic. Am Nat 101: 431-466

8.4 Functional Significance of Daily Cycles in Sexual Behavior of the Male Golden Hamster

G.A. Eskes[1]

One hypothesis about the function of circadian rhythms (CR's) is that they provide an endogenous temporal program for adaptive synchronization of the physiology and behavior of an animal to its fluctuating external world (Enright 1970, Daan 1981). For example, the circadian control of *Drosophila* eclosion presumably restricts the emergence of the adult fly to the generally humid conditions around dawn that permit successful eclosion (Pittendrigh 1958). Likewise, an endogenous clock in the honeybee allows it to synchronize its daily visits to a particular flower species at the time when pollen or nectar is available (Renner 1960). There is, however, little empirical evidence to support this hypothesis for the functional role of rhythmicity in mammals.

The adaptive significance of rhythmicity has been measured by correlating the occurrence of a behavioral rhythm with a variable, such as survival, that is critical to individual fitness (i.e., the number of viable offspring an organism can produce) (Daan and Tinbergen 1979, see Chap. 8.1, this Vol.). An alternative method for assessing the contribution of CR's to fitness would be to examine the negative consequences that follow the elimination of normal circadian organization. If intact circadian rhythmicity is of functional significance to an animal, then disruptions of rhythmicity should have detrimental effects. The influence of the loss of rhythmicity on fitness would provide evidence for the functional role of a CR in the ecology of the organism.

The suprachiasmatic nuclei (SCN) are critical for the generation of all physiological and behavioral rhythms that have been studied in rodents (Rusak and Zucker 1979, see Chap. 3, this Vol.). However, despite severe disruptions of rhythmicity after SCN damage, most parameters studied in the laboratory remain within normal limits without obvious detriment to the health of the animal. Total daily water intake, proportion of paradoxical sleep to total sleep, total sleep time, mean level of plasma corticosterone, and average daily body temperature in SCN-lesioned rats were similar to values for control animals (Ibuka et al. 1977, Stephan and Nunez 1977, Stephan and Zucker 1972).

The lack of functional consequences after SCN lesions must be interpreted with caution, pending studies in more natural environments. Arrhythmicity might have severe detrimental effects in a natural environment in which there are fluctuations in climatic factors and rhythms in the availability of resources (e.g., food, mates) and in the risk of predation. For example, intact male hamsters housed in a simulated burrow environment confined their activity outside a continuously dark burrow to the dark phase of

1 Dept. of Psychology University of California Berkeley, California, USA 94720

Vertebrate Circadian Systems
Ed. by J. Aschoff, S. Daan, and G. Groos

the external light-dark (LD) cycle, while a SCN-lesioned male emerged at irregular intervals and was active outside the burrow at all phases of the LD cycle. This absence of rhythmicity presumably would have negative consequences by increasing the risk of exposure to diurnal predators (Rusak 1978).

Another example of a potential functional deficit in SCN-lesioned animals is in relation to temporally coded learning. The quality of performance on a passive avoidance task depends on the interval between training and testing, and is independent of phase of the LD cycle at the time of training (Holloway and Wansley 1973). Animals exhibit better retention at 12 and 24 h following training and at circadian multiples thereof. This passive avoidance rhythm may have important functional consequences for the organism (Rusak 1978). If an animal were to escape from a dangerous encounter with a predator or conspecific at a particular place at one time of day, it would be adaptive to avoid that locus. However, given the rhythmic behavior of other organisms, it would be more beneficial to avoid an area only at those times when the probability of further encounters is highest, i.e., every 24 h, and to take advantage of the resources at that location at other times. SCN lesions eliminated the variation in performance, and lesioned animals performed equally well at all test intervals following training (Stephan and Kovacevic 1978, Wansley and Holloway 1975). Without an internal temporal program guided by the SCN to optimally coordinate behavior to the environment, an animal might suffer from an inability to fully exploit available resources.

Study of the circadian mechanisms underlying reproductive behavior in rodents may provide an illustration of the functional value of the internal temporal control of physiology and behavior. Successful mating involves both physiological readiness and behavioral synchrony in the two sexes. The precise timing of the proestrous surge of luteinizing hormone in female rats and hamsters is controlled by a circadian clock that also ensures the onset of behavioral receptivity around dusk, a time when both sexes are active and male rats show an increase in copulatory readiness (Beach and Levinson 1949, Beach et al. 1976, Zucker and Carmichael 1981). Circadian synchronization of the events underlying mating is likely to increase the probability of successful reproduction. For example, fertilization of blinded female rats housed with blinded male rats for 6-11 days only occurred when the extrapolated free running activity phases of the male and female coincided (Richter 1970).

Coordination of the sexes by CR's entrained to environmental cycles would seem to be a valuable adaptation in hamsters. Available evidence suggests that hamsters are fossorial and solitary; females are generally intolerant of males except during the brief period of estrus (Goldman and Swanson 1975, Murphy 1971, Payne and Swanson 1972). Circadian regulation of copulatory behavior in hamsters may influence fitness by ensuring the coincidence of male and female sexual behaviors and by increasing the probability of successful reproduction. Male hamsters show diurnal rhythms of sexual behavior; they are quicker to initiate copulation and to ejaculate in the dark (D) phase than in the light (L) phase of the light-dark cycle. The rhythm depends on the integrity of the SCN; destruction of the SCN eliminates the normal circadian rhythm of sexual behavior without affecting the mean level of copulation per se (Table 1). There were no significant differences between the overall scores of control and SCN animals (Eskes 1981). By eliminating CR's, SCN lesions could abolish the coincidence of male and female readiness to copulate and thereby reduce reproductive success. A male hamster that is equal-

Table 1. Sexual performance of male hamsters tested at different times of day

Group	N	Time of day[a] 1500	2100	0300	0900
		Intromission Latency			
Control	10	64.0[b]	71.0	90.0	147.0[c,e]
		(26.0)	(20.0)	(38.0)	(32.0)
SCN-lesioned	6	132.0	81.0	94.0	72.0
		(71.0)	(27.0)	(29.0)	(24.0)
		Ejaculation Latency			
Control	10	73.0	63.0	92.0	154.0[c,d]
		(18.0)	(10.0)	(33.0)	(37.0)
SCN-lesioned	6	117.0	100.0	70.0	74.0
		(52.0)	(39.0)	(38.0)	(24.0)

a LD 14 : 10 (L on 0100-1500 h).
b The percent deviation from the mean score of all test times was calculated for each animal; table entries represent the median (and semi-interquartile range) of each group
c Significantly different from 1500 h (p <.01), Wilcoxon test.
d Significantly different from 2100 h (p <.01), Wilcoxon test.
e Significantly different from 2100 h (p <.05), Wilcoxon test.

ly likely to attempt copulation at all times of day may also waste time and energy seeking unavailable mates and suffer from aggressive encounters after locating an unreceptive female.

To test the effects of the loss of CR's on a male hamster's ability to reproduce successfully, SCN-lesioned males were housed overnight with a female in estrus. Under these conditions, arrhythmic sexual behavior did not prevent the male from fertilizing a female; SCN-lesioned males were able to sire normal-sized litters (Eskes 1981).

These results again emphasize the problem of interpreting the consequences of circadian disorganization when animals are maintained in an artificial laboratory environment. Successful fertilization of a female hamster normally may require at least (1) location of a receptive mate and initiation of copulation; (2) coordination of the pattern of mating behavior to stimulate the neuroendocrine events underlying pregnancy; and (3) the ability to compete with other males to maximize the number of one's offspring sired in cases where females may mate with more than one male. In many species, male-male competition is an essential determinant of mating success and some female rodents are reported to mate with multiple males (Birdsall and Nash 1973, Henken and Sherman 1981, Wilson 1975). The functional deficits of arrhythmicity may be evident in a competitive environment in which temporal constraints are important. I decided to study the effects of circadian disorganization on the male hamster's ability to mate in a competitive experimental situation. The preliminary study to be described here measured the reproductive success of arrhythmic males in a one-female, two-male mating test. When two males are tested together, a SCN-lesioned male that does not show a consistent rise in copulatory performance during the limited daily phase of female receptivity would probably be at a disadvantage relative to an intact male in terms of reproductive success.

Thirty male hamsters screened for sexual behavior were divided into four groups balanced for the number of ejaculations each male obtained in the last screening test: Sham-operated controls, to be tested during their dark phase (CD) (1) or light phase (CL) (2); and SCN-lesioned hamsters, to be tested during their dark phase (SD) (3) or light phase (SL) (4). After surgery the males were placed in one of two LD 14 : 10 cycles 12 h out of phase with each other, with lights on at 1730 h (CD, SD) or at 0530 h (CL, SL).

Males were tested under dim red light in pairs consisting of: CD/CL; CD/SD; SD/SL; and CL/SL. Testing of each pair began at 1330 h and consisted of three parts. In part 1, the proceptive behavior of an ovariectomized receptive female (behavior considered to represent the female's attraction to the male) was measured by the method of Beach et al. (1976). In part 2, the males were placed together in a large mating test arena for 30 min while male-male interactions (investigation and agonistic encounters) were recorded. The female was introduced to the mating arena in part 3 and the mating behaviors of each male (mounts, intromissions, and ejaculations) were recorded. The males were allowed to mate to satiety. Details of this experiment are reported elsewhere (Eskes 1981).

Eleven animals sustained complete or nearly complete ablation of the SCN; the running-wheel activity of these animals (recorded after the mating tests were concluded) lacked normal circadian organization under constant dim lighting conditions. The results of this preliminary study of 11 pairs of animals are sufficiently consistent to identify trends and to encourage further testing.

Analysis of the results of parts 1 and 2 revealed that any differences in the mating performance between males of each pair are independent of changes in female behavior or male aggressiveness. There were no significant correlations between female preference or male aggression scores and various measures of mating behavior. Figure 1 presents the differences in scores between males of each pair for mount and intromission latency (ML, IL; the time elapsed from the start of the test to the first mount or intromission), ejaculation latency (EL; the time elapsed from the first intromission to the following ejaculation), and ejaculation frequency. Intact control males tested during D were quicker to begin copulating and to ejaculate than their counterparts tested in the L phase: ML, IL, and EL scores were shorter in CD than CL animals. CD males were also likely to ejaculate more frequently than CL males.

The performance of intact males during D also was better than that of arrhythmic (SCN) males tested during D. CD males were quicker to initiate copulation and to ejaculate than SD animals and usually ejaculated more frequently. The deficit in mating ability of SD animals probably reflects a lesion-induced loss of daily rhythms in copulatory behavior and not a general decline in mating ability. SCN-lesioned males do not show a consistent rise in copulatory performance during D, but their average daily copulatory behavior is similar to that of control animals (Eskes 1981).
The loss of diurnal rhythmicity in copulatory performance is also reflected in the absence of normal L/D differences in performance in SCN-lesioned hamsters (SD/SL pairs). While SL males showed shorter latencies to copulation, there were no corresponding differences in the number of ejaculations.

Since mating performance of SCN animals is quite variable at all daily phases, it is sometimes quite good in the L phase. However, performance of the SL males was not consistently better than that of the CL males. The addition of more pairs to this

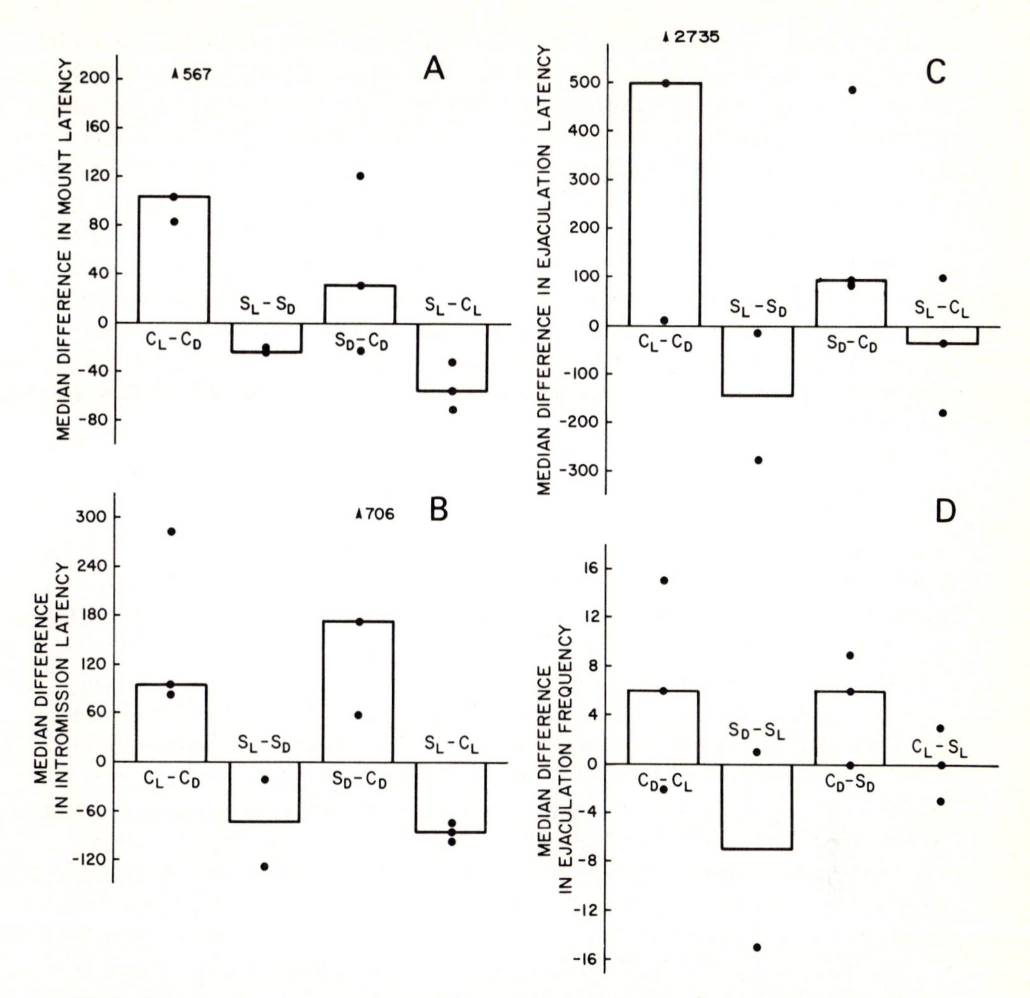

Fig. 1. Median differences in mount **(A)**, intromission **(B)** and ejaculation **(C)** latencies (s) and ejaculation frequency of two males tested simultaneously with one receptive female. Males were at different circadian phases during testing: *CD* and *SD* indicate control and SCN-lesioned animals tested at 6 h after the beginning of the D phase; *CL* and *SL* indicate control and SCN animals tested at 8 h after the beginning of their L phase. Each *dot* represents the difference in score between one pair as indicated on the abscissa

group might demonstrate that the variable performance of SCN hamsters is, on average, superior to the consistently poor performance of intact animals in L.

The difference in performance between CD/CL and CD/SD animals is important in view of some of the requirements already listed for successful reproduction:

Location of a Receptive Mate and Initiation of Copulation. Rapid initiation of mating is critical since female hamsters are receptive for only a limited period of time. Sexual receptivity is confined to the night of proestrus, beginning shortly before D onset and several hours before ovulation; if unmated, females may remain receptive throughout D (Beach et al. 1976, Carter and Schein 1971). However, the interval available for successful fertilization is probably much shorter due to the time needed for sperm penetration of ova (Yanagimachi 1966) and the deterioration of unfertilized ova after ovulation (Yamamoto and Ingalls 1972). The length of the receptive period may also be reduced by mating since females show a rapid decline in receptivity once mating occurs (Carter and Schein 1971).

Coordination of the Mating Behavior Pattern to Stimulate the Neuroendocrine Events of Pregnancy. The stimulus requirements for the initiation of pregnancy in female hamsters have not been studied in detail. It appears that the copulatory stimuli delivered by more than four ejaculations are needed to ensure successful induction of pregnancy in female hamsters (Lanier et al. 1975). Thus multiple ejaculations by the male hamster are critical to the successful fertilization of the female.

The Ability to Compete with Other Males to Maximize the Number of One's Offspring. Prolonged copulation may serve to inhibit female receptively and subsequent mating by other males (Carter and Schein 1971). The results of two-male one-female mating tests also indicate that the number of ejaculations may determine the number of offspring sired by one male (Lanier et al. 1979). The coat color of pups born to females mated with two male rats that were permitted to copulate to satiety indicated that the percent of total ejaculations by a single male was highly correlated with the proportion of pups sired by that male. Preliminary work with hamsters showed similar results (Dewsbury, personal comm., Lanier et al. 1979).

The daily rhythm of copulation may increase the probability of successful reproduction in male hamsters in several ways. Circadian mechanisms in males and females appear to coordinate sexual readiness; males are quicker to initiate copulation during the short period in D when females are receptive than at other times. Once copulation has begun, rhythmic males are likely to achieve a greater number of ejaculations during D. The stimulation provided by multiple ejaculations helps to initiate pregnancy and to decrease female receptivity, thereby reducing opportunities for other males. Multiple ejaculations by one male also maximize the proportion of offspring sired by that male if the female mates more than once.

Animals with SCN damage and disrupted copulatory rhythms appear to be at a selective disadvantage in competing with rhythmic animals. They are slower to begin mating and achieve fewer ejaculations once copulation has begun. These differences in the artificial laboratory environment are likely to be translated into major deficits in the natural milieu. This paradigm may serve to demonstrate the functional advantage of the circadian organization of reproductive behavior in hamsters.

Acknowledgments. I am grateful to Serge Daan, Gerard Groos and Benjamin Rusak for their helpful comments on the manuscript. Research was supported by Grant HD-02982 from NICHHD to Irving Zucker and by a NSERC of Canada Grant to Benjamin Rusak.

References

Beach FA, Levinson G (1949) Diurnal variations in the mating behavior of male rats. Proc Soc Exp Biol Med 72: 78-80

Beach FA, Stern B, Carmichael M, Ranson E (1976) Comparisons of sexual receptivity and proceptivity in female hamsters. Behav Biol 18: 473-487

Birdsall DA, Nash D (1973) Occurrence of successful multiple insemination of females in natural populations of deer mice *(Peromyscus maniculatus)*. Evolution 27: 106-110

Carter CS, Schein MW (1971) Sexual receptivity and exhaustion in the female golden hamster. Horm Behav 2: 191-200

Daan S (1981) Adaptive daily strategies in behaviour. In: Aschoff J (ed) Handbook of behavioural neurobiology, vol V. Plenum Press, New York, pp 275-298

Daan S, Tinbergen J (1979) Young guillemots *(Uria lomvia)* leaving their arctic breeding cliffs: a daily rhythm in numbers and risk. Ardea 67: 97-100

Enright JT (1970) Ecological aspects of endogenous rhythmicity. Annu Rev Ecol Syst 1: 221-238

Eskes GA (1981) Neural regulation of daily and seasonal cycles of sexual behavior in the male golden hamster. Unpubl Doct Diss, Univ California, Berkeley

Goldman L, Swanson HH (1975) Population control in confined colonies of golden hamsters (*Mesocricetus auratus* Waterhouse). Z Tierpsychol 37: 225-236

Henken J, Sherman PW (1981) Multiple paternity in Belding's ground squirrel litters. Science 212: 351-353

Holloway FA, Wansley R (1973) Multiphasic retention deficits at periodic intervals after passive-avoidance learning. Science 180: 208-210

Ibuka N, Inouye ST, Kawamura H (1977) Analysis of sleep-wakefulness rhythms in male rats after suprachiasmatic nucleus lesions and ocular enucleation. Brain Res 122: 33-47

Lanier DL, Estep DQ, Dewsbury DA (1975) Copulatory behavior of golden hamsters: effects on pregnancy. Physiol Behav 15: 209-212

Lanier DL, Estep DQ, Dewsbury DA (1979) Role of prolonged copulatory behavior in facilitating reproductive success in a competitive mating situation in laboratory rats. J Comp Physiol Psychol 93: 781-792

Murphy MR (1971) Natural history of the Syrian golden hamster - a reconnaissance expedition. Am Zool 11: 632

Payne AP, Swanson HH (1972) The effect of sex hormones on the aggressive behaviour of the female golden hamster *(Mesocricetus auratus* Waterhouse). Anim Behav 20: 782-787

Pittendrigh CS (1958) Adaptation, natural selection, and behavior. In: Roe A, Simpson GG (eds) Behavior and evolution. Yale Univ Press, New Haven, pp 390-416

Renner M (1960) The contribution of the honeybee to the study of time-sense and astronomical orientation. Cold Spring Harbor Symp Quant Biol 25: 361-367

Richter CP (1970) Dependence of successful mating in rats on functioning of the 24-hour clocks of the male and female. Comm Behav Biol 5: 1-5

Rusak B (1978) Neural mechanisms controlling behavioural rhythms in mammals. Abst Anim Behav Soc Ann Mtg, Seattle, Washington

Rusak B, Zucker I (1979) Neural regulation of circadian rhythms. Physiol Rev 59: 449-526

Stephan FK, Kovacevic NS (1978) Multiple retention deficit in passive avoidance in rats is eliminated by suprachiasmatic lesions. Behav Biol 22: 456-462

Stephan FK, Nunez A (1977) Elimination of the circadian rhythms in drinking, activity, sleep and temperature by isolation of the suprachiasmatic nuclei. Behav Biol 20: 1-16

Stephan FK, Zucker I (1972) Circadian rhythms in drinking behavior and locomotor activity of rats are eliminated by hypothalamic lesions. Proc Natl Acad Sci USA 69: 1583-1586

Wansley RA, Holloway FA (1975) Lesions of the suprachiasmatic nucleus (hypothalamus) alter the normal oscillation of retention performance in the rat. Soc Neurosci Abstr 1: 522

Wilson EO (1975) Sociobiology. The new synthesis. Harvard Univ Press, Cambridge

Yamamoto M, Ingalls TH (1972) Delayed fertilization and chromosome anomalies in the hamster embryo. Science 176: 518-521

Yanagimachi R (1966) Time and process of sperm penetration into hamster ova in vivo and in vitro. J Reprod Fertil 11: 359-370

Zucker I, Carmichael MS (1981) Circadian rhythms, brain peptides, and reproduction. In: Martin JB, Reichlin S, Bick KL (eds) Neurosecretion and brain peptides. Raven Press, New York, p 459-473

Subject Index